NEURAL STEM CELLS:
Development and Transplantation

Cover illustration is a micrograph of a differentiating cortical neurosphere immunostained with antibodies for nestin (green) and propidium iodide (red); see Chapter 3 by Nakano and Kornblum.

NEURAL STEM CELLS:
Development and Transplantation

edited by

Jane E. Bottenstein
*Marine Biomedical Institute
and
Department of Human Biological
Chemistry and Genetics
University of Texas Medical Branch
Galveston, TX*

KLUWER ACADEMIC PUBLISHERS
Boston / Dordrecht / London

Distributors for North, Central and South America:
Kluwer Academic Publishers
101 Philip Drive
Assinippi Park
Norwell, Massachusetts 02061 USA
Telephone (781) 871-6600
Fax (781) 681-9045
E-Mail: kluwer@wkap.com

Distributors for all other countries:
Kluwer Academic Publishers Group
Post Office Box 322
3300 AH Dordrecht, THE NETHERLANDS
Telephone 31 786 576 000
Fax 31 786 576 254
E-Mail: services@wkap.nl

 Electronic Services < http://www.wkap.nl>

QP
356
.25
.N465
2003

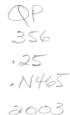

Library of Congress Cataloging-in-Publication Data

A C.I.P. Catalogue record for this book is available
from the Library of Congress.

NEURAL STEM CELLS: Development and Transplantation edited by Jane E. Bottenstein
ISBN 1-4020-7588-X

Printed on acid-free paper.

Printed in the United States of America.

*The Publisher offers discounts on this book for course use and bulk purchases.
For further information, send email to <Melissa Ramondetta@wkap.com> .*

PREFACE

During development of the central nervous system, multiple types of neurons and glial cells ultimately arise from self-renewing pluripotent embryonic stem cells. Little is known about the regulation of their differentiation into multipotent neural stem cells and their subsequent progeny. Neural stem cells are a topic of intense interest at the moment for two major reasons. First, they provide models for neural development that are easily manipulated and analyzed in vitro. Second, they are candidates for cellular and gene therapy of many intractable neurological disorders, e.g., Parkinson's disease, multiple sclerosis, spinal cord injury, and others. The availability of human neural stem cells is bringing us even closer to achieving the goal of effective cellular and molecular therapy for focal as well as disseminated neurological syndromes. Although there are numerous technical and ethical/legal problems yet to resolve, the enormous potential of this field of research has driven its exponential growth. This volume will be particularly useful for students, basic scientists, and clinicians in the academic or industrial sectors who have an interest in understanding neural development and/or its application to repairing the nervous system. In addition, it will provide vital information to those interested in the ethical/legal issues.

The current work on neural stem cells was preceded by studies of embryonal carcinoma (or teratocarcinoma) cells. This earlier work showed that these stem cells of the blastocyst stage could produce derivatives of all three germ layers (ectoderm, mesoderm and endoderm; Martin & Evans, 1975; McBurney, 1976) and could produce neuronal cells in vitro (Darmon et al., 1981). The latter study was the first to use serum-free N2 medium (Bottenstein and Sato, 1979) to generate large numbers of neurons from these pluripotent stem cells and to show a default mechanism of neural specification without a feeder layer, formation of embryoid bodies, or the presence of inducers, e.g., retinoic acid. These and other findings suggested that understanding some aspects of early neural development could indeed be derived from studying stem cells in vitro. It is interesting to review the comments of an NIH study section that evaluated a grant application I submitted in 1983 in which I proposed using the clonal 1003 mouse embryonal carcinoma cell line we described in Darmon et al. (1981) to isolate a neural stem cell line, generate monoclonal antibodies against different stages of differentiation to produce additional lineage markers, and determine the environmental signals that would induce neurotransmitter phenotypes other than cholinergic. I provided data showing I could obtain >95% postmitotic neurons that synthesized high levels of acetylcholine (but not tyrosine hydroxylase, serotonin, or glutamic acid decarboxylase) and exhibited both regenerative responses and delayed rectification using patch clamp techniques (Bottenstein, 1985). No GFAP-positive astrocytes were generated. The critique stated that "one's confidence in such studies is shaken by the ease (amply demonstrated by experiments carried out by the principal investigator) with which neuronal characteristics can be changed by various manipulations of the culture environment" and "there was some skepticism whether this kind of phenomenology in culture can provide basic insight into the problem of differentiation." These statements were not prescient of where this field is now. There was also resistance during this time to using these cells as models of normal development due to their tumorigenic origin, even though it had been shown that transplanted teratocarcinoma cells could participate in normal development and integrate into the host (Brinster, 1974; Mintz & Illmensee, 1975) and 1003 cell cultures, after neural differentiation occurs, contain no undifferentiated stem cells and are unable to form tumors in nude mice (Darmon et al., 1982).

The use of serum-free N2 medium (Bottenstein and Sato, 1979) has been of great benefit in identifying neural stem/progenitor cells in vitro and permitting their differentiation. In addition to its widespread use for neural cultures in general, it made possible our initial findings with the 1003 pluripotent stem cells, the discovery of oligodendrocyte progenitor cells by Raff et al. (1983), and is extensively used by investigators in the neural stem cell field. The addition of epidermal growth factor to N2 medium permitted the expansion and detection of mouse embryonic and adult neural stem cells first described by Reynolds et al. (1992) and Reynolds and Weiss (1992), respectively.

A seminal discovery was the identification of neural stem cells in the adult mouse (Reynolds and Weiss, 1992), which suggested that generation of new neurons and oligodendrocytes in vivo might be possible after development was complete, contrary to the extant view that this did not occur. The activation and generation of new astrocytes at injury sites is well known and can inhibit the repair process. This needs to be considered in transplant studies. Current studies are only now addressing the issue of stimulation of endogenous stem cells to produce the desired neural progeny to participate in the repair process.

It is now clear that both embryonic and neural stem cell lines as well as embryonic and adult sources of neural stem cells provide an expandable source of neurons and glia that can be used for studies of neural development and for cellular transplants that may be able to affect repair of nervous system injuries or disorders. Five major issues require further study in this field. First is the absence of a library of stage-specific markers to

identify the lineage position of embryonic and neural stem cells more precisely. Some markers have been described but many more need to be
identified. Second is the need to better understand the differentiation potential of cells derived from different sites in the nervous system and at
different stages of neural development. Third is the need to discover additional regulators of differentiation into specific phenotypes, e.g., notably
cholinergic neurons and myelinating oligodendrocytes, and their molecular mechanisms of action. Fourth is the problem of immunological rejection of transplants. One solution is the use of autologous transplants. The
fifth is to formulate standardized culture methods for maintaining and handling neural stem cells before and during experiments and transplantations.
This will require optimization of and consensus on a variety of parameters,
including different culture media (basal medium and supplements) for proliferation and differentiation protocols, passage technique, and acceptable
passage numbers for specific purposes. Standardization is essential for
replicating the findings of various investigators in this field, for comparing
data from different investigators, and in order to consistently produce desired differentiated phenotypes. Currently, there are multiple methods being used and this complicates analysis of experimental data and can result
in variability in the repertoire of differentiated progeny.

The range of topics covered in this volume is wide and the authors
were carefully selected for their expertise in the various subfields. I asked
them to share their view of the "state of the art", its present limitations, and
future perspectives. The book begins with a chapter on stem cells as models for neural development and neurological disorders to provide a context
for the subsequent chapters. This is followed by a discussion of stem cell
lineage and fate determination and subsequently a related chapter on stage-
specific and cell fate markers. Traditional sources and properties of embryonic and neural stem cells are described as well as alternative
transdifferentiated ones. Methods of purification of neural stem cells from
heterogeneous tissue sources and their clonal analyses are included. The
generation and properties of rodent and human embryonic and neural stem
cell lines and their use in research and repair paradigms is covered. The
following chapters discuss the regulation of survival, proliferation, and differentiation of neural stem cells and specifics regarding culture methods.
The final five chapters address the use of neural stem cells for cellular and
gene therapy. Two of these review the various animal transplantation studies and one discusses the exciting new topic of stimulation of endogenous
neural stem cells. This is followed by a discussion of cellular therapy in
humans directed at repairing injuries and diseases in the central nervous
system. The final chapter reviews methods of regulating and modifying

neural stem cell/progenitor gene expression for multiple purposes that include human gene therapy.

In summary, although we may be at the early stages of understanding neural stem cell lineage, differentiation, and transplant potential, the goals of this area of research are clear and the interest level is very high. The new fields of bioinformatics, genomics, and proteomics and their associated techniques should provide further insights that will result in exciting new information about early neural development and its clinical application to humans with developmental, metabolic, immunological, degenerative, aging, traumatic, or ischemic disorders of genetic or epigenetic origin. While the intractability of many neurological disorders drives the clinical side of this field, caution is imperative and success will depend on the findings of the basic scientists and their wise application by clinicians.

I would like to commend my Editorial Assistant Pat Gazzoli for her outstanding skill with graphics and page layout programs, excellent attention to detail, long hours spent in compiling this volume, and diplomatic interface with the various contributors. Her untiring efforts are gratefully acknowledged. Thanks are also extended to Jennifer LaScala for help with the CD included with this book.

Jane E. Bottenstein, Ph.D.
jebotten@utmb.edu

References

Bottenstein JE (1985) Growth and differentiation of neural cells in defined media. In: Cell Culture in the Neurosciences (Bottenstein JE, Sato G, eds), pp 3-43. New York: Plenum.

Bottenstein JE, Sato GH (1979) Growth of a rat neuroblastoma cell line in serum-free supplemented medium. Proc Natl Acad Sci U S A 76: 514-517.

Brinster RL (1974) The effect of cells transferred into the mouse blastocyst on subsequent development. J Exp Med 140: 1049-1056.

Darmon M, Bottenstein J, Sato G (1981) Neural differentiation following culture of embryonal carcinoma cells in a serum-free defined medium. Dev Biol 85: 463-473.

Darmon M, Sato G, Stallcup W, Pittman, Q (1982) Control of differentiation pathways by the extracellular environment in an embryonal carcinoma cell line. In: Growth of Cells in Hormonally Defined Media, Book B (Sato G, Pardee A, Sirbasku D, eds), pp 997-1006. Cold Spring Harbor: Cold Spring Harbor Laboratory.

Martin GR, Evans MJ (1975) Differentiation of clonal lines of teratocarcinoma cells: formation of embryoid bodies in vitro. Proc Natl Acad Sci U S A 72: 1441-1445.

McBurney MW (1976) Clonal lines of teratocarcinoma cells in vitro: differentiation and cytogenetic characteristics. J Cell Physiol 89: 441-455.

Mintz B, Illmensee K (1975) Normal genetically mosaic mice produced from malignant teratocarcinoma cells. Proc Natl Acad Sci U S A 72: 3585-3589.

Raff MC, Miller RH, Noble MD (1983) A glial progenitor cell that develops in vitro into an astrocyte or an oligodendrocyte depending on culture medium. Nature 303: 390-396.

Reynolds BA, Weiss S (1992) Generation of neurons and astrocytes from isolated cells of the adult mammalian central nervous system [see comments]. Science 255: 1707-1710.

Reynolds BA, Tetzlaff W, Weiss S (1992) A multipotent EGF-responsive striatal embryonic progenitor cell produces neurons and astrocytes. J Neurosci 12: 4565-4574.

Contents

Contributors

Paola Arlotta
Program in Neuroscience
Harvard Medical School
Edwards-Wellman 4
Edwards 410A
Massachusetts General Hospital
50 Blossom Street
Boston, MA 02114
Email: paola_arlotta@hms.harvard.edu

Carlos Bueno
Center of Molecular Biology
Severo Ochoa
Laboratory CX-450
Autonomous University of Madrid
Cantoblanco, 28049-Madrid, Spain
Email: cbueno@cbm.uam.es

Alexandra Capela
Stem Cells, Inc.
3155 Porter Drive
Palo Alto, CA 94304-1213
Email: capela@stemcellsinc.com

Jinhui Chen
Program in Neuroscience
Harvard Medical School
Edwards-Wellman 4, Edwards 410A
Massachusetts General Hospital
50 Blossom Street
Boston, MA 02114
Email: jinhui_chen@hmsarvard.edu

Cyndy D. Davis
Saneron CCEL Therapeutics, Inc.
USF Center for Entrepreneurship
13101 Telecom Drive, Suite 105
Temple Terrace, FL 33617
Email: cdh@saneron-ccel.com

Ian Duncan
University of Wisconsin – Madison
School of Veterinary Medicine
Department of Medical Sciences
2015 Linden Drive
Madison, WI 53706-1102
Email: duncani@svm.vetmed.wisc.edu

Thomas B. Freeman
University of South Florida
Tampa General Hospital
4 Columbia Drive, Suite 730
Tampa, FL 33606
Email: tfreeman@hsc.usf.edu

Ryan Fryer
Abbott Laboratories
100 Abbott Park Road
Abbott Park, IL 60064-3500
Ryan.Fryer@abbott.com

Daniel J. Guillaume
Stem Cell Research Program
Waisman Center, Rm T613
University of Wisconsin
1500 Highland Avenue
Madison, WI 53705
Email: guillaume@Waisman.Wisc.Edu

Steven R. Gullans
Brigham and Women's Hospital
Harvard Institutes of Medicine
77 Avenue Louis Pasteur
BWH Renal Div
Boston, MA 02115
Email: sgullans@rics.bwh.harvard.edu

Yoichi Kondo
University of Wisconsin – Madison
School of Veterinary Medicine
Department of Medical Sciences
2015 Linden Drive
Madison, WI 53706-1102
Email: kondoy@svm.vetmed.wisc.edu

Ichiro Nakano
Crump Institute for Molecular Imaging
UCLA School of Medicine
Department of Pharmacology
Box 951770
700 Westwood Plaza, 1423 CIMI
Los Angeles, CA 90095-1770
Email: inakano@mednet.ucla.edu

Harley I. Kornblum
Department of Molecular & Medical Pharma-
cology
University of California Los Angeles
1126 CIMI
700 Westwood Plaza
Los Angeles, CA 90095
Email: hkornblum@mednet.ucla.edu

Mahesh Lachyankar
Abbott Bioresearch Center
100 Research Drive
Worcester, MA 01605
Email: Mahesh.Lachyankar@abbott.com

Isabel Liste
Center of Molecular Biology Severo Ochoa
Laboratory CX-450
Autonomous University of Madrid
Cantoblanco, 28049-Madrid
Spain
Email: iliste@cbm.uam.es

Rick Livesey
Wellcome Trust/Cancer Research
UK Institute of Cancer & Develop. Biol.
University of Cambridge
Tennis Court Road
Cambridge, CB2 1QR
Email: rick@welc.cam.ac.uk

Jeffrey Macklis
Program in Neuroscience
Harvard Medical School
Division of Neuroscience, Children's Hospital
320 Longwood Avenue, Enders 354
Boston, MA 02115
Email : jeffrey.macklis@tch.harvard.edu

Alberto Martínez-Serrano
Center of Molecular Biology
Severo Ochoa Laboratory CX-450
Autonomous University of Madrid
Cantoblanco, 28049-Madrid, Spain
Email: amserrano@cbm.uam.es

Sanjay S. P. Magavi
Harvard Medical School
Edwards-Wellman 4, Edwards 410A
Massachusetts General Hospital
50 Blossom Street
Boston, MA 02114
Email: sanjay_magavi@hms.harvard.edu

Eva Mezey
National Institutes of Health (NINDS)
Building 36/3D-06
Covent Drive MSC 4157
Bethesda, MD 220892
Email: mezey@codon.nih.gov

Beatriz Navarro
Center of Molecular Biology
Severo Ochoa Laboratory CX-450
Autonomous University of Madrid
Cantoblanco, 28049-Madrid, Spain
Email: bnavarro@cbm.uam.es

Mary B. Newman
University of South Florida
College of Medicine MDC78
Department of Neurosurgery
Tampa, FL 33612
Email: mnewman@hsc.usf.edu

Larysa Halyna Pevny
Neuroscience Research Center
Department of Genetics
University of North Carolina at Chapel Hill
103 Mason Farm Road
Chapel Hill, NC 27599
Email: larysa_pevny@med.unc.edu

K. Sue O'Shea
University of Michigan
Department of Cell & Developmental Biology
4748 Med Sci II, Rm 0616
Ann Arbor, MI 48109
Email: oshea@umich.edu

Sabhi Rahman
Wellcome Trust/Cancer Research UK Institute
Department of Biochemistry
University of Cambridge,
Tennis Court Road,
Cambridge, CB2 1QR, UK.
Email: sr339@cam.ac.uk

Mahendra Rao
National Institutes of Health
Laboratory of Neurosciences
Gerontology Research Center
Room 4-B-17
5600 Nathan Shock Drive
Baltimore, MD 21224-6825
Email: raomah@grc.nia.nih.gov

Stephen N. Sansom
Wellcome Trust/Cancer Research UK Institute
Department of Biochemistry
University of Cambridge,
Tennis Court Road,
Cambridge, CB2 1QR, UK.
Email: sns27@hermes.cam.ac.uk

Paul Sanberg
Center for Aging-Neuroscience
University of South Florida
College of Medicine
12901 Bruce B. Downs Boulevard
Tampa, FL 33612
Email: psanberg@hsc.usf.edu

Evan Snyder
Professor & Director, Stem Cell Program
The Burnham Institute
10901 North Torrey Pines Road
La Jolla, CA 92037
E-mail: esnyder@burnham.org

Lorenz Studer
Laboratory of Stem Cell & Tumor Biology
Memorial Sloan–Kettering Cancer Center
New York, NY 10021
Email: studerl@mskcc.org

Stanley Tamaki
Stem Cells, Inc.
3155 Porter Drive
Palo Alto, CA 94304-1213
Stan.tamaki@stemcellsinc.co

Uruporn Thammongkol
Wellcome Trust/Cancer Research UK Institute
Department of Biochemistry
University of Cambridge,
Tennis Court Road,
Cambridge, CB2 1QR, UK.
Email: ut204@cam.ac.uk

Mark Tomishima
Laboratory of Stem Cell and Tumor Biology
Memorial Sloan-Kettering Cancer Center
New York, NY 10021
Email: tomishim@mskcc.org

Nobuko Uchida
Stem Cells, Inc.
3155 Porter Drive
Palo Alto, CA 94304-1213
Email: nobuko.uchida@stemcellsinc.com

Ana Villa
Center of Molecular Biology
Severo Ochoa Laboratory CX-450
Autonomous University of Madrid
Cantoblanco, 28049-Madrid, Spain
Email: anavilla@cbm.uam.es

Ping Wu
Marine Biomedical Institute and
Department of Anatomy & Neurosciences
University of Texas Medical Branch
Galveston, TX 77555-1043
Email: piwu@utmb.edu

Weidong Xiao
302G Abramson Research Center
Children's Hospital of Philadelphia
3516 Civic Center Blvd.
Philadelphia, PA 19104
Email: wxiao@mail.med.upenn.edu

Su Chun Zhang
Stem Cell Research Program
Waisman Center, Rm T613
University of Wisconsin
1500 Highland Avenue
Madison, WI 53705
Email: zhang@Waisman.Wisc.Edu

Chapter 1

Neural Stem Cell Models of Development and Disease

K. Sue O'Shea

EARLY DEVELOPMENT OF THE CNS

With extrusion of the second polar body from the fertilized zygote, the anterior–posterior axis of the mouse embryo is established (Gardner, 2001). From that point on, regionalization and differentiation of the embryo is established by cyclic expression of signaling molecules, growth factors and the extracellular matrix molecules that create and maintain gradients of these critical factors. Surprisingly, many are reutilized during development and differentiation of the embryo, and some are reexpressed following injury. At gastrulation, cells from the epiblast delaminate from the ectoderm and migrate to form endoderm and mesoderm, establishing the three-layered embryo. A posterior signaling center, the node, secretes molecules such as noggin and chordin that induce the embryonic ectoderm to form neural ectoderm by inhibiting the interaction of bone morphogenetic protein-4 (BMP-4) with its receptor (e.g., Harland, 2000). A second signaling center in the anterior visceral endoderm (AVE) of mammalian embryos produces additional signaling factors, including cerberus, Dickkopf1 and Otx2 that induce the formation of anterior (forebrain) structures (e.g., Perea-Gomez et al., 2001).

Once induced, neuroepithelial cells lengthen and the neural plate begins to undergo a series of morphogenetic shaping changes that result in the transformation of the sheet of neuroepithelial cells into a closed neural tube (e.g. Colas and Schoenwolf, 2001, Figure 1). As the the primitive streak regresses into the tail bud, newly formed mesenchymal cells aggregate to form a cord of cells that canalizes, extending the spinal cord into the lumbo-sacral region; a process termed secondary neurulation. Further differentiation of the CNS relies on anterior–posterior signals from regional signaling centers such as the isthmus and the forebrain organizer; with dorsal-ventral patterning controlled by gradients of ventral signals from the notochord and dorsalizing signals from the roof plate and surface ectoderm.

From: *Neural Stem Cells: Development and Transplantation*
Editor: Jane E. Bottenstein © 2003 Kluwer Academic Publishers, Norwell, MA

STEM CELLS OF THE NERVOUS SYSTEM: LEXICON

After neurulation is complete, the primitive nervous system is composed of a single-layered cylinder of cells, the neural tube, that gives rise to all the differentiated derivatives of the adult brain and spinal cord (Figure 1). Although many of the neuroepithelial cells are mitotically active, there are already regions that are restricted in their developmental potential, such as the roof plate and the floor plate (Figure 1) that contain glial-like cells (Silver, 1994). Stem cells are present throughout the development of the nervous system, remaining into maturity in many species including humans (Kukekov et al., 1999), although their precise locations are unresolved. The most simple definition of a stem cell would include the ability to both self-renew and to differentiate into multiple derivatives, in the case of the nervous system into neurons, astrocytes and oligodendroglia. By this definition, cells of the early neural tube are clearly stem cells, as most are tripotential (Mujtaba et al., 1999), and have been termed neuroepithelial stem cells (NEP) (Kalyani et al., 1997; Rao, 1999) to reflect their early origin (Figure 2).

NEPs proliferate in response to fibroblast growth factor (FGF2) exposure *in vitro*, and can be identified by their expression of fibroblast growth factor receptor (FGFr) 4, Frizzled 9 and Sox2 (Cai et al., 2002). During development, the number of NEP is reduced as they form neuronal and glial restricted precursors (Mayer-Proschel et al., 1997) that can be identified by their expression of ps-NCAM and A2B5 (Cai et al., 2002). As differentiating cells migrate radially from the lumen of the neural tube, the neuroepithelium begins to stratify, and precursor cells begin to express receptors for additional growth and differentiation factors. Neuronal and glial precursors typically have some, albeit less, mitotic capability. The precise mechanisms by which NEPs become determined to form glial precursors (GPs), cells destined to form neurons (NP; neuronal precursors), or precursors of the peripheral nervous system (neural crest stem cells) are unknown, but likely involve multiple mechanisms that may include cell-cell interactions, environmental conditions, and alterations in cell cycle characteristics.

There appears to be a gradual restriction in developmental options by precursor cells, both positive [bHLH genes including neurogenin, MATH, MASH, Olig1,2 (Lee, 1997; Sasai, 1998)] and negative regulators (ID genes; Hollnagel et al., 1999) combine with local signaling molecules and growth factors to specify cell fate. Clearly, instruction of a cell to a particular fate can be interpreted as inhibition of another fate, e.g., proneural genes inhibit gliogenesis (Kageyama et al., 1995; Tomita et al., 2000).

During fetal development (after approximately E13.5 in the mouse embryo) as NEPs become diminished in number, neural stem cells resident in the

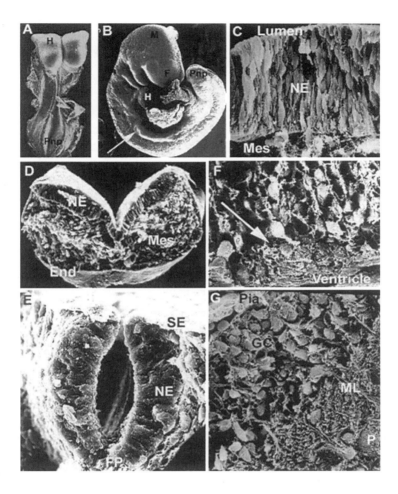

Figure 1. Scanning electron micrographs illustrating patterns of early CNS development.
A. Dorsal aspect of an early somite staged E8 mouse embryo. The neural folds are beginning to fuse in the midline in the future cervical region. The neural folds in the headfold region (H) and posterior neuropore (Pnp) are elevating but are not yet fused. **B.** Side view of an E9 embryo. The neural tube is closed throughout its extent except at the posterior neuropore (Pnp). F=forebrain, M=midbrain, H=heart, arrow indicates forelimb bud. **C.** High magnification view of a fracture through the E9 neuroepithelium. Dividing cells are present near the lumen, with neuroepithelial processes contacting both luminal and basal surfaces. NE=neuroepithelium, Mes=mesenchyme. **D.** Fracture through the midbrain neural folds of an E8.5 embryo illustrating the relationship of neuroepithelium (NE), underlying mesenchyme (Mes) and endoderm (End). **E.** Fracture through the closing neural folds in the future spinal cord region. SE=surface ectoderm, NE=neuroepithelium, FP=floorplate. **F.** Fracture through the E16 SVZ/ependymal zone illustrating small pockets of dividing cells (arrow) near the ventricle. **G.** Fracture through P8 cerebellar cortex illustrating the presence of dividing granule cells (GC) near the pial surface, the forming molecular layer (ML) above the large Purkinje cell (P) bodies.

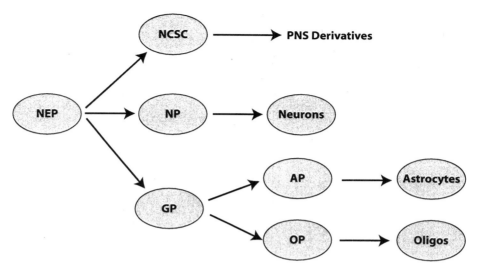

Figure 2. Sequential differentiation of stem cells. (after Rao, 1999). Neuroepithelial progenitor cells (NEP) differentiate to neural crest stem cells (NCSC) that form peripheral structures; to neuronal progenitors (NP) that differentiate into neurons; to glial progenitors (GP) that form astrocyte progenitors (AP) and oligodendrocyte progenitors (OP).

neuroepithelium begin to express epidermal growth factor (EGF) receptors, and respond both *in vivo* and *in vitro* to high (50 ng/ml) levels of EGF by proliferating (Rao, 1999). At late fetal as well as adult stages of development, forebrain neuroepithelial cells grown in suspension culture as neurospheres, contain a population of neural stem cells that respond to both EGF and to FGF (Reynolds and Weiss, 1992), unlike NEPs that respond only to FGF and lack the EGF receptor (Kalyani et al., 1999). Thus, during fetal development, the NEP population is decreased in number, forming precursors committed to a particular lineage, and neural stem cells acquire the ability to respond to EGF, a characteristic that is retained into adulthood (Okano et al., 1996). Surprisingly little is known about the transition between fetal and adult stem cells, other than that they respond to similar growth factors. In fact, the assumption that the adult stem cell must express primitive features, i.e., be located in a fetal microenvironment and express genes typical of primitive cells, may have significantly misled investigators, as it is only now being recognized that adult neural stem cells may have characteristics of mature glial cells (Alvarez-Buylla et al., 2001).

It has been known for some time that dividing cells are present in the adult central nervous system (CNS, Altman and Das, 1965). Whether this is a fundamental property of all neurons and glia that is repressed by environmental factors late in development, or whether remnants of embryonic structures containing stem cells are sequestered in the adult brain remains to be determined. This is an area of considerable research effort, with the goal of both understanding how to expand specific stem cell and precursor populations following injury or degeneration, as well as to elucidate the basis of and source of cells involved in tumor formation. Although it is clear that stem cells are present throughout the development of the nervous system, the lack of sufficient markers makes it difficult to precisely identify them, particularly in adult CNS. It has been argued based on their growth factor responsiveness, that the forebrain periventricular region contains a mixed population of both rapidly dividing cells destined to form olfactory neurons and a "true" stem cell population. Identification of a stem cell population based on its *in vitro* growth factor responsiveness may result in erroneous conclusions since cells are deprived of their normal cell-cell interactions, and may express new receptors and behaviors not normally observed *in vivo*. Conversely, it has been argued that this environment simply allows cells to express their entire repertoire of behaviors. Unfortunately, *in vitro* assays have often identified the presence of stem cells retrospectively and likely contain mixed populations. Whether the initial starting population is mixed or simply contains different developmental stages of the same cell type is not known.

Adult neural stem cells share important similarities and significant differences with other tissue stem cells. Like tissue stem cells, neural stem cells exhibit surprising plasticity and can differentiate into a wide number of derivatives when grown in particular culture environments (e.g., Bjornson et al., 1999), or when aggregated with cells of the early blastocyst (Clarke et al., 2000). The fact that adult neural stem cells and adult hematopoietic stem cells express similar growth factor receptors (Ivanova et al., 2002; Parati et al., 2002) argues more that there are conserved signaling pathways for differentiation, rather than explaining phenotypic plasticity of these cells. Many, but not all, tissue stem cells are characterized by long cell cycle times, remaining largely quiescent (G0). However, neural stem cells isolated from the lateral ventricles of the adult nervous system produce tens of thousands of new neurons daily (Lois and Alvarez-Buylla, 1994), turning over every 12-28 days (Craig et al., 1999). It is possible that this specialized region is a relic rather than a "true" tissue stem cell field, and that "real" CNS stem cells are those that are resident in other locations in the CNS that may have much longer cell cycle times. However, cell cycle time per se should not be a primary determinant of "stemness" as populations of stem cells vary considerably in their cell cycling times. What may be more relevant is that with cell division there is limited nuclear reprogramming that may contribute significantly to the phenotypic lability of stem cells.

SITES OF NEUROGENESIS AND GLIOGENESIS

During embryonic development, the ventricular zone is relatively wide and actively producing neurepithelial stem cells and progenitors (Jacobson, 1991). It has been estimated that in the mouse embryo, neuroepithelial stem cells undergo 10-12 cell divisions (Takahashi et al., 1994), with neuronal precursers born first, followed by glial precursors (Rao, 1999; Alvarez-Buylla et al., 2001; Sommer and Rao, 2002). The peak period of neurogenesis occurs over days E9.5-15.5, while major rounds of gliogenesis take place in the late prenatal to early postnatal period (Altman, 1966), although glial progenitors are born as early as E12. In the spinal cord, the pattern of differentiation is relatively simple, with the ventricular zone (VZ) gradually thinning to form the ependymal layer, stratifying neuronal and glial precursors forming the intermediate zone. With continued differentiation, neuronal cells in the intermediate zone make connections with each other and send axons into the cell sparse marginal zone. These axons eventually are myelinated by oligodendrocytes, forming the marginal zone of the mature spinal cord (Figure 3).

In the brain, the pattern is much more complex. After E14 in the mouse embryo, the cortical layers begin to stratify and the ventricular zone (that contains both neuronal and glial precursor cells as well as undifferentiated stem cells (Raff et al., 1983; Turner and Cepko, 1987; Luskin, 1993; Sakakibara and Okano, 1997) gradually thins to form the ependymal layer (Caviness and Takahashi, 1995). The subventricular zone (SVZ) immediately beneath the VZ becomes the source of proliferating stem cells and precursors (both neuronal and glial). Neurons migrate radially and tangentially to the pial surface forming the characteristic inside out organization of the cerebral cortex (Figure 3). The SVZ is particularly prominent in the ganglionic eminences (Altman and Bayer, 1985), generating cells for basal ganglia, diencephalon, and cortex (Garcia-Verdugo et al., 1998; Caviness and Takahashi, 1995). Progenitors from the SVZ migrate to various cortical regions from P0 to P30, after which time they are restricted to the rostral migratory stream (Peretto et al., 1999).

Postnatal Neurogenesis and Gliogenesis

Neurogenesis is largely complete prenatally (by E17.5 in the mouse) except in regions such as the neonatal external granule cell layer in the cerebellum where cell division followed by inward migration of granule cells to their adult location in the granule cell layer is completed in the third postnatal week (Figure 1G; Altman and Bayer, 1985). In the adult brain, active neurogenesis is largely restricted to two regions, one in the granular layer of the adult dentate gyrus of the hippocampus, a region involved in learning and memory, in a strip

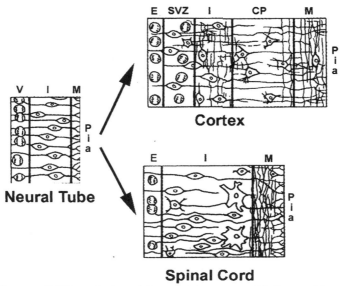

Figure 3. Pattern of differentiation of the primitive neuroepithelium (left) to more mature cortex and spinal cord (after Jacobson, 1991). In the cortex, the intermediate zone will thicken as white matter grows, the cortical plate will contain both neuroblasts and glioblasts; while in the spinal cord the ventricular zone will eventually thin, forming the ependymal zone. Lumen is oriented toward the left, pia to the right. V=ventricular zone, I=intermediate zone, M=marginal layer, E=ependymal layer, SVZ=subventricular zone, CP=cortical plate.

of cells known as the subgranular zone. Although neurogenesis significantly increases neuronal number and size of the hippocampus over time (Bayer et al., 1982), surprisingly little is known regarding the stem cell properties of these cells. Hippocampal progenitors respond to exogenous growth factors by forming multipotent neurospheres (Gage et al., 1998); they proliferate throughout life (Kuhn et al., 1996) and in response to injury (Parent et al., 1997). Recent evidence suggests that the hippocampal stem cell (like the SVZ astrocyte below) may have astrocyte characteristics (Seri et al., 2001), although whether it forms de novo or from SVZ stem cells is not yet resolved. Hippocampal astrocytes (unlike astrocytes from the spinal cord) are capable not only of stimulating proliferation of hippocampal neural stem cells, but also control cell fate decisions. Young but not older astrocytes stimulate proliferation and neuronal differentiation, while co-culture of hippocampal stem cells with hippocampal neurons produces oligodendrocytes (Song et al., 2002). These studies and others (below) suggest a primary role for the astrocyte in both controlling proliferation of neural stem cells and determining their neuronal vs glial fate, and they invoke the possibility that the NSC may be an astroglial cell.

The other region of active adult neurogenesis is the subventricular zone (SVZ) aka the subependymal layer (SEL) in the forebrain, which is also thought to be the major source of glial cells (Levison and Goldman, 1993). Gliogenesis begins on about E12 with the formation of glial progenitor cells then astrocytes, oligodendroglia (Timsit et al., 1995) and continues throughout life. After birth, the SVZ is depleted in most regions of the neuraxis, remaining in the anterior forebrain along the lateral ventricles, where cells proliferate throughout life (Goldman et al., 1997; Tropepe et al., 1997). The SVZ contains several identi- fied cell types including: ependymal cells and astrocytes (B cells), that make contact with the ventricular surface by sending processes between ependymal cells. The fact that astrocyte processes contact the CSF has complicated many lineage studies when tracers are placed into the cebrospinal fluid (CSF) (Johansson et al., 1999), as both ependyma and astrocytes are labeled. How- ever, ependymal cells can reenter the cell cycle and differentiate into glial and neuronal cells. The astrocytes also surround immature precursors – transit am- plifying cells (C cells), as well as the majority of the cells [migrating neuroblasts (A cells)] that are already organized into chains (Doetsch et al., 1997,1999; Garcia-Verdugo et al., 1998; Luskin et al., 1998). The neuroblasts migrate in the rostral migratory stream to the olfactory bulb where they differentiate into interneurons (Luskin, 1993; Lois and Alvarez-Buylla, 1994).

Astrocytes (B cells) in the SVZ are mitotically active and appear to be the source of the rapidly expanding population of progenitors (C cells), that give rise to stem cells in multipotent neurospheres (Doetsch et al., 1999). It is clear that many "glial" cells have considerably more plasticity that previously antici- pated, as it now appears that fetal radial glial cells that span the neuroepithelial wall and serve as a substratum for radial migration of neuroblasts, may dedif- ferentiate to form B cells when migration is complete, and thus are a source of neurons as well as astrocytes in the adult brain (Lim and Alvarez-Buylla, 2002; Parnavelas and Nadarajah, 2001). The transient radial glial cell may therefore be the missing link between embryonic and adult neural stem cells (Alvarez- Buylla et al., 2001).

A terminological difficulty is that authors variously refer to the embry- onic SVZ as being maintained in the adult brain. To avoid confusion, the Boul- der Committee (1970) recommended that the remnant of the embryonic SVZ that persists into adult life in the forebrain ventricles be called the subependymal layer (SEL) to recognize that it is not present along the entire neuraxis and that in the adult it has a unique organization and cell composition. Since processes from astrocytes (B cells) contact the ventricular fluid, it has been argued that the layer is not anatomically entirely "beneath" the ependymal layer, and the term SVZ continues to be used as do hybrid terms such as the subependymal zone (SEZ). The term "adult SVZ" will be employed here.

Glial cells, the most numerous cell type in the adult CNS, begin to differentiate from VZ progenitor cells (NEPs) at approximately E12-E13 in the rat (Rao et al., 1998), slightly later than neuronal progenitors. Noble and Mayer-Proshel (2002) have proposed that the glial precursor (GP) is ancestoral to all CNS glia, giving rise to the astrocyte progenitor (AP) and the O2A/OP (oligodendocyte-type 2 astrocyte/oligodendrocyte progenitor) by a process of sequential lineage restriction. GPs are present from E12 through P2 and express A2B5 and nestin, but do not express PDGFRα. Despite the surprisingly early origin of GPs, the VZ is more likely to give rise to NPs than GPs, while the later SVZ is more likely to produce glial cells either *in vitro* or when growth factors are infused intraventricularly. Treatment of GPs with BMPs or EGF promotes the differentiation of astrocytes from both spinal cord and from the SVZ (Fok-Seang and Miller, 1994). Consistent with this observation, astrocytes differentiate from E12.5 through postnatal development (Richardson et al., 1997) in the dorsal neuroepithelium where BMP levels are high. Although there are not particularly good markers for these cells, they express nestin and S-100, but do not express A2B5, and are PDGFRα-negative (Rao, 1999). A repressor of transcription (N-CoR) appears to play a critical role in astrocyte differentiation. In N-CoR null mice, self-renewal is impaired and NSCs differentiate prematurely to astrocytes, suggesting that N-CoR is a repressor of astrocyte fate (Hermanson et al., 2002)

Because there are better markers for OP than AP cells, there is considerably more information about their differentiation. The default differentiation of the OP appears to be to oligodendrocytes, but OPs can also form type 2 astrocytes *in vitro*. OPs form in ventral regions of the neuraxis, both ventral spinal cord as well as ventral rhombencephalon (e.g., Chandross et al., 1999) where gradients of secreted factors from the notochord and floor plate such as SHH are thought to influence their differentiation (Mekki-Dauriac et al., 2002) via expression of bHLH lineage control genes including Olig1, 2, 3 (Fu et al., 2002; Takebayashi et al., 2002). Misexpression of SHH in dorsal portions of the neuroepithelium is sufficient to cause ectopic differentiation of oligodendrocytes in this location (Orentas and Miller, 1996). From their origin as clusters along the neuraxis (Thomas et al., 2000), OPs migrate extensively, undergoing multiple rounds of division as they migrate, before differentiating at their final adult locations (Timsit et al., 1995; Chari and Blakemore, 2002). OPs express PDGFRα and respond to PDGF by dividing. There is a gradient of differentiation of these cells, with OPs differentiating earlier in the spinal cord than in the cortex, although myelination follows a rostral to caudal pattern within regions.

Interestingly, an internal clock that measures time rather than the number of cell divisions appears to control differentiation of OPs to oligodendrocytes (Ibarolla et al., 1996). Whether grown *in vivo* or *in vitro*, oligodendrocytes differentiate from OPs at precisely the same time (Abney et al., 1981). This stereotyped timing of differentiation can be overridden *in vitro* by thyroid hormone and factors produced by type 1 astrocytes that are potent inducers of oligodendrocyte differentiation. The presence of GPs, OPs, and APs at the same stage of differentiation, in combination with *in vitro* studies that have demonstrated the ability of GPs, but not APs or OPs to give rise to all three differentiated glial cell types, has led to the suggested lineage relationship between them. In addition, the rapidly dividing, rapidly migrating OPs characteristic of fetal staged embryos are thought to mature directly into the slowly dividing OPs present in the white matter of the adult brain (Armstrong et al., 1992; Scolding et al., 1998).

Outside the hippocampal granular zone and SVZ, scattered proliferating cells do not typically form neurons, but form glial cells in the adult brain (Gage 2000; Horner et al., 2000; Kornack and Rakic, 2001; Rakic, 2002). However pluripotent cells have been obtained from these regions considered to be non neurogenic, including the adult spinal cord (Gage et al., 1995; Weiss, 1999), septum, striatum (Palmer et al., 1995), the white matter (Palmer et al., 1999), and the cortex (Marmur et al., 1998; Magavi et al., 2000). These studies suggest that the plasticity characteristic of early development may be gradually inhibited rather than being entirely lost during differentiation. It also is possible that local factors act to restrict developmental plasticity (as would be appropriate to maintain the pristine organization of the adult brain), rather than there being an intrinsic genetic program that restricts potential. Wounding studies have indicated that many growth factors and extracellular matrix proteins characteristic of the early CNS can be reexpressed, and may contribute to abortive process outgrowth and glial hypertrophy that follow injury (Brodkey et al., 1995). It has been argued that if the adult environment (growth factor depleted) simply inhibits differentiation potential, removing neurons and glial cells from repression by placing them into cell culture should allow the reexpression of the developmental program (Gage, 2000). Clearly, the situation is more complex, and it is likely that cell-intrinsic variables and the environment of the cell both produce and maintain the progressive restriction in fate that characterizes the adult CNS.

INTRINSIC CONTROL OF STEM CELL BEHAVIOR: CELL DIVISION, PROGRAMMED CELL DEATH, QUIESCENCE

Shortly after neural tube closure is complete, the neuroepithelium is pseudostratified (Figure 1), and except during the M-phase, cells maintain contact with both the luminal and basal surfaces of the epithelium. Neuroepithelial cells undergo characteristic rounds of cell division at the lumen, then enter G1, elongate and reestablish contact with the basal surface of the neuroepithelium. During S-phase they move to the outer levels of the neuroepithelium, then in G2, the nucleus moves toward the ventricular surface where mitosis occurs, and the cycle begins again (Figure 4). Initially, cell cycle time is short (7-11h, Jacobson, 1991), increasing later in development to 18-24h (Takahashi et al., 1994; Garcia-Verdugo et al., 1998). Still later, stem cells have cell cycle times of many days (Morshead et al., 1998), and a subset appears to remain in G0 as quiescent GFAP+ astrocyte like stem cells (Alvarez-Buylla et al., 2001).

There has been considerable interest in this process of interkinetic nuclear migration (Sauer, 1935), as it appears that the orientation of the mitotic spindle may determine whether a particular cell division generates two equivalent daughter cells (spindle parallel to the lumen; proliferation; see Figure 4) or a committed precursor and a stem cell (spindle oriented perpendicular to the lumen; differentiation; see Figure 4). During early developmental stages, symmetrical cell division predominates (Caviness and Takahashi, 1995; Rakic, 1995). As the neuroepithelium grows, the pattern switches to asymmetrical cell division to produce differentiated progeny (Chenn and McConnell, 1995; Reznikov and van der Kooy, 1995; Cai et al., 1997). It has been reported that in differentiative divisions there is an extended G1, while in symmetrical or proliferative divisions G1 is relatively short. (Lukaszewicz et al., 2002), implicating direct control of cell cycle in differentiation. This is similar to the pluripotent embryonic stem (ES) cell, that has a very attenuated G1 (Savatier et al., 1996). In fact, changes in the regulation of proliferation e.g., conditional mutation of tumor suppressor genes such as PTEN produce enlarged abnormal brain structures characterized by overproduction of NEPs and progenitor cells (Groszer et al., 2001). Control of cell division as a mechanism to control differentiation is an area of considerable interest. In the case of neural stem cells, growth factors that stimulate cell division such as FGF2, EGF, IGF1 override differentiation promoting factors and must be removed from culture medium before differentiation can occur (Grinspan et al., 2000; Lillien and Raphael, 2000). Without EGF and FGF2, cells withdraw from the cell cycle and then differentiate. Consistent with this, both chemical inhibitors of cell division (LoPresti et al., 1992) and in-

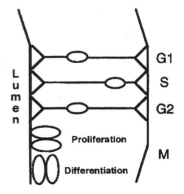

Figure 4. Interkinetic nuclear migration (after Frisen et al., 1998).

hibitors of cell cycle progression (Parker et al., 1995; Poluha et al., 1996) can promote differentiation.

The issue of whether there is a genetic program, i.e., a set number of rounds of division that NEP cells undergo, has been studied extensively. When precursor cells are explanted into a cell culture environment that stimulates cell division, they are capable of long term self renewal (Chenn and McConnell, 1995), suggesting the lack of a pre-existing program. *In vitro*, glioblasts went through a long series of symmetrical cell divisions while neuroblasts underwent predominantly asymmetric cell divisions, reinforcing the idea that the pattern of division may be cell-intrinsic, rather than environmentally controlled (Qian et al., 1998). Each clone generated different numbers of progeny, however, suggesting a considerable degree of heterogeneity in the starting population. It has also been shown that *in vitro* more differentiated precursors (particularly oligodendrocyte precursors), unlike true stem cells, may stop responding to growth factors, and withdraw from cycle, i.e., reach their Hayflick (1968) limit.

There has been considerable interest in understanding the mechanisms (both genetic and epigenetic) that control symmetrical cell divisions vs. the asymmetrical generation of progenitor cells. It should be pointed out that the cleavage plane model is extremely simplistic, and many have commented that it fails to account for the size and complexity of the mammalian CNS. In fact, two daughter cells produced by a "proliferative" division may find themselves in very different microenvironments and form different cell types. However, understanding the mechanisms involved could have important implications for expanding endogenous stem cell populations following trauma or disease, and also possibly suggest therapies for malignancies. In asymmetrical cell division, there is unequal division of cytoplasmic components, particularly of the Numb and Prospero proteins (and associated cytoplasmic complexes), producing different cell fates (Zhong et al., 1996). In the drosophila embryo where it has been most extensively studied, Numb and Prospero proteins segregate to the neuroblast where Prospero is thought to inhibit cell division and control transcription of neuronal genes and Numb to control neuronal fate by suppressing the activity of the integral membrane protein Notch (Wai et al., 1999).

The Notch pathway is of particular importance in the development of many tissues because of its role in controlling cell-cell interactions by lateral inhibition. In the nervous system, the Notch receptor Delta1 is expressed in the outer portion of the VZ (Henrique et al., 1995), while Notch1 is expressed in the VZ (Weinmaster et al., 1991, 1992; Henrique et al., 1995; Williams et al., 1995). In unequal divisions, the daughter cell that receives higher levels of Notch continues to divide (Chenn and McConnell, 1995) suppressing neurogenesis, while inhibition of Delta1 promotes neuronal differentiation (Austin et al., 1995). Consistent with this, loss of Notch activity results in embryos in which neuroepithelial cells remain progenitor-like with expansion of the VZ (Kopan et al., 1996). Overexpression of a dominant active Notch3 in stem cells using transgenesis-produced embryos that died by E12.5, and were characterized by neuroepithelial overgrowth, and increases in both cell number and mitotic rate (Lardelli et al., 1996). Activation of Notch suppresses neurogenesis but not gliogenesis (Nye et al., 1994), and there is evidence that Notch promotes a glial identity (Gaiano et al., 2000).

The Hes genes (homologues of the Drosophila bHLH gene Hairy/Enhancer of Split that act as repressors of neurogenic genes), also function as negative regulators of neural fate in the mammalian nervous system (Ishibashi et al., 1994; Nakamura et al., 2000). Hes1 is downstream of Notch, so knock-out of either Notch or of Hes1, stimulate early differentiation of neuronal progenitors (Ishibashi et al., 1995). Overexpression of Hes1 on the other hand, maintains the progenitor state and blocks neuronal differentiation via its inhibitory actions on neurogenic genes of the bHLH family of transcription factors, Mash1, neurogenin, neuroDs (Ishibashi et al., 1994). The Hes family members (Hes1-7) are expressed widely in the developing nervous system (Sasai et al., 1992; Takebayashi et al., 1995; Wu et al., 2003), where they may play an important role in controlling the timing of progenitor differentiation. For example, Hes1 and Hes5 are expressed in glial lineages and are downregulated with differentiation. Over-expression of Hes1 in glial precursors (but not neuroepithelial progenitors) promotes differentiation of astrocytes at the expense of oligodendrocytes (Wu et al., 2003). These results and others (Ohtsuka et al., 2001), suggest that later in development the Hes genes play a role in controlling precursor differentiation in a stage- and cell type-specific manner.

Another class of transcription factors act as dominant-negative inhibitors of differentiation. The Id genes (Inhibitors of differentiation) encode a family of proteins that lack the basic DNA binding region (i.e., HLH; Benezra et al., 1990). Id proteins heterodimerize with bHLH proteins, preventing them from binding to DNA, thereby inhibiting gene expression. Id genes are involved in many fundamental developmental processes from cell division to differentiation to apoptosis (reviews: Norton et al., 1998; Israel et al., 1999; Norton, 2000; Tzeng, 2003). There are four Id family members that are expressed in dynamic patterns during development (Kee and Bronner-Fraser, 2001), particularly in the nervous system, where they function as negative regulators of neural differentiation (Sun et al., 1991; Neuman et al., 1993). In general, expression of the Id genes inhibits cell type specific differentiation. For example, overexpression of Id2 inhibits neuronal differentiation and promotes apoptosis in cortical progenitors (Toma et al., 2000). Interestingly, Id2 regulates Rb activity (Lasorella et al., 1996), and Id4 regulates the expression of BRCA1 (Beger et al., 2001). Id genes are upregulated according to astrocyte tumor grade (Vandeputte et al.,

2002), and are downregulated following injury to the nervous system (Kabos et al., 2002). This family, like the Hes family appears to play a critical role in controlling the transition between precursor cell and differentiated neurons and glia. Ultimately, the dynamic balance between the expression of positive and negative regulators of differentiation may determine the ability of growth factors and signaling molecules to promote lineage specific differentiation (Mehler, 2002).

Cell number is ultimately controlled by cell division, differentiation and inhibition of differentiation, and by programmed cell death within the nervous system. Initially, it was thought that there was relatively little cell death during early stages of development; programmed cell death (PCD) has been reported to peak at E14-E15 and then decline (Blaschke et al., 1996). However estimation of the precise numbers of dying cells depends on the sensitivity of the technique employed. PCD appears to occur earlier than previously thought (during neurulation) and may involve classic neurotrophic factors as well as other classes of signaling molecules, particularly the bone morphognetic proteins (BMPs), to regulate early stem cell number (de la Rosa and De Pablo, 2000). A number of genes and gene products are known to affect PCD. One, the Rb protein has been shown to be essential for determining when neuronal precursors exit the cell cycle, as embryos with targeted mutations in Rb, are characterized at E12.5 by ectopic mitoses and massive cell death, possibly due to their inability to undergo terminal mitosis (Slack et al., 1998). Interestingly, embryos double null for Rb and Id2 survive and have no apparent CNS defects (Lasorella et al., 2000).

Many of the genes involved in the cell death pathway are expressed in the nervous system during development, and when deleted produced surprisingly severe CNS effects (e.g., Gillmore et al., 2000). For example, both caspase 3 (Pompeiano, 2000; Roth, 2000) and caspase 9 (Hakem et al., 1998; Kuida et al., 1998) are expressed in the ventricular zone. Activation of caspase 3 promotes PCD of progenitor cells; blocking its activity prevents cell death. Embryos null for either caspase exhibit regional hyperplasia of cerebellum, striatum, cortex, hippocampus and retina, with both an increase in proliferation and a decrease in cell death. Mutation of the anti-apoptotic gene Bclx or its inhibitor BAX have no effect on VZ cell number, and act instead to control cell death in postmitotic neurons (Motoyama et al., 1995; White et al., 1998; Roth et al., 2000). Clearly choice of quiescence, death, division, or differentiation of early progenitors and precursors will have drastic effects on the size and organization of various regions of the CNS (see Sommer and Rao, 2002). This aspect of stem cell behavior is just beginning to be studied, largely as a way to expand stem cell number.

EXTRINSIC CONTROL OF STEM CELL BEHAVIOR:
GROWTH FACTORS AND SIGNALING MOLECULES

One obvious way to identify candidate molecules that might influence the survival, proliferation, differentiation or quiescence of NSC is to examine the pattern of expression of growth and differentiation factors and their receptors during CNS development and then correlate expression patterns over time with cell behavior. Factors expressed and bound to VZ zone cells would be expected to affect proliferation rather than differentiation, while those associated with deeper layers might promote differentiation. Once the factors are identified, targeted gene deletion can be used to elicidate growth factor/stem cell relationships.

In fact, understanding of survival, growth and differentiation factors has progressed largely by adding candidates to tissue explants containing NSCs. When grown in suspension culture as neurospheres (Reynolds and Weiss, 1992), stem cells proliferate in response particularly to EGF and FGF. When plated on adhesive substrata in medium without mitogens, NSCs differentiate into neurons, oligodendrocyes and astrocytes (Reynolds et al., 1992; Morshead et al., 1994). Many growth factor effects appear to be determined by (1) the stage of differentiation at the time of exposure, (2) culture conditions (including cell-cell contact, cell-substratum interactions, cell density, serum content), and (3) the concentration of the tested factors (Qian et al., 1997; Palmer et al., 1995). Neurosphere culture has been useful in defining the effects of growth factors on NSC behavior (survival, proliferation, differentiation) as well as patterns of gene expression and lineage segregation during differentiation, with the caveat that the starting population is often heterogeneous and only a fraction of the cells in a neurosphere may be stem cells.

Growth Factors

Fibroblast growth factor (FGF)

FGF expression in the nervous system is developmentally regulated (Emoto et al., 1989; Wanaka et al., 1991; Nurcombe et al., 1993; Patstone et al., 1993; Peters et al., 1993). Of the more than 20 identified FGFs, FGF2 is expressed early in development in the VZ, with expression later in the SVZ and remaining into adulthood (Emoto et al., 1989); FGF1 is expressed slightly later (Fu et al., 1991). Since NEPs themselves express FGF1 and FGF2, in high density cultures NEPs do not require supplemental FGF (Kalyani et al., 1999). Of the approximately 10 identified FGF receptors, receptors 1-4 are differentially expressed by neuroepithelial cells. FGFR1-3 are expressed in the VZ (Peters et al., 1993) by early precursors (Reimers et al., 2001), while FGFR4 is uniquely expressed by NEPs (Kalyani et al., 1999).

Consistent with the pattern of receptor expression, cells of the neural tube are responsive to FGF2 as early as E8.5 in the rat embryo (Tropepe et al., 1999), and progenitors proliferate in response to exogenous FGF2 (Qian et al., 1997; Kalyani et al., 1999; Tropepe et al., 1999). FGF2 also stimulates proliferation of later differentiating NSCs and when it is removed from the medium, both neurons and glial cells differentiate (Kilpatrick and Bartlett, 1995; Cicciolini and Svendsen, 1998), suggesting that FGF may keep cells in the cell cycle (Palmer et al., 1995), actively repressing differentiation (Kessler et al., 2002). Recent evidence suggests that FGF2 exposure may decrease the amount of time a stem cell spends in G1, and thereby increase the number of proliferative divisions (Lukaszewicz et al., 2002).

The effects of FGF on NSCs appear to depend on a number of variables including: the stage of differentiation, cell-type, as well as FGF concentration. NEPs, GPs and NPs all require FGF; low levels of FGF appear to promote survival, and higher levels are mitogenic (Kalyani et al., 1999). At low doses, FGF acts as a survival factor for post-mitotic neurons (Walicke and Baird, 1988), while for cortical stem cells, low levels of FGF promote neuronal differentiation and higher levels promote glial differentiation (Qian et al., 1997). Intraventricular infusion of FGF2 has been shown to increase cortical size, as well as neuronal number (Vaccarino et al., 1999). FGF2 exposure may help later stem cells acquire the ability to form neurons (competency), as when cells from regions of the adult brain that normally produce glia are exposed to FGF2, they can produce neurons (Palmer et al., 1999). High levels of FGF2 *in vitro* appear to inhibit oligodendrocyte differentiation of NSCs (McKinnon et al., 1990), although EGF expanded neurospheres preferentially differentiate into oligodendrocytes at the expense of neurons when treated with FGF2 (Reimers et al., 2001). It has also been observed that stem cells near the ventricles require only EGF, while those located farther from the ventricles may need both FGF and EGF (Weiss, 1999). The requirements of neural stem cells for EGF and FGF are clearly stage-dependent. Early NEPs and precursors require FGF, while neurospheres derived later in development require EGF, although FGF may combine with EGF to promote cell division (Vescovi et al., 1993). However, NSCs in neurospheres are a mixed population of cells that express unique combinations of FGFRs, and have slightly different growth factor requirements consistent with their degree of lineage specification.

FGF1 and 2 receptor null mice die at very early stages of development (Yamaguchi et al., 1992; Deng et al., 1994; Ciruna et al., 1997), while FGF2 null mice have reductions in the size of the VZ and in the number of cortical neurons and glial cells (Ortega et al., 1998; Vaccarino et al., 1999; Raballo et al., 2000). Although it is clear that FGF2 can be mitogenic for both VZ and SVZ populations, these data combine to support a role for FGF signaling in maintaining the proliferative capabilities of progenitors early in embryonic development. Later they have a role in the expansion of more committed precursors, with a bias toward neuronal rather than glial differentiation.

Epidermal Growth Factor (EGF)

EGF-responsive cells emerge in the neural tube over E12.5-E13.5, in agreement with the observations that NSCs derived from fetal and adult, but not embryonic spinal cord, require EGF (Reynolds and Weiss, 1992). EGF is expressed in germinal zones late in embryonic development, remaining in this region into adulthood (Kaser et al., 1992; Kornblum et al., 1997). Cells of the primitive VZ do not express the EGF receptor, and NEP cells do not respond to EGF (Burrows et al., 1997; Kornblum et al., 1997). The EGF receptor is expressed later by neural progenitors and oligodendrocytes and by subsets of astrocytes (Kalyani et al., 1999) and is expressed in the SVZ during fetal and postnatal stages of development at a time and position to influence perinatal to adult gliogenesis (Kaser et al., 1992; Burrows et al., 1997; Kornblum et al., 1997). Neural stem cells of the early SVZ do not respond to EGF, but EGF does promote proliferation of SVZ cells at later developmental stages (Kilpatrick and Bartlett, 1993, 1995), and infusion of EGF into the lateral ventricles of adults increased cell division in the SVZ (Craig et al., 1996; Kuhn et al., 1997). Like EGF, TGFα (which activates the EGF receptor) is not expressed until later in development, from E13 to adult (Kornblum et al., 1997).

Overexpression of the EGFRs can induce a glial fate in NSCs (Burrows et al., 1997), but blocking EGFR signaling does not block astrocyte differentiation (Zhu et al., 1999a). It has been shown that exposure of early NSCs to FGF2 induces expression of the EGFR (Ciccolini and Svendsen, 1998), suggesting a method to transition between FGF- and EGF-responsive populations (Martens et al., 2000).

The EGF receptor null mouse is characterized by impaired cortical growth, with delayed migration of neuroblasts to the intermediate zone, and germinal zones are abnormally thickened (Sibilia and Wagner, 1995; Kornblum et al., 1998). There is later cortical atrophy (Miettinen et al., 1995; Threadgill et al., 1995) and abnormal differentiation of astrocytes (Sibilia and Wagner, 1995; Kornblum et al., 1998; Sibilia et al., 1998). Mice null for TGFα similarily show impaired proliferation of progenitors in the SVZ (Tropepe et al., 1997).

Overall, it is clear that multipotent cells that respond to EGF differentiate later than those that respond to FGF2. EGF promotes division and differentiation of astrocytes but does not act to restrict the developmental options of stem cells (Johe et al., 1996). EGF-responsive cells may be a single lineage at different developmental stages, may be composed of related lineages (Tropepe et al., 1999), or may represent a heterogeneous starting population of cells (Bonfanti et al., 2002). As more markers for specific populations of cells become available, it should be possible to better discriminate these options.

Platelet Derived Growth Factor (PDGF)

PDGF-A,B are expressed in the VZ at E11.5 (Hutchins and Jefferson, 1992), although the PDGF alpha receptor has been reported to be present on cell processes at the ventricular surface as early as E8.5 in the mouse embryo (Andrae et al., 2001). PDGFRα is expressed by GPs on E12 (Kalyani et al., 1999) while PDGFRβ is expressed later at E14 in the rat SVZ (Sasahara et al., 1992). Later OPs, but not astrocytes, express PDGFRα and divide in response to PDGF exposure (Kalyani et al., 1999). PDGFRα is expressed by differentiating precursors; however, very low levels of PDGFRβ are detected in uncommitted cells, expression increasing with differentiation (Erlandsson et al., 2001). *In vitro*, PDGF has been shown to stimulate cell division and maintain neuronal precursor proliferation, thus expanding the pool of immature neurons (Erlandsson et al., 2001), rather than playing an instructive role in their differentiation (Johe et al., 1996). Similarly, exposure of human fetal neurospheres to PDGF increased neuronal survival rather than promoting differentiation (Caldwell et al., 2001). Consistent with the pattern of receptor expression, PDGF does not stimulate proliferation or differentiation of NEPs, but both neuronal and glial precursors differentiate in response to PDGF (Pringle et al., 1992; Johe et al., 1996; Valenzuela et al., 1997; Williams et al., 1997; Rao et al., 1998). Interestingly, EGF dependent neurospheres express PDGFRs suggesting that they may be primed to differentiate.

Considerable evidence suggests a role for PDGF in oligodendrocyte differentiation (Ibarrola et al., 1996), and expression of PDGFα is a useful marker for oligodendrocyte precursors (Kalyani et al., 1999; Spassky et al., 2001), as OPs in the ventral neuroepithelium (the region of oligodendrocyte differentiation) *in vivo* are PDGFRα positive (Fu et al., 2002). Targeted inactivation of genes encoding the PDGFs have demonstrated a more severe CNS effect of PDGFA deletion than of PDGFB. There is decreased proliferation of glial precursors, fewer oligodendrocytes, and tremor suggesting defective myelination (Fruttiger et al., 1999).

Dai et al. (2001) targeted ectopic PDGFB expression to astrocytes using the glial fibrillary acidic protein (GFAP) promoter and to progenitor cells using the nestin promoter. Ectopic PDGFB increased astrocyte proliferation and caused them to become precursor-like cells. Misexpression in progenitor cells (via the nestin promoter) produced oligodendrogliomas, suggesting that these tumors may be composed of GPs blocked in their ability to differentiate. Since GFAP$^+$ SVZ astrocytes have NSC properties, it is not really surprising that they can form neural progenitors.

Brain Derived Neurotrophic Factor (BDNF)

BDNF mRNA can be detected at gastrulation (Yao et al., 1994), becoming restricted later in development to populations of neurons and neurogenic placodes (Hallbook et al., 1993). The neurotrophin receptors TrkA (NGF), B (BDNF, NT4), and C (NT3) are expressed early in neural progenitor cells, with some cells expressing multiple receptors (Lachyankar et al., 1997). Addition of BDNF to cultures of neurospheres obtained from E15 mouse brain promotes differentiation of multipolar neurons (Lachyankar et al., 1997). Exogenous BDNF increased neuronal survival and differentiation of hippocampal stem cells *in vitro* (Shetty and Turner, 1998, 1999), while human neural progenitors exposed to BDNF increased secretion of dopamine up to 41% (Riaz et al., 2002).

In vivo, infusion of BDNF into the lateral ventricle on E16 increased mitotic activity in the SVZ and increased the number of neurons in the cortical plate (Fukumitsu et al., 1998). When BDNF was infused later into the lateral ventricle of the adult rat there was increased mitotic activity and neuronal number in the olfactory bulb (Zigova et al., 1998). Using adenoviral infection via the lateral ventricle of adult rats, Benraiss et al. (2001) demonstrated that BDNF increased the number of neurons in both neurogenic zones in regions that normally do not form neurons. NSCs transfected to secrete BDNF exhibited improved survival, while antisense treatment decreased survival of cerebellar neurons and was lethal to hippocampal stem cells (Rubio et al., 1999).

Consistent with these observations, BDNF null mice are viable and have gait anomalies and abnormalities of neural patterning, particularly of granule cell neurons in the cerebellar cortex (Schwartz et al., 1997). BDNF affects neuronal precursors by a mechanism that appears to involve both survival and later differentiation.

Neurotrophin 3 (NT3)

NT3 is expressed in the ventricular zone by E13 in the rat (Fukumitsu et al., 1998), in the embryonic cortex (Maisonpierre et al., 1990), and by proliferating granule cells in neonatal cerebellum (Katoh-Semba et al., 2000). Its high affinity receptor TrkC is expressed in the VZ on E14-16 (Tessarollo et al., 1993; Ghosh and Greenberg, 1995). The catalytic isoform of the TrkC receptor (TrkC

K, kinase-containing) is present both in mitotic and postmitotic cells, while the noncatalytic TrkC NC2 isoform was expressed as neural stem cells exited the cell cycle forming neuronal and glial progenitors, suggesting it may be involved in both neuronal and glial differentiation (Menn et al., 2000).

Somewhat surprisingly, exposure of E15 mouse neurospheres to NT3 produced bipolar neurons as well as astrocytes (Lachvanakar et al., 1997). NT3 has been shown to promote differentiation of NSCs (Lukaszewicz et al., 2002) by increasing the length of G1 and promoting differentiation via decreasing cyclin D2. Intraventricular infusion of NT3, like BDNF, increased the number of neurons in the cortical plate four days later (Fukumitsu et al., 1998), and both NT3 and NT4 increased neuronal survival rather than neuronal commitment in human neurospheres (Caldwell et al., 2001).

The SVZ is small in both the NT3$^{-/-}$ and the TrkC null mouse, with decreases in the number of neurons, OPs and astrocytes, leading to the suggestion that in addition to its widely accepted role in promoting neuronal survival and differentiation, NT3 may also play a previously unsuspected role in glial cell development (Kahn et al., 1999).

Ciliary neurotrophic factor (CNTF), interleukin-6 (IL-6), leukemia inhibitory factor (LIF), and oncostatin M (OSM)

These factors signal through either homodimers of a common receptor subunit gp130, or heterodimers of gp130 and either LIFR, OSMR, or CNTFR. Because LIF maintains the proliferative capability and stem cell properties of embryonic stem (ES) cells (Smith et al., 1988) there has been considerable interest in the ability of cytokines that signal through the gp130 pathway to maintain neural stem cells. LIF mRNA is present in the spinal cord as early as E12 (Murphy et al., 1993), and the LIFR is expressed at the earliest stages of neurogenesis (Murphy et al., 1993; Bartlett et al., 1998), later in the SVZ (Shimazaki et al., 2001). Treatment of E10 neuroepithelial cells with LIF promoted neuronal differentiation, but did not support their survival. LIF also promoted astrocyte differentiation of both E10 progenitors (Richards et al., 1996) and later NSCs (Rajan and McKay, 1998).

Animals null for either LIF (Stewart et al., 1992) or CNTF (Masu et al., 1993) are born without striking CNS abnormalities, although there is mild loss of motor neurons in the adult spinal cord, and crossing of the null animals produces only an accelerated loss of motor neurons (Sendtner et al., 1996). Not surprisingly, receptor null animals (LIFR, gp130, or CNTFR) have the most severe, neonatal lethal, phenotypes. LIFR$^{-/-}$ animals lack astrocytes (Bartlett et al., 1998) and have fewer adult NSCs (due to decreased cell proliferation). The heterozygous null$^{(+/-)}$ mouse has fewer neural stem cells and a reduction in the number of tyrosine hydroxylase-positive neurons in the olfactory bulb (Shimizaki et al., 2001). Analysis of the LIFR$^{-/-}$ animals suggests that signaling through the LIFR promotes astrocyte differentiation and inhibits neuronal differentiation (Bartlett et al., 1998), possibly by its action on early precursors. Later, signaling through the LIFR may be required for long term self-renewal of NSCs (Shimizaki et al., 2001).

Surprisingly similar observations have been made regarding the development of gp130$^{-/-}$ mice. It was expected that gp130 null animals would exhibit a more severe phenotype than individual receptor null animals, since all cytokine signaling through this complex would be abrogated. There is a severe depletion of astrocytes in mature brain of these animals, as well as cell death restricted to selected populations of motor and sensory neurons (Nakashima et al., 1999). Interestingly, these defects develop during late phases of development. Neuronal number is normal on E14.5, but there is significant neuronal loss by E18.5, suggesting that signaling through this receptor is required not for early neuronal differentiation but is necessary to maintain differentiating neurons through the phase of normal programmed cell death.

CNTF promotes the differentiation and survival of astrocytes, oligodendrocytes, and certain types of neurons (Johe et al., 1996). Although CNTF is expressed at low levels during embryonic development, the CNTF receptor is provocatively associated with proliferating regions of the CNS (DeChiara et al., 1995). It is expressed in the VZ of the developing neuroepithelium at E11 (Ip et al., 1993), later in the fetal SVZ, and remains in the adult forebrain SVZ (Shimazaki et al., 2001). Intraventricular infusion of CNTF in the adult promotes proliferation of NSCs and enhances differentiation of astrocyte progenitors, thus altering the commitment of NSCs to APs (Shimiazaki et al., 2001). These authors also suggest that CNTF enhances self-renewal not survival and keeps cells in the mitotic cycle, inhibiting their differentiation to GPs.

Null mutation of the CNTFR has demonstrated a requirement for signaling through this receptor in astrocyte differentiation, although oligodendrocytes were also affected (DeChiara et al., 1995). NSC derived from early embryos (E15) express the CNTFr, and CNTF treatment promotes the differentiation of bipolar neurons as well as astrocytes from these early NSC (Lachyankar et al., 1997). Later in development CNTF promotes the differentiation of both OP (Whittemore et al., 1999), and oligodendrocyte progenitors derived from cerebral cortex to oligodendrocytes (Marmur et al., 1998).

In NSCs derived from human embryonic brain, exposure to LIF or to CNTF supported astrocyte survival, and increased neuronal number two fold, supporting a role of these growth factors in astroglial and neuronal fate (Galli et al., 2000), or simply in NSC proliferation.

Overall, signaling through this receptor complex appears to particularly affect the diferentiation of glial precursors to astrocytes and oligodendrocytes, but the early expression of these receptors and subtle effects on later neuronal differentiation suggest a broader role, or possibly existance of additional ligands.

Insulin like growth factors (IGFs)

Insulin, IGF1 and IGF2 are widely expressed during CNS development, with IGF expressed in the ventricular zone and in the SVZ (Garcia-Seguro et al., 1991; Bondy et al., 1993; Kar et al., 1993; Aberg et al., 2000). Undifferentiated progenitors from the SVZ (Arsenijevic and Weiss, 1998) and the striatum (Arsenijevic et al., 2001) express the IGF1R. IGF1 has been shown to maintain human fetal forebrain neurospheres (Chalmers-Redman et al., 1997), and the absence of IGF1 abrogates the proliferation of striatal stem cells in response to EGF or to FGF2 (Arsenijevic et al., 2001). Virtually no neurons differentiated from NSCs without IGF1, while addition of IGF1 to EGF-expanded neurospheres increased up to 40-fold the number of neurons obtained from these cultures (Arsenijevic and Weiss, 1998). When IGF1 was administered peripherally, there was a significant increase in proliferation of neuronal progenitors in the dentate subgranular zone, with no effect on astrocyte number (Aberg et al., 2000).

IGF1 and IGF2 null mice are characterized by overall growth deficiency (Liu et al., 1993). Like IGFR null animals, there are abnormalities of CNS development including an overall reduction in CNS size and lack of particular striatal neurons (Baker et al., 1993; Liu et al., 1993; Beck et al., 1995). The IGF1 null mouse is characterized by hypomyelination, reduction in the number of both axons and oligodendrocytes, with significant decreases in the size of the dentate gyrus granule cell layer, although the brain is grossly morphologically normal. There is a decrease in the number of both oligodendrocyte progenitors and oligodendrocytes in developing, but not adult, brain (Ye et al., 2002), and the size of white matter tracts is decreased due to a reduction in the number of axons and oligodendrocytes. IGF2 was upregulated in the CNS of these animals and may have protected them from oligodendrocyte depletion. In addition to promoting neuronal differentiation, IGF1 appears to play an important role in later axon growth and in myelination.

Signaling Molecules

Considerable research has identified combinations of genes (members of bHLH families Ngn, NeuroDs, Math, Mash, Olig1,2,3) and signaling molecules that produce regional specialization within the neural tube. Gradients of bone morphogenetic proteins (BMPs) and Sonic hedgehog (SHH) specify dorsal/ventral fates respectively (Jessell, 2000), while signaling molecules such as fibroblast growth factors (Crossley et al., 1996) and retinoic acid (Hogan et al., 1992) determine anterior-posterior characteristics. Many of these signaling molecules are re-utilized during differentiation, so that understanding the specific role of a signaling molecule (like a growth factor) requires the precise knowledge of developmental stage and detailed identification of target cells. Given the large number of signaling molecules that are known to participate in early development, this review will focus on two families, the BMPs and SHH. Factors that influence anterior-posterior identity will be considered briefly in the section below on regional characteristics.

Bone Morphogenetic Protein (BMP)

At induction on E6, the embryonic ectoderm is extremely senssistive to gradients of BMPs. Controlled diffusion of inhibitors (noggin and chordin) from the node produces regions of low BMP activity that differentiate into neural ectoderm (Harland, 2000). The transient signaling center in the anterior visceral endoderm (AVE) produces additional BMP antagonists that induce anterior head structures (Perea-Gomez et al., 2001). Antagonism of BMP signals also play an important role in formation of the secondary neural tube (Goldman et al., 2000).

BMPs (Gross et al., 1996), BMP receptors (DeWulf et al., 1995; Soderstrom et al., 1996; Zhang et al., 1998), and inhibitors such as noggin (Valenzuela et al., 1995) are expressed throughout development, remaining in the adult in discrete CNS regions (Mabie et al., 1999). Their receptors are also expressed early; BMPR-1a is expressed throughout the neuroepithelium, followed at E9 by expression of BMPR-1b in the dorsal neural tube (Panchision et al., 2001). In one of the best examples of BMP/inhibitor interaction, in the late SVZ, C (amplifying cells) and B cells (astrocytes) express BMP2,4 while noggin is produced by ependymal cells (Lim et al., 2000) where it may play an essential role in influencing NSCs (that have an innate bias toward astrocytes) to differentiate into neurons. When noggin producing ependymal cells were killed by neuraminidase treatment, there was compensatory proliferation and differentiation of SVZ astrocytes to form a glial scar (Grondona et al., 1996). BMP misexpression in the ependyma decreased proliferation of SVZ progenitors and abolished neuroblast differentiation from the SVZ, while noggin expression in the ependyma increased neuronal differentiation. Grafts of SVZ cells to the striatum did not produce neurons unless adenoviral noggin was delivered, resulting in extensive neuronal differentiation of the graft (Lim et al., 2000). Thus, like other signaling centers in the nervous system, the interaction between BMP produced by B/C cells and noggin from ependyma may precisely control the differentiation of these cells and may explain the bias of the SVZ toward astrocyte production.

In the neural tube, TGFβ superfamily members, particularly BMPs are expressed at high levels in dorsal regions and at low levels in ventral zones (Liem et al., 1997). SHH is expressed in the ventral region of the neural tube where it antagonizes BMPs, partially by promoting expression of the BMP antagonist noggin (Hirsinger et al., 1997), thereby controlling dorsal/ventral patterning of the neural tube (Roelink, 1996; Goodrich and Scott, 1998). As proliferative cells of the neuroepithelium gradually increase their expression of EGFR, BMP4 antagonizes this differentiation, inhibiting EGFR expression and responsiveness, and FGF2 also antagonizes the effects of BMP4 (Lillien and Raphael, 2000). Depending on developmental stage, BMPs promote differentiation by inhibiting neuroepithelial cell proliferation, cells exit the cell cycle and begin to express lineage-specific genes. Ligand binding to the BMPR-1a dorsalizes cells and determines early dorsal patterning in the neural tube, while increasing proliferation of neural precursors both *in vivo* and *in vitro*. Consistent with its later expression, BMPR-1b promotes apoptosis early and promotes differentiation of precursors slightly later in gestation (Panchision et al., 2001).

There is strong stage and regional specificity in these effects. Cells from late embryonic SVZ exposed to BMPs produce astrocytes (Gross et al., 1996), while BMP treatment of neurospheres derived from earlier stages produce oligodendrocytes (Rogister et al., 1999) and neurons (Li et al., 1998; Zhu et al., 1999b). BMP treatment of OP can produce astrocytes (Grinspan et al., 2000), while high levels of BMP inhibit oligodendrocyte differentiation (Mekki-Dauriac et al., 2002), and promote differentiation of cerebellar granule neurons in the BMP-rich region near the rhombic lip (Alder et al., 1999).

Gene targeting has demonstrated that double knockout of the BMP inhibitors noggin and chordin produce embryos with defects of anterior and posterior neuraxis development (Bachiller et al., 2000). Consistent with its early, widespread expression, targeted deletion of BMPR-1a is prenatal lethal at gastrulation (Mishina et al., 1995), unlike gene-targeted BMPR-1b embryos that do not appear to have a CNS phenotype (Yi et al., 2000). Overexpression of BMPR1a under the control of the nestin enhancer (to target expression to progenitors) resulted in excessive proliferation and induction of a dorsal phenotype in the neuroepithelium, while overexpression of BMPR-1b induced apoptosis in early gestation and later accelerated the terminal differentiation of precursors by promoting cell cycle arrest (Panchision et al., 2001). Overexpression of noggin and to a lesser extent chordin produced rapid differentiation of neurons from totipotent ES cells (Gratsch and O'Shea, 2002).

Sonic Hedgehog (SHH)

BMPs and SHH are expressed at opposite (dorsal/ventral respectively) regions of the early neural tube, and consistent with this non-overlapping distribution they have opposite effects on the proliferation and differentiation of NSCs (Goodrich and Scott, 1998; Dutton et al., 1999; Zhu et al., 1999b). In early development, SHH promotes neuronal (Goodrich and Scott, 1998) and oligodendrocyte differentiation (Orentas and Miller, 1996; Pringle et al., 1996; Zhu et al., 1999b; Mekki-Dauriac et al., 2002). The SHH null embryo dies at E14, so little has been determined about its role in regional cell fate specifiction, although SHH null embryos have anterior neural induction defects, and holoprosencephaly (Chiang et al., 1996). Injection of SHH into the adult rat striatum increases signaling in the SVZ suggesting that as in the embryo, this pathway may be active in the adult CNS (Charytoniuk et al., 2002).

The diversity of cell types required in the adult nervous system may ultimately be controlled by gradients of signaling molecules and growth factors. It appears that intersections of growth factor/signaling molecules may produce many types of neurons and glial cells e.g., intersections of SHH and FGFs have been shown to promote dopaminergic and serotonergic neuronal differentiation (Ye et al., 1998). Interpretation of the effects of growth factors and signaling molecules on NSCs therefore requires detailed information on developmental stage and region of isolation from the neuraxis.

REGIONALIZATION AND THE STEM CELL NICHE

It is somewhat ironic that the concept of the stem cell niche as a region enriched in growth factors, extracellular matrix (ECM) and cytokines that creates an incubator for stem cells, originated in the hematopoietic field (Schofield, 1978), where the niche within the marrow of long bones is perhaps the most difficult to study. In the case of the adult CNS, it has been accepted for some time that once cell proliferation, migration and synaptic connections are established, the adult brain becomes largely growth factor and ECM depleted (Scheffler et al., 1999). That adult cells retain the ability to express many of the components of the rich embryonic environment is clear from injury studies (Brodkey et al., 1995), where re-expression of embryonic ECM and signaling molecules can promote untargeted process outgrowth and cell division characteristic of gliosis (McKeon et al., 1999). As well as maintaining the niche, cell surface carbohydrates are critical to successful integration of stem cells following transplantation. When explants of neural stem cells have been enzymatically disaggregated without a recovery period to allow reexpression of cell surface constituents, transplantation has been less successful (Olsson et al., 1998).

Regions of the adult nervous system that contain neural stem cells such as the hippocampus and adult SVZ continue to express astrocyte-associated (Gates et al., 1995) extracellular matrix molecules including tenascin-c, thrombospondin and chondroitin sulfate proteoglycan (Miragall et al., 1990; Jankovski and Sotelo, 1996; Thomas et al., 1996). These ECM molecules bind and present growth factors to create gradients of signaling factors and cytokines that control cell migration, division and differentiation. Growth factors and extracellular matrix molecules modulate differentiation of embryonic stem cells (Czyz and Wobus, 2001), and growth factors control adhesion of neural stem cells to their extracellular matrix (Kinashi and Springer, 1994). Others, such as tenascin-c appear to play a role in delimiting boundary fields during development (Faissner and Steindler, 1995). The perseverance of embryonic ECM molecules and continued cell division in regions containing stem cells has lead to the suggestion that stem cell zones are persistently immature (Whittemore et al., 1999). While it is tempting to attribute "control" of stem cell behavior to these molecules, many are rapidly turned over (produced by the cells themselves), and others mark, rather than determine, stem cell fate. Since there appears to be considerable heterogeneity in neural stem cell populations, with rapidly dividing populations as well as quiescent NSCs and EGF-responsive and FGF-responsive populations, different microenvironmental niches may exist side by side within stem cell zones. Supporting cells, whether hematopoietic stroma or astrocytes in the CNS, play an important role in creating the stem cell microenvironment, as they are the source of many ECM molecules and growth factors, and form partitions or boundaries throughout the nervous system: barrel-field boundaries, roof plate, and astrocyte processes that cordon off damaged areas.

In the search for factors common to the stem cell niche of multiple organs, it has often been assumed that similar ECM and signaling molecules will be present (Scheffler et al., 1999), that stem cells should exhibit a "primitive morphology", and that there should be similar gene expression profiles of stem cells throughout the body (Ivanova et al., 2002; Parati et al., 2002). One signaling system common to several stem cell niches, is BMP-noggin antagonism. In drosophila, the BMP homologue dpp determines the differentiation pattern of male and female germ cells, high levels of dpp favor proliferation and can lead to tumor formation, while low levels of dpp cause differentiation (Xie and Spradling, 1998). A similar niche for neural stem cells appears to exist late in CNS development with noggin from ependymal cells antagonizing BMP2,4 from astrocytes to control precursor fate in the SVZ (Lim et al., 2000). Consistent with this, expression of noggin in other stem cell zones such as the bone marrow stroma induces neuronal differentiation from hematopoietic stem cells (Gratsch et al., 2001).

In addition to producing/maintaining an enriched environment, ECM molecules also play an important role in promoting and channeling neural stem cell migration. In the telencephalon, polysialylated NCAM supports the migration of neuroblasts in the rostral migratory stream (RMS) to the olfactory bulb, and promotes neuronal differentiation of NSC *in vitro* (Amoureaux et al., 2000). The receptor tyrosine kinases of the Eph family and their transmembrane associated ephrin ligands have also been implicated in chain migration and proliferation of stem cells in the adult SVZ (Conover et al., 2000). Following intraventricular infusion of the ectodomain of the EphB2 receptor or ephrin-B2 ligand, cell proliferation was increased and cell migration to the olfactory bulb was interrupted. Slit proteins also appear to be involved in restricting neuronal migration to the RMS (Hu, 1999). Although ECM is often considered to be permissive rather than instructive in cell behavior, gene targeting experiments have often suggested a more active role. For example, targeted deletion of tenascin-c specifically inhibits proliferation and migration of neural stem cells, resulting in depletion of OPs (Garcion et al., 2001).

Ultimately, the combination of molecular markers of stem cells, targeted deletion of putative stem cell genes and growth and extracellular matrix factors that maintain stem cell identity will determine their precise roles in stem cell behavior.

Regional differentiation along the neural tube occurs with induction, when the secondary signaling center, the anterior visceral endoderm (AVE), secretes unique BMP inhibitors and forebrain inducers including cerberus, Dkk1 and Otx2. Slightly later, the entire neuraxis is divided into a metameric pattern beginning with the prosomer (prosencephalon; Rubenstein et al., 1994), with additional regional characteristics determined by the Hox gene code (Lumsden and Krumlauf, 1996). Fibroblast growth factors and retinoic acid have a strong posteriorizing effect on the nervous system, although it is not yet known if they can similarly affect regional gene expression patterns of NSCs. It appears that stem cells develop regional characteristics relatively early in development, and are able to maintain these regional markers through multiple rounds of division *in vitro* (Robel et al., 1995; Nakagawa et al., 1996; Zappone et al., 2000; Hitoshi et al., 2002; Ostenfeld et al., 2002). For example, expression of Sox2 by telencephalic NSCs is maintained over many divisions *in vitro* (Zappone et al., 2000). Consistent with their *in vivo* behavior, forebrain NSCs have the ability to produce more neurons *in vitro* than stem cells from hindbrain (Hitoshi et al., 2002; Ostenfeld et al., 2002). NSC derived prior to E10.5 are capable of both heterotypic and heterochronic integration, but that ability is largely lost after E10.5 (Olsson et al., 1997; Brustle et al., 1999), similar to the observations of Zigova et al. (1996) that heterochronically transplanted VZ cells often fail to exhibit normal migration in the absence of radial glial cells later in development. Thus regional patterning appears to be established quite early. To develop appropriate cells for transplantation it may be possible to simply maintain cells for extended periods *in vitro* with the hope that they may undergo some nuclear reprogramming with cell division; alternatively, region appropriate cells will need to be developed.

The ultimate test of regional identity is to transplant stem cells into heterotypic or heterochronic environments to determine if they differentiate into region-specific cells, or retain their original identity (memory). Perhaps the most extreme examples of plasticity following heterotypic transplantation, are the ability of NSCs to form blood cells (Bjornson et al., 1999) and embryonic tissues when aggregated with pre-implantation stage embryos (Clarke et al., 2000).

Transplanted neural stem cells can integrate and differentiate *in vivo*, but in many cases they retain their original pattern of regional gene expression (Na et al., 1998), are restricted in their neurotransmitter phenotype (Nishino et al., 2000), or may not be able to respond to extracellular cues (Sheen et al., 1999; Prestoz et al., 2001). Integration may be more complete when NSCs are grafted to neurogenic compared with non-neurogenic zones, as SVZ stem cells transplanted to SVZ produced new neurons (Herrera et al., 1999), but when transplanted to the cortex they differentiated into astrocytes (Doetsch and Alvarez-Buylla, 1996). Transplated hippocampal stem cells integrate within the rostral migratory stream and form interneurons of the olfactory bulb, although hippocampal NSC, but not stem cells derived from cerebellum or midbrain, form hippocampal pyramidal neurons when transplanted to the neurogenic zone of the hippocampus (Shetty and Turner, 1998,1999). These studies suggest that cell fate may be controlled by the local microenvironment of the cell, and that the regional microenvironment may be more directive than previously suspected. While there appears to be considerable plasticity in fate restriction of neural stem cells, stem cells isolated early in development are better able to differentiate into region appropriate phenotypes specified by their new environment, while cells derived later in differentiation integrate less well, even when placed in an embryonic environment (Sheen et al., 1999). The genetic events that specify this restriction remain to be determined.

In addition to its ability to affect regional patterning, the teratogen/morphogen retinoic acid has been shown to play a role in CNS differentiation (Balkan et al., 1992; LaMantia et al., 1993). It has been suggested that more "plastic" (i.e., stem cell containing) CNS zones might be retinoic acid sensitive. Cells present in the SVZ are in fact sensitive to retinoic acid, and scattered cells throughout the CNS express retinoic acid-responsive elements, suggesting that they too may exhibit unexpected plasticity (Thomson Haskell et al., 2002).

If it is only the environment that represses stem cell fate, then removing cells from inhibition should be sufficient for them to reexpress stem cell properties. That appears to be the case during early development; however sequential rounds of growth factor, ECM, and cytokine exposure limit stem cell plasticity, except in very few regions of the cortex. How the stem cell microenvironment both protects stem cells from exposure to growth and differentiation factors like the primordial germ cell niche, and at the same time produces the unique niche remains to be determined. Understanding the role of the unique regional environment may eventually allow expansion of NSCs for therapeutic uses.

TOOLS

Transgenic technologies are increasingly being employed to develop tools to study NSC behavior. Initially a constitutively active promoter such as beta-actin was used to express a marker, usually beta-galactosidase, throughout the embryo/adult, so that marked cells could be identified after transplantation to a non-transgenic recipient (e.g., Mujtaba et al., 2002). Transgenic mice have also been developed in which a cell type-restricted promoter is used with the temperature sensitive SV40 large T antigen (Kilty et al., 1999), to develop NSC-like cell lines. This "knock in" approach has been used extensively to monitor promoter activity, and has enabled the study of expression of putative lineage restricted genes during development. Markers such as the fluorophores eGFP, eYFP, DsRed, sometimes in combination with an antibiotic resistance gene such as neomycin phosphotransferase, have also been driven by CNS restricted promoters. Incorporation of fluorescent markers and antibiotic resistance cassettes allows cells to be dissociated and enriched using flow cytometry, and/or growth in high levels of antibiotics to select specific populations of cells. As more is known about lineage-restricted genes, this approach will allow the "purification" of NSCs from populations that emerge at sequential stages of development or that have unique regional characteristics or neurotransmitter phenotypes. Once sorted, cells can be studied for their growth factor responsiveness, their pattern of gene expression, and be monitored for their ability to integrate in the CNS at various axial levels, or in region-specific or stage-specific manners. This approach would also allow disease genes to be targeted to specific populations of cells, e.g., HIV1 to astrocytes (Goudreau et al., 1996), or to replace disease genes (Readhead et al., 1987).

Genes restricted to specific neural cell types have been used to target markers to specific cells of the developing CNS. Using the murine 2'3'-cyclic nucleotide 3'-phosphodiesterase (CNP) promoter to drive expression of β-geo it was possible to monitor oligodendrocyte differentiation (Chandross et al., 1999; Belachew et al., 2001), and to select OPs in the presence of antibiotic (Chandross et al., 1999) or by flow sorting for eGFP expression (Belachew et al., 2001).

The myelin basic protein (MBP) promoter has also been used to target gene expression to oligodendrocytes, but because it is expressed late in differentiation, it is not as useful to target stem cells or oligodendrocyte progenitors. The MPB promoter has been used to create models of dysmyelination by expressing MHCI genes in oligodendrocytes (Turnley et al., 1991), to express growth factors (Ma et al., 1995), cytokines (Taupin et al., 1997), transcription factors (Jensen et al., 1998a), and oncogenes in these cells (Jensen et al., 1998b).

Promoters from "astrocyte-specific" genes such as the glial fibrillary acidic protein have also been used to generate astrocyte restricted genetic modifications. However, the glial fibrillary acidic protein promoter, thought to be restricted to astrocytes, directs surprisingly widespread expression of cre recombinase to many neurons, ependyma and glial cells (Zhuo et al., 2001), possibly because the GFAP$^+$ SVZ stem cell can produce these cell types. The murine CMV immediate early gene has also been employed to direct astrocyte-specific gene expression in adult transgenic mice (Li et al., 2001).

A number of investigators have used the second intronic enhancer of the nestin gene to drive gene expression to neural stem-like cells (Lendahl, 1997; Yamaguchi et al., 2000; Kawaguchi et al., 2001; Keyoung et al., 2001). Nestin is broadly expressed in progenitor cells within the nervous system and is upregulated following injury, then expressed in astrocytes (Lendahl et al., 1990), although it marks later precursors as well as progenitor cells (Cai et al., 2002). Nestin-eGFP selected cells differentiated into multiple cell types consistent with a NSC phenotype, including dopaminergic neurons, that successfully replaced damaged cells in a Parkinson's Disease model (Sawamoto et al., 2001a). These authors have also used the Tα1 α-tubulin promoter to express eYFP in neuronal progenitor cells, and these mice were then crossed with the nestin-eGFP mice to create animals in which progenitors express both eGFP and eYPF (Sawamoto et al., 2001b)! Using an adenovirus, human fetal neural stem cells have been transduced to express the pan-neuroepithelial marker Musashi1 driving eGFP or the nestin enhancer to express eGFP. Sorted cells were multipotent, and could be passaged, expanded, and grafted successfully to E17 and P2 rat forebrain (Keyoung et al., 2001).

As regional patterning is better understood, and better molecular markers that uniquely identify NSCs are developed (e.g., Cai et al., 2002), it will be possible to develop highly specific populations of stem cells using selection techniques. Identification of genes such as Sox2, that are expressed in the blastocyst, and later identify neural stem cells during embryonic development and in the adult nervous system (Rao and Pevny, 2003, Chapter 4 in this volume), will significantly expedite this process. The mouse Dach1 gene is highly expressed in neurons of the cortical plate, in the hippocampus, olfactory bulb, and SVZ, and a forebrain enhancer D6 is active in adult neural stem cells (Machon et al., 2002). Other transcriptional regulators have been examined for their ability to target marker genes to developing stem cell populations. BF1 is expressed in telencephalon, otic vesicle, mid-hindbrain junction (Hebert and McConnell, 2000), and Emx1 initiates expression at E9.5 in the telencephalon and CA1-3 and dentate gyrus of the hippocampus. (Guo et al., 2000). These and other transcription factors (Josephson et al., 1998); neural RNA binding proteins such as Musashi1 (Sakakibara and Okano, 1997), regionally expressed genes such as Emx2 (Galli et al., 2002), and possibly cell surface markers (CD133, Uchida et al., 2000) should make it possible to obtain region-specified, stage-specified NSCs. Once identified, these cells will provide stem cell populations for expansion and transplantation, studies of growth factor responsiveness, and gene expression patterns.

STEM CELLS AS MODELS OF DEVELOPMENT AND DISEASE

Stem cells have tremendous potential to replace cells damaged by disease or trauma (Lim et al., 2002), to replace missing enzymes e.g., β-glucuronidase replacement for MPSIII (Buchet et al., 2002), for growth factor delivery (e.g., NGF or BDNF; Martinez-Serrano and Bjorklund, 1996). As cellular vectors, they have the considerable advantage over fibroblasts (Fisher and Ray, 1994), or even astrocytes (e.g., Fitoussi et al., 1998), that they can integrate into host CNS and differentiate in a region- and cell-type appropriate manner, while providing growth factors or specific missing cells to the region (e.g., Lacorazza et al., 1996; Flax et al., 1998; Brustle et al., 1999). Neural stem cells have also been immortalized using oncogenes to create "neural stem cell lines" for these studies (e.g., Auerbach et al., 2000; Tate et al., 2002).

Stem cells, both neural stem cells and the earlier embryonic stem cells, also provide an opportunity to study the effects of genes, growth factors, and signaling molecules on lineage segregation from an uncommitted progenitor. They also form an important source of models of CNS disease. Thus, normal stem cells can be exposed to growth and differentiation factors (above), stem cells can be derived from genetically modified animals, or NSCs can be transfected to express mutant genes and their proteins *in vitro*.

Three basic approaches have been employed. In the first, the gene of interest is expressed in transgenic animals either constitutively, in a particular cell type or in the stem cell population, using promoters described above (in TOOLS), then NSCs are derived in neurospheres. This approach has the advantage of being relatively simple, and allows the effects of a particular gene to be studied both *in vivo* and when cells are explanted in tissue culture. Further, the time and pattern of expression of the promoter is known and potentially marked. Alternatively, NSCs are derived from normal growth factor expanded neurospheres, then cells are transfected *in vitro* to express markers, disease genes, or signaling molecules. Both approaches can produce targeted cell lines, but since the constructs will integrate randomly (and several copies may integrate), expression will be influenced by unknown enhancers and repressors, although insulator sequences can be included in the construct to lessen this problem. Transfection of cultured cells often results in variable expression levels, transfection is transient, and efficiency is often low (10-20% of cells). These models provide the opportunity to study the cellular processes involved in differentiation or in the degenerative phenotype, and are particularly useful in producing a "gain of function" phenotype. This approach may fail or produce erroneous results if the gene is not expressed normally in neural stem cells, or if the *in vivo* effect is due to another cell population and the effect on the NSC is secondary. The lack of complexity provided by the cell culture system is therefore both helpful and too simplistic (Sipione and Cattaneo, 2002).

The third approach is to derive neural stem cell lines from transgenic animals that have been engineered by homologous recombination to express a genetic mutation or a selectable marker. The advantage is that the targeting construct is stably integrated, so that endogenous enhancer and repressor elements are in their native configuration. In addition, the ability to remove the foreign DNA using Cre-Lox technology reduces the concern of genetic instability.

Thus, neural stem cells carrying a particular genetic mutation, or with targeted overexpression of a signaling molecule or growth factor, can be developed, expanded and stored for future studies of genetic or pharmacological intervention. These cells could also represent a long term source of transgenic protein.

Development and disease models

The numerous studies described above regarding growth factor and signaling molecule mediated differentiation of neural stem cells attest to the value of using NSC as a model of lineage segregation during differentiation; they may also be useful models of disease. Neurodegenerative conditions may result from an early disruption of neural development that over time develops into full blown disease or predisposes cells to future injury (Gokhan and Mehler, 2001; Mehler and Gokhan, 2001). Since both NSC and ES cells normally express many genes associated with neurological diseases during their differentiation, they may be a good model of early degenerative events. For example, during their differentiation, NSCs express both huntingtin, and presenilin 1, while ES cells express α-synuclein, ataxin and frataxin (Gokhan and Mehler, 2001). Gene targeting has revealed a requirement for many of these genes surprisingly early in development. For example, embryos in which the huntingtin gene has been deleted die at E7.5 (Nasir et al., 1995), deletion of the presenilin gene is also prenatal lethal (Shen et al., 1997). Subtle alterations in genes essential for induction, neural patterning, or regional specification may therefore produce a genetic predisposition to injury and disease. Thus, study of gene targeted neural stem cells or embryonic stem cells plus neurotoxic compounds might sugest novel pathways to disease, early diagnosis and intervention strategies, and produce a starting cell population to test pharmacological interventions.

Despite the potential of this approach, relatively few studies have used neural stem cells as model systems to study mechanisms involved in normal development or in disease. Neural stem cell differentiation has largely been carried out by exposing neurospheres to soluble factors. Targeted misexpression of developmental control genes is another way of interfering with local gene expression, and has been widely employed in submammalian species. In the case of noggin misexpression, BMP4 signaling can be abrogated in a precise niche (Kulessa et al., 2000). In another recent study, the nestin promoter was employed to express the reelin gene in wild type and reelin mutant brain (Magdaleno et al., 2002). Contrary to some predictions, overexpression of reelin in the normal CNS did not induce patterning defects, while expression in the mutant background reversed some, but not all, of the cell migration defects characteristic of the reelin mutants.

Neurodegenerative conditions

Study of neurodegenerative diseases using stem cell models is particularly appropriate when the disease mutation initiates a cell-intrinsic degenerative process. This is the case in Huntington's disease (HD), where the HD mutation is an expanded CAG repeat that increases a polyglutamine segment in the huntingtin protein. To study this process, neural stem cells can either be obtained from mutant (gene targeted) mice or normal stem cells can be transfected to express the mutant protein to better understand the cascade of events that produce cellular injury and to examine potential pharmacological interventions.

NSCs derived from the R6/2 transgenic mouse which has more than 150 CAG repeats knocked into exon one of the human huntingtin gene, were passaged, frozen, and reconstituted into "cell lines" that when differentiated as neurospheres expressed the expanded polyglutamine tract. When these cells were differentiated, mainly astrocytes and a few neurons formed (Chu-LaGraff et al., 2001). The authors attribute this to the requirement of EGF to expand neurospheres, biasing differentiation. This could be an effect of the mutation itself and suggests that other growth factors might be employed in the expansion phase. In another study of HD, striatal cell lines were obtained from the

HdhQ111 mouse that expresses expanded CAG repeats under the control of the temperature-sensitive SV40 large T antigen. In these studies, there was no preferential differentiation to a neuronal or glial phenotype, although cells expressing mutant Htt were characterized by longer cell cycle times, and toxicity that may involve specific stress pathways (Trettel et al., 2000). Conditionally immortalized striatal cells were also transfected with either wild type Htt or constructs to express the expanded CAG repeats to study normal and mutant protein function in these cells. Mutant Htt induced apoptosis in striatal neurons by a caspase-dependent mechanism, while wild type Htt inhibited caspase 3 activation (Rigamonti et al., 2000). In an embryonic stem cell based model, the neuronal differentiation capacity of ES cells containing expanded polyglutamine tracts was significantly impaired (Lorincz et al., 2001). Although mice null for huntingtin die at E7.5, null embyonic stem cells appear to differentiate normally (Metzler et al., 1999).

Tumors

Neural stem cells may be the source of transformed cells, particularly primitive tumors such as glioma or neuroblastoma (Wernig and Brustle, 2002). NSC have been shown to be the target of chemical mutagens that produce gliomas in animal models (Kleihues et al., 1979), and the presence of multiple lineages, e.g., neuronal cells in gliomas, would also be explained if the cell of origin was a multipotent progenitor (Igatova et al., 2002). Many CNS tumors express genes typical of primitive stem cells including: BMP receptors (Yamada et al., 1996), bHLH genes including Olig1,2 (Hoang-Xuan et al., 2002), Notch receptors Jagged or Delta (Igatova et al., 2002), and Id genes (Hasskarl and Munger, 2002). Mutation of cell cycle control genes, and those involved in programmed cell death also produce proliferation phenotypes (above). Tumor stem cells may be genetically defective in their ability to respond to growth and differentiation factors (Ignatova et al., 2002) or common signaling pathways may be dysregulated (Bachoo et al., 2002), producing a population of stem cells that is unable to undergo terminal differentiation.

When nestin was used to drive overexpression of a constitutively active EGFR to progenitors, glial tumors were induced more efficiently than when expressed in astrocytes by the GFAP promoter (Holland et al., 1998), possibly because nestin expressing cells were less mature and therefore more permissive to tumorigenesis (Bachoo et al., 2002). However inactivation of the tumor suppressor gene Ink4a/ARF with EGFR activation in either neural stem cells or astrocytes induced high grade gliomas, astrocyte "dedifferentiation", with some neuronal elements present in the tumors. These studies suggest that both astrocytes and NSCs may serve as the cell of origin for these highly lethal tumors.

In addition to providing models to examine the behavior of transformed cells, NSCs have been used to deliver chemotherapeutic agents to tumors. It has been possible to overexpress IL4 (Aboody et al., 2000) or cytosine deaminase (Benedetti et al., 2000) in neural stem cells which were then injected and showed surprising trophism to tumors, reducing tumor size.

Understanding the cell biology and molecular biological characteristics of neural stem cells offers much for developmental biologists, as well as those who would use this population of cells in proteomic and drug testing experiments, or to target cells, growth factors, or genes to the nervous system and possibly elsewhere. As we learn more about their unique niche, and restricted patterns of gene expression, it will be possible to develop improved tools for more precise monitoring and engineering of these amazing cells.

Acknowledgements

Supported in part by NS-39438 and assistance from Dr. TE Gratsch and Ms. A Dohring.

REFERENCES

Aberg MA, Aberg ND, Hedbacker H, Oscarsson J, Eriksson PS (2000) Peripheral infusion of IGF-I selectively induces neurogenesis in the adult rat hippocampus. J Neurosci 20: 2896-2903.

Abney ER, Bartlett PP, Raff MC (1981) Astrocytes, ependymal cells, and oligodendrocytes develop on schedule in dissociated cell cultures of embryonic rat brain. Dev Biol 83: 301-310.

Aboody KS, Brown A, Rainov NG (2000) Neural stem cells display extensive tropism for pathology in adult brain: evidence from intracranial gliomas. Proc Natl Acad Sci USA 97: 12846-12851.

Alder J, Lee KJ, Jessell TM, Hatten ME (1999) Generation of cerebellar granule neurons *in vivo* by transplantation of BMP-treated neural progenitor cells. Nat Neurosci 2: 535-540.

Altman J (1966) Proliferation and migration of undifferentiated precursor cells in the rat during postnatal gliogenesis. Exp Neurol 16: 263-278.

Altman J, Bayer SA (1985) Embryonic development of the rat cerebellum. I. Delineation of the cerebellar primordium and early cell movements. J Comp Neurol 231: 1-26.

Altman J, Das GD (1965) Autoradiographic and histological evidence of postnatal hippocampal neurogenesis in rats. J Comp Neurol 124: 319-335.

Alvarez-Buylla A, Garcia-Verdugo JM, Tramontin AD (2001) A unified hypothesis on the lineage of neural stem cells. Nat Rev Neurosci 2: 287-293.

Amoureux MC, Cunningham BA, Edelman GM, Crossin KL (2000) N-CAM binding inhibits the proliferation of hippocampal progenitor cells and promotes their differentiation to a neuronal phenotype. J Neurosci 20: 3631-3640.

Andrae J, Hansson I, Afink GB, Nister M (2001) Platelet-derived growth factor receptor-alpha in ventricular zone cells and in developing neurons. Mol Cell Neurosci 17: 1001-1013.

Armstrong RC, Dorn HH, Kufta CV, Friedman E, Dubois-Dalcq ME (1992) Pre-oligodendro-cytes from adult human CNS. J Neurosci 12: 1538-1547.

Arsenijevic Y, Weiss S (1998) Insulin-like growth factor-I is a differentiation factor for postmitotic CNS stem cell-derived neuronal precursors: distinct actions from those of brain-derived neurotrophic factor. J Neurosci 18: 2118-2128.

Arsenijevic Y, Weiss S, Schneider B, Aebischer P (2001) Insulin-like growth factor-I is necessary for neural stem cell proliferation and demonstrates distinct actions of epidermal growth factor and fibroblast growth factor-2. J Neurosci 21: 7194-7202.

Auerbach JM, Eiden MV, McKay RD (2000) Transplanted CNS stem cells form functional synapses *in vivo*. Eur J Neurosci 12: 1696-1704.

Austin CP, Feldman DE, Ida JA Jr, Cepko CL (1995) Vertebrate retinal ganglion cells are selected from competent progenitors by the action of Notch. Develop 121: 3637-3650.

Bachiller D, Klingensmith J, Kemp C, Belo JA, Anderson RM, May SR, McMahon JA, McMahon AP, Harland RM, Rossant J, De Robertis EM (2000) The organizer factors Chordin and Noggin are required for mouse forebrain development. Nature 403: 658-661.

Bachoo RM, Maher EA, Ligon KL, Sharpless NE, Chan SS, You MJ, Tang Y, DeFrances J, Stover E, Weissleder R, Rowitch DH, Louis DN, DePinho RA (2002) Epidermal growth factor receptor and Ink4a/Arf: convergent mechanisms governing terminal differentiation and transformation along the neural stem cell to astrocyte axis. Cancer Cell 1: 269-277.

Baker J, Liu JP, Robertson EJ, Efstratiadis A (1993) Role of insulin-like growth factors in embryonic and postnatal growth. Cell 75: 73-82.

Balkan W, Colbert M, Bock C, Linney E (1992) Transgenic indicator mice for studying activated retinoic acid receptors during development. Proc Natl Acad Sci USA 89: 3347-3351.

Bartlett PF, Brooker GJ, Faux CH, Dutton R, Murphy M, Turnley A, Kilpatrick TJ (1998) Regulation of neural stem cell differentiation in the forebrain. Immunol Cell Biol 76: 414-418.

Bayer SA, Yackel JW, Puri PS (1982) Neurons in the rat dentate gyrus granular layer substantially increase during juvenile and adult life. Science 216: 890-892.

Beck KD, Powell-Braxton L, Widmer HR, Valverde J, Hefti F (1995) Igf1 gene disruption results in reduced brain size, CNS hypomyelination, and loss of hippocampal granule and striatal parvalbumin-containing neurons. Neuron 14: 717-730.

Beger C, Pierce LN, Kruger M, Marcusson EG, Robbins JM, Welsch P, Welch PJ, Welte K, King MC, Barber JR, Wong-Staal F (2001) Identification of Id4 as a regulator of BRCA1 expression by using a ribozyme-library-based inverse genomics approach. Proc Natl Acad Sci USA 98: 130-135.

Belachew S, Yuan X, Gallo V (2001) Unraveling oligodendrocyte origin and function by cell-specific transgenesis. Dev Neurosci 23: 287-298.

Benedetti S, Pirola B, Pollo B, Magrassi L, Bruzzone MG, Rigamonti D, Galli R, Selleri S, DiMeco F, DeFraja C, Vescovi A, Cattaneo E, Finocchiaro G (2000) Gene therapy of experimental brain tumors using neural progenitor cells. Nat Med 6: 447-450.

Benezra R, Davis RL, Lassar A, Tapscott S, Thayer M, Lockshon D, Weintraub H (1990) Id: a negative regulator of helix-loop-helix DNA binding proteins. Control of terminal myogenic differentiation. Ann NY Acad Sci 599: 1-11.

Benraiss A, Chmielnicki E, Lerner K, Roh D, Goldman SA (2001) Adenoviral brain-derived neurotrophic factor induces both neostriatal and olfactory neuronal recruitment from endogenous progenitor cells in the adult forebrain. J Neurosci 21: 6718-6731.

Bjornson CR, Rietze RL, Reynolds BA, Magli MC, Vescovi AL (1999) Turning brain into blood: a hematopoietic fate adopted by adult neural stem cells *in vivo*. Science 283: 534-537.

Blaschke AJ, Staley K, Chun J (1996) Widespread programmed cell death in proliferative and postmitotic regions of the fetal cerebral cortex. Develop 122: 1165-1174.

Bondy C, Lee WH (1993) Correlation between insulin-like growth factor (IGF)-binding protein 5 and IGF-I gene expression during brain development. J Neurosci 13: 5092-5104.

Bonfanti L, Gritti A, Galli R, Vescovi Al (2002) Multipotent stem cells in the adult central nervous system. In: Stem Cells and CNS Development. (Rao MS, ed), pp 49-70: Totowa, NJ: Humana Press.

Boulder Committee (1970) Embryonic vertebrate central nervous system: revised terminology. Anat Rec 166: 257-261.

Brodkey JA, Laywell ED, O'Brien TF, Faissner A, Stefansson K, Dorries HU, Schachner M, Steindler DA (1995) Focal brain injury and upregulation of a developmentally regulated extracellular matrix protein. J Neurosurg 82: 106-112.

Brüstle O, Jones KN, Learish RD, Karram K, Choudhary K, Wiestler OD, Duncan ID, McKay RD (1999) Embryonic stem cell-derived glial precursors: a source of myelinating transplants. Science 285: 754-756.

Buchet D, Serguera C, Zennou V, Charneau P, Mallet J (2002) Long-term expression of b-glucuronidase by genetically modified human neural progenitor cells grafted into the mouse central nervous system. Mol Cell Neurosci 19: 389-401.

Burrows RC, Wancio D, Levitt P, Lillien L (1997) Response diversity and the timing of progenitor cell maturation are regulated by developmental changes in EGFR expression in the cortex. Neuron 19: 251-267.

Cai J, Wu Y, Mirua T, Pierce JL, Lucero MT, Albertine KH, Spangrude GJ, Rao MS (2002) Properties of a fetal multipotent neural stem cell (NEP cell). Dev Biol 251: 221-240.

Cai L, Hayes NL, Nowakowski RS (1997) Local homogeneity of cell cycle length in developing mouse cortex. J Neurosci 17: 2079-2087.

Caldwell MA, He X, Wilkie N, Pollack S, Marshall G, Wafford KA, Svendsen CN (2001) Growth factors regulate the survival and fate of cells derived from human neurospheres. Nature Biotech 19: 475-479.

Caviness VS Jr, Takahashi T (1995) Proliferative events in the cerebral ventricular zone. Brain Develop 17: 159-163.

Chalmers-Redman RM, Priestley T, Kemp JA, Fine A (1997) *In vitro* propagation and inducible differentiation of multipotential progenitor cells from human fetal brain. Neurosci 76: 1121-1128.

Chandross KJ, Cohen RI, Paras P Jr, Gravel M, Braun PE, Hudson LD (1999) Identification and characterization of early glial progenitors using a transgenic selection strategy. J Neurosci 19: 759-774.

Chari DM, Blakemore WF (2002) Efficient recolonisation of progenitor-depleted areas of the CNS by adult oligodendrocyte progenitor cells. Glia 37: 307-313.

Charytoniuk D, Traiffort E, Hantraye P, Hermel JM, Galdes A, Ruat M (2002) Intrastriatal sonic hedgehog injection increases Patched transcript levels in the adult rat subventricular zone. Eur J Neurosci 16: 2351-2357.

Chenn A, McConnell SK (1995) Cleavage orientation and the asymmetric inheritance of Notch1 immunoreactivity in mammalian neurogenesis. Cell 82: 631-641.

Chiang C, Litingtung Y, Lee E, Young KE, Corden JL, Westphal H, Beachy PA (1996) Cyclopia and defective axial patterning in mice lacking Sonic hedgehog gene function. Nature 383: 407-413.

Chu-LaGraff Q, Kang X, Messer A (2001) Expression of the Huntington's disease transgene in neural stem cell cultures from R6/2 transgenic mice. Brain Res Bull 56: 307-312.

Ciccolini F, Svendsen CN (1998) Fibroblast growth factor 2 (FGF-2) promotes acquisition of epidermal growth factor (EGF) responsiveness in mouse striatal precursor cells: identification of neural precursors responding to both EGF and FGF-2. J Neurosci 18: 7869-7880.

Ciruna BG, Schwartz L, Harpal K, Yamaguchi TP, Rossant J (1997) Chimeric analysis of fibro-blast growth factor receptor-1 (Fgfr1) function: A role for FGFR1 in morphogenetic movement through the primitive streak. Develop 124: 2829-2841.

Clarke DL, Johansson CB, Wilbertz J, Veress B, Nilsson E, Karlstrom H, Lendahl U, Frisen J (2000) Generalized potential of adult neural stem cells. Science 288: 1660-1663.

Colas JF, Schoenwolf GC (2001) Towards a cellular and molecular understanding of neurulation. Dev Dyn 221: 117-145.

Conover JC, Doetsch F, Garcia-Verdugo JM, Gale NW, Yancopoulos GD, Alvarez-Buylla A (2000) Disruption of Eph/ephrin signaling affects migration and proliferation in the adult subventricular zone. Nat Neurosci 3: 1091-1097.

Craig CG, Tropepe V, Morshead CM, Reynolds BA, Weiss S, van der Kooy D (1996) *In vivo* growth factor expansion of endogeneous subependymal neural precursor cell populations in the adult mouse brain. J Neurosci 16: 2649-2658.

Craig CG, D'sa R, Morshead CM, Roach A, van der Kooy D (1999) Migrational analysis of the constitutively proliferating subependyma population in adult mouse forebrain. Neurosci 93: 1197-1206.

Crossley PH, Martinez S, Martin GR (1996) Midbrain development induced by FGF8 in the chick embryo. Nature 380: 66-68.

Czyz J, Wobus A (2001) Embryonic stem cell differentiation: the role of extracellular factors. Differentiation 68: 167-174.

Dai C, Celestino JC, Okada Y, Louis DN, Fuller GN, Holland EC (2001) PDGF autocrine stimulation dedifferentiates cultured astrocytes and induces oligodendrogliomas and oligoastrocytomas from neural progenitors and astrocytes *in vivo*. Genes Develop 15: 1913-1925.

DeChiara TM, Vejsada R, Poueymirou WT, Acheson A, Suri C, Conover JC, Friedman B, McClain J, Pan L, Stahl N, Ip NY, Kato A, Yancopoulos GD (1995) Mice lacking the CNTF receptor, unlike mice lacking CNTF, exhibit profound motor neuron deficits at birth. Cell 83: 313-322.

de la Rosa EJ, de Pablo F (2000) Cell death in early neural development: beyond the neurotrophic theory. TINS 23: 454-458.

Deng CX, Wynshaw-Boris A, Shen MM, Daugherty C, Ornitz DM, Leder P (1994) Murine FGFR-1 is required for early postimplantation growth and axial organization. Genes Dev 8: 3045-3057.

Dewulf N, Verschueren K, Lonnoy O, Moren A, Grimsby S, van de Spiegle K, Miyazono K, Huylebroeck D, ten Dijke P (1995) Distinct spatial and temporal expression patterns of two type I receptors for bone morphogenetic proteins during mouse embryogenesis. Endo 136: 2652-2663.

Ding H, Guha A (2001) Mouse astrocytoma models: embryonic stem cell mediated transgenesis. J Neurooncol 53: 289-296.

Doetsch F, Alvarez-Buylla A (1996) Network of tangential pathways for neuronal migration in adult mammalian brain. Proc Natl Acad Sci USA 93: 14895-14900.

Doetsch F, Caille I, Lim DA, Garcia-Verdugo JM, Alvarez-Buylla A (1999) Subventricular zone astrocytes are neural stem cells in the adult mammalian brain. Cell 97: 703-716.

Doetsch F, Garcia-Verdugo JM, Alvarez-Buylla A (1997) Cellular composition and three - dimensional organization of the subventricular germinal zone in the adult mammalian brain. J Neurosci 17: 5046-5061.

Dutton R, Yamada T, Turnley A, Bartlett PF, Murphy M (1999) Sonic hedgehog promotes neuronal differentiation of murine spinal cord precursors and collaborates with neurotrophin 3 to induce Islet - 1. J Neurosci 19: 2601-2608.

Emoto N, Gonzalez AM, Walicke PA, Wada E, Simmons DM, Shimasaki S, Baird A (1989) Basic fibroblast growth factor (FGF) in the central nervous system: identification of specific loci of basic FGF expression in the rat brain. Growth Factors 2: 21-29.

Erlandsson A, Enarsson M, Forsberg-Nilsson K (2001) Immature neurons from CNS stem cells proliferate in response to platelet - derived growth factor. J Neurosci 21: 3483-3491.

Faissner A, Steindler D (1995) Boundaries and inhibitory molecules in developing neural tissues. Glia 13: 233-254.

Fitoussi N, Sotnik-Barkai I, Tornatore C, Herzberg U, Yadid G (1998) Dopamine turnover and metabolism in the striatum of parkinsonian rats grafted with genetically-modified human astrocytes. Neurosci 85: 405-413.

Flax JD, Aurora S, Yang C, Simonin C, Wills AM, Billinghurst LL, Jendoubi M, Sidman RL, Wolfe JH, Kim SU, Snyder EY (1998) Engraftable human neural stem cells respond to developmental cues, replace neurons, and express foreign genes. Nat Biotech 16: 1033-1039.

Fok-Seang J, Miller RH (1994) Distribution and differentiation of A2B5+ glial precursors in the developing rat spinal cord. J Neurosci Res 37: 219-235.

Frisen J, Johansson CB, Lothian C, Lendahl U (1998) Central nervous system stem cells in the embryo and adult. Cell Mol Life Sci 54: 935-945.

Fruttiger M, Karlsson L, Hall AC, Abramsson A, Calver AR, Bostrom H, Willetts K, Bertold CH, Heath JK, Betsholtz C, Richardson WD (1999) Defective oligodendrocyte development and severe hypomyelination in PDGF-A knockout mice. Develop 126: 457-467.

Fu H, Qi Y, Tan M, Cai J, Takebayashi H, Nakafuku M, Richardson W, Qiu M (2002) Dual origin of spinal oligodendrocyte progenitors and evidence for the cooperative role of Olig2 and Nkx2.2 in the control of oligodendrocyte differentiation. Develop 129: 681-693.

Fu YM, Spirito P, Yu ZX, Biro S, Sasse J, Lei J, Ferrans VJ, Epstein SE, Casscells W (1991) Acidic fibroblast growth factor in the developing rat embryo. J Cell Biol 114: 1261-1273.

Fukumitsu H, Furukawa Y, Tsukaka M, Kinukawa H, Nitta A, Nomoto H, Mima T, Furukawa S (1998) Simultaneous expression of brain-derived neurotrophic factor and neurotrophin-3 in Cajal-Retzius, subplate and ventricular progenitor cells during early development stages of the rat cerebral cortex. Neurosci 84: 115-127.

Gage FH (2000) Mammalian neural stem cells. Science 287: 1433-1438.

Gage FH, Coates PW, Palmer TD, Kuhn HG, Fisher LJ, Suhonen JO, Peterson DA, Suhr ST, Ray J (1995) Survival and differentiation of adult neural precursor cells transplanted to adult brain. Proc Natl Acad Sci USA 92: 11879-11883.

Gage FH, Kempermann G, Palmer TD, Peterson DA, Ray J (1998) Multipotent progenitor cells in the adult dentate gyrus. J Neurobiol 36: 249-266.

Gaiano N, Nye JS, Fishell G (2000) Radial glial identity is promoted by Notch1 signaling in the murine forebrain. Neuron 26: 395-404.

Galli R, Pagano SF, Gritti A, Vescovi AL (2000) Regulation of neuronal differentiation in human CNS stem cell progeny by leukemia inhibitory factor. Dev Neurosci 22: 86-95.

Galli R, Fiocco R, De Filippis L, Muzio L, Gritti A, Mercurio S, Broccoli V, Pellegrini M, Mallamaci A, Vescovi AL (2002) Emx2 regulates the proliferation of stem cells of the adult mammalian central nervous system. Develop 129: 1633-1644.

Garcia-Segura LM, Perez J, Pons S, Rejas MT, Torres-Aleman I (1991) Localization of insulin like growth factor (IGF-I)-like immunoreactivity in the developing and adult rat brain. Brain Res 560: 167-174.

Garcia-Verdugo JM, Doetsch F, Wichterle H, Lim DA, Alvarez-Buylla A (1998) Architecture and cell types of the adult subventricular zone: in search of the stem cells. J Neurobiol 36: 234-248.

Garcion E, Faissner A, ffrench-Constant C (2001) Knockout mice reveal a contribution of the extracellular matrix molecule tenascin-C to neural precursor proliferation and migration. Develop 128: 2485-2496.

Gardner RL (2001) Specification of embryonic axes begins before cleavage in normal mouse development. Develop 128: 839-847.

Gates MA, Thomas LB, Howard EM, Laywell ED, Sajin B, Faissner A, Gotz B, Silver J, Steindler DA (1995) Cell and molecular analysis of the developing and adult mouse subventricular zone of the cerebral hemispheres. J Comp Neurol 361: 249-266.

Ghosh A, Greenberg ME (1995) Distinct roles for bFGF and NT-3 in the regulation of cortical neurogenesis. Neuron 15: 89-103.

Gilmore EC, Nowakowski RS, Caviness VS Jr, Herrup K. (2000) Cell birth, cell death, cell diversity and DNA breaks: how do they all fit together? TINS 23: 100-105.

Gloster A, El-Bizri H, Bamji SX, Rogers D, Miller FD (1999) Early induction of Ta1 a-tubulin transcription in neurons of the developing nervous system. J Comp Neurol 405: 45-60.

Gökan S, Mehler MF (2001) Basic and clinical neuroscience applications of embryonic stem cells. Anat Rec 265: 142-156.

Goldman DC, Martin GR, Tam PP (2000) Fate and function of the ventral ectodermal ridge during mouse tail development. Develop 127: 2113-2123.

Goldman SA, Kirschenbaum B, Harrison-Restelli C, Thaler HT (1997) Neuronal precursors of the rat subependymal zone persist into senescence, with no decline in spatial extent or response to BDNF. J Neurobiol 32: 554-566.

Goodrich LV, Scott MP (1998) Hedgehog and patched in neural development and disease. Neuron 21: 1243-1257.

Goudreau G, Carpenter S, Beaulieu N, Jolicoeur P (1996) Vacuolar myelopathy in transgenic mice expressing human immunodeficiency virus type 1 proteins under the regulation of the myelin basic protein gene promoter. Nat Med 2: 655-661.

Gratsch TE, Fuller J, Long MW, O'Shea KS (2001) Noggin and chordin induce neural gene expression in bone marrow stromal cells. Soc Neurosci Abstr 27: 24.12.

Gratsch TE, O'Shea KS (2002) Noggin and chordin have distinct activities in promoting lineage commitment of mouse embryonic stem (ES) cells. Dev Biol 245: 83-94.

Grinspan JB, Edell E, Carpio DF, Beesley JS, Lavy L, Pleasure D, Golden JA (2000) Stage-specific effects of bone morphogenetic proteins on the oligodendrocyte lineage. J Neurobiol 43: 1-17.

Grondona JM, Perez-Martin M, Cifuentes M, Perez J, Jimenez AJ, Perez-Figares JM, Fernandez-Llebrez P (1996) Ependymal denudation, aqueductal obliteration and hydrocephalus after a single injection of neuraminidase into the lateral ventricle of adult rats. J Neuropathol Exp Neurol 55: 999-1008.

Gross RE, Mehler MF, Mabie PC, Zang Z, Santschi L, Kessler JA (1996) Bone morphogenetic proteins promote astroglial lineage commitment by mammalian subventricular zone progenitor cells. Neuron 17: 595-606.

Groszer M, Erickson R, Scripture-Adams DD, Lesche R, Trumpp A, Zack JA, Kornblum HI, Liu X, Wu H (2001) Negative regulation of neural stem/progenitor cell proliferation by the Pten tumor suppressor gene *in vivo*. Science 294: 2186-2189.

Guo H, Hong S, Jin XL, Chen RS, Avasthi PP, Tu YT, Ivanco TL, Li Y (2000) Specificity and efficiency of Cre-mediated recombination in EMX1-Cre knock-in mice. Biochem Biophys Res Commun 273: 661-665.

Hakem R et al (1998) Differential requirement for caspase 9 in apoptotic pathways *in vivo*. Cell 94: 339-352.

Hallbook F, Ibanez CF, Ebendal T, Persson H (1993) Cellular localization of brain-derived neu-rotrophic factor and neurotrophin-3 mRNA expression in the early chicken embyo. Eur J Neurosci 5: 1-14.

Harland R (2000) Neural induction. Curr Opin Genet Dev 10: 357-362.

Hasskarl J, Munger K (2002) Id proteins—tumor markers or oncogenes? Cancer Biol Ther 1: 91-96.

Hayflick L (1968) Human cells and aging. Sci Am 218: 32-37.

Hebert JM, McConnell SK (2000) Targeting of cre to the Foxg1 (BF-1) locus mediates loxP recombination in the telencephalon and other developing head structures. Dev Biol 222: 296-306.

Henrique D, Adam J, Myat A, Chitnis A, Lewis J, Ish-Horowicz D (1995) Expression of a Delta homologue in prospective neurons in the chick. Nature 375: 787-790.

Hermanson O, Jepsen K, Rosenfeld MG (2002) N-CoR controls differentiation of neural stem cells into astrocytes. Nature 419: 934-939.

Herrera DG, Garcia-Verdugo JM, Alvarez-Buylla A (1999) Adult-derived neural precursors trans-planted into multiple regions in the adult brain. Ann Neurol 46: 867-877.

Hirsinger E, Duprez D, Jouve C, Malapert P, Cooke J, Pourquie O (1997) Noggin acts down-stream of Wnt and Sonic Hedgehog to antagonize BMP4 in avian somite patterning. Develop 124: 4605-4614.

Hitoshi S, Tropepe V, Ekker M, van der Kooy D (2002) Neural stem cell lineages are regionally specified, but not committed, within distinct compartments of the developing brain. Develop 129: 233-244.

Hogan BL, Thaller C, Eichele G (1992) Evidence that Hensen's node is a site of retinoic acid synthesis. Nature 359: 237-241.

Holland EC, Hively WP, DePinho RA, Varmus HE (1998) A constitutively active epidermal growth factor receptor cooperates with disruption of G1 cell-cycle arrest pathways to induce glioma-like lesions in mice. Genes Develop 12: 3675-3685.

Hollnagel A, Oehlmann V, Heymer J, Ruther U, Nordheim A (1999) Id genes are direct targets of bone morphogenetic protein induction in embryonic stem cells. J Biol Chem 274: 19838-19845.

Horner P et al. (2000) Proliferation and differentiation of progenitor cells throughout the intact adult rat spinal cord. J Neurosci 20: 2218-2228.

Hu H (1999) Chemorepulsion of neuronal migration by Slit2 in the developing mammalian fore-brain. Neuron 23: 703-711.

Hutchins JB, Jefferson VE (1992) Developmental distribution of platelet-derived growth factor in the mouse central nervous system. Brain Res Dev Brain Res 67: 121-135.

Ibarrola N, Mayer-Proschel M, Rodriguez-Pena A, Noble M (1996) Evidence for the existence of at least two timing mechanisms that contribute to oligodendrocyte generation *in vitro*. Dev Biol 180: 1-21.

Ignatova TN, Kukekov VG, Laywell ED, Suslov ON, Vrionis FD, Steindler DA (2002) Human cortical glial tumors contain neural stem-like cells expressing astroglial and neuronal markers *in vitro*. Glia 39: 193-206.

Ip NY, McClain J, Barrezueta NX, Aldrich TH, Pan L, Li Y, Wiegand SJ, Friedman B, Davis S, Yancopoulos GD (1993) The alpha component of the CNTF receptor is required for signaling and defines potential CNTF targets in the adult and during development. Neuron 10: 89-102.

Ishibashi M, Moriyoshi K, Sasai Y, Shiota K, Nakanishi S, Kageyama R (1994) Persistent ex-pression of helix-loop-helix factor HES-1 prevents mammalian neural differentiation in the central nervous system. EMBO J 13: 1799-1805.

Ishibashi M, Ang SL, Shiota K, Nakanishi S, Kageyama R, Guillemot F (1995) Targeted disrup-tion of mammalian hairy and Enhancer of split homolog-1 (HES-1) leads to upregulation of neural helix-loop-helix factors, premature neurogenesis, and severe neural tube defects. Genes Dev 9: 3136-3148.

Israel MA, Hernandez MC, Florio M, Andres-Barquin PJ, Mantani A, Carter JH, Julin CM (1999) Id gene expression as a key mediator of tumor cell biology. Cancer Res 59: 1726s-1730s.

Ivanova NB, Dimos JT, Schaniel C, Hackney JA, Moore KA, Lemischka IR (2002) A stem cell molecular signature. Science 298: 601-604.

Jacobson M (1991) Developmental Neurobiology, 3rd ed., NY: Plenum Press.

Jankovski A, Sotelo C (1996) Subventricular zone-olfactory bulb migratory pathway in the adult mouse: cellular composition and specificity as determined by heterochronic and heterotopic transplantation. J Comp Neurol 371: 376-396.

Jensen NA, Pedersen KM, Celis JE, West MJ (1998a) Neurological disturbances, premature lethality, and central myelination deficiency in transgenic mice overexpressing the homeo domain transcription factor Oct-6. J Clin Invest 101: 1292-1299.

Jensen NA, Pedersen KM, Celis JE, West MJ (1998b) Failure of central nervous system myelination in MBP/c-myc transgenic mice: evidence for c-myc cytotoxicity. Oncogene 16: 2123-2129.

Jessell TM (2000) Neuronal specification in the spinal cord: inductive signals and transcriptional codes. Nat Rev Genet 1: 20-29.

Johansson CB, Momma S, Clarke DL, Risling M, Lendahl U, Frisen J (1999) Identification of a neural stem cell in the adult mammalian central nervous system. Cell 96: 25-34.

Johe KK, Hazel TG, Muller T, Dugich-Djordjevic MM, McKay RD (1996) Single factors direct the differentiation of stem cells from the fetal and adult central nervous system. Genes Dev 10: 3129-3140.

Josephson R, Müller T, Pickel J, Okabe S, Reynolds K, Turner PA, Zimmer A, McKay RD (1998) POU transcription factors control expression of CNS stem cell-specific genes. Develop 125: 3087-3100.

Kabos P, Kabosova A, Neuman T (2002) Neuronal injury affects expression of helix-loop-helix transcription factors. NeuroReport 13: 2385-2388.

Kageyama R, Sasai Y, Akazawa C, Ishibashi M, Takebayashi K, Shimizu C, Tomita K. Nakanishi S (1995) Regulation of mammalian neural development by helix-loop-helix transcription factors. Crit Rev Neurobiol 9: 177-188.

Kahn MA, Kumar S, Liebl D, Chang R, Parada LF, De Vellis J (1999) Mice lacking NT-3, and its receptor TrkC, exhibit profound deficiencies in CNS glial cells. Glia 26: 153-165.

Kalyani A, Hobson K, Rao MS (1997) Neuroepithelial stem cells from the embryonic spinal cord: isolation, characterization, and clonal analysis. Dev Biol 186: 202-223.

Kalyani AJ, Mujtaba T, Rao MS (1999) Expression of EGF receptor and FGF receptor isoforms during neuroepithelial stem cell differentiation. J Neurobiol 38: 207-224.

Kar S, Chabot JG, Quirion R (1993) Quantitative autoradiographic localization of (^{125}I)insulin-like growth factor I, (^{125}I)insulin-like growth factor II, and (^{125}I)insulin receptor binding sites in developing and adult rat brain. J Comp Neurol 333: 375-397.

Kaser MR, Lakshmanan J, Fisher DA (1992) Comparison between epidermal growth factor, transforming growth factor-alpha and EGF receptor levels in regions of adult rat brain. Brain Res Mol Brain Res 16: 316-322.

Katoh-Semba R, Takeuchi IK, Semba R, Kato K (2000) Neurotrophin-3 controls proliferation of granular precursors as well as survival of mature granule neurons in the developing rat cerebellum. J Neurochem 74: 1923-1930.

Kawaguchi A, Miyata T, Sawamoto K, Takashita N, Murayama A, Akamatsu W, Ogawa M, Okabe M, Tano Y, Goldman SA, Okano H (2001) Nestin-EGFP transgenic mice: visualization of the self-renewal and multipotency of CNS stem cells. Mol Cell Neurosci 17: 259-273.

Kee Y, Bronner-Fraser M (2001) Id4 expression and its relationship to other Id genes during avian embryonic development. Mech Dev 109: 341-345.

Kessler JA, Mehler MF, Mabie PC (2002) Multipotent stem cells in the nervous system. In: Stem Cells and CNS Development (Rao MS, ed), pp 31-48. Totowa, NJ: Humana Press.

Keyoung HM, Roy NS, Benraiss A, Louissant A Jr, Suzuki A, Hashimoto M, Rashbaum WK, Okano H, Goldman SA (2001) High-yield selection and extraction of two promoter-defined phenotypes of neural stem cells from the fetal human brain. Nat Biotech 19: 843-850.

Kilpatrick TJ, Bartlett PF (1993) Cloning and growth of multipotential neural precursors: requirements for proliferation and differentiation. Neuron 10: 255-265.

Kilpatrick TJ, Bartlett PF (1995) Cloned multipotential precursors from the mouse cerebrum require FGF-2, whereas glial restricted precursors are stimulated with either FGF-2 or EGF. J Neurosci 15: 3653-3661.

Kilty IC, Barraclough R, Schmidt G, Rudland PS (1999) Isolation of a potential neural stem cell line from the internal capsule of an adult transgenic rat brain. J Neurochem 73: 1859-1870.

Kinashi T, Springer TA (1994) Steel factor and c-kit regulate cell-matrix adhesion. Blood 83: 1033-1038.

Kleihues P, Lantos PL, Magee PN (1976) Chemical carcinogenesis in the nervous system. Int Rev Exp Pathol 15: 153-232.

Kopan R, Schroeter EH, Weintraub H, Nye JS (1996) Signal transduction by activated mNotch: importance of proteolytic processing and its regulation by the extracellular domain. Proc Natl Acad Sci USA 93: 1683-1688.

Kornack DR, Rakic P (2001) Cell proliferation without neurogenesis in adult primate neocortex. Science 294: 2127-2130.

Kornblum HI, Hussain RJ, Bronstein JM, Gall CM, Lee DC, Seroogy KB (1997) Prenatal ontogeny of the epidermal growth factor receptor and its ligand, transforming growth factor alpha, in the rat brain. J Comp Neurol 380: 243-261.

Kornblum HI, Hussain R, Wiesen J, Miettienen P, Zurcher SD, Chow K, Derynck R, Werb Z (1998) Abnormal astrocyte development and neuronal death in mice lacking the epidermal growth factor receptor. J Neurosci Res 53: 697-717.

Kuhn HG, Dickinson-Anson H, Gage FH (1996) Neurogenesis in the dentate gyrus of the adult rat: age related decrease of neural progenitor proliferation. J Neurosci 16: 2027-2033.

Kuhn HG, Winkler J, Kempermann G, Thal LJ, Gage FH (1997) Epidermal growth factor and fibroblast growth factor-2 have different effects on neural progenitors in the adult rat brain. J Neurosci 17: 5820-5829.

Kuida K, et al (1998) Reduced apoptosis and cytochrome c-mediated caspase activation in mice lacking caspase 9. Cell 94: 325-337.

Kukekov VG, Laywell ED, Suslov O, Davies K, Scheffler B, Thomas LB, O'Brien TF, Kusakabe M, Steindler DA (1999) Multipotent stem/progenitor cells with similar properties arise from two neurogenic regions of adult human brain. Exp Neurol 156: 333-344.

Kulessa H, Turk G, Hogan BL (2000) Inhibition of Bmp signaling affects growth and differentiation in the anagen hair follicle. EMBO J 19: 6664-6674.

Lachyankar MB, Condon PJ, Quesenberry PJ, Litofsky NS, Recht LD, Ross AH (1997) Embryonic precursor cells that express Trk receptors: induction of different cell fates by NGF, BDNF, NT-3 and CNTF. Exp Neurol 144: 350-360.

Lacorazza HD, Flax JD, Snyder EY, Jendoubi M (1996) Expression of human b-hexosaminidase a-subunit gene (the gene defect of Tay-Sachs disease) in mouse brains upon engraftment of transduced progenitor cells. Nature Med 2: 424-429.

LaMantia AS, Colbert MC, Linney E (1993) Retinoic acid induction and regional differentiation prefigure olfactory pathway formation in the mammalian forebrain. Neuron 10: 1035-1048.

Lardelli M, Williams R, Mitsiadis T, Lendahl U (1996) Expression of the Notch 3 intracellular domain in mouse central nervous system progenitor cells is lethal and leads to disturbed neural tube development. Mech Dev 59: 177-190.

Lasorella A, Iavarone A, Israel MA (1996) Id2 specifically alters regulation of the cell cycle by tumor suppressor proteins. Mol Cell Biol 16: 2570-2578.

Lasorella A, Noseda M, Beyna M, Iavarone A (2000) Id2 is a retinoblastoma protein target and mediates signalling by Myc oncoproteins. Nature 407: 592-598.

Lee JE (1997) Basic helix-loop-helix genes in neural development. Curr Opin Neurobiol 7: 13-20.

Lendahl U (1997) Transgenic analysis of central nervous system development and regeneration. Acta Anaesthesiol Scand S110: 116-118.

Lendahl U, Zimmerman LB, McKay RD (1990) CNS stem cells express a new class of intermediate filament protein. Cell 60: 585-595.

Levison SW, Goldman JE (1993) Both oligodendrocytes and astrocytes develop from progenitors in the subventricular zone of the postnatal rat forebrain. Neuron 10: 201-212.

Li RY, Baba S, Kosugi I, Arai Y, Kawasaki H, Shinmura Y, Sakakibara SI, Okano H, Tsutsui Y (2001) Activation of murine cytomegalovirus immediate-early promoter in cerebral ventricular zone and glial progenitor cells in transgenic mice. Glia 35: 41-52.

Li W, Cogswell CA, LoTurco JJ (1998) Neuronal differentiation of precursors in the neocortical ventricular zone is triggered by BMP. J Neurosci 18: 8853-8862.

Liem KF Jr, Tremml G, Jessell TM (1997) A role for the roof plate and its resident TGFb-related proteins in neuronal patterning in the dorsal spinal cord. Cell 91: 127-138.

Lillien L, Raphael H (2000) BMP and FGF regulate the development of EGF-responsive neural progenitor cells. Develop 127: 4993-5005.

Lim DA, Alvarez-Buylla A (1999) Interaction between astrocytes and adult subventricular zone precursors stimulates neurogenesis. Proc Natl Acad Sci USA 96: 7526-7531.

Lim DA, Alvarez-Buylla A (2002) Glial characteristics of adult subventricular zone stem cells. In: Stem Cells and CNS Development (Rao MS, ed), pp 71-92. Totowa, NJ: Humana Press.

Lim DA, Flames N, Collado L, Herrera DG (2002) Investigating the use of primary adult subventricular zone neural precursor cells for neuronal replacement therapies. Brain Res Bull 57: 759-764.

Lim DA, Tramontin AD, Trevejo JM, Herera DG, Garcia-Verdugo JM, Alvarez-Buylla A (2000) Noggin antagonizes BMP signaling to create a niche for adult neurogenesis. Neuron 28: 713-726.

Liu JP, Baker J, Perkins AS, Robertson EJ, Efstratiadis A (1993) Mice carrying null mutations of the genes encoding insulin-like growth factor I (Igf-1) and type 1 IGF receptor (Igf1r). Cell 75: 59-72.

Lois C, Alvarez-Buylla A (1994) Long-distance neuronal migration in the adult mammalian brain. Science 264: 1145-1148.

LoPresti P, Poluha W, Poluha DK, Drinkwater E, Ross AH (1992) Neuronal differentiation triggered by blocking cell proliferation. Cell Growth Differ 3: 627-635.

Lorincz MT, O'Shea KS, Dettlof PD, Albin RL (2001) Neuronally differentiated ES cells as a Huntington's Disease model. Soc Neurosci Abstr 27: 99.10.

Lukaszewicz A, Savatier P, Cortay V, Kennedy H, Dehay C (2002) Contrasting effects of basic fibroblast growth factor and neurotrophin 3 on cell cycle kinetics of mouse cortical stem cells. J Neurosci 22: 6610-6622.

Lumsden A, Krumlauf R (1996) Patterning the vertebrate neuraxis. Science 274: 1109-1115.

Luskin MB (1993) Restricted proliferation and migration of postnatally generated neurons derived from the forebrain subventricular zone. Neuron 11: 173-189.

Luskin MB, Pearlman AL, Sanes JR (1988) Cell lineage in the cerebral cortex of the mouse studied *in vivo* and *in vitro* with a recombinant retrovirus. Neuron 1: 635-647.

Ma W, Ribeiro-da-Silva A, Noel G, Julien JP, Cuello AC (1995) Ectopic substance P and calcitonin gene-related peptide imunoreactive fibres in the spinal cord of transgenic mice over-expressing nerve growth factor. Eur J Neurosci 7: 2021-2035.

Mabie PC, Mehler MF, Kessler JA (1999) Multiple roles of bone morphogenetic protein signaling in the regulation of cortical cell number and phenotype. J Neurosci 19: 7077-7088.

Machon O, van den Bout CJ, Backman M, Rosok O, Caubit X, Fromm SH, Geronimo B, Krauss S (2002) Forebrain-specific promoter/enhancer D6 derived from the mouse Dach1 gene controls expression in neural stem cells. Neurosci 112: 951-966.

Magavi SS, Leavitt BR, Macklis JD (2000) Induction of neurogenesis in the neocortex of adult mice. Nature 405: 951-955.

Magdaleno S, Keshvara L, Curran T (2002) Rescue of ataxia and preplate splitting by ectopic expression of Reelin in reeler mice. Neuron 33: 573-586.

Maisonpierre PC, Belluscio L, Friedman B, Alderson RF, Wiegand SJ, Furth ME, Lindsay RM, Yancopoulos GD (1990) NT-3, BDNF, and NGF in the developing rat nervous system: parallel as well as reciprocal patterns of expression. Neuron 5: 501-509.

Marmur R, Kessler JA, Zhu G, Gokhan S, Mehler MF (1998) Differentiation of oligodendroglial progenitors derived from cortical multipotent cells requires extrinsic signals including activation of gp130/LIFbeta receptors. J Neurosci 18: 9800-9811.

Martens DJ, Tropepe V, and van der Kooy D (2000) Separate proliferation kinetics of fibroblast growth factor-responsive and epidermal growth factor-responsive neural stem cells within the embryonic forebrain germinal zone. J Neurosci 20: 1085-1095.

Martinez-Serrano A, Bjorklund A (1996) Protection of the neostriatum against excitotoxic damage by neurotrophin-producing, genetically modified neural stem cells. J Neurosci 16: 4604-4616.

Masu Y, Wolf E, Holtmann B, Sendtner M, Brem G, Thoenen H (1993) Disruption of the CNTF gene results in motor neuron degeneration. Nature 365: 27-32.

Mayer-Proschel M, Kalyani AJ, Mujtaba T, Rao MS (1997) Isolation of lineage-restricted neuronal precursors from multipotent neuroepithelial stem cells. Neuron 19: 773-785.

McKeon RJ, Schreiber RC, Rudge JS, Silver J (1991) Reduction of neurite outgrowth in a model of glial scaring following CNS injury is correlated with the expression of inhibitory molecules on reactive astrocytes. J Neurosci 11: 3398-3411.

McKinnon RD, Matsui T, Dubois-Dalcq M, Aarsonson SA. (1990) FGF modulates the PDGF-driven pathway of oligodendrocyte development. Neuron 5: 603-614.

Mehler MF (2002) Mechanisms regulating lineage diversity during mammalian cerebral cortical neurogenesis and gliogenesis. Results Probl Cell Differ 39: 27-52.

Mehler MF (2002) Regional forebrain patterning and neural subtype specification: implications for cerebral cortical functional connectivity and the pathogenesis of neurodegenerative diseases. Results Probl Cell Differ 39: 157-178.

Mehler MF, Gökhan S (2001) Developmental mechanisms in the pathogenesis of neurodegenerative diseases. Prog Neurobiol 63: 337-363.

Mekki-Dauriac S, Agius E, Kan P, Cochard P (2002) Bone morphogenetic proteins negatively control oligodendrocyte precursor specification in chick spinal cord. Develop 129: 5117-5130.

Menn B, Timsit S, Represa A, Mateos S, Calothy G, Lamballe F (2000) Spatiotemporal expression of noncatalytic TrkC NC2 isoform during early and late CNS neurogenesis: a comparative study with TrkC catalytic and p75NTR receptors Eur J Neurosci 12: 3211-3223.

Metzler M, Chen N, Helgason CD, Graham RK, Nichol K, McCutcheon K, Nasir J, Humphries RK, Raymond LA, Hayden MR (1999) Life without huntingtin: normal differentiation into functional neurons. J Neurochem 72: 1009-1018.

Miettinen PJ, Berger JE, Meneses J, Phung Y, Pedersen RA, Werb Z, Derynck R (1995) Epithelial immaturity and multiorgan failure in mice lacking epidermal growth factor receptor. Nature 376: 337-341.

Miragall F, Kadmon G, Faissner A, Antonicek H, Schachner M (1990) Retention of J1/tenascin and the polysialylated form of the neural cell adhesion molecule (N-CAM) in the adult olfactory bulb. J Neurocytol 19: 899-914.

Mishina Y, Suzuki A, Ueno N, Behringer RR (1995) Bmpr encodes a type I bone morphogenetic protein receptor that is essential for gastrulation during mouse embryogenesis. Genes Dev 9: 3027-3037.

Morshead CM, Reynolds BA, Craig CG, McBurney MW, Staines WA, Morassutti D, Weiss S, van der Kooy D (1994) Neural stem cells in the adult mammalian forebrain: a relatively quiescent subpopulation of subependymal cells. Neuron 13: 1071-1082.

Morshead CM, Craig CG, van der Kooy D(1998) *In vivo* clonal analyses reveal the properties of endogenous neural stem cell proliferation in the adult mammalian forebrain. Develop 125: 2251-2261.

Motoyama N, Wang F, Roth KA, Sawa H, Nakayama K, Negishi I, Senju S, Zhang Q, Fujii S (1995) Massive cell death of immature hematopoietic cells and neurons in Bcl-X-deficient mice. Science 267: 1506-1510.

Mujtaba T, Han SS, Fischer I, Sandgren EP, Rao MS (2002) Stable expression of the alkaline phosphatase marker gene by neural cells in culture and after transplantation into the CNS using cells derived from a transgenic rat. Exp Neurol 174: 48-57.

Mujtaba T, Piper DR, Kalyani A, Groves AK, Lucero MT, Rao MS (1999) Lineage - restricted neural precursors can be isolated from both the mouse neural tube and cultured ES cells. Dev Biol 214: 113-127.

Murphy M, Reid K, Brown MA, Bartlett PF (1993) Involvement of leukemia inhibitory factor and nerve growth factor in the development of dorsal root ganglion neurons. Develop 117: 1173-1182.

Na E, McCarthy M, Neyt C, Lai E, Fishell G (1998) Telencephalic progenitors maintain antero-posterior identities cell autonomously. Curr Biol 8: 987-990.

Nakagawa Y, Kaneko T, Ogura T, Suzuki T, Torii M, Kaibuchi K, Arai K, Nakamura S, Nakafuku M (1996) Roles of cell-autonomous mechanisms for differential expression of region-specific transcription factors in neuroepithelial cells. Develop 122: 2449-2464.

Nakamura Y, Sakakibara S, Miyata T, Ogawa M, Shimazaki T, Weiss S, Kageyama R, Okano H (2000) The bHLH gene Hes1 as a repressor of the neuronal commitment of CNS stem cells. J Neurosci 20: 283-293.

Nakashima K, Wiese S, Yanagisawa M, Arakawa H, Kimura N, Hisatsune T, Yoshida K, Kishimoto T, Sendtner M, Taga T (1999) Developmental requirement of gp130 signaling in neuronal survival and astrocyte differentiation. J Neurosci 19: 5429-5434.

Nasir J, Floresco SB, O'Kusky JR, Diewert VM, Richman JM, Zeisler J, Borowski A, Marth JD, Phillips AG, Hayden MR (1995) Targeted disruption of the Huntington's disease gene results in embryonic lethality and behavioral morphological changes in heterozygotes. Cell 81: 811-823.

Neuman T, Keen A, Zuber MX, Kristjansson GI, Gruss P, Nornes HO (1993) Neuronal expression of regulatory helix-loop-helix factor Id2 gene in mouse. Dev Biol 160: 186-195.

Nishino H, Hida H, Takei N, Kumazaki M, Nakajima K, Baba H (2000) Mesencephalic neural stem (progenitor) cells develop to dopaminergic neurons more strongly in dopamine-depleted striatum than in intact striatum. Exp Neurol 164: 209-214.

Noble M, Mayer-Proschel M (2002) Glial restricted precursors. In: Stem Cells and CNS Development (Rao MS, ed), pp 123-151. Totowa, NJ: Humana Press.

Norton JD (2000) Id helix-loop-helix proteins in cell growth, differentiation and tumorigenesis. J Cell Sci 113: 3897-3905.

Norton JD, Deed RW, Craggs G, Sabilitzky F (1998) Id helix-loop-helix proteins in cell growth and differentiation. Trends Cell Biol 8: 58-65.

Nurcombe V, Ford MD, Wildschut JA, Bartlett PF (1993) Developmental regulation of neural response to FGF-1 and FGF-2 by heparan sulfate proteoglycan. Science 260: 103-106.

Nye JS, Kopan R, Axel R (1994) An activated Notch suppresses neurogenesis and myogenesis but not gliogenesis in mammalian cells. Develop 120: 2421-2430.

Ohtsuka T, Sakamoto M, Guillemot F, Kageyama R (2001) Roles of the basic helix-loop-helix genes Hes1 and Hes 5 in expansion of neural stem cells of the developing brain. J Biol Chem 276: 30467-30474.

Okano HJ, Pfaff DW, Gibbs RB (1996) Expression of EGFR, p75NGFR, and PSTAIR (cdc2)-like immunoreactivity by proliferating cells in the adult rat hippocampal formation and forebrain. Dev Neurosci 18: 199-209.

Olsson M, Bjerregaard K, Winkler C, Gates M, Bjorklund A, Campbell K (1998) Incorporation of mouse neural progenitors transplated into the rat embryonic forebrain is developmentally regulated and dependent on regional and adhesive properties. Eur J Neurosci 10: 71-85.

Olsson M, Campbell K, Turnbull DH (1997) Specification of mouse telencephalic and mid-hindbrain progenitors following heterotopic ultrasound guided embryonic transplantation. Neuron 19: 761-772.

Orentas DM, Miller RH (1996) The origin of spinal cord oligodendrocytes is dependent on local influences from the notochord. Dev Biol 177: 43-53.

Ortega S, Ittmann M, Tsang SH, Ehrlich M, Basilico C (1998) Neuronal defects and delayed wound healing in mice lacking fibroblast growth factor 2. Proc Natl Acad Sci USA 95: 5672-5677.

Ostenfeld T, Joly E, Tai YT, Peters A, Caldwell M, Jauniaux E, Svendsen CN (2002) Regional specification of rodent and human neurospheres. Dev Brain Res 134: 43-55.

Palmer TD, Markakis EA, Willhoite AR, Safar F, Gage FH (1999) Fibroblast growth factor-2 activates a latent neurogenic program in neural stem cells from diverse regions of the adult CNS. J Neurosci 19: 8487-8497.

Palmer TD, Ray J, Gage FH (1995) FGF-2-responsive neuronal progenitors reside in proliferative and quiescent regions of the adult rodent brain. Mol Cell Neurosci 6: 474-486.

Panchision DM, Pickel JM, Studer L, Lee SH, Turner PA, Hazel TG, McKay RD (2001) Sequential actions of BMP receptors control neural precursor cell production and fate. Genes Develop 15: 2094-2110.

Parati EA, Be ZA, Ponti D, de Grazia U, Corsini E, Cova L, Sala S, Colombo A, Alessandri G, Pagano SF (2002) Human neural stem cells express extra-neural markers. Brain Res 925: 213-221.

Parent JM, Yu TW, Leibowitz RT, Geschwind DH, Sloviter RS, Lowenstein DH (1997) Dentate granule cell neurogenesis is increased by seizures and contributes to aberrant network reorganization in the adult rat hippocampus. J Neurosci 17: 3727-3738.

Parker SB, Eichele G, Zhang P, Rawls A, Sands AT, Bradley A, Olson EN, Harper JW, Elledge SJ (1995) p53-independent expression of p21 Cip1 in muscle and other terminally differentiating cells. Science 267: 1024-1027.

Parnavelas JG, Nadarajah B (2001) Radial glial cells: are they really glia? Neuron 31: 881-884.

Patstone G, Pasquale EB, Maher PA (1993) Different members of the fibroblast growth factor receptor family are specific to distinct cell types in the developing chicken embryo. Dev Biol 155: 107-123.

Perea-Gomez A, Rhinn M, Ang SL (2001) Role of the anterior visceral endoderm in restriciting posterior signals in the mouse embryo. Int J Dev Biol 45: 311-320.

Peretto P, Merighi A, Fasolo A, Bonfanti L (1999) The subependymal layer in rodents: a site of structural plasticity and cell migration in the adult mammalian brain. Brain Res Bull 49: 221-243.

Peters K, Ornitz D, Werner S, Williams L (1993) Unique expression pattern of the FGF receptor 3 gene during mouse organogenesis. Dev Biol 155: 423-430.

Poluha W, Poluha DK, Chang B, Crosbie NE, Schonhoff CM, Kilpatrick DL, Ross AH (1996) The cyclin-dependent kinase inhibitor p21^{WAF1} is required for survival of differentiating neuroblastoma cells. Mol Cell Biol 16: 1335-1341.

Pompeiano M, Blaschke AJ, Flavell RA, Srinivasan A, Chun J (2000) Decreased apoptosis in proliferative and postmitotic regions of the caspase 3-deficient embryonic central nervous system. J Comp Neurol 423: 1-12.

Prestoz L, Relvas JB, Hopkins K, Patel S, Sowinski P, Price J, ffrench-Constant C (2001) Association between integrin-dependent migration capacity of neural stem cells *in vitro* and anatomical repair following transplantation. Mol Cell Neurosci 18: 473-484.

Pringle NP, Mudhar HS, Collarini EJ, Richardson WD (1992) PDGF receptors in the rat CNS: during late neurogenesis, PDGF alpha-receptor expression appears to be restricted to glial cells of the oligodendrocyte lineage. Develop 115: 535-551.

Pringle NP, Yu WP, Guthrie S, Roelink H, Lumsden A, Peterson AC, Richardson WD (1996) Determination of neuroepithelial cell fate: induction of the oligodendrocyte lineage by ventral midline cells and sonic hedgehog. Dev Biol 177: 30-42.

Qian X, Davis AA, Goderie SK, Temple S (1997) FGF2 concentration regulates the generation of neurons and glia from multipotent cortical stem cells. Neuron 18: 81-93.

Qian X, Goderie SK, Shen Q, Stern JH, Temple S (1998) Intrinsic programs of patterned cell lineages in isolated vertebrate CNS ventricular zone cells. Develop 125: 3143-3152.

Raballo R, Rhee J, Lyn-Cook R, Leckman JF, Schwartz ML, Vaccarino FM (2000) Basic fibroblast growth factor (Fgf2) is necessary for cell proliferation and neurogenesis in the developing cerebral cortex. J Neurosci 20: 12-23.

Raff MC, Miller RH, Noble M (1983) A glial progenitor cell that develops *in vitro* into an astrocyte or an oligodendrocyte depending on culture medium. Nature 303: 390-396.

Rajan P, McKay RD (1998) Multiple routes to astrocytic differentiation in the CNS. J Neurosci 18: 3620-3629.

Rakic P (1995) A small step for the cell, a giant leap for mankind: a hypothesis of neocortical expansion during evolution. TINS 8: 383-388.

Rakic P (2002) Adult neurogenesis in mammals: an identity crisis. J Neurosci 22: 614-618.

Rao MS (1999) Multipotent and restricted precursors in the central nervous system. Anat Rec 257: 137-148.

Rao M, Noble M, Mayer-Proschel M (1998) A tripotential glial precursor cell is present in the developing spinal cord. Proc Natl Acad Sci USA 95: 3996-4001.

Readhead C, Popko B, Takahashi N, Shine HD, Saavedra RA, Sidman RL, Hood L (1987) Expression of a myelin basic protein gene in transgenic shiverer mice: correction of the dysmyelinating phenotype. Cell 48: 703-712.

Reimers D, Lopez-Toledano MA, Mason I, Cuevas P, Redondo C, Herranz AS, Lobo MVT, Bazan E (2001) Developmental expression of fibroblast growth factor (FGF) receptors in neural stem cell progeny. Modulation of neuronal and glial lineages by basic FGF treatment. Neurol Res 23: 612-621.

Reynolds BA, Tetzlaff W, Weiss S (1992) A multipotent EGF-responsive striatal embryonic progenitor cell produces neurons and astrocytes. J Neurosci 12: 4565-4574.

Reynolds BA, Weiss S (1992) Generation of neurons and astrocytes from isolated cells of the adult mamalian central nervous system. Science 255: 1707-1710.

Reznikov K, van der Kooy D (1995) Variability and partial synchrony of the cell cycle in the germinal zone of the early embryonic cerebral cortex. J Comp Neurol 360: 536-554.

Riaz SS, Jauniaux E, Stern GM, Bradford HF (2002) The controlled conversion of human neural progenitor cells derived from foetal ventral mesencephalon into dopaminergic neurons *in vitro*. Brain Res Dev Brain Res 136: 27-34.

Richards LJ, Kilpatrick TJ, Dutton R, Tan SS, Gearing DP, Bartlett PF, Murphy M (1996) Leukaemia inhibitory factor or related factors promote the differentiation of neuronal and astrocytic precursors within the developing murine spinal cord. Eur J Neurosci 8: 291-299.

Richardson WD, Pringle NP, Yu WP, Hall AC (1997) Origins of spinal cord oligodendrocytes: possible developmental and evolutionary relationships with motor neurons. Dev Neurosci 19: 58-68.

Rigamonti D, Bauer JH, De-Fraja C, Conti L, Sipione S, Sciorati C, Clementi E, Hackam A, Hayden MR, Li Y, Cooper JK, Ross CA, Govoni S, Vincenz C, Cattaneo E (2000) Wild-type huntingtin protects from apoptosis upstream of caspase-3. J Neurosci 20: 3705-3713.

Robel L, Ding M, James AJ, Lin X, Simeone A, Leckman JF, Vaccarino FM (1995) Fibroblast growth factor 2 increases Otx2 expression in precursor cells from mamalian telencephalon. J Neurosci 15: 7879-7891.

Roelink H (1996) Tripartite signaling of pattern: interacions between Hedgehogs, BMPs and Wnts in the control of vertebrate development. Cur Opin Neurobiol 6: 33-40.

Rogister B, Ben-Hur T, Dubois-Dalcq M (1999) From neural stem cells to myelinating oligodendrocytes. Mol Cell Neurosci 14: 287-300.

Roth KA, Kuan C, Haydar TF, D'Sa-Eipper C, Shindler KS, Zheng TS, Kuida K, Flavell RA, Rakic P (2000) Epistatic and independent functions of caspase-3 and Bcl-X(L) in developmental programmed cell death. Proc Natl Acad Sci USA 97: 466-471.

Rubenstein JLR, Martinez S, Shimamura K, Puelles L (1994) The embryonic vertebrate forebrains : the prosomeric model. Science 266: 578-580.

Rubio F, Kokaia Z, Arco A, Garcia-Simon M, Snyder E, Lindvall O, Satrustegui J, Martinez-Serrano A (1999) BDNF gene transfer to the mamalian brain using CNS-derived neural precursors. Gene Ther 6: 1851-1866.

Sakakibara SI, Okano H (1997) Expression of neural RNA-binding proteins in the postnatal CNS: implications of their roles in neuronal and glial cell development. J Neurosci 17: 8300-8312.

Sasahara A, Kott J, Sasahara M, Raines E, Ross R, Westrum L (1992) Platelet-derived growth factor B-chain-like imunoreactivity in the developing and adult brain. Dev Brain Res 68: 41-53.

Sasai Y, Kageyama R, Tagawa Y, Shigemoto R, Nakanishi S (1992) Two mammalian helix-loop-helix factors structurally related to Drosophila hairy and Enhancer of split. Genes Dev 6: 2620-2634.

Sasai Y (1998) Identifying the missing links: genes that connect neural induction and primary neurogenesis in vertebrate embryos. Neuron 21: 455-458.

Sauer FC (1935) Mitosis in the neural tube. J Comp Neurol 62: 377-405.

Savatier P, Lapillonne H, van Grunsven LA, Rudkin BB, Samarut J (1996) Withdrawal of differentiation inhibitory activity/leukemia inhibitory factor up-regulates D-type cyclins and cyclin-dependent kinase inhibitors in mouse embryonic stem cells. Oncogene 12: 309-322.

Sawamoto K, Nakao N, Kakishita K, Ogawa Y, Toyama Y, Yamamoto A, Yamaguchi M, Mori K, Goldman SA, Itakura T, Okano H (2001a) Generation of dopaminergic neurons in the adult brain from mesencephalic precursor cells labeled with a nestin-GFP transgene. J Neurosci 21: 3895-3903.

Sawamoto K, Yamamoto A, Kawaguchi A, Yamaguchi M, Mori K, Goldman SA, Okano H (2001b) Direct isolation of committed neuronal progenitor cells from transgenic mice coexpressing spectrally distinct fluorescent proteins regulated by stage-specific neural promoters. J Neurosci Res 65: 220-227.

Scheffler B, Horn M, Blumcke I, Laywell ED, Coomes D, Kukekov VG, Steindler DA (1999) Marrow-mindedness: a perspective on neuropoiesis. TINS 22: 348-357.

Schofield R (1978) The relationship between the spleen colony-forming cell and the haemopoietic stem cell. Blood Cells 4: 7-25.

Scolding N, Franklin R, Stevens S, Heldin CH, Compston A, Newcombe J (1998) Oligodendrocyte progenitors are present in the normal adult human CNS and in the lesions of multiple sclerosis. Brain 121: 2221-2228.

Schwartz PM, Borghesani PR, Levy RL, Pomeroy SL, Segal RA (1997) Abnormal cerebellar development and foliation in BDNF -/- mice reveals a role for neurotrophins in CNS patterning. Neuron 19: 269-281.

Sendtner M, Gotz R, Holtmann B, Escary JL, Masu Y, Carroll P, Wolf E, Brem G, Brulet P, Thoenen H (1996) Cryptic physiological trophic support of motoneurons by LIF revealed by double gene targeting of CNTF and LIF. Curr Biol 6: 686-694.

Seri B, Garcia-Verdugo JM, McEwen BS, Alvarez-Buylla AA (2001) Astrocytes give rise to new neurons in the adult mammalian hippocampus. J Neurosci 21: 7153-7160.

Sheen VL, Arnold MW, Wang Y, Macklis JD (1999) Neural precursor differentiation following transplantation into neocortex is dependent on intrinsic developmental state and receptor competence. Exp Neurol 158: 47-62.

Shen J, Bronson RT, Chen DF, Xia W, Selkoe DJ, Tonegawa S (1997) Skeletal and CNS defects in presenilin 1 deficient mice. Cell 89: 629-639.

Shetty AK, Turner DA (1998) *In vitro* survival and differentiation of neurons derived from epidermal growth factor-responsive postnatal hippocampal stem cells: inducing effects of brain-derived neurotrophic factor. J Neurobiol 35: 395-425.

Shetty AK, Turner DA (1999) Neurite outgrowth from progeny of epidermal growth factor-responsive hippocampal stem cells is significantly less robust than from fetal hippocampal cells following grafting onto organotypic hippocampal slice cultures: effect of brain-derived neurotrophic factor. J Neurobiol 38: 391-413.

Shimazaki T, Shingo T, Weiss S (2001) The ciliary neurotrophic factor/leukemia inhibitory factor/gp130 receptor complex operates in the maintenance of mammalian forebrain neural stem cells. J Neurosci 21: 7642-7653.

Sibilia M, Steinbach JP, Stingl L, Aguzzi A, Wagner EF (1998) A strain-independent postnatal neurodegeneration in mice lacking the EGF receptor. EMBO J 17: 719-731.

Sibilia M, Wagner EF (1995) Strain-dependent epithelial defects in mice lacking the EGF receptor. Science 269: 234-238.

Silver J (1994) Inhibitory molecules in development and regeneration. J Neurol 242 (S): S22-S24.

Sipione S, Cattaneo E (2002) Modeling brain pathologies using neural stem cells. Methods Mol Biol 198: 245-262. (Zigova T, Sanberg PR, Sanchez-Ramos JR, eds). Totowa, NJ: Humana Press.

Slack RS, El-Bizri H, Wong J, Belliveau DJ, Miller FD (1998) A critical temporal requirement for the retinoblastoma protein family during neuronal differentiation. J Cell Biol 140: 1497-1509.

Smith AG, Heath JK, Donaldson DD, Wong GG, Moreau J, Stahl M, Rogers D (1988) Inhibition of pluripotential embryonic stem cell differentiation by purified polypeptides. Nature 336: 688-690.

Soderstrom S, Bengtsson H, Ebendal T (1996) Expression of serine/threonine kinase receptors including the bone morphogenetic factor type II receptor in the developing and adult rat brain. Cell Tiss Res 286: 269-279.

Sommer L, Rao M (2002) Neural stem cells and regulation of cell number. Prog Neurobiol 66: 1-18.

Song H, Stevens CF, Gage FH (2002) Astroglia induce neurogenesis from adult neural stem cells. Nature 417: 39-44.

Spassky N, Heydon K, Mangatal A, Jankovski A, Olivier C, Oueraud-Lessaux F, Goujet- Zalc C, Thomas JL, Zalc B (2001) Sonic hedgehog-dependent emergence of oligodendrocytes in the telencephalon: evidence for a source of oligodendrocytes in the olfactory bulb that is independent of PDGFR a signaling. Develop 128: 4993-5004.

Stewart CL, Kaspar P, Brunet LJ, Bhatt H, Gadi I, Kontgen F, Abbondanzo SJ (1992) Blastocyst implantation depends on maternal expression of leukaemia inhibitory factor. Nature 359: 76-79.

Takahashi T, Nowakowski RS, Caviness VS Jr (1994) Mode of cell proliferation in the developing mouse neocortex. Proc Natl Acad Sci USA 91: 375-379.

Takebayashi H, Ohtsuki T, Uchida T, Kawamoto S, Okubo K, Ikenaka K, Takeichi M, Chisaka O, Nabeshima Y (2002) Non-overlapping expression of Olig3 and Olig2 in the embryonic neural tube. Mech Dev 113: 169-174.

Takebayashi K, Akazawa C, Nakanishi S, Kageyama R (1995) Structure and promoter analysis of the gene encoding the mouse helix-loop-helix factor HES-5. Identification of the neural precursor cell-specific promoter element. J Biol Chem 270: 1342-1349.

Tate BA, Bower KA, Snyder EY (2002) Transplant therapy. In: Stem Cells and CNS Development (Rao MS, ed), pp 291-305. Totowa, NJ: Humana Press.

Taupin V, Renno T, Bourbonniere L, Peterson AC, Rodriguez M, Owens T (1997) Increased severity of experimental autoimmune encephalomylitis, chronic macrophage/microglial reactivity and demyelination in transgenic mice producing tumor necrosis factor-alpha in the central nervous system. Eur J Immunol 27: 905-913.

Tessarollo L, Tsoulfas P, Martin-Zanca D, Gilbert DJ, Jenkins NA, Copeland NG, Parada LF (1993) trkC, a receptor for neurotrophin-3 is widely expressed in the developing nervous system and in non-neuronal tissues. Develop 118: 463-475.

Thomas J-L, Spassky N, Perez-Villegas EM, Olivier C, Cobos I, Goujet-Zalc C, Martinez S, Zalc B (2000) Spatiotemporal development of oligodendrocytes in the embryonic brain. J Neurosci Res 59: 471-476.

Thomas LB, Gates MA, Steindler DA (1996) Young neurons from the adult subependymal zone proliferate and migrate along an astrocyte, extracellular matrix-rich pathway. Glia 17: 1-14.

Thompson Haskell G, Maynard TM, Shatzmiller RA, Lamantia AS (2002) Retinoic acid signaling at sites of plasticity in the mature central nervous system. J Comp Neurol 452: 228-241.

Threadgill DW, Dlugosz AA, Hansen LA, Tennenbaum T, Lichti U, Yee D, LaMantia C, Mourton T, Herrup K, Harris RC, Barnard JA, Yuspa SH, Coffey RJ, Magnuson T (1995) Targeted disruption of mouse EGF receptor: effect of genetic background on mutant phenotype. Science 269: 230-234.

Timsit S, Martinez S, Allinquant B, Peyron F, Puelles L, Zalc B (1995) Oligodendrocytes originate in a restricted zone of the embryonic ventral neural tube defined by DM-20 mRNA expression. J Neurosci 15: 1012-1024.

Toma JG, el-Bizri H, Barnabe-Heider F, Aloyz R, Miller RD (2000) Evidence that helix-loop-helix proteins collaborate with retinoblastoma tumor suppressor protein to regulate cortical neurogenesis. J Neurosci 20: 7648-7656.

Tomita K, Moriyoshi K, Nakanishi S, Guillemot F, Kageyama R (2000) Mammalian achaete-scute and atonal homologs regulate neuronal versus glial fate determination in the central nervous system. EMBO J 19: 5460-5472.

Trettel F, Rigamonti D, Hilditch-Maguire P, Wheeler VC, Sharp AH, Persichetti F, Cattaneo E, MacDonald ME (2000) Dominant phenotypes produced by the HD mutation in STHdh^{Q111}striatal cells. Hum Mol Genet 9: 2799-2809.

Tropepe V, Craig CG, Morshead CM, van der Kooy D (1997) Transforming growth factor-a null and senescent mice show decreased neural progenitor cell proliferation in the forebrain subependyma. J Neurosci 17: 7850-7859.

Tropepe V, Sibilia M, Ciruna BG, Rossant J, Wagner EF, van der Kooy D (1999) Distinct neural stem cells proliferate in response to EGF and FGF in the developing mouse telencephalon. Dev Biol 208: 166-188.

Turner DL and Cepko CL (1987) A common progenitor for neurons and glia persists in rat retina late in development. Nature 328: 131-136.

Turnley AM, Morahan G, Okano H, Bernard O, Mikoshiba K, Allison J, Bartlett PF, Miller JF (1991) Dysmyelination in transgenic mice resulting from expression of class I histocompatibility molecules in oligodendrocytes. Nature 353: 566-569.

Tzeng SF (2003) Inhibitors of DNA binding in neural cell proliferation and differentiation. Neurochem Res 28: 45-52.

Uchida N, Buck DW, He D, Reitsma MJ, Masek M, Phan TV, Tsukamoto AS, Gage FH, Weissman IL (2000) Direct isolation of human central nervous system stem cells. Proc Natl Acad Sci USA 97: 14720-14725.

Vaccarino FM, Schwartz ML, Raballo R, Nilsen J, Rhee J, Zhou M, Doetschman T, Coffin JD, Wyland JJ, Hung YT (1999) Changes in cerebral cortex size are governed by fibroblast growth factor during embryogenesis. Nature Neurosci 2: 246-253.

Valenzuela DM, Economides AN, Rojas E, Lamb TM, Nunez L, Jones P, Lp NY, Espinosa R 3rd, Brannan CI, Gilbert DJ (1995) Identification of mammalian noggin and its expression in the adult nervous system. J Neurosci 15: 6077-6084.

Valenzuela CF, Kazlauskas A, Weiner JL (1997) Roles of platelet derived growth factor in the developing and mature nervous systems. Brain Res Rev 24: 77-89.

Vandeputte DAA, Troost D, Leenstra S, Ijlst-Keizers H, Ramkema M, Boxch DA, Baas F, Das NK, Aronica E (2002) Expression and distribution of Id helix-loop-helix proteins in human astrocytic tumors. Glia 38: 329-338.

Vescovi AL, Reynolds BA, Fraser DD, Weiss S (1993). bFGF regulates the proliferative fate of unipotent (neuronal) and bipotent (neuronal/astroglial) EGF-generated CNS progenitor cells. Neuron 11: 951-966.

Wai P, Truong B, Bhat KM (1999) Cell division genes promote asymmetric interaction between numb and notch in the drosophila CNS. Develop 126: 2759-2770.

Walicke PA, Baird A (1988) Trophic effects of fibroblast growth factor on neural tissue. Prog Brain Res 78: 333-338.

Wanaka A, Milbrandt J, Johnson EM (1991) Expression of FGF receptor gene in rat development Develop 111: 455-468.

Weinmaster G, Roberts VJ, Lemke G (1991) A homolog of Drosophila Notch expressed during mammalian development. Develop 113: 199-205.

Weinmaster G, Roberts VJ, Lemke G (1992) Notch2: a second mammalian Notch gene. Develop 116: 931-941.

Weiss S (1999) Pathways for neural stem cell biology and repair. Nat Biotechnol 17: 850-851.

Weiss S, Dunne C, Hewson J, Wohl C, Wheatley M, Peterson AC, Reynolds BA (1996) Multipotent CNS stem cells are present in the adult mammalian spinal cord and ventricular neuroaxis. J Neurosci 16: 7599-7609.

Wernig M, Brüstle O (2002) Fifty ways to make a neuron: shifts in stem cell hierarchy and their implications for neuropathology and CNS repair. J Neuropathol Exp Neurol 61: 101-110.

White FA, Keller-Peck CR, Knudson CM, Korsmeyer SJ, Snider WP (1998) Widespread elimination of naturally occurring neuronal death in Bax-deficient mice. J Neurosci 18: 1428-1439.

Whittemore SR, Morassutti DJ, Walters WM, Liu RH, Magnuson DSK (1999) Mitogen and substrate differentially affect the lineage restriction of adult rat subventricular zone neural precursor cell populations. Exp Cell Res 252: 75-95.

Williams BP, Park JK, Alberta JA, Muhlebach SG, Hwang GY, Roberts TM, Stiles CD (1997) A PDGF-regulated immediate early gene response initiates neuronal differentiation in ventricular zone progenitor cells. Neuron 18: 553-562.

Williams R, Lendahl U, Lardelli M (1995) Complementary and combinatorial patterns of Notch gene family expression during early mouse development. Mech Dev 53: 357-368.

Wu Y, Levine EM, Rao MS (2003) Hes1 but not Hes5 regulates an astrocyte versus oligodendrocyte fate choice in glial restricted precursors. Dev Dyn 226: 675-689.

Xie T, Spradling AC (1998) decapentaplegic is essential for the maintenance and division of germline stem cells in the Drosophila ovary. Cell 94: 251-260.

Yamada N, Kato M, ten Dijke P, Yamashita H, Sampath TK, Heldin CH, Miyazono K, Funa K (1996) Bone morphogenetic protein type IB receptor is progressively expressed in malignant glioma tumors. Br J Cancer 73: 624-629.

Yamaguchi TP, Conlon RA, Rossant J (1992) Expression of the fibroblast growth factor receptor FGFR-1/flg during gastrulation and segmentation in the mouse embryo. Dev Biol 152: 75-88.

Yao L, Zhang D, Bernd P (1994) The onset of neurotrophin and trk mRNA expression in early embryonic tissues of the quail. Dev Biol 165: 727-730.

Yamaguchi M, Saito H, Suzuki M, Mori K (2000) Visualization of neurogenesis in the central nervous system using nestin promoter-GFP transgenic mice. Neuroreport 11: 1991-1996.

Ye P, Li L, Richards RG, DiAugustine RP, D'Ercole AJ (2002) Myelination is altered in insulin-like growth factor-I null mutant mice. J Neurosci 22: 6041-6051.

Ye W, Shimamura K, Rubenstein JLR, Hynes MA, Rosenthal A (1998) FGF and Shh signals control dopaminergic and serotonergic cell fate in the anterior neural plate. Cell 93: 755-766.

Yi SE, Daluiski A, Pederson R, Rosen V, Lyons KM (2000) The type I BMP receptor BMPRIB is required for chondrogenesis in the mouse limb. Develop 127: 621-630.

Zappone MV, Galli R, Catena R, Meani N, De Biasi S, Mattei E, Tiveron C, Vescovi AL, Lovell-Badge R, Ottolenghi S, Nicolis SK (2000) Sox2 regulatory sequences direct expression of a (beta)-geo transgene to telencephalic neural stem cells and precursors of the mouse embryo, revealing regionalization of gene expression in CNS stem cells. Develop 127: 2367-2382.

Zhang SC, Lundberg C, Lipsitz D, O'Connor LT, Duncan ID (1998) Generation of oligodendroglial progenitors from neural stem cells. J Neurocytol 27: 475-489.

Zhong W, Feder JN, Jiang MM, Jan LY, Jan YN (1996) Asymmetric localization of a mammalian numb homolog during mouse cortical neurogenesis. Neuron 17: 43-53.

Zhu G, Mehler MF, Mabie PC, Kessler JA (1999a) Developmental changes in neural progenitor cell lineage commitment do not depend on epidermal growth factor receptor signaling. J Neurosci Res 59: 312-320.

Zhu G, Mehler MF, Zhao J, Yu Yung S, Kessler JA (1999b) Sonic hedgehog and BMP2 expert opposing actions on proliferation and differentiation of embryonic neural progenitor cells. Dev Biol 215: 118-129.

Zhuo L, Theis M, Alvarez-Maya I, Brenner M, Willecke K, Messing A (2001) hGFAP-cre transgenic mice for manipulation of glial and neuronal function *in vivo*. genesis 31: 85-94.

Zigova T, Betarbet R, Soteres BJ, Brock S, Bakay RA, Luskin MB (1996) A comparison of the patterns of migration and the destinations of homotopically transplanted neonatal subventricular zone cells and heterotopically transplanted telencephalic ventricular zone cells. Dev Biol 73: 459-474.

Zigova T, Pencea V, Wiegand SJ, Luskin MB (1998) Intraventricular administration of BDNF increases the number of newly generated neurons in the adult olfactory bulb. Mol Cell Neurosci 11: 234-245.

Chapter 2

Neural Stem and Progenitor Cells: Lineage and Cell Fate Determination

Stephen N. Sansom, Sabhi Rahman, Uruporn Thammongkol and Frederick J. Livesey

INTRODUCTION

Neurogenesis, the process by which postmitotic neurons are generated from pools of mitotic progenitor cells, is a highly regulated process in all organisms studied (Edlund and Jessell, 1999; Livesey and Cepko, 2001). Different types of neurons are produced in a temporal sequence that is conserved in different species, and different types of neurons are produced in different parts of the nervous system (Cepko et al., 1996). Discrete phenotypes or identities are assigned to the postmitotic progeny of neural progenitor cells through a process of cell fate determination. To a significant degree, the fates of those progeny are decided within the mitotic progenitor cell before it divides. Thus, progenitor cells have an integrative function whereby they combine extrinsic information in the form of extracellular signals with information intrinsic to the cell to decide the fates of their daughter cells, as will be discussed in more detail below.

Given the emergence of findings in recent years illustrating the conservation of mechanisms controlling neural cell fate determination in vertebrate and invertebrate development, it is unlikely that alternative mechanisms are acting in adult neural stem cells. Therefore, an understanding of the cellular and molecular mechanisms involved in this process during development will be of direct benefit to efforts to exploit neural stem cells for therapeutic uses. The developmental biology of neural cell fate determination can be broadly divided into a series of processes: the induction or appearance of neurogenic tissue(s), that is tissue containing neural stem and progenitor cells; the division of this tissue into distinct territories or regions that go on to form different components of the adult nervous system; and the ordered production of region-specific neurons within each territory. Several striking recent studies have clearly shown that this process can be recapitulated *in vitro*, generating particular classes of neurons from embryonic stem (ES) cells through a series of discrete steps aimed

From: *Neural Stem Cells: Development and Transplantation*
Editor: Jane E. Bottenstein © 2003 Kluwer Academic Publishers, Norwell, MA

at guiding cells through each stage in this process (Kim et al., 2002; Wichterle et al., 2002).

In contrast with our increasing understanding of lineage and cell fate determination during neural development, much less is known of the origins, lineage choices and cell fate determination mechanisms operating in adult neural stem cells (NSCs). Therefore, this chapter will discuss what is known of neural lineage and cell fate determination mechanisms in both vertebrates and invertebrates, comparing this with what is known in adult NSCs.

THE ORIGINS OF NEURAL LINEAGES DURING EMBRYONIC DEVELOPMENT

The neural lineages of *Drosophila*

In the fruit fly, *Drosophila*, the neural lineages of the central nervous system (CNS) originate from bilaterally symmetrical neurogenic regions on either side of the embryonic ventral midline. This ventral neurogenic region will give rise to the ventral nerve cord (Bate, 1976; Bossing et al., 1996). The neuroectodermal cells of this region have the potential to become either epidermal or neural progenitor cells (Doe and Goodman, 1985). Initially this ventral neurogenic region is characterized by the broad expression of a group of genes that confer a neural fate, referred to as the proneural genes and described in detail below (Alonso and Cabrera, 1988; Cabrera, 1990; Cabrera and Alonso, 1991; Cabrera et al., 1987; Jimenez and Campos-Ortega, 1990; Martin-Bermudo et al., 1991). Not all cells within the neurogenic region become neural progenitors. Instead, only some nonadjacent cells continue to express proneural genes and proneural gene expression in the surrounding cells is lost (Cabrera, 1990; Cabrera et al., 1987).

Developmental origins of the vertebrate nervous system

In vertebrates the central nervous system derives from a neural plate that is induced in the dorsal ectoderm. The induction of this neuroectoderm has been most intensively studied in *Xenopus* and it is understood to take place according to a process known as the default model (Wilson and Hemmati-Brivanlou, 1997). In the default model, neural fate represents the default state of the ectoderm of the early embryo, that is normally repressed by factors of the bone morphogenetic protein (BMP) family. In *Xenopus*, neural induction is achieved by the secretion of BMP inhibitors, including chordin and noggin, from the organizer (Harland, 2000). The overexpression of Bmp2/4 prevents neural induction and promotes the formation of ectoderm, whilst the ectopic expression of the BMP inhibitors promotes neural induction at the expense of the ectodermal fate (Wilson and Edlund, 2001). The vertebrate neural plate, like the neurogenic regions of the *Drosophila* embryo, is characterised, by the expression of proneural genes (Blader et al., 1997; Henrique et al., 1997; Ma et al., 1996).

Notch signaling is responsible for the selection of neural progenitor cells

In the neuroectoderm of *Drosophila* lateral inhibition through the notch signaling pathway is responsible for the process by which the broad initial expression of the proneural genes becomes restricted to the subset of progenitor cells which will give rise to the nervous system (Artavanis-Tsakonas et al., 1999; Chitnis and Kintner, 1996; Heitzler and Simpson, 1991; Lewis, 1998). Although the overexpression of notch or its ligand delta in the neuroectoderm does not affect neurogenesis, the ectopic activation of notch signaling prevents the formation of neural progenitor cells, while the inactivation of notch sigalling results in the generation of ectopic neural progenitors (Hartenstein and Posakony, 1990; Heitzler and Simpson, 1991; Lehmann et al., 1983; Lieber et al., 1993; Nakao and Campos-Ortega, 1996; Parks and Muskavitch, 1993; Rebay et al., 1993; Seugnet et al., 1997; Struhl et al., 1993).

A situation comparable to that in *Drosophila* is found in the neural plates of *Xenopus* and zebrafish where lateral inhibition by notch signaling also restricts proneural gene expression and neural fate to a subset of cells (Blader et al., 1997; Chitnis et al., 1995; Chitnis and Kintner, 1996; Ma et al., 1996). In both flies and vertebrates, the lateral inhibition of neural fate in the prospective neural territories is due to an upregulation of the notch ligand, delta, by the proneural genes in cells fated to become neural precursors (Cau et al., 2002; Chitnis and Kintner, 1996; Fode et al., 1998; Lewis, 1998; Ma et al., 1999; Perron and Harris, 2000a). The notch target genes, hairy and enhancer of split E(spl) in *Drosophila*, and the vertebrate hairy and enhancer of split homolog (Hes), hairy and enhancer of split related (Her), and enhancer of split related (Esr) are all repressors of proneural gene expression (Casarosa et al., 1999; Cau et al., 2002; Chitnis and Kintner, 1996; Fode et al., 1998; Heitzler et al., 1996; Hinz et al., 1994; Kunisch et al., 1994; Ma et al., 1998).

The precise mechanism by which notch signaling selects neural progenitor cells is still unclear. In *Drosophila*, notch and its ligand, delta, are initially expressed evenly in the neuroectoderm (Fehon et al., 1991; Kooh et al., 1993) but, unlike other cells of the neuroectoderm, prospective neural precursors do not express the notch target gene E(spl) (Baker et al., 1996; Dokucu et al., 1996; Jennings et al., 1994). It is therefore thought that the cells of the *Drosophila* neuroectodem are specified as neural progenitors by a spatial inactivation of the notch pathway (Baker, 2000).

ROLES OF PRONEURAL GENES IN VERTEBRATE AND INVERTEBRATE NEUROGENESIS

Proneural gene families in flies and vertebrates

The proneural genes have been clearly implicated in neurogenesis in both flies and vertebrates [for review see Bertrand et al. (2002)]. Proneural genes are both necessary and sufficient to initiate the development of neuronal lineages and to promote the generation of progenitors which are committed to neuronal differentiation. Proneural genes are transcription factors which contain a basic helix-loop-helix (bHLH) domain which confers dimerization and DNA binding properties (Murre et al., 1989). The proneural genes were originally identified in *Drosophila* in the early 1970s as a complex of genes involved in the early stages of neural development (Garcia-Bellido, 1979; Ghysen and Dambly-Chaudiere, 1988).

Two classes of proneural genes are known in *Drosophila*. The achaete-scute (asc) family consists of four genes: achaete, scute, lethal of scute and asense (Gonzalez et al., 1989; Villares and Cabrera, 1987). The second, the atonal (ato) family, has three members, atonal, amos and cato (Goulding et al., 2000a,b; Huang et al., 2000b; Jarman et al., 1993). In vertebrates there are several families of proneural genes named according to their homology with those in *Drosophila*: these are the achaete-scute homologs (ath), the atonal homologs (ath), and the atonal-related (atr) gene families (Guillemot, 1999; Lee, 1997). The vertebrate ash family consists of four members, ash1-4, which are prefixed in vertebrates by the first letter of the species name such that: ash1 in mice is Mash1, in *Xenopus* is Xasth1, and in zebrafish is Zash1. The vertebrate ath gene family is larger but only two of its members are considered true orthologs of the *Drosophila* ato genes (these are Math1 and Math5 in mice). Examples of the vertebrate atonal-related families are the NeuroD, Neurogenin and Olig gene families (Hassan and Bellen, 2000; Lee, 1997). These family relations are based on the presence of specific residues within the bHLH domain.

How do proneural genes function?

Proneural genes function by binding to DNA as heterodimers with the ubiquitously expressed bHLH 'E' proteins: E2A, HEB and E2-2 in vertebrates, and daughterless (da) in *Drosophila* (Cabrera and Alonso, 1991; Johnson et al.,

1992; Massari and Murre, 2000). The bHLH domain of the proneural genes contains a stretch of ten DNA binding residues, of which nine are conserved between all proneural genes (Bertrand et al., 2002; Chien et al., 1996). These conserved DNA binding residues recognise the E-box (CANNT) promoter element. Most proneural genes function as activators of target gene transcription, with the exception of Olig2, which is a repressor (Cabrera and Alonso, 1991; Johnson et al., 1992; Mizuguchi et al., 2001; Novitch et al., 2001).

Repression of proneural function can be achieved by disruption of their heterodimerisation with the ubiquitous E proteins. The *Drosophila* extra macrochaetae (emc) and vertebrate inhibitor of differentiation (Id) genes possess bHLH domains but lack DNA binding motifs, and are thought to compete with proneural proteins for E proteins, thus inhibiting proneural gene function (Cabrera and Alonso, 1991; Campuzano, 2001; Yokota, 2001). The *Drosophila* hairy and enhancer of split (Espl), and the vertebrate hairy and enhancer of split homolog (Hes), hairy and enhancer of split related (Her), and enhancer of split related (Esr) genes are transcriptional repressors of proneural genes and are also thought to repress proneural function by the disruption of heterodimer formation (Davis and Turner, 2001; Kageyama and Nakanishi, 1997).

Expression of the proneural genes

In *Drosophila*, expression of the proneural genes begins in the quiescent cells (the cells of the neuroectoderm are not actively cycling at this stage) of the neuroectoderm which are competent to adopt both epidermal and neural fates. Proneural gene expression begins in clusters of cells in the neuroectoderm, which reflect the later distribution of neural progenitor cells in the peripheral and central nervous systems (Campuzano and Modolell, 1992). The refinement of proneural gene expression by notch signaling results in the selection and delamination of neural progenitors from the neuroectoderm (Jan and Jan, 1994; Jimenez and Campos-Ortega, 1990). In vertebrates, proneural genes are first expressed in the neural plate, the cells of which, in contrast to those of the neuroectoderm of *Drosophila*, are actively cycling and have already been specified for a neural fate. Proneural gene expression thus acts in combination with notch signaling to specify the the formation of neuronal progenitor cells in the neural plate that possess a limited mitotic potential (Casarosa et al., 1999; Cau et al., 2002; Fode et al., 1998; Fode et al., 2000; Horton et al., 1999; Ma et al., 1998; Ma et al., 1999).

Proneural genes are responsible for the specification of neural progenitor cells

In *Drosophila* a major role of the proneural genes is to promote the specification of neural progenitors in both the peripheral nervous system (PNS) and central nervous system (CNS). Mutations that disrupt proneural gene function in *Drosophila* result in a reduction in the numbers of neural progenitors generated, whereas the overexpression of proneural genes results in the ectopic formations of neural progenitors (Dominguez and Campuzano, 1993; Jimenez and Campos-Ortega, 1990; Rodriguez et al., 1990). In vertebrates, the ash, atoh and ngn genes have a proneural role which is similar to the role of their *Drosophila* homologues. The loss of neural progenitors in vertebrate models mutant for proneural gene function is correlated with premature astrocyte generation, and there is evidence that proneural genes promote the neuronal fate and repress the glial fate in vertebrates (Casarosa et al., 1999; Cau et al., 2002; Fode et al., 1998; Guillemot and Joyner, 1993; Horton et al., 1999; Ma et al., 1998; Ma et al., 1999; Scardigli et al., 2001).

Other vertebrate proneural genes, for example NeuroD and Math3/NeuroM, have characteristics more similar to those of neural differentiation genes, but are also implicated in dictating a neuronal rather than a glial cell fate choice in some regions (Morrow et al., 1999; Tomita et al., 2000). As in *Drosophila*, the over-expression of many vertebrate proneural genes has the opposite effect to loss of function studies, promoting neuronal differentiation (Blader et al., 1997; Ma et al., 1996; Mizuguchi et al., 2001). However, direct evidence for the proneural function of some vertebrate proneural genes is lacking. For example, Math1 and Math5 are involved in specifying neuronal identity, but do not seem to have a proneural function (Bermingham et al., 1999; Gowan et al., 2001; Hassan and Bellen, 2000). Mutational studies in the mouse have only established classical proneural function for a few genes, including Mash1, Ngn1 and Ngn2. Furthermore, the known vertebrate proneural genes do not account for the generation of all the known neural lineages (Fode et al., 1998; Ma et al., 1998; Ma et al., 1999; Sommer et al., 1995). There are therefore many similarities, but also clear differences in the roles of the proneural genes in progenitor cell selection in vertebrates and flies.

The role of proneural genes in neuronal differentiation

After selection, neural progenitor cells further upregulate proneural gene expression before becoming commited to differentiation (Culi and Modolell, 1998; Kintner, 2002; Koyano-Nakagawa et al., 1999; Vaessin et al., 1994). Positive feedback loops serve to maintain and upregulate proneural gene expression in prospective progenitor cells. For example, the transcription factors senseless in *Drosophila* and Xcoe2 and Hes6 in vertebrates are induced by proneural genes and upregulate proneural gene expression (Bae et al., 2000; Dubois et al., 1998; Koyano-Nakagawa et al., 2000; Nolo et al., 2000). Some proneural genes are subject to autoregulation, such as the vertebrate atonal homolog, Math1, and conversely other vertebrate proneural genes are known not to autoregulate such as Mash1 and Ngn1 (Guillemot et al., 1993; Helms et al., 2000; Nieto et al., 2001; Sun et al., 1998; Van Doren et al., 1992).

Whilst the proneural genes have a role in the promotion of neural fate, proneural gene expression in neural progenitors is transient. In vertebrates, proneural genes are downregulated before progenitors exit the proliferative zone of the neural tube and begin to differentiate (Ben-Arie et al., 1996; Gradwohl et al., 1996; Ma et al., 1998). In *Drosophila*, proneural genes are downregulated before progenitors start to generate the sense organs of the PNS and the ganglion mother cells of the CNS (Cubas et al., 1991; Jarman et al., 1993; Skeath and Carroll, 1991). Proneural genes therefore function to confer a neural fate by switching on downstream genes, known as the neuronal differentiation genes.

Many neuronal differentiation genes possess bHLH domains, and are related to the proneural genes, and this has given rise to the idea that, as is the situation in muscle differentiation, cascades of different bHLH genes are responsible for neural cell fate determination and differentiation (Jan and Jan, 1993; Kintner, 2002; Lee, 1997; Weintraub, 1993). bHLH neuronal differentiation genes are expressed later than the proneural genes, are under the transcriptional control of proneural genes, and can promote neuronal differentiation if ectopically expressed. In the fly, Asense is a direct transcriptional target of Achaete and Scute and is involved in sense organ differentiation (Dominguez and Campuzano, 1993; Jarman et al., 1993). In vertebrates, bHLH genes of the NeuroD family are downstream of the neurogenins (Fode et al., 1998; Huang et al., 2000a; Ma et al., 1998) and have the characteristics of neural differentiation genes (Farah et al., 2000; Lee et al., 1995; Liu et al., 2000a; Miyata et al., 1999; Olson et al., 2001; Schwab et al., 2000). Because proneural genes and neuronal differentiation genes are structurally related, it is plausible that their distinct functions may be due to the different times at which they are expressed. This possibility has not been fully investigated, although there is evidence that several proneural genes control differentiation steps in neuronal lineages.

Proneural genes have a role in the specification of neuronal identity

As well as their role in progenitor selection, a role for proneural genes in the specification of neuronal identity has emerged. Proneural genes are often expressed in restricted progenitor domains that will give rise to particular types of neurons. In the dorsal vertebrate spinal cord, Math1, Ngn1 and Mash1 are expressed in distinct dorsoventral progenitor domains that produce distinct types of interneurons (Gowan et al., 2001). Mutant analysis in the mouse has shown that Math1 and Ngn1 are necessary for the correct specification of some neural progenitor domains, further linking proneural gene expression to neural cell fate determination. (Bermingham et al., 2001; Gowan et al., 2001). In *Drosophila*, loss of function studies have shown that different types of proneural genes are involved both in the formation of different types of sense organs (Huang et al., 2000b; Jarman et al., 1993; Jarman et al., 1994), and in the formation of different types of neurons in the CNS (Parras et al., 1996; Skeath and Doe, 1996).

In vertebrates a well studied example of the role of the proneural genes in the specification of neuronal identity is the role of Mash1 in the generation of noradrenergic neurons. In the PNS, loss and gain of function experiments have shown that Mash1 acts together with the homeodomain protein Phox2b to induce the expression of Phox2a, a related homeobox gene, and dopamine b-hydroxylase (DBH), in the specification of noradrenergic neurons in the sympathetic ganglia (Goridis and Rohrer, 2002; Hirsch et al., 1998; Lo et al., 1998; Pattyn et al., 1999). By contrast, in the noradrenergic centres of the brain, Mash1 induces the expression of both Phox2b and Phox2a (Goridis and Brunet, 1999; Pattyn et al., 2000). Mash1 has also been implicated in the specification of other neuronal identities, for example, in the ventral forebrain Mash1 is expressed in domains which give rise γ-amino butyric acid (GABA) GABAergic neurons (Fode et al., 2000; Parras et al., 2002). The involvement of Mash1 in the specification of different kinds of neurons indicates that it must interact with regionally expressed factors that modify its specificity.

The vertebrate Neurogenin genes are also thought to be involved in the specification of neuronal identity. In the PNS a role has been established for the Neurogenins in the specification of sensory neurons, and in the CNS Ngn2 has been shown to cooperate with Olig2 in motor neuron induction (Lo et al., 2002). In the retina NeuroD and Math3 are necessary and sufficient for the generation of amacrine interneurons (Inoue et al., 2002; Morrow et al., 1999), whilst Math3 and Mash1 are involved in specifying bipolar fate (Hatakeyama et al., 2001). The specification of neuronal fate can therefore be carried out by non-proneural bHLH proteins, and is uncoupled from the selection of progenitors in some neural lineages.

Evidence from *Drosophila* indicates that different proneural genes regulate different target genes. For example, the gene cut, expressed in the progenitors of external sense organs, is induced by the asc genes but is repressed by atonal (Blochlinger et al., 1991; Jarman and Ahmed, 1998). Specificity in the regulation of target genes is thought to be conferred both by the different DNA binding properties of the different proneural genes, and by regionally expressed cofactors. Sequence analysis of E-box motifs has revealed that different proneural proteins recognise distinct E-box sequences (Bertrand et al., 2002; Chien et al., 1996). In *Drosophila*, the regionally expressed cofactors Pannier and Chip have been identified and shown to modulate Achaete/Scute-Daughterless mediated activation of achaete transcription (Ramain et al., 2000).

THE GENERATION OF NEURONAL TYPES: PROGENITOR REGIONALIZATION

A primary event in the construction of a nervous system is the division of the nascent CNS into a number of discrete territories or regions, typically by conferring distinct regional identities on neural progenitor cells. A further round of spatial patterning of progenitor cells then occurs within each region, as discussed below. As already described in both *Drosophila* and vertebrates, many neural progenitor cells acquire a regional identity during their initial induction or generation in the neuroectoderm and neural plate.

Regionalization of the ventral neuroectoderm in *Drosophila*

The ventral neurogenic region of the *Drosophila* embryo gives rise to thirty neuroblasts per hemisegment during neurogenesis. These neuroblasts delaminate from the neuroepithelium in five successive waves along the ventrodorsal and anterior-posterior axes in rows and columns in a stereotyped spatiotemporal pattern (Bate, 1976; Bossing et al., 1996; Doe, 1992; Hartenstein and Campos-Ortega, 1984; Schmidt et al., 1997). Upon formation, each neuroblast has a unique fate, and each neuroblast that arises in a particular position at a particular time during development always has the same fate (Bossing et al., 1996; Schmidt et al., 1997). The neural progenitors of the ventral neurogenic region thus have an identity and fate that is conferred in the neuroepithelium.

The initial dorsoventral patterning of neuroblasts is conferred by a set of transcription factors known as the columnar genes. These are the homeodomain proteins encoded by ventral nervous system defective (vnd), intermediate neuroblasts defective (ind), msh and dichaete (Buescher and Chia, 1997; Isshiki et al., 1997; McDonald et al., 1998; Skeath et al., 1994; Skeath et al., 1995). These factors are expressed in ventral to dorsal stripes within the ventral neurogenic region. The anteroposterior patterning of neuroblasts is understood to be under control of the segment polarity genes wingless (wg), gooseberry (gsb), patched (ptc), and hedgehog (hh) (Bhat, 1996; Bhat, 1999; Bhat and Schedl, 1997; Chu-LaGraff and Doe, 1993; Skeath et al., 1995). It is thought the segment polarity genes and the columnar genes function together with the proneural genes to confer regional identity during neural progenitor formation (Skeath and Thor, 2003). In this interpretation, the interaction of proneural proteins with the regionally expressed factors forms a code to specify the unique fate of every neuroblast.

Regionalization of the vertebrate neural tube

The primary regionalization event in the vertebrate CNS is its division into the broad territories of forebrain, midbrain, hindbrain and spinal cord. Within each territory, a fine-scale regionalization of neural progenitors takes place. In vertebrates the best understood example of fine-scale neural progenitor regionalisation is found in developing the neural tube. This structure is patterned dorsoventrally by molecules homologous to those responsible for patterning the ventral neurogenic region of the fly. The dorsoventral axis of the ventral half of the neural tube can be subdivided into five progenitor domains known as p0, p1, p2, MN and p3, based on differential gene expression (Briscoe et al., 2000). Each of these domains gives rise to a distinct class of neurons (for details, see Figure1). Domains p0-p3 give rise to V0-V3 interneurons, whilst the pMN domain gives rise to motor neurons (MNs) (Briscoe et al., 2000; Ericson et al., 1997a; Pierani et al., 2001; Sharma et al., 1998).

The five progenitor domains are initially specified by a gradient of the signaling molecule sonic hedgehog (SHH), secreted from the ventral floor plate (Ericson et al., 1996; Roelink et al., 1995). The progenitors of the neural tube are highy sensitive to the ambient concentration of SHH, and this results in the graded expression of a group of transcription factors (hereafter referred to as spinal cord TFs) by the neural tube progenitor cells (Briscoe et al., 2001; Briscoe et al., 2000; Ericson et al., 1996). Many of the spinal cord TFs possess homeodomains, although one is a bHLH factor (Olig2) (Lee and Pfaff, 2001). These spinal cord TFs can be divided into two classes: Class I factors (Pax6, Irx3, Dbx2, Dbx1, Pax3/7) are repressed by SHH signaling whereas Class II factors (Nkx2.2/2.9, Olig2, Nkx6.1, Nkx6.2) are induced by SHH (Briscoe et al., 2000).

Furthermore, Class I and II spinal cord TFs are antagonistic and downregulate the expression of one another, in a process known as cross regulation, which functions to establish sharp boundaries in gene expression (Briscoe et al., 2000; Ericson et al., 1997b; Mizuguchi et al., 2001; Novitch et al., 2001; Sander et al., 2000; Vallstedt et al., 2001). The expression of combinations of these spinal cord TFs defines five progenitor domains (Jessell, 2000). Functional studies have demonstrated that these transcription factors act in a combinatorial manner to specify distinct neural identities (Briscoe et al., 2000; Briscoe

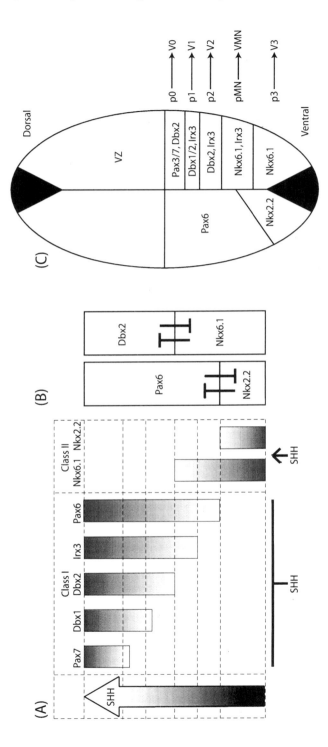

Figure 1. Regionalization of spinal cord progenitor cells. (A) Sonic hedgehog, produced by the floor plate, mediates the inhibition of expression of class I homeodomain proteins and the induction of expression of class II proteins within progenitor cells. The expression of each factor has differing sensitivities to SHH concentration. **(B)** Homeodomain proteins cross-repress one another at a common progenitor domain boundary. **(C)** Cross repression between class I and II proteins sets up five distinct progenitor domains (p0, p1, p2, pMN and p3). These progenitor domains have been shown to act in a combinatorial manner to specify distinct neural identities (V0, V1, V2 and V3 represent interneurons and MN denotes motor neurons) within the spinal cord. Ventricular zone (VZ). For detailed review, see Jessell, 2000.

et al., 1999; Ericson et al., 1997b; Mansouri and Gruss, 1998; Sander et al., 2000). Several of the spinal cord TFs are homologous to genes involved in the dorsoventral patterning of the *Drosophila* ventral neurogenic region: nkx2.2 is related to vnd, gsh-1/2 to ind and msx is related to msh(dr) (Cornell and Ohlen, 2000).

The intracellular mechanisms by which spinal cord TFs are expressed in response to the SHH gradient has not been fully elucidated. However members of the Gli/ci gene family, which is known to be downstream of SHH signaling in *Drosophila,* have been implicated in this process (Ding et al., 1998; Litingtung and Chiang, 2000; Matise et al., 1998). It is also known that SHH is able to induce the expression of target genes via Gli/ci independent mechanisms (Krishnan et al., 1997).

How do spinal cord TFs function in the specification of neuronal identity?

Spinal cord TFs, contrary to expectation, are understood to act by the repression of their target genes. Eight of the eleven spinal cord TFs possess a engrailed homology (eh1) domain, conserved with the engrailed repressor (Muhr et al., 2001; Smith and Jaynes, 1996). This domain is understood to interact with the Groucho-TLE (Gro/TLE) corepressors, which are broadly expressed in the developing neural tube (Allen and Walsh, 1999; Muhr et al., 2001). The Gro/TLE repressors are thought to mediate gene regulation by promoting interaction with histone deacetylases to modulate chromatin structure or possibly by more direct interaction with the transcription machinery (Chen et al., 1999; Edmondson and Roth, 1998; Edmondson et al., 1996; Lee and Pfaff, 2001). It is possible that additional corepressors function with spinal cord TFs in the developing neural tube.

Spinal cord TFs are thus understood to act to confer neural fate by the negative regulation of their target genes. Differential target gene expression in the five progenitor regions might be achieved by the presence of different transcription factor binding sites in the promoters of different target genes (Lee and Pfaff, 2001). In this model the downstream target genes of the spinal cord TFs are initially broadly expressed, but become restricted to permissive progenitor domains. Examples of targets of spinal cord TFs are evx1 in V0 cells, en1 in V1 cells, Lhx3/4 and Chx10 in V2 cells, MNR2/HB0 and Isl1/2 in motor neurons and Sim1 in V3 interneurons (Lee and Pfaff, 2001). Expression of these downstream genes begins in the spinal cord neurons when they exit the cell cycle, and is responsible for conferring specific neuronal phenotypes (Lee and Pfaff, 2001). Studies of mouse mutant models of the spinal cord TF target genes have revealed that whilst some of these genes dictate all aspects of cell identity, others only act to specify certain features of cell fate, such as axon guidance properties (Matise and Joyner, 1997; Moran-Rivard et al., 2001; Saueressig et al., 1999).

The rostrocaudal patterning of neural progenitors in the spinal cord is not as well understood as their dorsoventral patterning. Although most kinds of neuron are represented at the different segmental levels of the spinal cord, strikingly some classes of motor neurons are not (Jessell, 2000). Grafting experiments have shown that the initial rostrocaudal patterning of the neural tube is carried out by interactions with the paraxial mesoderm (Appel et al., 1995; Ensini et al., 1998; Itasaki et al., 1996; Lance-Jones et al., 2001; Lumsden and Krumlauf, 1996). Although the signals involved in this patterning are still being elucidated, signaling by retinoic acid is known to be important (Muhr et al., 1999).

A class of genes understood to be important for the rostrocaudal patterning of neurons in the spinal cord are the classical Homeobox (Hox) genes. There are four Hox gene clusters: a, b, c, and d in vertebrates, which are related to the homeotic genes of *Drosophila*. In vertebrates, Hox genes have a well established role in axial patterning (Burke et al., 1995). Members of the Hox-c and Hox-d gene clusters are expressed at different rostrocaudal levels of the spinal cord and there is evidence that they are necessary for the specification of some classes of neuron in the developing neural tube, indicating that Hox genes play a role in the rostrocaudal regionalisation of the nervous system (Belting et al., 1998; de la Cruz et al., 1999; Keynes and Krumlauf, 1994).

Progenitor cells within a given domain are multipotent

An important feature of the progenitor domains in the developing spinal cord is that whilst they may be defined by the expression of regional factors and proneural genes, these domains are not resticted to the generation of a single type of neuron. In the pMN domain of the ventral vertebrate spinal cord progenitors are known to undergo a switch from motor neuron production to oligodendrocyte generation over time (Lu et al., 2000; Pringle et al., 1998; Richardson et al., 1997; Soula et al., 2001; Zhou et al., 2001; Zhou et al., 2000). In addition to temporal changes it is possible that there is heterogeneity within individual progenitor domains.

CONTROL OF THE TEMPORAL ORDER OF NEUROGENESIS

During neurogenesis, multipotent neural progenitors give rise to a series of differentiated neurons (Bossing et al., 1996; Schmidt et al., 1997). The known mechanisms by which neural progenitors gain a spatial identity, through the activity of regionalised factors and the proneural genes have been discussed. However, the generation of the individual neurons of a specific neural lineage over time presents a new problem. How does a neural progenitor give rise to a series of neurons with distinct identities over time? A key mechanism used in all nervous systems studied is asymmetric division of stem and progenitor cells.

Asymmetric cell division in *Drosophila*

In *Drosophila* both the CNS progenitors, neuroblasts (NBs), and the PNS progenitors, the sensory organ precursors (SOPs) undergo a series of asymmetric divisions in order to generate characteristic lineages of neurons and glia (Bossing et al., 1996; Gho et al., 1999; Reddy and Rodrigues, 1999; Schmid et al., 1999; Schmidt et al., 1997). After delaminating from the neuroectoderm, *Drosophila* neuroblasts undergo a series of apical/basal orientated asymmetric divisions (Figure 2). These asymmetric divisions give rise to a smaller daughter cell, the ganglion mother cell (GMC), which buds off from the dorsal/lateral cortex of the neuroblast. GMCs then divide terminally to give rise to two neurons or glia.

The asymmetric division of neuroblasts requires the asymmetric localisation of cell fate determinants and the correct orientation of the mitotic spindle in order for the proper segregation of cell fate determinants to the GMC daughter cell. The polarity of neuroblasts is established by an apical protein complex consisting of Bazooka, DaPKC and DmPar6, which also mediates polarity in the epithelium (Petronczki and Knoblich, 2001; Schober et al., 1999; Wodarz et al., 2000; Wodarz et al., 1999). The cell-fate determinants Prospero, prospero mRNA and Numb, and the adapter molecules that help to localise them which are Miranda, Staufen, and Partner of numb, respectively, form a basal crescent within the neuroblast (Broadus et al., 1998; Hirata et al., 1995; Ikeshima-Kataoka et al., 1997; Knoblich et al., 1995; Li et al., 1997; Lu et al., 1998; Rhyu et al., 1994; Schuldt et al., 1998; Shen et al., 1997; Spana and Doe, 1995). This basal crescent, which overlies the basal spindle pole of mitotic NBs, segregates to the GMC daughter cell. The neural lineages of the peripheral nervous system are generated by the SOPs which undergo a series of asymmetric cell divisions to give rise to four different cell types which together constitute an external sense organ (Bodmer et al., 1989).

Asymmetric cell division in vertebrates

In vertebrates, the asymmetric cell division of neural progenitors has been reported in the cortex and in the retina of the rat (Cayouette et al., 2001; Chenn and McConnell, 1995). In the mammalian cerebral cortex, neural progenitors of

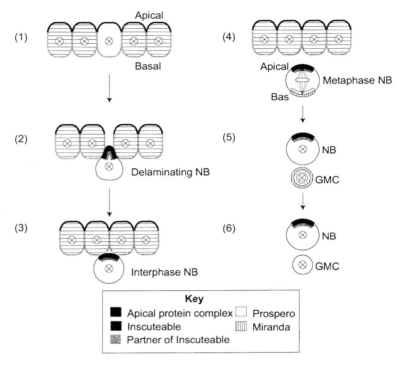

Figure 2: The origin and asymmetric division of *Drosophila* neuroblasts. (**1**) The neuroblast (NB) is specified in the neuroectoderm. Polarity is determined in the neuroblast and the epithelium by a conserved apical protein complex, consisting of Bazooka, DmPar6, and DaPkC. Partner of Inscuteable (Pins) is cortically localised (**2-3**) In the interphase delaminating neuroblast Inscuteable expression begins and Inscuteable binds to Bazooka, localising Pins apically. (**4**) In the delaminated neuroblast a basal crescent forms consisting of Miranda, which binds to Prospero and Staufen, and Partner of numb, which binds Numb (not shown). (**5**) The NB divides asymmetrically producing a smaller daughter cell, the GMC and regenerating the NB. The basal crescent of the NB is segregated to the GMC. (**6**) In the GMC, Miranda is rapidly degraded and Prospero translocates to the nucleus where it specifies a neural cell fate in this lineage.

the ventricular zone (VZ) give rise to the outer radial layers, which are composed of differentiated neurons which posses distinct identities. This process of layer formation in the cerebral cortex is known to involve the asymmetric division of neural progenitor cells in the VZ followed by the outward migration of postmitotic neurons. It is thought the formation of the different radial layers is due to changes in, and restriction of progenitor cell competence in the VZ over time (Desai and McConnell, 2000).

The mouse homologue of the *Drosophila* cell fate determinant Numb, m-Numb, is known to be involved in the asymmetric divisions of the progenitors of the cerebral cortex (Zhong et al., 1996) and retina (Cayouette et al., 2001), and is capable of rescuing numb mutant flies (Zhong et al., 1996). In cultures of cortical progenitor cells, m-Numb has been shown to preferentially localise to the postmitotic cell in progenitor-neuron divisions, and, in comparison m-Numb inheritance is unbiased in progenitor-progenitor divisions (Shen et al., 2002). Recent video microscopy analysis of retinal progenitor explants has demonstrated that the asymmetric inheritance of Numb between two retinal daughter cells promotes a different fate for each daughters, whereas the symmetric inheritance of Numb tend to leads to the same fate for both daughter cells (Cayouette and Raff, 2003). These observations indicate that there is conservation of Numb function in the developing nervous systems of *Drosophila* and vertebrates, and that the generation of neural lineages by a series of asymmetric divisions is a common feature of neurogenesis.

Temporal aspects of neural cell fate determination in the *Drosophila* retina

In the *Drosophila* retina, neural differentiation is initiated in a posterior to anterior wave which sweeps across the retina, following the morphogenetic furrow (Ready et al., 1976; Tomlinson and Ready, 1987). In front of the morphogenetic furrow differentiation is inhibited by notch signaling via hairy (Brown et al., 1995). Behind the furrow, photoreceptor differentiation occurs in an invariant sequence, starting with the differentiation of single precisely spaced 'founder' R8 photoreceptor cells (Jarman et al., 1994; Tomlinson and Ready, 1987). The R8 photoreceptor founder cells each contribute to one ommatidium, of which there are 750 in the adult retina (Ready et al., 1976). The founder cells are recruited by a process of lateral inhibition involving notch signaling in which the broad initial expression of the proneural gene, atonal, becomes restricted to the prospective R8 photoreceptor cells (Frankfort and Mardon, 2002).

Unlike the generation of the CNS and sensory organs, the different retinal neurons in *Drosophila* do not arise by the asymmetric division of a progenitor cell, but are understood to be specified over time by bursts of epidemeral growth factor (EGF) signaling, which have been shown to be necessary and sufficient for the generation of the different photoreceptor neurons in the retina (Freeman, 1997). It has been suggested that the ability of the retinal precursor cells to respond to EGF signaling changes over time, either due to intrinsic or extrinsic factors, such that the age of the retinal cell exposed to the EGF signal determines the type of neuron born. In this model, temporal intrinsic information is integrated with an extrinsic signal to determine the type of neuron generated (Freeman, 1997). In this situation where the same extra cellular signal is used to stimulate each wave of differentiation, the changing internal configuration of the neural precursor is likely to be crucial to determine the cell fate it will adopt.

Temporal aspects of neural cell fate determination in the vertebrate retina

In the vertebrate retina, the initial steps of neural cell fate determination are remarkably similar to those in *Drosophila*. The first-born neurons are retinal ganglion cells, and their production requires the expression of the atonal homolog, ath5, which as in the *Drosophila* retina is induced by sonic hedgehog and opposed by a gradient of notch signaling (Neumann and Nuesslein-Volhard, 2000). In the vertebrate retina, six types of neurons and one type of glial cell are generated during development. The order in which these cell types appear is invariant across vertebrate species, with the retinal ganglion cells being produced first and rods, bipolar and Müller glial cells last (Carter-Dawson and LaVail, 1979; Cepko et al., 1996a; LaVail et al., 1991; Stiemke and Hollyfield, 1995; Young, 1985). Unlike in *Drosophila*, the neurons of the vertebrate retina are generated from a pool of actively cycling neural progenitor cells. Neurogenesis in the vertebrate retina is characterised by several features. Firstly retinal progenitors are multipotent, and can generate more than one or two cell types (Holt et al., 1988; Turner and Cepko, 1987; Turner et al., 1990; Wetts and Fraser, 1988). Secondly, despite the conserved birth order, there is an overlap in the generation of different retinal cell types (LaVail et al., 1991; Stiemke and Hollyfield, 1995; Young, 1985).

Vertebrate retinal progenitor cells are only able to give rise to certain subsets of cell types at different stages in development (Austin et al., 1995; Belliveau and Cepko, 1999; Belliveau et al., 2000). It has been shown that whilst extrinsic signals can regulate the proportion of different cell types being made, they cannot alter the range of cell types generated. Combined with the multipotency of retinal progenitor cells, these observations led to the proposal of a competence model for retinal development (Figure 3), which suggested that retinal progenitor cells pass through a series of intrinsically determined configurations, or competence states, in each of which they are able to give rise to only a subset of cell types in response to extracellular signals (Cepko et al., 1996a).

Potentially, competence states might be determined by chromatin modulation, transcriptional states, gene expression profiles, translational regulation, protein accumulation/degradation, and by post-translational protein modification (Livesey and Cepko, 2001). There is some evidence for a transcriptional or translational control of progenitor competence. Two markers have been identified which show heterogeneity of expression in retinal progenitors, syntaxin-1a and VC1.1 (Alexiades and Cepko, 1997). Retinal progenitors also display a changing response to mitogens over time, and the level of epidermal growth factor receptor (EGFR) expression is known to change over time in the retina (Lillien, 1995; Lillien and Cepko, 1992). In addition, the cyclin kinase inhibi-

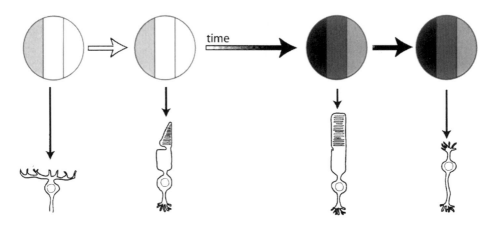

Figure 3. The competence model of neural development. Progenitors have been proposed to undergo changes in competence to generate different cell types during the course of development (Cepko et al., 1996). Evidence exists for at least two distinct competence states in the developing vertebrate retina: an early progenitor competence state and a late state. Little is currently known of how many competence states exist, how they are defined at the molecular level and how progenitor cells shift between competence states, the latter is illustrated by the middle arrow between the two competence states shown.

tors (CKI's) p27 and p57 which regulate cell cycle exit are expressed in different subsets of progenitors (Dyer and Cepko, 2000b). It has also been suggested that the level of p27^{xic1}, another cyclin kinase inhibitor, increases over time in retinal progenitors, and that its accumulation over a certain level is responsible for driving the formation of the final retinal cell type, the Müller glial cell (Ohnuma et al., 1999).

The mechanisms that govern the switch between different progenitor states or competences are unknown. It is possible that intrinsic factors, extrinsic factors or a combination of the two are responsible for changes in progenitor competence. Several types of retinal neuron are known to produce signals that negatively feedback on the retinal progenitor cells, regulating the types of neurons which they can generate (Bermingham et al., 1999; Reh and Tully, 1986; Waid and McLoon, 1998). A complicating factor in understanding retinal progenitor competence is the existence of heterogenous progenitor states at any one time (Brown et al., 1998; Dyer and Cepko, 2000a; Jasoni and Reh, 1996; Levine et al., 2000; Lillien and Cepko, 1992). This progenitor heterogeneity raises the complicating possibility that functionally different subsets of retinal progenitors exist, and that each subset of progenitors may generate only a selection of retinal cell types.

A temporal identity for neural progenitors and their progeny

In the asymmetrically dividing neuroblasts of the *Drosophila* embryo a sequentially expressed group of transcription factors encoded by hunchback, kruppel, castor, pdm and grainyhead has recently been reported (Brody and Odenwald, 2000; Isshiki et al., 2001). While the expression of these factors occurs as a temporal sequence in neuroblasts, the daughter ganglion mother cells (GMCs) that they give rise to maintain the expression of the transcription factor expressed in the mother neuroblast at the time that they were born. Hunchback and Kruppel have been shown to confer birth order specificity on many neuroblast lineages regardless of whether these lineages result in a neuronal or glial cell fate. How is this cascade of transcription factors regulated? Misexpression studies indicate that hunchback activates the expression of kruppel, and that Kruppel activates the expression of castor (Isshiki et al., 2001). Hunchback and Kruppel are also known to repress the expression of the next plus one gene in the sequence. Hunchback represses castor and Kruppel represses pdm. Interestingly the overexpression of hunchback has recently been shown to reset the sequential expression of these transcription factors (Pearson and Doe, 2003).

The expression of this transcriptional cascade in many different neuroblast lineages, some of which give rise to neurons and some to glia, suggests that it is responsible for conferring a temporal, rather than absolute identity on each GMC as it is born. As yet no such transcriptional cascade has been identified in vertebrates, although given the similarity of the other steps of neural cell fate determination between flies and vertebrates, the existence of similar mechanisms would not be a surprise.

A role is emerging for a novel group of regulatory genes, the microRNAs, in this process. The nematode worm *C.elegans* homologue of hunchback, hbl-1, has recently been shown to control developmental time and to be regulated by the microRNA let7 (Abrahante et al., 2003; Lin et al., 2003). Regulatory sites exist in the *Drosophila* hunchback 3'-untranslated region for the homologous *Drosophila* microRNAs and it is therefore likely that it too is temporally regulated in this way. Regulation by microRNA genes may therefore offer a novel mechanism for the temporal control of neurogenesis, in conjunction with the temporal transcription factor sequence outlined above.

Cell fate determination within a given competence state

Within an intrinsically defined progenitor competence state, cell fate has been shown to be influenced by extrinsic factors, for example by feedback inhibition from postmitotic neurons (Belliveau and Cepko, 1999; Reh and Tully,

1986; Waid and McLoon, 1998). Such a feedback mechanism has been shown to act on progenitor cells before M phase in order to affect daughter cell fate (Belliveau and Cepko, 1999) and a similar mechanism has been proposed for the developing neocortex (Desai and McConnell, 2000). It is noteworthy that extrinsic factors can also act to determine or respecify the fate of postmitotic cells, at least *in vitro*. For example, it is known that cilary neurotrophic factor (CNTF) and leukaemia inhibitory factor (LIF) can cause cells destined to become rods to adopt aspects of the bipolar cell phenotype (Ezzeddine et al., 1997).

Notch signaling is known to be involved in the differentiation of neurons and glia of the vertebrate retina and developing forebrain. However it is unclear whether notch signaling has a permissive or instructive role in these processes (Livesey and Cepko, 2001). In the neural crest, transient notch signaling is instructive in switching neural crest progenitors to neurogenesis, and then to gliogenesis (Morrison et al., 2000). In the late retina, notch acts to signal the transition between neurogenesis and gliogenesis (Furukawa et al., 2000), as it also does in the developing forebrain (Gaiano et al., 2000). Notch signaling is therefore likely to be important in regulating cell fate determination in vertebrates.

NEURAL CELL FATE DETERMINATION ELSEWHERE IN THE VERTEBRATE NERVOUS SYSTEM

In other regions of the developing vertebrate nervous system, cell fate determination is similar to the situation in the retina. Both the progenitor cells of the cortex and spinal cord are multipotent (Briscoe et al., 1999; Leber et al., 1990). In the developing cortex, as in the retina, progenitors give rise to neurons before generating glial cell types (Morrison et al., 2000; Qian et al., 2000b). Cortical progenitors progress through phases reminiscent of the competence states of retinal progenitors, in which they are competent to produce cells of a given laminar fate (Morrison et al., 2000; Qian et al., 2000b). However, unlike the situation in the retina, cortical neural progenitor cells are capable of generating later, but not earlier cell types upon heterochronic transplantation (Desai and McConnell, 2000; McConnell, 1988). This has lead to the concept of the progressive restriction model in cortical cell fate determination (Desai and McConnell, 2000).

Both cortical and spinal cord progenitors can respond to extrinsic factors that regulate their cell fate choices. Cortical progenitors are competent to respond to extrinsic signals until late S/early G2 in the cell cycle (McConnell, 1988), in agreement with the finding that retinal progenitors make cell fate choices prior to M phase (Belliveau and Cepko, 1999). As in the retina, feedback signaling is used as a mechanism of neural cell fate determination in the spinal cord where postmitotic motor neurons induce the genesis of interneurons (Pfaff et al., 1996). Heterogeneity also appears to be a conserved feature of vertebrate neurogenesis, as different populations of spinal cord progenitors can be distinguished by the expression of different transcription factors (Briscoe et al., 2000).

Vertebrate neural cell fate determination has many striking parallels with cell fate determination in *Drosophila*. As in the fly, precursor cells give rise to different types of neurons over time, implying that changes in competence (or cell fate potential) are common mechanisms of cell fate determination in both invertebrates and vertebrates. Other conserved features of neural cell fate determination are feedback inhibition, multipotency and progenitor heterogeneity. Whilst much still remains to be understood about neural cell fate determination, our existing knowledge and paradigms suggest that with the advent of genomics technologies we might soon have a much deeper understanding of key aspects such as the changes in neural progenitor cell fate potential over time.

TERMINAL DIFFERENTIATION PROGRAMS

When a postmitotic neuron has been specified to assume a particular fate, it must then terminally differentiate to realize that fate. This process is understood to involve locking a cell into a terminal transcriptional program. The transcriptional networks involved in terminal differentiation are only just beginning to be understood in vertebrates, and several transcription factors have been identified which are specific for different neurons. For example, in the retina the transcription factors Brn3a-c are specific to ganglion cells, Crx is specific to photoreceptors, and each is required for the full phenotypic differentiation of their respective neurons (Furukawa et al., 1997; Furukawa et al., 1999; Gan et al., 1999; Liu et al., 2000b). As discussed above, combinations of homeodomain-containing transcription factors control survival and differentiation of many classes of spinal cord neurons (for review, see (Jessell, 2000)).

Several different transcription factors regulating terminal differentiation have been identified in the worm *C.elegans*, including the Chx10 homologue ceh-10 and several different LIM-domain containing transcription factors (Altun-Gultekin et al., 2001; Hobert and Westphal, 2000). Single transcription factors have been shown to control distinct aspects of neuronal phenotypes, the most striking example being definition of dendritic arbor morphology in *Drosophila* neurons by different levels of expression of the cut transcription factor (Grueber et al., 2003). It is thought that terminal transcription programs are likely to involve autoregulatory loops to maintain the cell specific transcriptional program.

ORIGINS AND LINEAGE RELATIONSHIPS OF ADULT NEURAL STEM CELLS

A key issue for neural stem cell biology, discussed in more detail elsewhere in this book, is the origin of neural stem cells and their lineage relationships to the progenitor cells present during embryonic development. As has been well described, neurosphere-forming cells can be isolated from diverse regions of the adult CNS, including the lateral wall of the lateral ventricle, the hippocampus and the ciliary margin of the retina (for review, see Kuhn and Svendsen, 1999; Temple, 2001; and elsewhere in this volume). The precise location of the primary neural stem cell within the lateral wall of the lateral ventricle has been studied in some detail (Doetsch et al., 1999; Johansson et al., 1999), with a consensus emerging that a subpopulation of SVZ astrocytes are true stem cells. However, there is still much to know about the developmental origins of these cells.

Firstly, it is not clear when neural stem cells are generated during development. Using the classic neurosphere formation assay, typically it is not possible to harvest stem cells until relatively late in development. However, using the alternative approach of adherent clone generation, rather than sphere formation, for identifying NSCs suggests that stem cells are present from the earliest stages of neural development (Qian et al., 2000a). Is it possible that NSCs are a late generated cell type, or that key NSC properties are not acquired by endogenous cycling neural progenitor and stem cells until late in development? This latter possibility raises the other pivotal question of which cells are NSCs related to and derived from. Are NSCs actually astrocytes, or a distinct subpopulation of astrocyte-like cells, generated late in development? There are striking similarities between astrocytes and progenitor cells in terms of gene expression, as noted by several authors (Fischer and Reh, 2001; Furukawa et al., 2000), and astrocytes are typically the last-born cell type. Radial glia within the developing cerebral cortex have been demonstrated to not only give rise to astrocytes late in development, but are also neurogenic progenitor cells earlier in development (Campbell and Gotz, 2002; Hartfuss et al., 2001; Malatesta et al., 2000). So are astrocytes, or a subpopulation thereof, effectively quiescent late stage neural progenitor cells?

If neural stem cells are late stage progenitor cells, this would not be compatible with multipotency, as it has been demonstrated in both the retina and spinal cord that late progenitor cells are only capable of generating late-born cell types. The potential of NSCs isolated from different regions of the CNS has not been analysed in detail, particularly with respect to the temporal order of normal neurogenesis. One exception is the ciliary marginal zone of the retina, where the stem cells within this region in amphibia and fish are multipotent, and give rise to mitotic progenitor cells to go on to produce all of the cells of the adult retina (Perron and Harris, 2000b; Perron et al., 1998).

Thus, it would appear that NSCs are not simply the functional equivalent of late stage progenitor cells. One alternative possibility for the origin of NSCs is that there is a population of neural stem cells distinct from the much larger population of mitotic neural progenitor cells, and that adult NSCs are the direct descendants of these stem cells (Temple, 2001). If this is the case, the ability to generate neurospheres in culture may be a property that is lacking in early embryonic NSCs, but which has been gained by adult NSCs.

CELL FATE DETERMINATION IN ADULT NEURAL STEM CELLS

In contrast with developing neural stem and progenitor cells, little is known of the basic mechanisms regulating cell fate decision-making in adult NSCs. Most strategies for inducing the neural differentiation of cultured NSCs rely on growth factor withdrawal or retinoid exposure, and assay the relative degree of neuronal and glial differentiation within the culture system, rather than the diversity of neuronal cell types (Kuhn and Svendsen, 1999). Thus much interest has focused on the efficiency of neuron generation from neural stem cells, rather than on the mechanisms controlling neurogenesis and cell fate determination. However, it is clear that NSCs express many of the same markers as developing neural progenitor cells, and do use the delta-notch pathway to regulate neural differentiation (Morrison et al., 2000). It is likely, therefore, that similar or identical cellular mechanisms and key genes are redeployed in NSCs for the generation of discrete types of neurons.

With regard to the developmental potential of NSCs, transplantation of NSCs into the adult CNS results in differentiation into regionally appropriate cell types after an initial culture period of weeks to months, typically as neurospheres (Englund et al., 2002; Fricker et al., 1999). However, the degree of neurogenesis depends on transplantation into one of the areas of adult neurogenesis, such as the SVZ. NSCs that are transplanted outside of these regions generate mostly glial cells (Wu et al., 2002). Time spent in culture prior to transplantation appears to have striking effects on the potential and/or competence of NSCs, suggesting that some form of reprogramming or dedifferentiation may take place. Human NSCs are capable of generating region-specific neuronal progeny when transplanted into the adult rat CNS following a series of *in vitro* treatments, including growth factor exposure (Wu et al., 2002). In contrast, *in vitro* studies indicate that NSCs isolated from different regions of the forebrain are intrinsically different in terms of gene expression and responses to mitogens (Parmar et al., 2002).

The ability of tissue culture and growth factor exposure to alter the potential of NSCs appears to be a general phenomenon. Peripheral nerve stem cells demonstrate a striking difference in potential when transplanted immediately after harvesting, compared to transplantation following a period of time in culture (Bixby et al., 2002; Kruger et al., 2002; White et al., 2001). Such a change in potential is reminiscent of the change in potential of oligodendrocyte precursor cells *in vitro* from being dedicated to the production of oligodendrocytes to the ability to form astrocytes and neurons, again through a series of growth factor treatments (Kondo and Raff, 2000).

Those studies have important implications for our understanding of the biology underlying progenitor and stem cell potential and competence, and also for the practical aspects of manipulating these cells for replacement therapies. Several reports have been made of the plasticity of neural stem cells both within and outside the nervous system and in the developing chick embryo (Bjornson et al., 1999; Blau et al., 2001; Clarke et al., 2000; Vescovi et al., 2002). A number of alternative mechanisms have been proposed and demonstrated for this putative plasticity, including cell fusion (Terada et al., 2002; Wurmser and Gage, 2002; Ying et al., 2002). However, it is also clear that culture conditions can have marked effects on the potential of these cells, through unknown cellular mechanisms, a phenomenon that could possibly be harnessed for therapeutic benefits, but could also be of concern for experimental design and interpretation.

CONCLUSION: IS THERE A GENERAL BLUEPRINT FOR CONTROLLING NEURAL CELL FATE DETERMINATION FROM ES CELLS AND NS CELLS?

Recent findings on the generation of particular classes of neurons from ES cells have emphasized the importance of a thorough understanding of the mechanisms controlling cell fate determination during development (Kim et al., 2002; Wichterle et al., 2002). In both cases, specific types of neurons were generated from ES cells by first inducing neural differentiation in the ES cells, and then using the available knowledge of the extracellular signals and transcription factors required in the production of each cell type. In one case, overexpression of a transcription factor normally expressed in midbrain dopaminergic neuron progenitor cells was used to bias progeny towards that fate (Kim et al., 2002). In the other, ES cells were exposed to a series of treatments designed to mimic normal *in vivo* development, such that ES cells were first induced to form neurogenic tissue and then this tissue was exposed to signals that confer caudal, or spinal cord, fates. Lastly, this tissue was treated with a combination of factors known to induce motor neuron production and differentiation in the developing neural tube (Wichterle et al., 2002).

These studies have raised the possibility that there is a single framework for generating classes of neurons which can be applied to both ES cells and NSCs isolated from different regions of the CNS. However, they also highlight the potential difficulties of working with NSCs, as opposed to ES cells. In the case of ES cells, it is possible to prospectively walk cultured neural tissue down developmental pathways appropriate to the cell type one wishes to produce. This may not be the case for regionally-derived NSCs. As discussed above, we know little about the cellular status of these NSCs, especially after periods of time in culture. In particular, it is not clear if NSCs with increased potential after time spent in culture are the equivalent of developing neural progenitor cells of a distinct development stage, or whether this represents an independent pathway for neural development or differentiation. Understanding this aspect of NSC biology will be essential for efforts to manipulate NSCs to generate desired cell types at high efficiencies.

Finally, given the presence of endogenous stem cells in the adult mammalian CNS, it is also of interest to ask whether approaches that manipulate such stem cells are likely to be more successful for stimulating repair than the transplantation of exogenous stem cells (See Chapter 12 in this book). In this case, as in the manipulation of NSCs *in vitro*, an understanding of the competence and potential of these cells, allied to our increasing knowledge of the general mechanisms regulating neurogenesis, will be essential for the rational development of therapeutics aimed at stimulating those cells to repair the damaged CNS.

REFERENCES

Abrahante, J. E., Daul, A. L., Li, M., Volk, M. L., Tennessen, J. M., Miller, E. A., and Rougvie, A. E. (2003). The Caenorhabditis elegans hunchback-like Gene lin-57/hbl-1 Controls Developomental Time and Is Regulated by MicroRNAs. Developmental Cell 4, 625-637.

Alexiades, M. R., and Cepko, C. L. (1997). Subsets of retinal progenitors display temporally regulated and distinct biases in the fates of their progeny. Development 124, 1119-1131.

Allen, K. M., and Walsh, C. A. (1999). Genes that regulate neuronal migration in the cerebral cortex. Epilepsy Res 36, 143-154.

Alonso, M. C., and Cabrera, C. V. (1988). The achaete-scute gene complex of *Drosophila melanogaster* comprises four homologous genes. Embo J 7, 2585-2591.

Altun-Gultekin, Z., Andachi, Y., Tsalik, E. L., Pilgrim, D., Kohara, Y., and Hobert, O. (2001). A regulatory cascade of three homeobox genes, ceh-10, ttx-3 and ceh-23, controls cell fate specification of a defined interneuron class in C. elegans. Development 128, 1951-1969.

Appel, B., Korzh, V., Glasgow, E., Thor, S., Edlund, T., Dawid, I. B., and Eisen, J. S. (1995). Motoneuron fate specification revealed by patterned LIM homeobox gene expression in embryonic zebrafish. Development 121, 4117-4125.

Artavanis-Tsakonas, S., Rand, M. D., and Lake, R. J. (1999). Notch signaling: cell fate control and signal integration in development. Science 284, 770-776.

Austin, C. P., Feldman, D. E., Ida, J. A., and Cepko, C. L. (1995). Vertebrate retinal ganglion cells are selected from competent progenitors by the action of Notch. Development 121, 3637-3650.

Bae, S., Bessho, Y., Hojo, M., and Kageyama, R. (2000). The bHLH gene Hes6, an inhibitor of Hes1, promotes neuronal differentiation. Development 127, 2933-2943.

Baker, N. E. (2000). Notch signaling in the nervous system. Pieces still missing from the puzzle. Bioessays 22, 264-273.

Baker, N. E., Yu, S., and Han, D. (1996). Evolution of proneural atonal expression during distinct regulatory phases in the developing *Drosophila* eye. Curr Biol 6, 1290-1301.

Bate, C. M. (1976). Embryogenesis of an insect nervous system. I. A map of the thoracic and abdominal neuroblasts in Locusta migratoria. J Embryol Exp Morphol 35, 107-123.

Belliveau, M. J., and Cepko, C. L. (1999). Extrinsic and intrinsic factors control the genesis of amacrine and cone cells in the rat retina. Development, In press.

Belliveau, M. J., Young, T. L., and Cepko, C. L. (2000). Late retinal progenitor cells show intrinsic limitations in the production of cell types and the kinetics of opsin synthesis. J Neurosci 20, 2247-2254.

Belting, H. G., Shashikant, C. S., and Ruddle, F. H. (1998). Multiple phases of expression and regulation of mouse Hoxc8 during early embryogenesis. J Exp Zool 282, 196-222.

Ben-Arie, N., McCall, A. E., Berkman, S., Eichele, G., Bellen, H. J., and Zoghbi, H. Y. (1996). Evolutionary conservation of sequence and expression of the bHLH protein Atonal suggests a conserved role in neurogenesis. Hum Mol Genet 5, 1207-1216.

Bermingham, N. A., Hassan, B. A., Price, S. D., Vollrath, M. A., Ben-Arie, N., Eatock, R. A., Bellen, H. J., Lysakowski, A., and Zoghbi, H. Y. (1999). Math1: an essential gene for the generation of inner ear hair cells. Science 284, 1837-1841.

Bermingham, N. A., Hassan, B. A., Wang, V. Y., Fernandez, M., Banfi, S., Bellen, H. J., Fritzsch, B., and Zoghbi, H. Y. (2001). Proprioceptor pathway development is dependent on Math1. Neuron 30, 411-422.

Bertrand, N., Castro, D. S., and Guillemot, F. (2002). Proneural genes and the specification of neural cell types. Nat Rev Neurosci 3, 517-530.

Bhat, K. M. (1996). The patched signaling pathway mediates repression of gooseberry allowing neuroblast specification by wingless during *Drosophila* neurogenesis. Development 122, 2921-2932.

Bhat, K. M. (1999). Segment polarity genes in neuroblast formation and identity specification during *Drosophila* neurogenesis. Bioessays 21, 472-485.

Bhat, K. M., and Schedl, P. (1997). Requirement for engrailed and invected genes reveals novel regulatory interactions between engrailed/invected, patched, gooseberry and wingless during *Drosophila* neurogenesis. Development 124, 1675-1688.

Bixby, S., Kruger, G. M., Mosher, J. T., Joseph, N. M., and Morrison, S. J. (2002). Cell-intrinsic differences between stem cells from different regions of the peripheral nervous system regulate the generation of neural diversity. Neuron 35, 643-656.

Bjornson, C. R., Rietze, R. L., Reynolds, B. A., Magli, M. C., and Vescovi, A. L. (1999). Turning brain into blood: a hematopoietic fate adopted by adult neural stem cells *in vivo*. Science 283, 534-537.

Blader, P., Fischer, N., Gradwohl, G., Guillemont, F., and Strahle, U. (1997). The activity of neurogenin1 is controlled by local cues in the zebrafish embryo. Development 124, 4557-4569.

Blau, H. M., Brazelton, T. R., and Weimann, J. M. (2001). The evolving concept of a stem cell: entity or function? Cell 105, 829-841.

Blochlinger, K., Jan, L. Y., and Jan, Y. N. (1991). Transformation of sensory organ identity by ectopic expression of Cut in *Drosophila*. Genes Dev 5, 1124-1135.

Bodmer, R., Carretto, R., and Jan, Y. N. (1989). Neurogenesis of the peripheral nervous system in *Drosophila* embryos: DNA replication patterns and cell lineages. Neuron 3, 21-32.

Bossing, T., Udolph, G., Doe, C. Q., and Technau, G. M. (1996). The embryonic central nervous system lineages of *Drosophila* melanogaster. I. Neuroblast lineages derived from the ventral half of the neuroectoderm. Dev Biol 179, 41-64.

Briscoe, J., Chen, Y., Jessell, T. M., and Struhl, G. (2001). A hedgehog-insensitive form of patched provides evidence for direct long-range morphogen activity of sonic hedgehog in the neural tube. Mol Cell 7, 1279-1291.

Briscoe, J., Pierani, A., Jessell, T. M., and Ericson, J. (2000). A homeodomain protein code specifies progenitor cell identity and neuronal fate in the ventral neural tube. Cell 101, 435-445.

Briscoe, J., Sussel, L., Serup, P., Hartigan-O'Connor, D., Jessell, T. M., Rubenstein, J. L., and Ericson, J. (1999). Homeobox gene Nkx2.2 and specification of neuronal identity by graded Sonic hedgehog signalling. Nature 398, 622-627.

Broadus, J., Fuerstenberg, S., and Doe, C. Q. (1998). Staufen-dependent localization of prospero mRNA contributes to neuroblast daughter-cell fate. Nature 391, 792-795.

Brody, T., and Odenwald, W. F. (2000). Programmed transformations in neuroblast gene expression during *Drosophila* CNS lineage development. Dev Biol 226, 34-44.

Brown, N. L., Kanekar, S., Vetter, M. L., Tucker, P. K., Gemza, D. L., and Glaser, T. (1998). Math5 encodes a murine basic helix-loop-helix transcription factor expressed during early stages of retinal neurogenesis. Development 125, 4821-4833.

Brown, N. L., Sattler, C. A., Paddock, S. W., and Carroll, S. B. (1995). Hairy and emc negatively regulate morphogenetic furrow progression in the *Drosophila* eye. Cell 80, 879-887.

Buescher, M., and Chia, W. (1997). Mutations in lottchen cause cell fate transformations in both neuroblast and glioblast lineages in the *Drosophila* embryonic central nervous system. Development 124, 673-681.

Burke, A. C., Nelson, C. E., Morgan, B. A., and Tabin, C. (1995). Hox genes and the evolution of vertebrate axial morphology. Development 121, 333-346.

Cabrera, C. V. (1990). Lateral inhibition and cell fate during neurogenesis in *Drosophila*: the interactions between scute, Notch and Delta. Development 110, 733-742.

Cabrera, C. V., and Alonso, M. C. (1991). Transcriptional activation by heterodimers of the achaete-scute and daughterless gene products of *Drosophila*. Embo J 10, 2965-2973.

Cabrera, C. V., Martinez-Arias, A., and Bate, M. (1987). The expression of three members of the achaete-scute gene complex correlates with neuroblast segregation in *Drosophila*. Cell 50, 425-433.

Campbell, K., and Gotz, M. (2002). Radial glia: multi-purpose cells for vertebrate brain development. Trends Neurosci 25, 235-238.

Campuzano, S. (2001). Emc, a negative HLH regulator with multiple functions in *Drosophila* development. Oncogene 20, 8299-8307.

Campuzano, S., and Modolell, J. (1992). Patterning of the *Drosophila* nervous system: the achaete-scute gene complex. Trends Genet 8, 202-208.

Carter-Dawson, L. D., and LaVail, M. M. (1979). Rods and cones in the mouse retina. II. Autoradiographic analysis of cell generation using tritiated thymidine. J Comp Neurol 188, 263-272.

Casarosa, S., Fode, C., and Guillemot, F. (1999). Mash1 regulates neurogenesis in the ventral telencephalon. Development 126, 525-534.

Cau, E., Casarosa, S., and Guillemot, F. (2002). Mash1 and Ngn1 control distinct steps of determination and differentiation in the olfactory sensory neuron lineage. Development 129, 1871-1880.

Cayouette, M., and Raff, M. (2003). The orientation of cell division influences cell-fate choice in the developing mammalian retina. Development 130, 2329-2339.

Cayouette, M., Whitmore, A. V., Jeffery, G., and Raff, M. (2001). Asymmetric segregation of Numb in retinal development and the influence of the pigmented epithelium. J Neurosci 21, 5643-5651..

Cepko, C. L., Austin, C. P., Yang, X., Alexiades, M., and Ezzeddine, D. (1996). Cell fate determination in the vertebrate retina. Proc Natl Acad Sci USA 93, 589-595.

Chen, G., Fernandez, J., Mische, S., and Courey, A. J. (1999). A functional interaction between the histone deacetylase Rpd3 and the corepressor groucho in *Drosophila* development. Genes Dev 13, 2218-2230.

Chenn, A., and McConnell, S. K. (1995). Cleavage orientation and the asymmetric inheritance of Notch1 immunoreactivity in mammalian neurogenesis. Cell 82, 631-641.

Chien, C. T., Hsiao, C. D., Jan, L. Y., and Jan, Y. N. (1996). Neuronal type information encoded in the basic-helix-loop-helix domain of proneural genes. Proc Natl Acad Sci U S A 93, 13239-13244.

Chitnis, A., Henrique, D., Lewis, J., Ish-Horowicz, D., and Kintner, C. (1995). Primary neurogenesis in *Xenopus* embryos regulated by a homologue of the *Drosophila* neurogenic gene Delta. Nature 375, 761-766.

Chitnis, A., and Kintner, C. (1996). Sensitivity of proneural genes to lateral inhibition affects the pattern of primary neurons in *Xenopus* embryos. Development 122, 2295-2301.

Chu-LaGraff, Q., and Doe, C. Q. (1993). Neuroblast specification and formation regulated by wingless in the *Drosophila* CNS. Science 261, 1594-1597.

Clarke, D. L., Johansson, C. B., Wilbertz, J., Veress, B., Nilsson, E., Karlstrom, H., Lendahl, U., and Frisen, J. (2000). Generalized potential of adult neural stem cells. Science 288, 1660-1663.

Cornell, R. A., and Ohlen, T. V. (2000). Vnd/nkx, ind/gsh, and msh/msx: conserved regulators of dorsoventral neural patterning? Curr Opin Neurobiol 10, 63-71.

Cubas, P., de Celis, J. F., Campuzano, S., and Modolell, J. (1991). Proneural clusters of achaete-scute expression and the generation of sensory organs in the *Drosophila* imaginal wing disc. Genes Dev 5, 996-1008.

Culi, J., and Modolell, J. (1998). Proneural gene self-stimulation in neural precursors: an essential mechanism for sense organ development that is regulated by Notch signaling. Genes Dev 12, 2036-2047.

Davis, R. L., and Turner, D. L. (2001). Vertebrate hairy and Enhancer of split related proteins: transcriptional repressors regulating cellular differentiation and embryonic patterning. Oncogene 20, 8342-8357.

de la Cruz, C. C., Der-Avakian, A., Spyropoulos, D. D., Tieu, D. D., and Carpenter, E. M. (1999). Targeted disruption of Hoxd9 and Hoxd10 alters locomotor behavior, vertebral identity, and peripheral nervous system development. Dev Biol 216, 595-610.

Desai, A. R., and McConnell, S. K. (2000). Progressive restriction in fate potential by neural progenitors during cerebral cortical development. Development 127, 2863-2872.

Ding, Q., Motoyama, J., Gasca, S., Mo, R., Sasaki, H., Rossant, J., and Hui, C. C. (1998). Diminished Sonic hedgehog signaling and lack of floor plate differentiation in Gli2 mutant mice. Development 125, 2533-2543.

Doe, C. Q. (1992). Molecular markers for identified neuroblasts and ganglion mother cells in the *Drosophila* central nervous system. Development 116, 855-863.

Doe, C. Q., and Goodman, C. S. (1985). Early events in insect neurogenesis. I. Development and segmental differences in the pattern of neuronal precursor cells. Dev Biol 111, 193-205.

Doetsch, F., Caille, I., Lim, D. A., Garcia-Verdugo, J. M., and Alvarez-Buylla, A. (1999). Subventricular zone astrocytes are neural stem cells in the adult mammalian brain. Cell 97, 703-716.

Dokucu, M. E., Zipursky, S. L., and Cagan, R. L. (1996). Atonal, rough and the resolution of proneural clusters in the developing *Drosophila* retina. Development 122, 4139-4147.

Dominguez, M., and Campuzano, S. (1993). asense, a member of the *Drosophila* achaete-scute complex, is a proneural and neural differentiation gene. Embo J 12, 2049-2060.

Dubois, L., Bally-Cuif, L., Crozatier, M., Moreau, J., Paquereau, L., and Vincent, A. (1998). XCoe2, a transcription factor of the Col/Olf-1/EBF family involved in the specification of primary neurons in *Xenopus*. Curr Biol 8, 199-209.

Dyer, M. A., and Cepko, C. L. (2000a). Control of Müller glial cell proliferation and activation following retinal injury. Nat Neurosci 3, 873-880.

Dyer, M. A., and Cepko, C. L. (2000b). p57(Kip2) regulates progenitor cell proliferation and amacrine interneuron development in the mouse retina. Development 127, 3593-3605.

Edlund, T., and Jessell, T. M. (1999). Progression from extrinsic to intrinsic signaling in cell fate specification: a view from the nervous system. Cell 96, 211-224.

Edmondson, D. G., and Roth, S. Y. (1998). Interactions of transcriptional regulators with histones. Methods 15, 355-364.

Edmondson, D. G., Smith, M. M., and Roth, S. Y. (1996). Repression domain of the yeast global repressor Tup1 interacts directly with histones H3 and H4. Genes Dev 10, 1247-1259.

Englund, U., Bjorklund, A., Wictorin, K., Lindvall, O., and Kokaia, M. (2002). Grafted neural stem cells develop into functional pyramidal neurons and integrate into host cortical circuitry. Proc Natl Acad Sci U S A 99, 17089-17094.

Ensini, M., Tsuchida, T. N., Belting, H. G., and Jessell, T. M. (1998). The control of rostrocaudal pattern in the developing spinal cord: specification of motor neuron subtype identity is initiated by signals from paraxial mesoderm. Development 125, 969-982.

Ericson, J., Briscoe, J., Rashbass, P., van Heyningen, V., and Jessell, T. M. (1997a). Graded sonic hedgehog signaling and the specification of cell fate in the ventral neural tube. Cold Spring Harb Symp Quant Biol 62, 451-466.

Ericson, J., Morton, S., Kawakami, A., Roelink, H., and Jessell, T. M. (1996). Two critical periods of Sonic Hedgehog signaling required for the specification of motor neuron identity. Cell 87, 661-673.

Ericson, J., Rashbass, P., Schedl, A., Brenner-Morton, S., Kawakami, A., van Heyningen, V., Jessell, T. M., and Briscoe, J. (1997b). Pax6 controls progenitor cell identity and neuronal fate in response to graded Shh signaling. Cell 90, 169-180.

Ezzeddine, Z. D., Yang, X., DeChiara, T., Yancopoulos, G., and Cepko, C. L. (1997). Postmitotic cells fated to become rod photoreceptors can be respecified by CNTF treatment of the retina. Development 124, 1055-1067.

Farah, M. H., Olson, J. M., Sucic, H. B., Hume, R. I., Tapscott, S. J., and Turner, D. L. (2000). Generation of neurons by transient expression of neural bHLH proteins in mammalian cells. Development 127, 693-702.

Fehon, R. G., Johansen, K., Rebay, I., and Artavanis-Tsakonas, S. (1991). Complex cellular and subcellular regulation of notch expression during embryonic and imaginal development of *Drosophila*: implications for notch function. J Cell Biol 113, 657-669.

Fischer, A. J., and Reh, T. A. (2001). Müller glia are a potential source of neural regeneration in the postnatal chicken retina. Nat Neurosci 4, 247-252.

Fode, C., Gradwohl, G., Morin, X., Dierich, A., LeMeur, M., Goridis, C., and Guillemot, F. (1998). The bHLH protein NEUROGENIN 2 is a determination factor for epibranchial placode-derived sensory neurons. Neuron 20, 483-494.

Fode, C., Ma, Q., Casarosa, S., Ang, S. L., Anderson, D. J., and Guillemot, F. (2000). A role for neural determination genes in specifying the dorsoventral identity of telencephalic neurons. Genes Dev 14, 67-80.

Frankfort, B. J., and Mardon, G. (2002). R8 development in the *Drosophila* eye: a paradigm for neural selection and differentiation. Development 129, 1295-1306.

Freeman, M. (1997). Cell determination strategies in the *Drosophila* eye. Development 124, 261-270.

Fricker, R. A., Carpenter, M. K., Winkler, C., Greco, C., Gates, M. A., and Bjorklund, A. (1999). Site-specific migration and neuronal differentiation of human neural progenitor cells after transplantation in the adult rat brain. J Neurosci 19, 5990-6005.

Furukawa, T., Morrow, E. M., and Cepko, C. L. (1997). Crx, a novel otx-like homeobox gene, shows photoreceptor-specific expression and regulates photoreceptor differentiation. Cell 91, 531-541.

Furukawa, T., Morrow, E. M., Li, T., Davis, F. C., and Cepko, C. L. (1999). Retinopathy and attenuated circadian entrainment in Crx-deficient mice. Nat Genet 23, 466-470.

Furukawa, T., Mukherjee, S., Bao, Z. Z., Morrow, E. M., and Cepko, C. L. (2000). rax, Hes1, and notch1 promote the formation of Müller glia by postnatal retinal progenitor cells. Neuron 26, 383-394.

Gaiano, N., Nye, J. S., and Fishell, G. (2000). Radial glial identity is promoted by Notch1 signaling in the murine forebrain. Neuron 26, 395-404.

Gan, L., Wang, S. W., Huang, Z., and Klein, W. H. (1999). POU domain factor Brn-3b is essential for retinal ganglion cell differentiation and survival but not for initial cell fate specification. Dev Biol 210, 469-480.

Garcia-Bellido, A. (1979). Genetic analysis of the achaete-scute system of *Drosophila* Melanogaster. Genetics 91, 491-520.

Gho, M., Bellaiche, Y., and Schweisguth, F. (1999). Revisiting the *Drosophila* microchaete lineage: a novel intrinsically asymmetric cell division generates a glial cell. Development 126, 3573-3584.

Ghysen, A., and Dambly-Chaudiere, C. (1988). From DNA to form: the achaete-scute complex. Genes Dev 2, 495-501.

Gonzalez, F., Romani, S., Cubas, P., Modolell, J., and Campuzano, S. (1989). Molecular analysis of the asense gene, a member of the achaete-scute complex of *Drosophila* melanogaster, and its novel role in optic lobe development. Embo J 8, 3553-3562.

Goridis, C., and Brunet, J. F. (1999). Transcriptional control of neurotransmitter phenotype. Curr Opin Neurobiol 9, 47-53.

Goridis, C., and Rohrer, H. (2002). Specification of catecholaminergic and serotonergic neurons. Nat Rev Neurosci 3, 531-541.

Goulding, S. E., White, N. M., and Jarman, A. P. (2000a). cato encodes a basic helix-loop-helix transcription factor implicated in the correct differentiation of *Drosophila* sense organs. Dev Biol 221, 120-131.

Goulding, S. E., zur Lage, P., and Jarman, A. P. (2000b). amos, a proneural gene for *Drosophila* olfactory sense organs that is regulated by lozenge. Neuron 25, 69-78.

Gowan, K., Helms, A. W., Hunsaker, T. L., Collisson, T., Ebert, P. J., Odom, R., and Johnson, J. E. (2001). Crossinhibitory activities of Ngn1 and Math1 allow specification of distinct dorsal interneurons. Neuron 31, 219-232.

Gradwohl, G., Fode, C., and Guillemot, F. (1996). Restricted expression of a novel murine atonal-related bHLH protein in undifferentiated neural precursors. Dev Biol 180, 227-241.

Grueber, W. B., Jan, L. Y., and Jan, Y. N. (2003). Different levels of the homeodomain protein cut regulate distinct dendrite branching patterns of *Drosophila* multidendritic neurons. Cell 112, 805-818.

Guillemot, F. (1999). Vertebrate bHLH genes and the determination of neuronal fates. Exp Cell Res 253, 357-364.

Guillemot, F., and Joyner, A. L. (1993). Dynamic expression of the murine Achaete-Scute homologue Mash-1 in the developing nervous system. Mechanisms of Development 42, 171-185.

Guillemot, F., Lo, L. C., Johnson, J. E., Auerbach, A., Anderson, D. J., and Joyner, A. L. (1993). Mammalian achaete-scute homolog 1 is required for the early development of olfactory and autonomic neurons. Cell 75, 463-476.

Harland, R. (2000). Neural induction. Curr Opin Genet Dev 10, 357-362.

Hartenstein, V., and Campos-Ortega, J. A. (1984). Early neurogenesis in wild type *Drosophila melanogaster*. Roux's Arch Dev Biol 193, 308-325.

Hartenstein, V., and Posakony, J. W. (1990). A dual function of the Notch gene in *Drosophila* sensillum development. Developmental Biology 142, 13-30.

Hartfuss, E., Galli, R., Heins, N., and Gotz, M. (2001). Characterization of CNS precursor subtypes and radial glia. Dev Biol 229, 15-30.

Hassan, B. A., and Bellen, H. J. (2000). Doing the MATH: is the mouse a good model for fly development? Genes Dev 14, 1852-1865.

Hatakeyama, J., Tomita, K., Inoue, T., and Kageyama, R. (2001). Roles of homeobox and bHLH genes in specification of a retinal cell type. Development 128, 1313-1322.

Heitzler, P., Bourouis, M., Ruel, L., Carteret, C., and Simpson, P. (1996). Genes of the Enhancer of split and achaete-scute complexes are required for a regulatory loop between Notch and Delta during lateral signalling in *Drosophila*. Development 122, 161-171.

Heitzler, P., and Simpson, P. (1991). The choice of cell fate in the epidermis of *Drosophila*. Cell 64, 1083-1092.

Helms, A. W., Abney, A. L., Ben-Arie, N., Zoghbi, H. Y., and Johnson, J. E. (2000). Autoregulation and multiple enhancers control Math1 expression in the developing nervous system. Development 127, 1185-1196.

Henrique, D., Tyler, D., Kintner, C., Heath, J., Lewis, J., Ish-Horowicz, D., and Storey, K. (1997). cash4, a novel achaete-scute homolog induced by Hensen's node during generation of the posterior nervous system. Genes Dev 11, 603-615.

Hinz, U., Giebel, B., and Campos-Ortega, J. A. (1994). The basic-helix-loop-helix domain of *Drosophila* lethal of scute protein is sufficient for proneural function and activates neurogenic genes. Cell 76, 77-87.

Hirata, J., Nakagoshi, H., Nabeshima, Y., and Matsuzaki, F. (1995). Asymmetric segregation of the homeodomain protein Prospero during *Drosophila* development. Nature 377, 627-630.

Hirsch, M. R., Tiveron, M. C., Guillemot, F., Brunet, J. F., and Goridis, C. (1998). Control of noradrenergic differentiation and Phox2a expression by MASH1 in the central and peripheral nervous system. Development 125, 599-608.

Hobert, O., and Westphal, H. (2000). Functions of LIM-homeobox genes. Trends Genet 16, 75-83.

Holt, C. E., Bertsch, T. W., Ellis, H. M., and Harris, W. A. (1988). Cellular determination in the *Xenopus* retina is independent of lineage and birth date. Neuron 1, 15-26.

Horton, S., Meredith, A., Richardson, J. A., and Johnson, J. E. (1999). Correct coordination of neuronal differentiation events in ventral forebrain requires the bHLH factor MASH1. Mol Cell Neurosci 14, 355-369.

Huang, H. P., Liu, M., El-Hodiri, H. M., Chu, K., Jamrich, M., and Tsai, M. J. (2000a). Regulation of the pancreatic islet-specific gene BETA2 (neuroD) by neurogenin 3. Mol Cell Biol 20, 3292-3307.

Huang, M. L., Hsu, C. H., and Chien, C. T. (2000b). The proneural gene amos promotes multiple dendritic neuron formation in the *Drosophila* peripheral nervous system. Neuron 25, 57-67.

Ikeshima-Kataoka, H., Skeath, J. B., Nabeshima, Y., Doe, C. Q., and Matsuzaki, F. (1997). Miranda directs Prospero to a daughter cell during *Drosophila* asymmetric divisions. Nature 390, 625-629.

Inoue, T., Hojo, M., Bessho, Y., Tano, Y., Lee, J. E., and Kageyama, R. (2002). Math3 and NeuroD regulate amacrine cell fate specification in the retina. Development 129, 831-842.

Isshiki, T., Pearson, B., Holbrook, S., and Doe, C. Q. (2001). *Drosophila* neuroblasts sequentially express transcription factors which specify the temporal identity of their neuronal progeny. Cell 106, 511-521.

Isshiki, T., Takeichi, M., and Nose, A. (1997). The role of the msh homeobox gene during *Drosophila* neurogenesis: implication for the dorsoventral specification of the neuroectoderm. Development 124, 3099-3109.

Itasaki, N., Sharpe, J., Morrison, A., and Krumlauf, R. (1996). Reprogramming Hox expression in the vertebrate hindbrain: influence of paraxial mesoderm and rhombomere transposition. Neuron 16, 487-500.

Jan, Y. N., and Jan, L. Y. (1993). HLH proteins, fly neurogenesis, and vertebrate myogenesis. Cell 75, 827-830.

Jan, Y. N., and Jan, L. Y. (1994). Genetic control of cell fate specification in *Drosophila* peripheral nervous system. Annu Rev Genet 28, 373-393.

Jarman, A. P., and Ahmed, I. (1998). The specificity of proneural genes in determining *Drosophila* sense organ identity. Mech Dev 76, 117-125.

Jarman, A. P., Grau, Y., Jan, L. Y., and Jan, Y. N. (1993). atonal is a proneural gene that directs chordotonal organ formation in the *Drosophila* peripheral nervous system. Cell 73, 1307-1321.

Jarman, A. P., Grell, E. H., Ackerman, L., Jan, L. Y., and Jan, Y. N. (1994). Atonal is the proneural gene for *Drosophila* photoreceptors. Nature 369, 398-400.

Jasoni, C. L., and Reh, T. A. (1996). Temporal and spatial pattern of MASH-1 expression in the developing rat retina demonstrates progenitor cell heterogeneity. J Comp Neurol 369, 319-327.

Jennings, B., Preiss, A., Delidakis, C., and Bray, S. (1994). The Notch signalling pathway is required for Enhancer of split bHLH protein expression during neurogenesis in the *Drosophila* embryo. Development 120, 3537-3548.

Jessell, T. M. (2000). Neuronal specification in the spinal cord: inductive signals and transcriptional codes. Nat Rev Genet 1, 20-29.

Jimenez, F., and Campos-Ortega, J. A. (1990). Defective neuroblast commitment in mutants of the achaete-scute complex and adjacent genes of D. melanogaster. Neuron 5, 81-89.

Johansson, C. B., Momma, S., Clarke, D. L., Risling, M., Lendahl, U., and Frisen, J. (1999). Identification of a neural stem cell in the adult mammalian central nervous system. Cell 96, 25-34.

Johnson, J. E., Birren, S. J., Saito, T., and Anderson, D. J. (1992). DNA binding and transcriptional regulatory activity of mammalian achaete-scute homologous (MASH) proteins revealed by interaction with a muscle-specific enhancer. Proc Natl Acad Sci U S A 89, 3596-3600.

Kageyama, R., and Nakanishi, S. (1997). Helix-loop-helix factors in growth and differentiation of the vertebrate nervous system. Curr Opin Genet Dev 7, 659-665.

Keynes, R., and Krumlauf, R. (1994). Hox genes and regionalization of the nervous system. Annu Rev Neurosci 17, 109-132.

Kim, J. H., Auerbach, J. M., Rodriguez-Gomez, J. A., Velasco, I., Gavin, D., Lumelsky, N., Lee, S. H., Nguyen, J., Sanchez-Pernaute, R., Bankiewicz, K., and McKay, R. (2002). Dopamine neurons derived from embryonic stem cells function in an animal model of Parkinson's disease. Nature 418, 50-56.

Kintner, C. (2002). Neurogenesis in embryos and in adult neural stem cells. J Neurosci 22, 639-643.

Knoblich, J. A., Jan, L. Y., and Jan, Y. N. (1995). Asymmetric segregation of Numb and Prospero during cell division. Nature 377, 624-627.

Kondo, T., and Raff, M. (2000). Oligodendrocyte precursor cells reprogrammed to become multipotential CNS stem cells. Science 289, 1754-1757.

Kooh, P. J., Fehon, R. G., and Muskavitch, M. A. (1993). Implications of dynamic patterns of Delta and Notch expression for cellular interactions during *Drosophila* development. Development 117, 493-507.

Koyano-Nakagawa, N., Kim, J., Anderson, D., and Kintner, C. (2000). Hes6 acts in a positive feedback loop with the neurogenins to promote neuronal differentiation. Development 127, 4203-4216.

Koyano-Nakagawa, N., Wettstein, D., and Kintner, C. (1999). Activation of *Xenopus* genes required for lateral inhibition and neuronal differentiation during primary neurogenesis. Mol Cell Neurosci 14, 327-339.

Krishnan, V., Pereira, F. A., Qiu, Y., Chen, C. H., Beachy, P. A., Tsai, S. Y., and Tsai, M. J. (1997). Mediation of Sonic hedgehog-induced expression of COUP-TFII by a protein phosphatase. Science 278, 1947-1950.

Kruger, G. M., Mosher, J. T., Bixby, S., Joseph, N., Iwashita, T., and Morrison, S. J. (2002). Neural crest stem cells persist in the adult gut but undergo changes in self-renewal, neuronal subtype potential, and factor responsiveness. Neuron 35, 657-669.

Kuhn, H. G., and Svendsen, C. N. (1999). Origins, functions, and potential of adult neural stem cells. Bioessays 21, 625-630.

Kunisch, M., Haenlin, M., and Campos-Ortega, J. A. (1994). Lateral inhibition mediated by the *Drosophila* neurogenic gene delta is enhanced by proneural proteins. Proc Natl Acad Sci U S A 91, 10139-10143.

Lance-Jones, C., Omelchenko, N., Bailis, A., Lynch, S., and Sharma, K. (2001). Hoxd10 induction and regionalization in the developing lumbosacral spinal cord. Development 128, 2255-2268.

LaVail, M. M., Rapaport, D. H., and Rakic, P. (1991). Cytogenesis in the monkey retina. J Comp Anat 309, 86-114.

Leber, S., Breedlove, S., and Sanes, J. (1990). Lineage, arrangement, and death of clonally related motoneurons in the chick spinal cord. J Neuroscience 10, 2451-2462.

Lee, J. E. (1997). Basic helix-loop-helix genes in neural development. Curr Opin Neurobiol 7, 13-20.

Lee, J. E., Hollenberg, S. M., Snider, L., Turner, D. L., Lipnick, N., and Weintraub, H. (1995). Conversion of *Xenopus* ectoderm into neurons by NeuroD, a basic helix-loop-helix protein. Science 268, 836-844.

Lee, S. K., and Pfaff, S. L. (2001). Transcriptional networks regulating neuronal identity in the developing spinal cord. Nat Neurosci 4 Suppl, 1183-1191.

Lehmann, R., Jimenez, F., Dietrich, U., and Campos-Ortega, J. A. (1983). On the phenotype and development of mutants of early neurogenesis in *Drosophila* Melanogaster. 192, 62-74.

Levine, E. M., Close, J., Fero, M., Ostrovsky, A., and Reh, T. A. (2000). p27(Kip1) regulates cell cycle withdrawal of late multipotent progenitor cells in the mammalian retina. Dev Biol 219, 299-314.

Lewis, J. (1998). Notch signalling and the control of cell fate choices in vertebrates. Semin Cell Dev Biol 9, 583-589.

Li, P., Yang, X., Wasser, M., Cai, Y., and Chia, W. (1997). Inscuteable and Staufen mediate asymmetric localization and segregation of prospero RNA during *Drosophila* neuroblast cell divisions. Cell 90, 437-447.

Lieber, T., Kidd, S., Alcamo, E., Corbin, V., and Young, M. W. (1993). Antineurogenic phenotypes induced by truncated Notch proteins indicate a role in signal transduction and may
point to a novel function for Notch in nuclei. Genes Dev 7, 1949-1965.

Lillien, L. (1995). Changes in retinal cell fate induced by overexpression of EGF receptor. Nature
377, 158-162.

Lillien, L., and Cepko, C. (1992). Control of proliferation in the retina: temporal changes in
responsiveness to FGF and TGF alpha. Development 115, 253-266.

Lin, S., Johnson, S. M., Abraham, M., Vella, M. C., Pasquinelli, A., Gamberi, C., Gottlieb, E.,
and Slack, F. (2003). The *C.elegans* hunchback homolog, hbl-1, Controls Temporal Patterning and Is a Probable MicroRNA Target. Developmental Cell 4, 639-650.

Litingtung, Y., and Chiang, C. (2000). Specification of ventral neuron types is mediated by an
antagonistic interaction between Shh and Gli3. Nat Neurosci 3, 979-985.

Liu, M., Pleasure, S. J., Collins, A. E., Noebels, J. L., Naya, F. J., Tsai, M. J., and Lowenstein, D.
H. (2000a). Loss of BETA2/NeuroD leads to malformation of the dentate gyrus and epilepsy.
Proc Natl Acad Sci U S A 97, 865-870.

Liu, W., Khare, S. L., Liang, X., Peters, M. A., Liu, X., Cepko, C. L., and Xiang, M. (2000b). All
Brn3 genes can promote retinal ganglion cell differentiation in the chick. Development 127,
3237-3247.

Livesey, F. J., and Cepko, C. L. (2001). Vertebrate neural cell-fate determination: lessons from
the retina. Nat Rev Neurosci 2, 109-118.

Lo, L., Dormand, E., Greenwood, A., and Anderson, D. J. (2002). Comparison of the generic
neuronal differentiation and neuron subtype specification functions of mammalian achaete-
scute and atonal homologs in cultured neural progenitor cells. Development 129, 1553-1567.

Lo, L., Tiveron, M. C., and Anderson, D. J. (1998). MASH1 activates expression of the paired
homeodomain transcription factor Phox2a, and couples pan-neuronal and subtype-specific
components of autonomic neuronal identity. Development 125, 609-620.

Lu, B., Rothenberg, M., Jan, L. Y., and Jan, Y. N. (1998). Partner of Numb colocalizes with
Numb during mitosis and directs Numb asymmetric localization in *Drosophila* neural and
muscle progenitors. Cell 95, 225-235.

Lu, Q. R., Yuk, D., Alberta, J. A., Zhu, Z., Pawlitzky, I., Chan, J., McMahon, A. P., Stiles, C. D.,
and Rowitch, D. H. (2000). Sonic hedgehog—regulated oligodendrocyte lineage genes encoding bHLH proteins in the mammalian central nervous system. Neuron 25, 317-329.

Lumsden, A., and Krumlauf, R. (1996). Patterning the Vertebrate Neuraxis. Science 274, 1109-
1115.

Ma, Q., Chen, Z., del Barco Barrantes, I., de la Pompa, J. L., and Anderson, D. J. (1998).
neurogenin1 is essential for the determination of neuronal precursors for proximal cranial
sensory ganglia. Neuron 20, 469-482.

Ma, Q., Fode, C., Guillemot, F., and Anderson, D. J. (1999). Neurogenin1 and neurogenin2
control two distinct waves of neurogenesis in developing dorsal root ganglia. Genes Dev 13,
1717-1728.

Ma, Q., Kintner, C., and Anderson, D. J. (1996). Identification of neurogenin, a vertebrate neuronal determination gene. Cell 87, 43-52.

Malatesta, P., Hartfuss, E., and Gotz, M. (2000). Isolation of radial glial cells by fluorescent-
activated cell sorting reveals a neuronal lineage. Development 127, 5253-5263.

Mansouri, A., and Gruss, P. (1998). Pax3 and Pax7 are expressed in commissural neurons and
restrict ventral neuronal identity in the spinal cord. Mech Dev 78, 171-178.

Martin-Bermudo, M. D., Martinez, C., Rodriguez, A., and Jimenez, F. (1991). Distribution and
function of the lethal of scute gene product during early neurogenesis in *Drosophila*. Development 113, 445-454.

Massari, M. E., and Murre, C. (2000). Helix-loop-helix proteins: regulators of transcription in eucaryotic organisms. Mol Cell Biol 20, 429-440.

Matise, M. P., Epstein, D. J., Park, H. L., Platt, K. A., and Joyner, A. L. (1998). Gli2 is required for induction of floor plate and adjacent cells, but not most ventral neurons in the mouse central nervous system. Development 125, 2759-2770.

Matise, M. P., and Joyner, A. L. (1997). Expression patterns of developmental control genes in normal and Engrailed-1 mutant mouse spinal cord reveal early diversity in developing interneurons. J Neurosci 17, 7805-7816.

McConnell, S. K. (1988). Fates of visual cortical neurons in the ferret after isochronic and heterochronic transplantation. J Neurosci 8, 945-974.

McDonald, J. A., Holbrook, S., Isshiki, T., Weiss, J., Doe, C. Q., and Mellerick, D. M. (1998). Dorsoventral patterning in the *Drosophila* central nervous system: the vnd homeobox gene specifies ventral column identity. Genes Dev 12, 3603-3612.

Miyata, T., Maeda, T., and Lee, J. E. (1999). NeuroD is required for differentiation of the granule cells in the cerebellum and hippocampus. Genes Dev 13, 1647-1652.

Mizuguchi, R., Sugimori, M., Takebayashi, H., Kosako, H., Nagao, M., Yoshida, S., Nabeshima, Y., Shimamura, K., and Nakafuku, M. (2001). Combinatorial roles of olig2 and neurogenin2 in the coordinated induction of pan-neuronal and subtype-specific properties of motoneurons. Neuron 31, 757-771.

Moran-Rivard, L., Kagawa, T., Saueressig, H., Gross, M. K., Burrill, J., and Goulding, M. (2001). Evx1 is a postmitotic determinant of v0 interneuron identity in the spinal cord. Neuron 29, 385-399.

Morrison, S. J., Perez, S. E., Qiao, Z., Verdi, J. M., Hicks, C., Weinmaster, G., and Anderson, D. J. (2000). Transient Notch activation initiates an irreversible switch from neurogenesis to gliogenesis by neural crest stem cells. Cell 101, 499-510.

Morrow, E. M., Furukawa, T., Lee, J. E., and Cepko, C. L. (1999). NeuroD regulates multiple functions in the developing neural retina in rodent. Development 126, 23-36.

Muhr, J., Andersson, E., Persson, M., Jessell, T. M., and Ericson, J. (2001). Groucho-mediated transcriptional repression establishes progenitor cell pattern and neuronal fate in the ventral neural tube. Cell 104, 861-873.

Muhr, J., Graziano, E., Wilson, S., Jessell, T. M., and Edlund, T. (1999). Convergent inductive signals specify midbrain, hindbrain, and spinal cord identity in gastrula stage chick embryos. Neuron 23, 689-702.

Murre, C., McCaw, P. S., Vaessin, H., Caudy, M., Jan, L. Y., Jan, Y. N., Cabrera, C. V., Buskin, J. N., Hauschka, S. D., Lassar, A. B., and et al. (1989). Interactions between heterologous helix-loop-helix proteins generate complexes that bind specifically to a common DNA sequence. Cell 58, 537-544.

Nakao, K., and Campos-Ortega, J. A. (1996). Persistent expression of genes of the enhancer of split complex suppresses neural development in *Drosophila*. Neuron 16, 275-286.

Neumann, C. J., and Nuesslein-Volhard, C. (2000). Patterning of the zebrafish retina by a wave of sonic hedgehog activity. Science 289, 2137-2139.

Nieto, M., Schuurmans, C., Britz, O., and Guillemot, F. (2001). Neural bHLH genes control the neuronal versus glial fate decision in cortical progenitors. Neuron 29, 401-413.

Nolo, R., Abbott, L. A., and Bellen, H. J. (2000). Senseless, a Zn finger transcription factor, is necessary and sufficient for sensory organ development in *Drosophila*. Cell 102, 349-362.

Novitch, B. G., Chen, A. I., and Jessell, T. M. (2001). Coordinate regulation of motor neuron subtype identity and pan-neuronal properties by the bHLH repressor Olig2. Neuron 31, 773-789.

Ohnuma, S., Philpott, A., Wang, K., Holt, C. E., and Harris, W. A. (1999). p27Xic1, a Cdk inhibitor, promotes the determination of glial cells in *Xenopus* retina. Cell 99, 499-510.

Olson, J. M., Asakura, A., Snider, L., Hawkes, R., Strand, A., Stoeck, J., Hallahan, A., Pritchard, J., and Tapscott, S. J. (2001). NeuroD2 is necessary for development and survival of central nervous system neurons. Dev Biol 234, 174-187.

Parks, A. L., and Muskavitch, M. A. (1993). Delta function is required for bristle organ determination and morphogenesis in *Drosophila*. Dev Biol 157, 484-496.

Parmar, M., Skogh, C., Bjorklund, A., and Campbell, K. (2002). Regional specification of neurosphere cultures derived from subregions of the embryonic telencephalon. Mol Cell Neurosci 21, 645-656.

Parras, C., Garcia-Alonso, L. A., Rodriguez, I., and Jimenez, F. (1996). Control of neural precursor specification by proneural proteins in the CNS of *Drosophila*. Embo J 15, 6394-6399.

Parras, C. M., Schuurmans, C., Scardigli, R., Kim, J., Anderson, D. J., and Guillemot, F. (2002). Divergent functions of the proneural genes Mash1 and Ngn2 in the specification of neuronal subtype identity. Genes Dev 16, 324-338.

Pattyn, A., Goridis, C., and Brunet, J. F. (2000). Specification of the central noradrenergic phenotype by the homeobox gene Phox2b. Mol Cell Neurosci 15, 235-243.

Pattyn, A., Morin, X., Cremer, H., Goridis, C., and Brunet, J. F. (1999). The homeobox gene Phox2b is essential for the development of autonomic neural crest derivatives. Nature 399, 366-370.

Pearson, B. J., and Doe, C. Q. (2003). What time is it? Controlling the neuroblast clock with the transcription factor hunchback. Abstract for 44th Annual *Drosophila* Research Conference.

Perron, M., and Harris, W. A. (2000a). Determination of vertebrate retinal progenitor cell fate by the Notch pathway and basic helix-loop-helix transcription factors. Cell Mol Life Sci 57, 215-223.

Perron, M., and Harris, W. A. (2000b). Retinal stem cells in vertebrates. Bioessays 22, 685-688.

Perron, M., Kanekar, S., Vetter, M. L., and Harris, W. A. (1998). The genetic sequence of retinal development in the ciliary margin of the *Xenopus* eye. Dev Biol 199, 185-200.

Petronczki, M., and Knoblich, J. A. (2001). DmPAR-6 directs epithelial polarity and asymmetric cell division of neuroblasts in *Drosophila*. Nat Cell Biol 3, 43-49.

Pfaff, S. L., Mendelsohn, M., Stewart, C. L., Edlund, T., and Jessell, T. M. (1996). Requirement for LIM homeobox gene Isl1 in motor neuron generation reveals a motor neuron-dependent step in interneuron differentiation. Cell 84, 309-320.

Pierani, A., Moran-Rivard, L., Sunshine, M. J., Littman, D. R., Goulding, M., and Jessell, T. M. (2001). Control of interneuron fate in the developing spinal cord by the progenitor homeodomain protein Dbx1. Neuron 29, 367-384.

Pringle, N. P., Guthrie, S., Lumsden, A., and Richardson, W. D. (1998). Dorsal spinal cord neuroepithelium generates astrocytes but not oligodendrocytes. Neuron 20, 883-893.

Qian, X., Shen, Q., Goderie, S. K., He, W., Capela, A., Davis, A. A., and Temple, S. (2000a). Timing of CNS cell generation: a programmed sequence of neuron and glial cell production from isolated murine cortical stem cells. Neuron 28, 69-80.

Qian, X., Shen, Q., Goderie, S. K., He, W., Capela, A., Davis, A. A., and Temple, S. (2000b). Timing of CNS cell generation: a programmed sequence of neuron and glial cell production from isolated murine cortical stem cells [In Process Citation]. Neuron 28, 69-80.

Ramain, P., Khechumian, R., Khechumian, K., Arbogast, N., Ackermann, C., and Heitzler, P. (2000). Interactions between chip and the achaete/scute-daughterless heterodimers are required for pannier-driven proneural patterning. Mol Cell 6, 781-790.

Ready, D. F., Hanson, T. E., and Benzer, S. (1976). Development of the *Drosophila* retina, a neurocrystalline lattice. Dev Biol 53, 217-240.

Rebay, I., Fehon, R. G., and Artavanis-Tsakonas, S. (1993). Specific truncations of *Drosophila* Notch define dominant activated and dominant negative forms of the receptor. Cell 74, 319-329.

Reddy, G. V., and Rodrigues, V. (1999). A glial cell arises from an additional division within the mechanosensory lineage during development of the microchaete on the *Drosophila* notum. Development 126, 4617-4622.

Reh, T. A., and Tully, T. (1986). Regulation of tyrosine hydroxylase-containing amacrine cell number in larval frog retina. Developmental Biology 114, 463-469.

Rhyu, M. S., Jan, L. Y., and Jan, Y. N. (1994). Asymmetric distribution of numb protein during division of the sensory organ precursor cell confers distinct fates to daughter cells. Cell 76, 477-491.

Richardson, W. D., Pringle, N. P., Yu, W. P., and Hall, A. C. (1997). Origins of spinal cord oligodendrocytes: possible developmental and evolutionary relationships with motor neurons. Dev Neurosci 19, 58-68.

Rodriguez, I., Hernandez, R., Modolell, J., and Ruiz-Gomez, M. (1990). Competence to develop sensory organs is temporally and spatially regulated in *Drosophila* epidermal primordia. Embo J 9, 3583-3592.

Roelink, H., Porter, J. A., Chiang, C., Tanabe, Y., Chang, D. T., Beachy, P. A., and Jessell, T. M. (1995). Floor plate and motor neuron induction by different concentrations of the amino-terminal cleavage product of sonic hedgehog autoproteolysis. Cell 81, 445-455.

Sander, M., Paydar, S., Ericson, J., Briscoe, J., Berber, E., German, M., Jessell, T. M., and Rubenstein, J. L. (2000). Ventral neural patterning by Nkx homeobox genes: Nkx6.1 controls somatic motor neuron and ventral interneuron fates. Genes Dev 14, 2134-2139.

Saueressig, H., Burrill, J., and Goulding, M. (1999). Engrailed-1 and netrin-1 regulate axon pathfinding by association interneurons that project to motor neurons. Development 126, 4201-4212.

Scardigli, R., Schuurmans, C., Gradwohl, G., and Guillemot, F. (2001). Crossregulation between Neurogenin2 and pathways specifying neuronal identity in the spinal cord. Neuron 31, 203-217.

Schmid, A., Chiba, A., and Doe, C. Q. (1999). Clonal analysis of *Drosophila* embryonic neuroblasts: neural cell types, axon projections and muscle targets. Development 126, 4653-4689.

Schmidt, H., Rickert, C., Bossing, T., Vef, O., Urban, J., and Technau, G. M. (1997). The embryonic central nervous system lineages of *Drosophila* melanogaster. II. Neuroblast lineages derived from the dorsal part of the neuroectoderm. Dev Biol 189, 186-204.

Schober, M., Schaefer, M., and Knoblich, J. A. (1999). Bazooka recruits Inscuteable to orient asymmetric cell divisions in *Drosophila* neuroblasts. Nature 402, 548-551.

Schuldt, A. J., Adams, J. H., Davidson, C. M., Micklem, D. R., Haseloff, J., St Johnston, D., and Brand, A. H. (1998). Miranda mediates asymmetric protein and RNA localization in the developing nervous system. Genes Dev 12, 1847-1857.

Schwab, M. H., Bartholomae, A., Heimrich, B., Feldmeyer, D., Druffel-Augustin, S., Goebbels, S., Naya, F. J., Zhao, S., Frotscher, M., Tsai, M. J., and Nave, K. A. (2000). Neuronal basic helix-loop-helix proteins (NEX and BETA2/Neuro D) regulate terminal granule cell differentiation in the hippocampus. J Neurosci 20, 3714-3724.

Seugnet, L., Simpson, P., and Haenlin, M. (1997). Transcriptional regulation of Notch and Delta: requirement for neuroblast segregation in *Drosophila*. Development 124, 2015-2025.

Sharma, K., Sheng, H. Z., Lettieri, K., Li, H., Karavanov, A., Potter, S., Westphal, H., and Pfaff, S. L. (1998). LIM homeodomain factors Lhx3 and Lhx4 assign subtype identities for motor neurons. Cell 95, 817-828.

Shen, C. P., Jan, L. Y., and Jan, Y. N. (1997). Miranda is required for the asymmetric localization of Prospero during mitosis in *Drosophila*. Cell 90, 449-458.

Shen, Q., Zhong, W., Jan, Y. N., and Temple, S. (2002). Asymmetric Numb distribution is critical for asymmetric cell division of mouse cerebral cortical stem cells and neuroblasts. Development 129, 4843-4853.

Skeath, J. B., and Carroll, S. B. (1991). Regulation of achaete-scute gene expression and sensory organ pattern formation in the *Drosophila* wing. Genes Dev 5, 984-995.

Skeath, J. B., and Doe, C. Q. (1996). The achaete-scute complex proneural genes contribute to neural precursor specification in the *Drosophila* CNS. Curr Biol 6, 1146-1152.

Skeath, J. B., Panganiban, G. F., and Carroll, S. B. (1994). The ventral nervous system defective gene controls proneural gene expression at two distinct steps during neuroblast formation in *Drosophila*. Development 120, 1517-1524.

Skeath, J. B., and Thor, S. (2003). Genetic control of *Drosophila* nerve cord development. Curr Opin Neurobiol 13, 8-15.

Skeath, J. B., Zhang, Y., Holmgren, R., Carroll, S. B., and Doe, C. Q. (1995). Specification of neuroblast identity in the *Drosophila* embryonic central nervous system by gooseberry-distal. Nature 376, 427-430.

Smith, S. T., and Jaynes, J. B. (1996). A conserved region of engrailed, shared among all en-, gsc-, Nk1-, Nk2- and msh-class homeoproteins, mediates active transcriptional repression *in vivo*. Development 122, 3141-3150.

Sommer, L., Shah, N., Rao, M., and Anderson, D. J. (1995). The cellular function of MASH1 in autonomic neurogenesis. Neuron 15, 1245-1258.

Soula, C., Danesin, C., Kan, P., Grob, M., Poncet, C., and Cochard, P. (2001). Distinct sites of origin of oligodendrocytes and somatic motoneurons in the chick spinal cord: oligodendrocytes arise from Nkx2.2-expressing progenitors by a Shh-dependent mechanism. Development 128, 1369-1379.

Spana, E. P., and Doe, C. Q. (1995). The prospero transcription factor is asymmetrically localized to the cell cortex during neuroblast mitosis in *Drosophila*. Development 121, 3187-3195.

Stiemke, M. M., and Hollyfield, J. G. (1995). Cell birthdays in *Xenopus* laevis retina. Differentiation 58, 189-193.

Struhl, G., Fitzgerald, K., and Greenwald, I. (1993). Intrinsic activity of the Lin-12 and Notch intracellular domains *in vivo*. Cell 74, 331-345.

Sun, Y., Jan, L. Y., and Jan, Y. N. (1998). Transcriptional regulation of atonal during development of the *Drosophila* peripheral nervous system. Development 125, 3731-3740.

Temple, S. (2001). The development of neural stem cells. Nature 414, 112-117.

Terada, N., Hamazaki, T., Oka, M., Hoki, M., Mastalerz, D. M., Nakano, Y., Meyer, E. M., Morel, L., Petersen, B. E., and Scott, E. W. (2002). Bone marrow cells adopt the phenotype of other cells by spontaneous cell fusion. Nature 416, 542-545.

Tomita, K., Moriyoshi, K., Nakanishi, S., Guillemot, F., and Kageyama, R. (2000). Mammalian achaete-scute and atonal homologs regulate neuronal versus glial fate determination in the central nervous system. Embo J 19, 5460-5472.

Tomlinson, A., and Ready, D. F. (1987). Neuronal differentiation in the *Drosophila* ommatidium. Devel Biol 120, 366-376.

Turner, D. L., and Cepko, C. L. (1987). A common progenitor for neurons and glia persists in rat retina late in development. Nature 328, 131-136.

Turner, D. L., Snyder, E. Y., and Cepko, C. L. (1990). Lineage-independent determination of cell type in the embryonic mouse retina. Neuron 4, 833-845.

Vaessin, H., Brand, M., Jan, L. Y., and Jan, Y. N. (1994). daughterless is essential for neuronal precursor differentiation but not for initiation of neuronal precursor formation in *Drosophila* embryo. Development 120, 935-945.

Vallstedt, A., Muhr, J., Pattyn, A., Pierani, A., Mendelsohn, M., Sander, M., Jessell, T. M., and Ericson, J. (2001). Different levels of repressor activity assign redundant and specific roles to Nkx6 genes in motor neuron and interneuron specification. Neuron 31, 743-755.

Van Doren, M., Powell, P. A., Pasternak, D., Singson, A., and Posakony, J. W. (1992). Spatial regulation of proneural gene activity: auto- and cross-activation of achaete is antagonized by extramacrochaetae. Genes Dev 6, 2592-2605.

Vescovi, A. L., Rietze, R., Magli, M. C., and Bjornson, C. (2002). Hematopoietic potential of neural stem cells. Nat Med 8, 535; author reply 536-537.

Villares, R., and Cabrera, C. V. (1987). The achaete-scute gene complex of D. melanogaster: conserved domains in a subset of genes required for neurogenesis and their homology to myc. Cell 50, 415-424.

Waid, D. K., and McLoon, S. C. (1998). Ganglion cells influence the fate of dividing retinal cells in culture. Development 125, 1059-1066.

Weintraub, H. (1993). The MyoD family and myogenesis: redundancy, networks, and thresholds. Cell 75, 1241-1244.

Wetts, R., and Fraser, S. E. (1988). Multipotent precursors can give rise to all major cell types of the frog retina. Science 239, 1142-1145.

White, P. M., Morrison, S. J., Orimoto, K., Kubu, C. J., Verdi, J. M., and Anderson, D. J. (2001). Neural crest stem cells undergo cell-intrinsic developmental changes in sensitivity to instructive differentiation signals. Neuron 29, 57-71.

Wichterle, H., Lieberam, I., Porter, J. A., and Jessell, T. M. (2002). Directed differentiation of embryonic stem cells into motor neurons. Cell 110, 385-397.

Wilson, P. A., and Hemmati-Brivanlou, A. (1997). Vertebrate neural induction: inducers, inhibitors, and a new synthesis. Neuron 18, 699-710.

Wilson, S. I., and Edlund, T. (2001). Neural induction: toward a unifying mechanism. Nat Neurosci 4 Suppl, 1161-1168.

Wodarz, A., Ramrath, A., Grimm, A., and Knust, E. (2000). *Drosophila* atypical protein kinase C associates with Bazooka and controls polarity of epithelia and neuroblasts. J Cell Biol 150, 1361-1374.

Wodarz, A., Ramrath, A., Kuchinke, U., and Knust, E. (1999). Bazooka provides an apical cue for Inscuteable localization in *Drosophila* neuroblasts. Nature 402, 544-547.

Wu, P., Tarasenko, Y. I., Gu, Y., Huang, L. Y., Coggeshall, R. E., and Yu, Y. (2002). Region-specific generation of cholinergic neurons from fetal human neural stem cells grafted in adult rat. Nat Neurosci 5, 1271-1278.

Wurmser, A. E., and Gage, F. H. (2002). Stem cells: cell fusion causes confusion. Nature 416, 485-487.

Ying, Q. L., Nichols, J., Evans, E. P., and Smith, A. G. (2002). Changing potency by spontaneous fusion. Nature 416, 545-548.

Yokota, Y. (2001). Id and development. Oncogene 20, 8290-8298.

Young, R. W. (1985). Cell differentiation in the retina of the mouse. The Anat Record 212, 199-205.

Zhong, W., Feder, J. N., Jiang, M. M., Jan, L. Y., and Jan, Y. N. (1996). Asymmetric localization of a mammalian numb homolog during mouse cortical neurogenesis. Neuron 17, 43-53.

Zhou, Q., Choi, G., and Anderson, D. J. (2001). The bHLH transcription factor Olig2 promotes oligodendrocyte differentiation in collaboration with Nkx2.2. Neuron 31, 791-807.

Zhou, Q., Wang, S., and Anderson, D. J. (2000). Identification of a novel family of oligodendrocyte lineage-specific basic helix-loop-helix transcription factors. Neuron 25, 331-343.

Chapter 3

Stage-Specific and Cell Fate Markers

Ichiro Nakano and Harley I. Kornblum

INTRODUCTION

For the purposes of this chapter, neural stem cells (NSCs) will be defined as cells capable of proliferating in symmetric and asymmetric fashions and that ultimately give rise to the three major cell types in the CNS: neurons, astrocytes and oligodendrocytes. Neural stem cells have the properties that they can be clonally and serially passaged (Reynolds and Weiss, 1992; Morshead et al., 1994; Gage et al., 1995). Neural stem cells are the subject of intense research and therapeutic interest. However, to date, NSC have been difficult to unambiguously identify *in vitro* or *in vivo* by their physical or molecular properties.

There are several different ways to propagate NSC *in vitro*. This fact alone has lead to a great deal of variation amongst results obtained for different studies (Reynolds and Weiss, 1992; Stemple and Anderson, 1992; Reynolds and Weiss, 1993; Vescovi et al., 1993; Gage et al., 1995; Gritti et al., 1995; Reynolds and Weiss, 1996; Weiss et al., 1996). Initially, NSC were propagated as floating balls of cells, termed neurospheres in the presence of epidermal growth factor (EGF; Reynolds and Weiss, 1992). Subsequently it has been shown that neurospheres can also be propagated in the presence of fibroblast growth factor, the most commonly used of which is basic FGF (FGF-2; Gritti et al., 1995; Kilpatrick and Bartlett, 1995; Ciccolini and Svendsen, 1998). Neurospheres can be grown from the brain at any developmental age beyond E8.5 in the mouse, although at these early stages NSC are responsive only to bFGF, while EGF-responsiveness comes later (Ciccolini and Svendsen, 1998). Numerous areas of the CNS have neurosphere-forming potential, from the olfactory bulb to the spinal cord, although the precise cells that give rise to neurospheres is not completely clear.

Neural stem cells have been propagated in monolayer cultures. This was initially accomplished with cells from the adult hippocampus, but has also been performed for NSCs from the embryonic neocortex (Gage et al., 1995; Palmer et al., 1995; Johe et al., 1996; Sun et al., 2001). These monolayer cultures use

From: *Neural Stem Cells: Development and Transplantation*
Editor: Jane E. Bottenstein © 2003 Kluwer Academic Publishers, Norwell, MA

different substrates: polyornithine/laminin for the hippocampus and polyornithine/fibronectin in the cortical stem cell cultures.

In addition to these methods of culture, NSCs can be derived from "other" cell types *in vitro*. Embryonic stem (ES) cells can be cultured in one of a variety of ways to produce cells with the characteristics of NSCs (Bain et al., 1995; Okabe et al., 1996; Brustle et al., 1997; Tropepe et al., 2001; Ying et al., 2003). Additionally, "traditional" astrocyte cultures can be driven to produce NSCs when propagated under specified conditions, as can oligodendrocyte precursors, and, possibly several other cell types (Kondo and Raff, 2000; Laywell et al., 2000; Imura et al., 2003).

Regardless of the method of culture, it must be emphasized that there are no conditions that give rise to a pure population of NSC. In each condition, a certain amount of spontaneous differentiation occurs and NSCs may exist simultaneously with a variety of other cell types. Thus, methods must be developed to distinguish neural stem cells from other cells contained within these cultures.

In vivo, NSCs are thought to reside within restricted regions, including the periventricular germinal epithelia, the dentate gyrus of the hippocampus and the region surrounding the central canal of the spinal cord (Lois and Alvarez-Buylla, 1993; Weiss et al., 1996; Laywell et al., 2000; Rietze et al., 2001; Seri et al., 2001). Recent observations have challenged these notions with evidence supporting the existence of neural stem cells in regions distant from these germinal zones, including the white matter of human cortex (Milosevic and Goldman, 2002). Regardless of the location of neural stem cells, the *in vivo* neural stem cell niche is a complex one and contains a variety of cell types. Additionally, many studies are testing implantation of NSCs as therapies for CNS disorders. These studies require that one is able to determine the phenotype of a cell in tissue sections following transplantation into the brain. Thus, molecular markers to label NSCs *in vivo* as well as to distinguish them from more differentiated cells are just as critically needed as for *in vitro* work.

Although the definition of NSC is relatively simple, this belies the complexity of the subject. One area that is of great current interest lies in the possible heterogeneity of NSCs, that is, that there may be many different "kinds" of neural stem cells. For instance, the cells isolated in the presence of FGF2 or EGF may have different properties in terms of their differentiation as well as cell cycle parameters (Kornblum et al., 1990; Vescovi et al., 1993; Weiss et al., 1996; Tropepe et al., 1999). Furthermore, different brain and spinal cord regions give rise to different classes of neurons and glia. It is possible that the NSCs existing in germinal regions of these different locales are different, as will be discussed below.

The complexity of NSC biology and cultures makes it imperative that markers be found to distinguish both between NSCs and other cells that may exist *in vitro* or *in vivo* as well as between potentially different NSCs. This chapter will focus on methods used to perform these tasks currently, as well as specific problems that need to be addressed in the future.

Molecular markers, simply put, are molecules that can be used to distinguish one cell type from another. There are several ways to take advantage of molecular markers. The most common method is through the use of immunocytochemistry. This method is relatively simple, but has the disadvantage of a high rate of nonspecific staining and the limitation of available antibodies. Another method used is in situ hybridization, which is highly sensitive and specific. However, this method is limited in use by its relative difficulty and the inability to label living cells.

Other methods exist to take advantage of molecular specificity. These include the use of promoter and/or enhancer regions to drive the cell-type specific production of a marker protein, such as green fluorescent protein or beta galactosidase (Price et al., 1987; Keyoung et al., 2001). Cells that would normally express the gene of interest would then express the marker. The genes can be introduced into cells in culture by standard methods or into animals using transgenic approaches (Kawaguchi et al., 2001). The difficulty here lies in the complexity of making the appropriate construct as well as promoter "leakiness" which will cause nonspecific production of the transgene.

In addition to molecular markers, other methods are used to track and label neural stem cells and their progeny *in vivo* and *in vitro*, that rely, not on molecular specificity, but the labeling of single cells and their progeny. Retroviral infection, at limiting dilutions, has been one method widely and successfully used to study and track neural stem cells (Price et al., 1987; Doetsch et al., 1999; Uchida et al., 2000; Milosevic and Goldman, 2002).

In the sections below, we provide examples of methods used to assay for cell types in NSC-containing cultures and *in vivo* with an attempt to place these methods in the context of NSC biology. We also describe other approaches to label NSCs and their progeny, as well as new approaches to identify markers and cell types. Needless to say, the lists and methods included here are incomplete and will continue to evolve.

MARKERS FOR DIFFERENTIATED CELL TYPES

An important aspect of the study of neural stem cells is the identification of their progeny, differentiated and committed cells. Multiple antibodies exist to label differentiated cells that are derived from neural stem cells. Because of a lack of completely specific markers, a combination of antibodies needs to be rationally developed to make identifications. A list of several of these markers is included in Table 1 (page 116). A few of the commonly used markers deserve specific comment and are discussed in each section. This list is by no means complete or exhaustive, but is meant to serve as a resource and provide examples.

General markers for neuronal identification

Neurons, when mature, possess characteristic and unambiguous morphological features. However, it is often difficult to clearly identify a cell as being a neuron *in vitro* or *in vivo*, when the neuron in question is not fully mature. Several antibodies exist to identify neurons at committed, postmitotic stages. Some of these are discussed below.

MAP2 (microtubule-associated protein2)

Both mRNA and protein are localized in dendrites, but not in all bodies, the developing brain, indicating that newly synthesized RNA is transported into dendrites in differentiating neurons for dendritic protein synthesis (Garner et al., 1988; Matus, 1991; Przyborski and Cambray-Deakin, 1995). MAP2 is known to have 3 isoforms. MAP2a appears first during the end of the second week of the postnatal period of mice, and MAP2b is present throughout brain development. Both of these isoforms have molecular weights of about 280kD. MAP2c, composed of several subunits of approximately 70kD each, is expressed at earlier stages of neuronal differentiation, compared to the other 2 isoforms, and is present largely during early embryonic development (Cassimeris and Spittle, 2001). This molecule disappears during brain maturation except in the retina, olfactory bulb, and cerebellum. Antibodies are available to selectively stain neurons in culture, but their use *in vivo* is limited.

Beta Tubulin III

Antibodies directed against beta tubulin III, one of which is the "TuJ1" antibody, are very frequently used to identify neurons in culture (Figure 1; Ferreira and Caceres, 1992; Menezes and Luskin, 1994). Microtubules consist principally of 2 soluble proteins, alpha- and beta-tubulin, each with a molecular weight of 55kD. In contrast to MAP2, mRNA for tubulin is localized exclusively in neuronal cell bodies, while its protein is present both in axons and in dendrites. Tubulin encodes an aminoterminal tetrapeptide sequence, which acts as the recognition element for autoregulated RNA instability, thus RNA does not correlate well with the expression of the protein, and biochemical measures

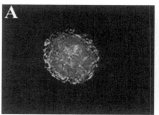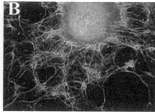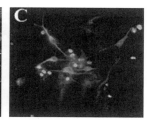

Figure 1. Immunocytochemistry of NSC-containing cultures (A) E17.5 cortical neurospheres with a supplement of bFGF were stained with antibodies for nestin (green) and propidium iodide (red). (B) Differentiating neurospheres were immunostained with antibodies for nestin (green) and propidium iodide (red). (C) Adherent progenitor cultures from E11.5 cerebral cortex were stained with antibodies directed against Beta tubulin III (TuJ1; green) and GFAP (red). Blue signals indicate nuclear staining by DAPI.

of Beta tubulin III mRNA, such as Northern blot or rt-PCR do not provide evidence of the extent to which neurons are present. The TuJ1 antibody will recognize neurons soon after differentiation, and, in some cases, even dividing neuroblasts (Luskin et al., 1997). *In vivo* staining is limited, with the best staining in embryonic tissue or isolated areas of neurogenesis, such as in the hippocampus and rostral migratory stream (Menezes and Luskin, 1994). This isoform of tubulin may also be present in highly proliferative cells, such as brain tumors like primitive neuroectodermal tumor (PNET) and astrocytoma (Katsetos et al., 1989; Katsetos et al., 2001).

Neurofilament

Neurofilament is one subclass of 5 cytoplastic intermediate filaments, which share a similar alpha-helical domain capable of forming coiled-coils (Katsetos et al., 1989; Katsetos et al., 2001). Neurofilament is composed of 3 neuron-specific proteins with molecular weights of 68kD (NFL), 125kD (NFM), and 200kD (NFH). The lower molecular weight of neurofilament is present in more immature neurons. A wide variety of commercially available antibodies exist and are most useful *in vitro* or in relatively early embryos *in vivo* (Gilad et al., 1989).

Doublecortin

Lissencephaly, which has the clinical manifestations of profound mental retardation and seizures, results from migrational arrest of virtually all cortical neurons short of their normal destinations and the development of the six layer formation in neocortex is disrupted (des Portes et al., 1998). One of the genes responsible for this disease is doublecortin. This gene is expressed in migrating neurons throughout the central and peripheral nervous systems exclusively in embryonic and neonatal brains (Hannan et al., 1999). The protein localization overlaps with that of beta tubulin III. However, conformational changes in the microtubules, such as depolymerization eliminate doublecortin staining. The function of this gene is in neuronal migration through the regulation of the organization and stability of microtubules. Antibodies raised against doublecortin are useful in detecting the presence of newly generated, migrating neurons *in vivo*.

Presenilin-1

Presenilin-1 is widely expressed in embryonic brain including neural progenitors, but is also found in adult brain, principally in postmitotic neurons (Busciglio et al., 1997; Wen et al., 2002). This gene is required for the release of the intracellular domain of Notch from the plasma membrane as well as for the processing of other cell surface proteins, including the amyloid precursor protein.

NeuN (neuronal nuclear antigen)

This somewhat inappropriately named antigen is expressed by the nuclei (Mullen et al., 1992) as well as the perinuclear cytoplasm of many neurons. Due to the limitations of many other neuronal antibodies *in vivo*, NeuN staining has been widely adopted as a benchmark for neuronal staining in transplantation and cell proliferation studies, especially using colocalization with bromodeoxyuridine staining to label neurons that have undergone DNA synthesis during a particular period of interest (Mullen et al., 1992).

Hu

A variety of "hu" antigens exist. The Hu (HuC) antigen is specifically expressed by most neurons and is useful for the identification of neurons *in vivo* and *in vitro* (Gultekin et al., 1998).

Identification of neurons at early stages of development

One particular area of intense interest lies in the early commitment of neurons from stem or other multipotent progenitor cells. A large diversity of cell types is observed by RNA expression of transcriptional marker genes during the formation of early neural tube and a number of genes are expressed by immmature or only recently "committed" neurons.

The transition from a proliferative neural precursor cell to a postmitotic neuron is a highly regulated step, which, in many instances, has been shown to involve a cascade of transcription factors that is triggered by proneural genes (Anderson, 1994; Bray, 2000). The events underlying this progression are easily visualized in the early neural tube, where they are spatially confined to distinct cell layers; the ventricular layer where neural precursors (including stem cells) divide, the intermediate layer where the first postmitotic cells can be identified, and the mantle layer where neuronal differentiation takes place (Anderson, 1994). This spatial organization allows for the use of in situ hybridization to distinguish amongst progenitors and committed cell types during embryonic periods. For example, Sox1, a stem cell-expressed gene is present solely in the ventricular layer and is downregulated prior to neuronal differentiation, while Math3/NeuroM, NeuroD and NKL (neuronal Kruppel-like) are expressed outside this layer and are present transiently in postmitotic cells, and superior cervical ganglion (SCG10) is expressed in differentiating neurons in the mantle layer (Perron et al., 1999; Cai et al., 2000; Buescher et al., 2002; Overton et al., 2002).

Proneural genes, such as Neurogenin1, promote neuronal differentiation, but in doing so also trigger the process of lateral inhibition; as a cell becomes a neuron it activates Notch signalling in neighboring cells, thereby inhibiting their differentiation (Ma et al., 1997; Mueller and Wullimann, 2003). This mechanism ensures that not all cells in the neuroepithelium differentiate simultaneously. Cells poised to become neurons express the Notch ligands Delta or Jagged, and expression of these genes in single cells marks a pivotal and conserved point in the neurogenesis pathway (Stump et al., 2002). As a consequence of Notch signaling (delivered by a Delta/Jagged-expressing cell), neighboring cells express repressors of neuronal differentiation that belong to the enhancer of split-hairy family of transcription factors, for example, Hes1 and Hes5 (Jarriault et al., 1998). Therefore, in situ hybridization for the Notch-activating ligands reveal committed cells within the neuroepithelium while identification of cells that have higher levels of receptor expression should reveal more uncommitted cells (Irvin et al., 2001). At later stages, however, the use of these genes to identify neurons or cells committed to become neurons can be more problematic. Transcription factors expressed by neurons may also be expressed by glia or glial-specific progenitors, while those expressed by neural stem cells may also be expressed by mature, differentiated cells present in the postnatal brain (Irvin et al., 2001).

In addition to the use of transcription factors as markers of immature neurons, there are factors important in regional identification of stem cells as well as neuronal subtypes which can be further used to distinguish amongst cells *in vivo*. In the ventral spinal cord, neurons of a specific subtype derive from a particular dorsoventral population of proliferating neural precursors that is referred to as a "progenitor domain"(Takahashi and Osumi, 2002). Each domain is characterized by the expression of a combination of homeodomain transcription factors (Dbx1, Dbx2, Irx3, Nkx2.2, Nkx6.1, Pax6 and Pax7), which is responsible for the specification of a particular neuronal subtype (Takahashi and Osumi, 2002). The neurons derived from these domains will continue to express at least some of the mRNAs characteristic of their region of origin. It is possible, therefore, that one can identify regionally distinct neuronal subtypes by establishing which combinations of transcription factors they express. Practically, this approach may be limited by the number of genes that one can simultaneously visualize using in situ hybridization or other methods.

In a manner analogous to the spinal cord, transcription factors regulating regional identification of neurons in the brain are now being elucidated and may also prove useful as molecular markers. For example, Dlx5 and 6 are expressed by GABAergic interneurons of the forebrain (He et al., 2001; Letinic et al., 2002). By using molecular markers for these factors, Anderson and colleagues were able to monitor the migration of these cells from the ventral neuroepithelium into neocortex, establishing this region of the germinal zone as the source of cortical GABAergic interneurons (Anderson et al., 2002).

Identification of oligodendrocytes

Like neurons, mature oligodendrocytes have a characteristic morphology and, additionally, synthesize myelin, which can be revealed by relatively simple lipid stains. However, *in vitro*, oligodendrocytes may not be easily distinguishable from other process-bearing cells. Some of the markers used to identify oligodendrocytes are described below.

PLP (proteolipid protein)

This protein is contained in CNS myelin and is expressed by maturing oligodendrocytes, but not at early stages of commitment or differentiation. DM20 is the other myelin proteolipid protein. Both PLP and DM20 are integral membrane proteins, which account for approximately half of the protein content of adult myelin. This gene encodes 5 hydrophobic domains that interact with the lipid bilayer as trans- and cis-membrane segments, and its transcript is found both in perinuclear and in peripheral processes, while another oligodendrocyte marker, MBP is highly localized in processes (Stoffel et al., 1984; Schliess and Stoffel, 1991).

MBP (myelin basic protein)

MBP is highly localized in processes and is another major constituent of myelin, although it appears *in vivo* at early stages of myelination (Sternberger et al., 1978)

GalC (galactocerebroside)

This cell surface protein is expressed by moderately mature oligodendrocytes (Raff et al., 1978), appearing after O4 and prior to MBP.

Claudin11 (oligodendrocyte-specific protein; OSP)

The central nervous system homolog of peripheral myelin protein-22, OSP, is a specific marker for oligodendrocytes in the brain and spinal cord. This gene is turned on later than Olig1/2, earlier than PLP/DM20, and persists throughout the oligodendrocyte maturations. As a member of the claudin family, this protein has a pivotal role in generating the paracellular physical barrier of tight junctions necessary for normail CNS function as well as spermatogenesis (Gow et al., 1999; Hellani et al., 2000).

CNPase (cyclic nucleotide phosphodiesterase)

Central nervous system myelin has high concentrations of a membrane-bound enzyme, CNPase, as does the outer segment of photoreceptors in the retina. Messenger RNA for this gene codes is predominantly localized both in perinuclear and in primary processes in oligodendrocytes. This gene codes for a microtubule-associated protein in promotes microtubule assembly (Dyer and Matthieu, 1994; Dyer et al., 1997).

O4

The O4 antigen is expressed by oligodendrocytes at the earliest stages of differentiation. O2A progenitors, which have the ability to differentiate into oligodendrocytes and type 2 astrocytes (though this cell type has not been clearly identified in the brain *in vivo*), are intermediate populations of O4-positive and galactocerebronide (GalC) negative cells (Trotter and Schachner, 1989; Baron et al., 1998). Selective anti-O4 antibodies are highly useful in discerning immature oligodendrocytes from immature neurons in mixed cultures, as both may have phase-bright nuclei and a bipolar morphology. O4 staining presists through maturation of oligodendrocytes.

Olig1/2

The transcription factors, Olig1 and 2 have a broad spectrum of RNA expression from multipotent progenitors to mature oligodendrocytes (Zhou and Anderson, 2002). These genes, like other bHLH type transcription factors, have major roles in cell fate determination. However, unlike other proneural genes that control neuronal versus glial commitment, these genes serve in both neuronal and glial subtype specification. Olig1 has roles in the development and maturation of oligodendrocytes, while Olig2 is required for oligodendrocyte and motor neuron specification in the spinal cord, a conclusion based on the study of targeted disruption of each gene (Zhou and Anderson, 2002). These genes are also expressed strongly in the neoplastic cells of oligodendrocytomas, contrasting to the absent or low expression in astrocytomas, which may suggest the origin from which these brain tumors are derived (Lu et al., 2001).

Identification of astrocytes and radial glia

Astrocytes are often stellate appearing *in vivo*, while *in vitro* they often are either flattened, polygonal cells or have long processes. They may be confused with neurons or with other cell types, such as fibroblasts or epithelial cells found in mixed cultures or cultures of embryonic stem cell origin. Some of the markers used to identify astrocytes *in vivo* and *in vitro* are described below. Radial glia are thought to be astrocyte precursors, but may act as stem cells (see below). *In vivo*, they are bipolar with long proocesses, extending from the ventricular to the pial surface. *In vitro*, they may possess similar morphology, and can be confused with a variety of cells, including immature neurons and oligodendrocytes.

GFAP (glial fibrillary acidic protein)

GFAP is an intermediate filament protein in brain, and is widely used as an astrocyte marker (Figure 1; Eng et al., 2000). However, GFAP expression is not limited to astrocytes. In late embryogenesis, radial glia turn on GFAP expression, and a subpopulation of GFAP expressing cells both in radial glia and in postnatal astrocytes behave as neural stem cells. Thus, while it appears that all astrocytes are immunoreactive for GFAP, not all cells that are GFAP positive are astrocytes (Lazzari and Franceschini, 2001; Parnavelas and Nadarajah, 2001). A number of good antibodies exist to stain for GFAP *in vitro* and *in vivo*. In situ hybridization is a convenient way to localize GFAP mRNA in cell bodies.

S100b

The beta-subunit of the S100 protein has a relatively high expression in astrocytes and some ependymal cells (Pfeiffer et al., 1992). A disulfide-bonded dimeric form of this protein induces axonal extension through glial-neuronal interaction (Ueda et al., 1996). Elevated levels of the protein are found in Down's syndrome (Seidl et al., 2001; Heizmann et al., 2002). The alpha sununit of S100 is also found in the brain, but the expression level is ten-fold lower than that of the beta subunit. Antibodies label astrocytes *in vitro* and *in vivo*, although some neurons have been reported to express S100b (Rickmann and Wolff, 1995; Yang et al., 1996).

GLAST (astrocyte-specific glutamate transporter)

GLAST is a member of a high-affinity sodium-dependent transporter moleculer family that regulates neurotransmitter concentrations at the excitatory glutaminergic synapses of the mammalian central nervous system. This gene is expressed in radial glia during developmental stages, and is also found in the substantia nigra, red nucleous, hippocampus, and in cerebral cortical layers in the mature brain (Shibata et al., 1997; Hartfuss et al., 2001).

Vimentin

The intermediate filament, vimentin, is expressed by proliferating astrocytes. It is also expressed by radial glia in the developing embryo as well as by mesenchymal cells (Lazzari and Franceschini, 2001; Parnavelas and Nadarajah, 2001). Antibodies to vimentin stain radial glial processes in most mammalian tissue. Other glia also express vimentin, including reactive astrocytes after injury, and oligodendrocyte progenitors (Lazzari and Franceschini, 2001; Parnavelas and Nadarajah, 2001).

MARKERS FOR STEM CELLS

Multiple kinds of stem cells exist

Stem cells can be isolated and propagated in numerous ways and from a variety of apparent cell types as described above. There is ample experimental evidence for the hypothesis that multiple cell types can function as stem cells *in vitro*, including radial glia and astrocytes, and some *in vivo* investigations support this idea as well (Gage et al., 1995; Johansson et al., 1999; Clarke et al., 2000). Numerous questions arise regarding the significance of these observations of multiple stem cell types. For example, it is not yet known how this multiplicity of stem cell-competent cells relates to the complexity of the central nervous system. Do these different stem cells give rise to different neuronal or glial subtypes? Does this stem cell competence reside in all radial glia or astrocytes, or is there a limitation? Does this diversity rely on spatial or temporal environment? A particularly interesting question is: what is the lineage relationship amongst these stem cells? In contrast to the classical model that suggested that neurons and glia are derived from two separate 'branches' of a lineage tree, neural stem cells may be contained within a continuum that forms the 'trunk' of a lineage tree (Berry and Rogers, 1965; Levitt et al., 1981; Temple and Raff, 1985; Price, 1987; Alvarez-Buylla and Garcia-Verdugo, 2002; Dietrich and Easterday, 2002; Gage, 2002). Depending on the time of development, cells within this trunk seem to have neuroepithelial, radial glial or astrocytic characteristics. Thus, molecular markers for cell lineages thought to be "committed", such as the astrocyte or radial glia may also be markers for a subpopulation of neural stem cells (Alvarez-Buylla and Garcia-Verdugo, 2002). However, not all radial glia serve as stem cells, nor do all astrocytes. According to this theory, therefore, a set of markers is required to identify universal stem cells independent of stages. Some of these issues are discussed below. However, this is a rapidly evolving area of research and many conclusions must be tentatively drawn.

General neural stem cell markers

NSC have been isolated from multiple regions and from multiple ages, and they share some morphological features, which are phase-bright round cell bodies with small processes. Although none of the markers are completely exclusive for NSC, some of them seem to be highly enriched in the stem cell populations. The markers used in these settings are described below.

Nestin

Nestin is an intermediate filament with an alpha helical domain. Nestin immuneoreactivity is present in the filamentous cytoskeletal network. It is expressed predominantly in stem/progenitor cells of the central nervous system, and upon terminal neural differentiation, this gene is downergulated and replaced by other intermediate filaments (Levison and Goldman, 1997; Doyle et al., 2001). It appears that all proliferating neural stem cell populations express nestin, however, nestin immunoreactivity does not verify that any particular cell is a stem cell. This gene is also known to be expressed in radial glia, reactive and immature astrocytes, and immature skeletal muscle (Zimmerman et al.,

1994; Sultana et al., 2000). Polyclonal and monoclonal antibodies exist which effectively label nestin under a wide variety of conditions *in vitro* and *in vivo* (Figure 1). The most commonly used antibody is the Rat401 antibody, available through the Developmental Studies Hybridoma Bank at the University of Iowa (Lendahl et al., 1990; Mokry and Nemecek, 1998).

Msi1 (homolog of drosophila musashi/Nrp-1)

Musashi1 (Msi1) is a neural RNA-binding protein, which contains 2 RNA recognition motifs (Sakakibara et al., 1996; Kaneko et al., 2000). Both RNA and protein are predominantly localized in fetal and adult brain, largely in neural stem/progenitor populations (Palm et al., 2000; Keyoung et al., 2001). Labeling for Msi1 is also found in some postmitotic neurons and astrocytes (Kaneko et al., 2000).

Sox1/2 (SRY-related HMG-box gene)

Both SOX1 and 2 have a single DNA-binding domain know as HMG box, resulting in the bending of DNA through large angles in a sequence specific manner. Messenger RNA for SOX1/2 is localized to the neural tube and is expressed at the earliest stages of neurulation(Pevny et al., 1998; Zappone et al., 2000; Cai et al., 2002). The use of antibodies directed against these genes or in situ hybridization probes to selectively label multipotent stem cells has not been validated.

Nucleostemin

Nucleostemin encodes two GTP-binding motifs. Immunocytochemistry revealed that it was found in the nucleoi of CNS stem cells, embryonic stem cells, and some cancer cell lines, accompanying with cell cycle progressions (Tsai and McKay, 2002). Analysis of distribution indicates this protein is expressed in the ventricular zone of embryonic spinal cord and in neuroepithelium and mesenchyme. Because nucleostemin has only been recently identified, the use of nucleostemin as a stem cell marker will need further study.

Notch1 (homolog of drosophila notch, 1)

Transcripts for Notch1 are expressed in many tissues both in fetal and in adult, but are most abundant in lymphoid tissues. In brain, this gene is highly expressed in the developing germinal zones, and can be co-expressed with CNTFRalpha, which suggests a link between Notch and CNTF signaling pathways (Chojnacki et al., 2003). Notch signaling has been implicated in neural stem cell maintenance but not their generation, and can also promote glial cell fate (Tanigaki et al., 2001; Lutolf et al., 2002). Notch1 is not a useful marker to distinguish stem from other cell types in culture or developing brain tissue, as it is expressed by immature neurons and glia. In adult CNS, Notch1 expression is likely to be largely confined to a proliferative population, likely to be stem cells (Irvin et al., 2001).

Markers for radial glia as neural stem cells

Are there any markers to identify stem cells among embryonic radial glia?

The seminal work of Rakic has demonstrated that radial glia act as a scaffold to support the migration of newly generated neurons from the embryonic germinal zone into the developing layers of the cortex, and until recently this was considered to be the major role for radial glial cells, before their postnatal transdifferentiation into astrocytes. In the mouse, radial glia can be identified with several markers, including RC2, brain lipid-binding protein (BLBP), vimentin, nestin, and GLAST (Yang et al., 1994; Parnavelas and Nadarajah, 2001; Sun et al., 2002). Radial glial cells begin to upregulate the astrocyte marker GFAP after the end of neurogenesis and begin to elaborate their ascending process and withdraw their descending processes. However, not all radial glia ultimately transform into mature astrocytes (Goldman et al., 1997; Parnavelas and Nadarajah, 2001). Factors such as ciliary neurotrophic factor, leukemia inhibitory factor, EGF receptor signaling, and members of the bone morphogenetic protein (BMP) family have been demonstrated to play a role in astrocyte differentiation from radial glia (Mabie et al., 1997; Zhu et al., 1999; Shimazaki et al., 2001).

Recent studies have questioned the simple view of radial glia functioning only as scaffolds and/or astrocyte precursors and suggest that these cells serve as multipotent progenitors (Hartfuss et al., 2001; Noctor et al., 2002). *In vivo* studies suggested that radial glial cells comprise the vast majority of precursor cells that give rise to neurons in the cortical germinal zone by using Ki-67, a marker of dividing cells, and nestin to label the precursor cell population in acutely dissociated cells derived from the cortex and ganglionic eminence of mouse embryonic brain (Noctor et al., 2001; Gotz et al., 2002). In the early embryonic brain until E12.5, the majority of precursor cells were found to express the radial glial marker RC2. As development progressed, the precursor cell population increasingly expressed the other radial glial markers BLBP and GLAST (Noctor et al., 2001). However, Feng et al. reported that most cells that expressed the radial glial cell marker BLBP did not incorporate bromodeoxyuridine (BrdU), and suggested that these cells played a conventional role in supporting the migration of new neurons, but not in the expansion of the neuronal precursor population (Feng and Heintz, 1995). This study implies that not all radial glial cells are neuronal precursor cells at least not actively dividing during neurogenesis, and there are no current markers to identify stem cells among embryonic radial glial populations.

Stem cells in postnatal/adult brain

Markers for different cell types in SVZ have been identified to elucidate stem cell specification in adult brain

Continuous neurogenesis occurs in discrete areas of adult brain subventricular zone (SVZ), and the inner granular layer of the dentate gyrus (Doetsch et al., 1997; van Praag et al., 2002). Within the SVZ, Doetsch et al. identified five cell types both by the expressions of markers using immunohistochemistry and by ultrastructure using electron microscopy (Doetsch et al., 1997; Doetsch et al., 1999). They showed that vitally labelled GFAP positive astrocytes (type B cell) can respond to EGF signaling *in vitro* to produce progeny that can be passaged and differentiated into neurons and glia, and that SVZ astrocytes can also behave as neural stem cells when cultured in calf serum, EGF and bFGF. There is, however, another conflicting theory regarding the cellular origin of neural stem cells in the adult brain, which suggests that adult neural stem cells are ependymal cells that express the intermediate filament protein nestin (Johansson et al., 1999; Clarke et al., 2000). As a complicating factor, further studies have indicated that stem cells propagated from the SVZ *in vitro*, are initially the type C GFAP negative cells (Doetsch et al., 2002). Further studies will need to resolve these controversies, although all are not completely mutually exclusive and it is possible that multiple cell types within the SVZ have stem cell characteristics.

Regional stem cell markers

As described above, many genes are expressed in the neural tube in a regional and time-specific manner. At least some of these genes are expressed by multipotent stem cells, themselves and may be markers for different populations of stem cells (Hitoshi et al., 2002). However, because the genes are expressed by the neuronal progeny of the NSCs in these regions, their expression may not be useful to distinguish NSCs from other cells within cultures from that region

MARKERS FOR COMMITTED PROGENITORS

The general theory of neural stem cell biology often calls for the existence of separate committed neuronal or glial progenitors in addition to NSC. However, experimental evidence validating this as a universal rule is lacking. Under some conditions, however, there do appear to be glial- and neuronal-restricted progenitors. Some markers used in these settings are described below.

A2B5

This surface antigen is expressed both by glial and neuronal progenitors (Eisenbarth et al., 1979; Fredman et al., 1984), although it is often used as a specific marker for O2A (oligodendrocyte/type 2 astrocyte) progenitor or oligodendroglial progenitor cells. Recent investigation using adult human subcortical white matter revealed that A2B5 positive fractions from FACS sorting contain multipotent progenitor cells, suggesting that a stem cell population expresses this antigen (Roy et al., 1999). In addition to progenitors, it appears that some relatively mature oligodendrocytes also express A2B5 (Roy et al., 1999).

PSA-NCAM (polysialylated neural cell adhesion molecule)

This molecule is expressed by a population of glial-restricted progenitors isolated from the postnatal brain. However, neuronal progenitors and, possibly, tripotent stem cells have also been reported to express this isoform (Doetsch et al., 1997; Ben-Hur et al., 1998). Some overlap is found with the neuronal markers TuJI and MAP2.

Beta tubulin III

As described above, this antigen is largely expressed by postmitotic neurons. However, within the rostral migratory stream, a population of cells expressing Beta tubulin III remains mitotically active (Luskin et al., 1997)..

METHODS FOR TRACKING CELLS LINEAGE

Tracking lineage using retroviral infection

In addition to the use of immunocytochemistry and in situ hybridization, other methods have been developed to track proliferation, migration, and differentiation *in vivo*. In classic studies, Luskin and colleagues used retroviruses encoding the enzyme beta-galactosidase at low titers to determine that neurons in the adult olfactory bulb were derived from the anterior portion of the subventricular zone—a clear demonstration of adult neurogenesis (Luskin and Boone, 1994; Betarbet et al., 1996; Smith and Luskin, 1998).

Retroviral lineage tracing was also used to demonstrate that radial glia were competent to produce neurons. For these studies, Noctor and colleagues injected a retrovirus that expressed green fluorescent protein (GFP) into the ventricle of E15-E16 rat embryos (Noctor et al., 2001; Noctor et al., 2002). Twenty-four hours after injection, cells with a radial glial morphology were infected by the virus and expressed high levels of GFP. These cells were also immunopositive for vimentin, consistent with a radial glial identity. They observed multiple radially arrayed clones three days after injection and each clone contained several cells that were immunopositive for neuronal tubulin and usually only one vimentin-positive radial glial cell. After that, they used time-lapse videomicroscopy and followed the development of these clones, and found that infected cortical radial glia divided asymmetrically to produce neuroblasts that migrated into the cortex, typically along the radial process of the same radial glial cell that produced them.

The use of viral constructs to isolate stem cells

There has been remarkable progress in the last few years in the prospective identification of NSCs. Goldman et al. sought to establish a means to identify, select, and observe (in real-time) live NSCs (Keyoung et al., 2001). They constructed a reporter gene using green fluorescent protein (GFP) placed under the transcriptional control of the neural-specific enhancer for the gene encoding nestin as a live-cell reporter of the neural progenitor phenotype. They found that this reporter gene cassette yielded progenitor cell-specific fluorescence in both fetal and adult human brain dissociates that had been transfected with nestin-enhanced green fluorescent protein (EGFP) plasmid DNA or infected with a recombinant adenovirus carrying this reporter gene cassette.

Based on a set of antigens and physical characteristics, cell sorting has been used for prospective identification of stem/ progenitor cells

The biological analysis of NSCs has lagged far behind that of hematopoetic stem cells (HSCs) due, in part, to the lack of available methodologies for the prospective identification or purification of NSC and the lack of *in vivo* repopulation assays. These capabilities that have proven to be of seminal value to the study of HSCs, in that HSCs can be isolated by the simultaneous use of several different phenotype and stage-selective surface markers (Uchida et al., 2000). Antibodies to Nestin, Musashi1, and Sox1 cannot be used for sorting living neural stem cells (NSCs) because these molecules are not cell surface antigens.

Uchida and her colleagues (see Chapter 7 in this volume) have succeeded in sorting live human fetal NSCs using the combination of antibodies to cell-surface antigens and fluorescence-activated cell sorting (Uchida et al., 2000). These human fetal NSCs were, phenotypically: $5F3^+$ ($CD133^+$), $5E12^+$, $CD34^-$, $CD45^-$, and $CD24^{-/lo}$. Single $CD133^{+,}$ $CD34^-/CD45^-$ sorted cells were used to initiated neurosphere cultures, and the progeny of clonogenic cells could differentiate into both neurons and glia. This approach has not been replicated in rodents, to date. In mice, Bartlett and colleagues have reported that one type of NSC present in adult mouse brains, which expresses only low levels of peanut aglutinin-binding and heat-stable antigen (HAS MCD4a) proteins, is found in both ependymal and subventricular zones and accounts for about 63% of the total NSC content (Rietze et al., 2001). Capela and Temple (2002) used another surface marker, LeX, a carbohydrate, which is expressed in embryonic pluripotent stem cells for the purpose of purifying adult SVZ stem cells (Capela and Temple, 2002). They determined that 4% of acutely isolated SVZ cells were LeX-positive and purified LeX-positive populations purified by FACS were found to have the characteristics of CNS stem cells using the neurosphere culture method. These experiments provide promising avenues for the isolation and identification of NSCs.

Transgenic mice are useful tools to examine cell-type specific expression and cell lineage

The characterization of transgenic animals where reporter genes, such as beta-galactosidase and green fluorscent protein (GFP) are placed under the control of cell-type specific promoters, is highly useful for tracking lineage *in vivo*. In particular, using GFP allows for the study of the dynamics of gene expression in real time, and places still further emphasis on understanding gene function in very specific cellular contexts. Transgenic mice expressing the nestin-EGFP reporter gene allow for the identification and enrichment of NSC (Kawaguchi et al., 2001). In transgenic mice, the nestin-EGFP-positive cells were identified and enriched via fluorescence-activated cell sorting, which revealed that EGFP expression correlated with both the mitotic index and the frequency of neurosphere formation *in vitro*. Similar techniques have been used to examine the process used by embryonic stem (ES) cells to acquire neural identity (Ying et al., 2003). A knock-in mouse was made by replacing the Sox1 gene with the coding region of GFP, and the conversion of ES cells into neuro-ectodermal progenitors was studied by FACS, where the GFP positive fraction yielded neurons and glia.

A transgenic mouse engineered to express the receptor for the avian leu-kosis virus was used to reveal that GFAP positive cells in the SVZ migrate through the rostral migratory pathway to the olfactory bulb and differentiate into interneurons (Doetsch et al., 1999). Another transgenic model using the GFAP promoter identified GFAP positive cells in the germinal zone as the pre-dominant source of multipotent neural stem cells in postnatal and adult but not in early embryonic brain (Imura et al., 2003).

In addition to the above, another transgenic model is currently being used to identify cell lineage through the use of Cre-lox technology. In these mice, the reporter gene is preceded by a "stop" signal which is, in turn, flanked by lox-p sequences. When these animals are crossed with animals expressing Cre-recombinase all cells that expressed the cell specific promoter, as well as all the progeny of these cells will be labeled by the reporter construct (Gorski et al., 2002).

The use of transgenic and other promoter-based methods to identify and sort cells, while highly significant and appealing, must be accompanied by cau-tionary statements. Not all promoters prove to be completely faithful in their cell and tissue-specific expression, which can lead to misleading results. Even the same promoter may be differentially active in different strains of mice.

APPROACHES FOR IDENTIFYING NEW MARKERS

Genomic and proteomic approaches

The complexity of neural stem cell biology and the heterogeneity of the brain and *in vitro* culture systems illustrate the need to identify more and better markers for neural stem cells and their committed progeny.

One approach to achieve this goal is through the use of DNA microarrays to identify genes expressed in the ventricular zone *in vivo* or in neural stem cells in cultures (Geschwind et al., 2001; Kornblum and Geschwind, 2001; Terskikh et al., 2001; Luo et al., 2002). Several investigators have applied genomic ap-proaches to neural stem cell cultures. In one set of studies, Geschwind et al. (2001) used a combination of representational difference analysis (a PCR-based genetic subtraction method), with custom cDNA microarrays, and downstream in situ hybridization to identify a number of potential marker candidates. A great deal of work will be needed, however, to determine which of the many genes identified by microarrays will truly serve as specific markers and under which conditions.

One potential drawback to the use of microarrays is that they will identify mRNA species of all types, while cell surface proteins would be the most useful markers, due to their accessibility in live cell cultures for immunostaining and for isolation of cells by FACS. Additionally, just because an mRNA species is expressed does not necessarily mean that the corresponding protein will also be expressed and available for staining with antibodies. For this reason, we have instituted a proteomics approach to the identification of cell surface proteins expressed by NSCs and other progenitors. In these studies, we are comparing the proteins present in the cell membrane fractions of proliferating neurospheres to those from differentiating cells (L. Shoemaker et al., unpublished results). Proteomics generally uses separation methods, such as 2-D gel electrophoresis or liquid chromatography, followed by the use of mass spectrometry to identify intact molecular weights or tryptic digests to obtain definitive protein identification. Proteomics technology and bioinformatics is rapidly evolving and will become more common place over the next several years.

Is there a neural stem cell molecular signature?

The availability of genomic and proteomic methodologies raises the possibility that nearly all of the genes or proteins synthesized within a cell can be identified simultaneously. It may be therefore possible to think of the identification of cell types by their molecular signature–the precise combination of genes and/or proteins that a specified cell type expresses. This approach has an obvious appeal, in that it does not rely on the use of one or even a handful of markers. However, there are also clear drawbacks, in addition to the technical challenge. Cells change the amount and types of genes that they express depending on numerous factors, including their position in the cell cycle, the degree of signaling from neighboring cells, their local metabolic environment and their lineage position. Thus, any true cellular molecular signature will need to be highly specific for not only a cell "type" but also a cellular condition and stage of differentiation. It is possible, however, that cells of a certain type will express a core, smaller set of genes under a variety of conditions. This set of genes and/or their proteins, may then serve to unambiguously identify particular cellular sets.

Stemness?

While the use of genomic and proteomic methodologies are promising for the identification of cell types, it must be emphasized that the notion of cell type, itself, is one of definitions. We have already seen that the term "neural stem cell" is likely to accompany a variety of phenotypically different cells.

However, it is possible that the core features of NSCs–self-renewal and the ability to yield neurons, astrocytes and oligodendrocytes–are accompanied by the expression of a core set of genes. An examination of those genes and proteins shared by multiple neural stem cell populations will prove interesting.

The concept of shared stem cell genes can be carried even beyond the NSC. All stem cells share some similar features: the ability to self-renew, to remain "undifferentiated" and to remain multipotent. It is possible that a small group of genes serves these common stem cell functions. We have found that several genes are shared by hematopoetic stem cells and NSCs when these cDNA libraries are highly subtracted to eliminate "housekeeping" and common structural genes (Geschwind et al., 2001; Terskikh et al., 2001). This concept has been extended to include embryonic stem cells (Ivanova et al., 2002; Ramalho-Santos et al., 2002). It remains to be determined, however, whether the genes identified in these studies truly serve unique stem cell functions or are shared for other reasons.

CONCLUSION

A great deal of progress has been made in the identification and use of markers in the study of neural progenitor and stem cells. The use of individual proteins or genes as highly specific markers, however, must always be made with a great deal of caution–no one antigen or gene is always likely to be only expressed by one specific cell type. Additionally, recent studies depicting the complexity and heterogeneity of neural stem cells will call for newer and better markers as well as new ways of thinking about cell identity.

Acknowledgements

We thank R. Erickson and R. Jackson for their contributions, and members of our laboratory for helpful comment on the manuscript. The investigators were supported by NIH grant MH65756 and DoE contract DE-FC03-02ER63420 during the preparation of this manuscript.

Table 1.   Molecular markers (including antigens and transcription factors) for identifying differentiated cell types as well as neural stem/ progenitor cells.

Antigen/Transcription Factor	Description
Neuronal Markers	
Beta tubulin III (TuJ1)	immature and mature neurons
MAP2	a: later than p14, b: mature, c: immature to mature
NeuN	mature neurons
Neurofilament	L: immature neuron, M and H: postmitotic neuron
TOAD64	newly postmitotic neurons
Presenilin	postmitotic neuron and other cells
Doublecortin	migrating neuron
Neuron specific-enolase	relatively immature and postmitotic neuron
SCG10 (superior cervical ganglion)	mature neuron, related to regeneration
Hu	neuron-specific RNA binding protein
GAP43 (growth associated protein)	growth cone, related to growth and regeneration
Transcription Factors	
Neurogenin1/2	cortical neuroblast
NeuroD	pyramidal neuron
Mash1	neuroblast to Interneuron
Math2	neuroblast
Tbr1	postmitotic neurons in cerebral cortex
Hoxa2,b2	rostral hindbrain
Phox2b	central and peripheral autonomic nervous system
Nkx2.2	neuroblast to interneuron
Pax6/7	neuroblast to pyramidal neuron
FGF8	isthmus
Wnt1/Lmx1b	isthmus
Dlx5/6	GABAergic interneuron
Synapse-related	
Synaptophysin	membrane of synaptic vesicle, endocrine cell
Synaptotagmin1	presynaptic vesicle
Synapsin1	membrane of synaptic vesicle
Neurabin	synapse in postmitotic neuron
SNAP(synaptosomal associated protein) 25	plasmamembrane
Synaptobrevin	synaptic vesicle
Syntaxin	plasmamembrane

Table 1. (cont.) Molecular markers (including antigens and transcription factors) for identifying differentiated cell types as well as neural stem/ progenitor cells.

Antigen/Transcription Factor	Description
Astrocyte Markers	
GFAP	astrocyte, radial glia, stem/progenitor cell
S100b	astrocyte, ependymal cell, some neuron
Vimentin	glial progenitor, radial glia, immature astrocyte
Nestin	immature astrocyte, stem cell, muscle
Radial Glial Markers	
BLBP (brain-lipid binding protein)	radial glia, multipotent progenitor
Vimentin	glial progenitor, radial glia, immature astrocyte
GFAP	radial glia, astrocyte, stem/progenitor cell
GLAST	radial glia to postmitotic astrocyte
RC1/2	radial glia
Nestin	immature astrocyte, stem cell, muscle
REST (RE1-silencing transcription factor)	nonneuronal progenitor
Oligodendrocyte Markers	
PLP/DM20	perinuclear and processes of mature oligodendrocyte (myelin)
CNPase	membrane of CNS myelin
Myelin-associated/ oligodendrocytic basic protein (MOBP)	oligodendrocyte (myelin)
GalC (galactocerebroside)	mature oligodendrocyte, not in O2A
Myelin basic protein (MBP)	processes of mature oligodendrocyte (myelin), bone marrow, immune system
Myelin associated glycoprotein (MAG)	oligodendrocyte (myelin)
Claudin11 / OSP	immature to mature oligodendrocyte
NOGO (neurite outgrowth inhibitor)	oligodendrocytes, endoplasmic reticulum of CNS myelin, not in Schwann cell
Transcription factors	
Olig1/2	oligodendrocyte lineage, some subtypes of neuron

Table 1. (cont.) Molecular markers (including antigens and transcription factors) for identifying differentiated cell types as well as neural stem/ progenitor cells.

Antigen/Transcription Factor	Description
Other cell types	
Ependymal/choroid plexus	
TTR	epithelial cell of choroid plexus
Noggin	ependymal and early neuron
S100b	astrocyte, ependymal cell, some neurons
Smooth muscle	
SMA	
Microglia	
lectin	
CD68	
CD11c	
Multipotent Stem/Progenitor Markers	
Nestin	immature astrocyte, all dividing neural stem cell, muscle
Msi1	immature progenitor, neural stem cell, some astrocyte
GFAP	radial glia, astrocyte, stem/progenitor cell
Notch1	immature cell, SVZ in adult brain, many tissues in embryo
gp130 (granulocyte glycoprotein)	many tissues in embryo
Nucleostemin	dividing stem cell
Sox1/2	neuroepithelium, neural stem cell, less in progenitor
Telomerase activity/TERT	embryonic, hematopoietic, and neural stem cell, some in progenitor
ABCG2 (BCRP1)	hematopoietic stem cell, myogenic progenitor, nestin positive pancreastic cell
EGFR	widely in embryo
FGFR	widely in embryo
Lex1	embryonic and adult neural stem cell, some differentiated cell type
CD133/CD34	hematopoietic and neural stem cell

Table 1. (cont.) Molecular markers (including antigens and transcription factors) for identifying differentiated cell types as well as neural stem/ progenitor cells.

Antigen/Transcription Factor	Description
Commited Progenitor Markers	
A2B5	oligodendrocyte progenitor, glial progenitor, early oligodendrocyte (not mature)
O4	immature oligodendrocyte progenitor, oligodendrocyte, Schwann cell
PSA-NCAM	cortical progenitor, ependymal cell, glial restricted progenitor, neuronal progenitors
Beta tubulin III (TuJ1)	immature neuroblast, mature neuron

REFERENCES

Alvarez-Buylla A, Garcia-Verdugo JM (2002) Neurogenesis in adult subventricular zone. J Neurosci 22:629-634.

Anderson DJ (1994) Stem cells and transcription factors in the development of the mammalian neural crest. Faseb J 8:707-713.

Anderson SA, Kaznowski CE, Horn C, Rubenstein JL, McConnell SK (2002) Distinct origins of neocortical projection neurons and interneurons *in vivo*. Cereb Cortex 12:702-709.

Bain G, Kitchens D, Yao M, Huettner JE, Gottlieb DI (1995) Embryonic stem cells express neuronal properties *in vitro*. Dev Biol 168:342-357.

Baron W, de Jonge JC, de Vries H, Hoekstra D (1998) Regulation of oligodendrocyte differentiation: protein kinase C activation prevents differentiation of O2A progenitor cells toward oligodendrocytes. Glia 22:121-129.

Ben-Hur T, Rogister B, Murray K, Rougon G, Dubois-Dalcq M (1998) Growth and fate of PSA-NCAM+ precursors of the postnatal brain. J Neurosci 18:5777-5788.

Berry M, Rogers AW (1965) The migration of neuroblasts in the developing cerebral cortex. J Anat 99:691–709.

Betarbet R, Zigova T, Bakay RA, Luskin MB (1996) Migration patterns of neonatal subventricular zone progenitor cells transplanted into the neonatal striatum. Cell Transplant 5:165-178.

Bray S (2000) Specificity and promiscuity among proneural proteins. Neuron 25:1-2.

Brustle O, Spiro AC, Karram K, Choudhary K, Okabe S, McKay RD (1997) *In vitro*-generated neural precursors participate in mammalian brain development. Proc Natl Acad Sci USA 94:14809-14814.

Buescher M, Hing FS, Chia W (2002) Formation of neuroblasts in the embryonic central nervous system of Drosophila melanogaster is controlled by SoxNeuro. Development 129:4193-4203.

Busciglio J, Hartmann H, Lorenzo A, Wong C, Baumann K, Sommer B, Staufenbiel M, Yankner BA (1997) Neuronal localization of presenilin-1 and association with amyloid plaques and neurofibrillary tangles in Alzheimer's disease. J Neurosci 17:5101-5107.

Cai J, Wu Y, Mirua T, Pierce JL, Lucero MT, Albertine KH, Spangrude GJ, Rao MS (2002) Properties of a fetal multipotent neural stem cell (NEP cell). Dev Biol 251:221-240.

Cai L, Morrow EM, Cepko CL (2000) Misexpression of basic helix-loop-helix genes in the murine cerebral cortex affects cell fate choices and neuronal survival. Development 127:3021-3030.

Capela A, Temple S (2002) LeX/ssea-1 is expressed by adult mouse CNS stem cells, identifying them as nonependymal. Neuron 35:865-875.

Cassimeris L, Spittle C (2001) Regulation of microtubule-associated proteins. Int Rev Cytol 210:163-226.

Chojnacki A, Shimazaki T, Gregg C, Weinmaster G, Weiss S (2003) Glycoprotein 130 signaling regulates Notch1 expression and activation in the self-renewal of mammalian forebrain neural stem cells. J Neurosci 23:1730-1741.

Ciccolini F, Svendsen CN (1998) Fibroblast growth factor 2 (FGF-2) promotes acquisition of epidermal growth factor (EGF) responsiveness in mouse striatal precursor cells: identification of neural precursors responding to both EGF and FGF-2. J Neurosci 18:7869-7880.

Clarke DL, Johansson CB, Wilbertz J, Veress B, Nilsson E, Karlstrom H, Lendahl U, Frisen J (2000) Generalized potential of adult neural stem cells. Science 288:1660-1663.

des Portes V, Pinard JM, Billuart P, Vinet MC, Koulakoff A, Carrie A, Gelot A, Dupuis E, Motte J, Berwald-Netter Y, Catala M, Kahn A, Beldjord C, Chelly J (1998) A novel CNS gene required for neuronal migration and involved in X-linked subcortical laminar heterotopia and lissencephaly syndrome. Cell 92:51-61.

Dietrich J, Easterday MC (2002) Developing concepts in neural stem cells. Trends Neurosci 25:129-131.

Doetsch F, Garcia-Verdugo JM, Alvarez-Buylla A (1997) Cellular composition and three-dimensional organization of the subventricular germinal zone in the adult mammalian brain. J Neurosci 17:5046-5061.

Doetsch F, Caille I, Lim DA, Garcia-Verdugo JM, Alvarez-Buylla A (1999) Subventricular zone astrocytes are neural stem cells in the adult mammalian brain. Cell 97:703-716.

Doetsch F, Petreanu L, Caille I, Garcia-Verdugo JM, Alvarez-Buylla A (2002) EGF converts transit-amplifying neurogenic precursors in the adult brain into multipotent stem cells. Neuron 36:1021-1034.

Doyle KL, Khan M, Cunningham AM (2001) Expression of the intermediate filament protein nestin by sustentacular cells in mature olfactory neuroepithelium. J Comp Neurol 437:186-195.

Dyer CA, Matthieu JM (1994) Antibodies to myelin/oligodendrocyte-specific protein and myelin/oligodendrocyte glycoprotein signal distinct changes in the organization of cultured oligodendroglial membrane sheets. J Neurochem 62:777-787.

Dyer CA, Phillbotte T, Wolf MK, Billings-Gagliardi S (1997) Regulation of cytoskeleton by myelin components: studies on shiverer oligodendrocytes carrying an Mbp transgene. Dev Neurosci 19:395-409.

Eng LF, Ghirnikar RS, Lee YL (2000) Glial fibrillary acidic protein: GFAP-thirty-one years (1969-2000). Neurochem Res 25:1439-1451.

Eisenbarth GS, Walsh FS, Nirenberg M (1979) Monoclonal antibody to a plasma membrane antigen of neurons. Proc Natl Acad Sci USA 76: 4913-4917.

Feng L, Heintz N (1995) Differentiating neurons activate transcription of the brain lipid-binding protein gene in radial glia through a novel regulatory element. Development 121:1719-1730.

Ferreira A, Caceres A (1992) Expression of the class III beta-tubulin isotype in developing neurons in culture. J Neurosci Res 32:516-529.

Fredman P, Magnani JL, Nirenberg M, Ginsburg V (1984) Monoclonal antibody A2B5 reacts with many gangliosides in neuronal tissue. Arch Biochem Biophys 233: 661-666.

Gage FH (2002) Neurogenesis in the adult brain. J Neurosci 22:612-613.

Gage FH, Coates PW, Palmer TD, Kuhn HG, Fisher LJ, Suhonen JO, Peterson DA, Suhr ST, Ray J (1995) Survival and differentiation of adult neuronal progenitor cells transplanted to the adult brain. Proc Natl Acad Sci USA 92:11879-11883.

Garner CC, Tucker RP, Matus A (1988) Selective localization of messenger RNA for cytoskeletal protein MAP2 in dendrites. Nature 336:674-677.

Geschwind DH, Ou J, Easterday MC, Dougherty JD, Jackson RL, Chen Z, Antoine H, Terskikh A, Weissman IL, Nelson SF, Kornblum HI (2001) A genetic analysis of neural progenitor differentiation. Neuron 29:325-339.

Gilad GM, Gilad VH, Dahl D (1989) Expression of neurofilament immunoreactivity in developing rat cerebellum *in vitro* and *in vivo*. Neurosci Lett 96:7-12.

Goldman JE, Zerlin M, Newman S, Zhang L, Gensert J (1997) Fate determination and migration of progenitors in the postnatal mammalian CNS. Dev Neurosci 19:42-48.

Gorski JA, Talley T, Qiu M, Puelles L, Rubenstein JL, Jones KR (2002) Cortical excitatory neurons and glia, but not GABAergic neurons, are produced in the Emx1-expressing lineage. J Neurosci 22:6309-6314.

Gotz M, Hartfuss E, Malatesta P (2002) Radial glial cells as neuronal precursors: a new perspective on the correlation of morphology and lineage restriction in the developing cerebral cortex of mice. Brain Res Bull 57:777-788.

Gow A, Southwood CM, Li JS, Pariali M, Riordan GP, Brodie SE, Danias J, Bronstein JM, Kachar B, Lazzarini RA (1999) CNS myelin and sertoli cell tight junction strands are absent in Osp/claudin-11 null mice. Cell 99:649-659.

Gritti A, Cova L, Parati EA, Galli R, Vescovi AL (1995) Basic fibroblast growth factor supports the proliferation of epidermal growth factor-generated neuronal precursor cells of the adult mouse CNS. Neurosci Lett 185:151-154.

Gultekin SH, Dalmau J, Graus Y, Posner JB, Rosenblum MK (1998) Anti-Hu immunolabeling as an index of neuronal differentiation in human brain tumors: a study of 112 central neuroepithelial neoplasms. Am J Surg Pathol 22:195-200.

Hannan AJ, Henke RC, Seeto GS, Capes-Davis A, Dunn J, Jeffrey PL (1999) Expression of doublecortin correlates with neuronal migration and pattern formation in diverse regions of the developing chick brain. J Neurosci Res 55:650-657.

Hartfuss E, Galli R, Heins N, Gotz M (2001) Characterization of CNS precursor subtypes and radial glia. Dev Biol 229:15-30.

He W, Ingraham C, Rising L, Goderie S, Temple S (2001) Multipotent stem cells from the mouse basal forebrain contribute GABAergic neurons and oligodendrocytes to the cerebral cortex during embryogenesis. J Neurosci 21:8854-8862.

Heizmann CW, Fritz G, Schafer BW (2002) S100 proteins: structure, functions and pathology. Front Biosci 7:d1356-1368.

Hellani A, Ji J, Mauduit C, Deschildre C, Tabone E, Benahmed M (2000) Developmental and hormonal regulation of the expression of oligodendrocyte-specific protein/claudin 11 in mouse testis. Endocrinology 141:3012-3019.

Hitoshi S, Tropepe V, Ekker M, van der Kooy D (2002) Neural stem cell lineages are regionally specified, but not committed, within distinct compartments of the developing brain. Development 129:233-244.

Imura T, Kornblum H, Sofroniew MV (2003) The Predominant Neural Stem Cell Isolated from Postnatal and Adult, but not Early Embryonic, Forebrain Expresses GFAP. J Neurol Sci In Press.

Irvin DK, Zurcher SD, Nguyen T, Weinmaster G, Kornblum HI (2001) Expression patterns of Notch1, Notch2, and Notch3 suggest multiple functional roles for the Notch-DSL signaling system during brain development. J Comp Neurol 436:167-181.

Ivanova NB, Dimos JT, Schaniel C, Hackney JA, Moore KA, Lemischka IR (2002) A stem cell molecular signature. Science 298:601-604.

Jarriault S, Le Bail O, Hirsinger E, Pourquie O, Logeat F, Strong CF, Brou C, Seidah NG, Isra l A (1998) Delta-1 activation of notch-1 signaling results in HES-1 transactivation. Mol Cell Biol 18:7423-7431.

Johansson CB, Momma S, Clarke DL, Risling M, Lendahl U, Frisen J (1999) Identification of a neural stem cell in the adult mammalian central nervous system. Cell 96:25-34.

Kaneko Y, Sakakibara S, Imai T, Suzuki A, Nakamura Y, Sawamoto K, Ogawa Y, Toyama Y, Miyata T, Okano H (2000) Musashi1: an evolutionally conserved marker for CNS progenitor cells including neural stem cells. Dev Neurosci 22:139-153.

Katsetos CD, Herman MM, Frankfurter A, Gass P, Collins VP, Walker CC, Rosemberg S, Barnard RO, Rubinstein LJ (1989) Cerebellar desmoplastic medulloblastomas. A further immunohistochemical characterization of the reticulin-free pale islands. Arch Pathol Lab Med 113:1019-1029.

Katsetos CD, Del Valle L, Geddes JF, Assimakopoulou M, Legido A, Boyd JC, Balin B, Parikh NA, Maraziotis T, de Chadarevian JP, Varakis JN, Matsas R, Spano A, Frankfurter A, Herman MM, Khalili K (2001) Aberrant localization of the neuronal class III beta-tubulin in astrocytomas. Arch Pathol Lab Med 125:613-624.

Kawaguchi A, Miyata T, Sawamoto K, Takashita N, Murayama A, Akamatsu W, Ogawa M, Okabe M, Tano Y, Goldman SA, Okano H (2001) Nestin-EGFP transgenic mice: visualization of the self-renewal and multipotency of CNS stem cells. Mol Cell Neurosci 17:259-273.

Keyoung HM, Roy NS, Benraiss A, Louissaint A, Jr., Suzuki A, Hashimoto M, Rashbaum WK, Okano H, Goldman SA (2001) High-yield selection and extraction of two promoter-defined phenotypes of neural stem cells from the fetal human brain. Nat Biotechnol 19:843-850.

Kilpatrick TJ, Bartlett PF (1995) Cloned multipotential precursors from the mouse cerebrum require FGF-2, whereas glial restricted precursors are stimulated with either FGF-2 or EGF. J Neurosci 15:3653-3661.

Kondo T, Raff M (2000) Oligodendrocyte precursor cells reprogrammed to become multipotential CNS stem cells. Science 289:1754-1757.

Kornblum H, Geschwind D (2001) The use of representational difference analysis and cDNA microarrays in neural repair research. Restor Neurol Neurosci 18:89-94.

Kornblum HI, Raymon HK, Morrison RS, Cavanaugh KP, Bradshaw RA, Leslie FM (1990) Epidermal growth factor and basic fibroblast growth factor: effects on an overlapping population of neocortical neurons *in vitro*. Brain Res 535:255-263.

Kukekov VG, Laywell ED, Suslov O, Davies K, Scheffler B, Thomas LB, O'Brien TF, Kusakabe M, Steindler DA (1999) Multipotent stem/progenitor cells with similar properties arise from two neurogenic regions of adult human brain. Exp Neurol 156:333-344.

Laywell ED, Rakic P, Kukekov VG, Holland EC, Steindler DA (2000) Identification of a multipotent astrocytic stem cell in the immature and adult mouse brain. Proc Natl Acad Sci USA 97:13883-13888.

Lazzari M, Franceschini V (2001) Glial fibrillary acidic protein and vimentin immunoreactivity of astroglial cells in the central nervous system of adult Podarcis sicula (Squamata, Lacertidae). J Anat 198:67-75.

Lendahl U, Zimmerman LB, McKay RD (1990) CNS stem cells express a new class of intermediate filament protein. Cell 60: 585-595.

Letinic K, Zoncu R, Rakic P (2002) Origin of GABAergic neurons in the human neocortex. Nature 417:645-649.

Levison SW, Goldman JE (1997) Multipotential and lineage restricted precursors coexist in the mammalian perinatal subventricular zone. J Neurosci Res 48:83-94.

Levitt P, Cooper ML, Rakic P (1981) Coexistence of neuronal and glial precursor cells in the cerebral ventricular zone of the fetal monkey: an ultrastructural immunoperoxidase analysis. J Neurosci 1:27-39.

Lois C, Alvarez-Buylla A (1993) Proliferating subventricular zone cells in the adult mammalian forebrain can differentiate into neurons and glia. Proc Natl Acad Sci U S A 90:2074-2077.

Lu QR, Park JK, Noll E, Chan JA, Alberta J, Yuk D, Alzamora MG, Louis DN, Stiles CD, Rowitch DH, Black PM (2001) Oligodendrocyte lineage genes (OLIG) as molecular markers for human glial brain tumors. Proc Natl Acad Sci U S A 98:10851-10856.

Luskin MB, Boone MS (1994) Rate and pattern of migration of lineally-related olfactory bulb interneurons generated postnatally in the subventricular zone of the rat. Chem Senses 19:695-714.

Luskin MB, Zigova T, Soteres BJ, Stewart RR (1997) Neuronal progenitor cells derived from the anterior subventricular zone of the neonatal rat forebrain continue to proliferate *in vitro* and express a neuronal phenotype. Mol Cell Neurosci 8:351-366.

Lutolf S, Radtke F, Aguet M, Suter U, Taylor V (2002) Notch1 is required for neuronal and glial differentiation in the cerebellum. Development 129:373-385.

Ma Q, Sommer L, Cserjesi P, Anderson DJ (1997) Mash1 and neurogenin1 expression patterns define complementary domains of neuroepithelium in the developing CNS and are correlated with regions expressing notch ligands. J Neurosci 17:3644-3652.

Mabie PC, Mehler MF, Marmur R, Papavasiliou A, Song Q, Kessler JA (1997) Bone morphogenetic proteins induce astroglial differentiation of oligodendroglial-astroglial progenitor cells. J Neurosci 17:4112-4120.

Matus A (1991) Microtubule-associated proteins and neuronal morphogenesis. J Cell Sci Suppl 15:61-67.

Menezes JR, Luskin MB (1994) Expression of neuron-specific tubulin defines a novel population in the proliferative layers of the developing telencephalon. J Neurosci 14:5399-5416.

Milosevic A, Goldman JE (2002) Progenitors in the postnatal cerebellar white matter are antigenically heterogeneous. J Comp Neurol 452:192-203.

Mokry J, Nemecek S (1998) Immunohistochemical detection of intermediate filament nestin. Acta Medica (Hradec Kralove) 41:73-80.

Morshead CM, Reynolds BA, Craig CG, McBurney MW, Staines WA, Morassutti D, Weiss S, van der Kooy D (1994) Neural stem cells in the adult mammalian forebrain: a relatively quiescent subpopulation of subependymal cells. Neuron 13:1071-1082.

Mueller T, Wullimann MF (2003) Anatomy of neurogenesis in the early zebrafish brain. Brain Res Dev Brain Res 140:137-155.

Mullen RJ, Buck CR, Smith AM (1992) NeuN, a neuronal specific nuclear protein in vertebrates. Development 116:201-211.

Noctor SC, Flint AC, Weissman TA, Dammerman RS, Kriegstein AR (2001) Neurons derived from radial glial cells establish radial units in neocortex. Nature 409:714-720.

Noctor SC, Flint AC, Weissman TA, Wong WS, Clinton BK, Kriegstein AR (2002) Dividing precursor cells of the embryonic cortical ventricular zone have morphological and molecular characteristics of radial glia. J Neurosci 22:3161-3173.

Okabe S, Forsberg-Nilsson K, Spiro AC, Segal M, McKay RD (1996) Development of neuronal precursor cells and functional postmitotic neurons from embryonic stem cells *in vitro*. Mech Dev 59:89-102.

Overton PM, Meadows LA, Urban J, Russell S (2002) Evidence for differential and redundant function of the Sox genes Dichaete and SoxN during CNS development in Drosophila. Development 129:4219-4228.

Palm K, Salin-Nordstrom T, Levesque MF, Neuman T (2000) Fetal and adult human CNS stem cells have similar molecular characteristics and developmental potential. Brain Res Mol Brain Res 78:192-195.

Palmer TD, Ray J, Gage FH (1995) FGF-2-responsive neuronal progenitors reside in proliferative and quiescent regions of the adult rodent brain. Mol Cell Neurosci 6:474-486.

Parnavelas JG, Nadarajah B (2001) Radial glial cells. are they really glia? Neuron 31:881-884.

Perron M, Opdecamp K, Butler K, Harris WA, Bellefroid EJ (1999) X-ngnr-1 and Xath3 promote ectopic expression of sensory neuron markers in the neurula ectoderm and have distinct inducing properties in the retina. Proc Natl Acad Sci U S A 96:14996-15001.

Pevny LH, Sockanathan S, Placzek M, Lovell-Badge R (1998) A role for SOX1 in neural determination. Development 125:1967-1978.

Pfeiffer B, Meyermann R, Hamprecht B (1992) Immunohistochemical co-localization of glycogen phosphorylase with the astroglial markers glial fibrillary acidic protein and S-100 protein in rat brain sections. Histochemistry 97:405-412.

Price J (1987) Retroviruses and the study of cell lineage. Development 101:409-419.

Price J, Turner D, Cepko C (1987) Lineage analysis in the vertebrate nervous system by retrovirus-mediated gene transfer. Proc Natl Acad Sci U S A 84:156-160.

Przyborski SA, Cambray-Deakin MA (1995) Developmental regulation of MAP2 variants during neuronal differentiation *in vitro*. Brain Res Dev Brain Res 89:187-201.

Raff MC, Mirsky R, Fields KL, Lisak RP, Dorfman SH, Silberberg DH, Gregson NA, Leibowitz S, Kennedy MC (1978) Galactocerebroside is a specific cell-surface antigenic marker for oligodendrocytes in culture. Nature 274: 813-816.

Ramalho-Santos M, Yoon S, Matsuzaki Y, Mulligan RC, Melton DA (2002) "Stemness": transcriptional profiling of embryonic and adult stem cells. Science 298:597-600.

Reynolds BA, Weiss S (1992a) Generation of neurons and astrocytes from isolated cells of the adult mammalian central nervous system [see comments]. Science 255:1707-1710.

Reynolds BA, Weiss S (1992b) Generation of neurons and astrocytes from isolated cells of the adult mammalian central nervous system. Science 255:1707-1710.

Reynolds BA, Weiss S (1993) Central nervous system growth and differentiation factors: clinical horizons—truth or dare? Curr Opin Biotechnol 4:734-738.

Reynolds BA, Weiss S (1996) Clonal and population analyses demonstrate that an EGF-responsive mammalian embryonic CNS precursor is a stem cell. Dev Biol 175:1-13.

Rickmann M, Wolff JR (1995) S100 protein expression in subpopulations of neurons of rat brain. Neuroscience 67:977-991.

Rietze RL, Valcanis H, Brooker GF, Thomas T, Voss AK, Bartlett PF (2001) Purification of a pluripotent neural stem cell from the adult mouse brain. Nature 412:736-739.

Roy NS, Wang S, Harrison-Restelli C, Benraiss A, Fraser RA, Gravel M, Braun PE, Goldman SA (1999) Identification, isolation, and promoter-defined separation of mitotic oligodendrocyte progenitor cells from the adult human subcortical white matter. J Neurosci 19:9986-9995.

Sakakibara S, Imai T, Hamaguchi K, Okabe M, Aruga J, Nakajima K, Yasutomi D, Nagata T, Kurihara Y, Uesugi S, Miyata T, Ogawa M, Mikoshiba K, Okano H (1996) Mouse-Musashi-1, a neural RNA-binding protein highly enriched in the mammalian CNS stem cell. Dev Biol 176:230-242.

Schliess F, Stoffel W (1991) Evolution of the myelin integral membrane proteins of the central nervous system. Biol Chem Hoppe Seyler 372:865-874.

Seidl R, Cairns N, Lubec G (2001) The brain in Down syndrome. J Neural Transm Suppl:247-261.

Seri B, Garcia-Verdugo JM, McEwen BS, Alvarez-Buylla A (2001) Astrocytes give rise to new neurons in the adult mammalian hippocampus. J Neurosci 21:7153-7160.

Shibata T, Yamada K, Watanabe M, Ikenaka K, Wada K, Tanaka K, Inoue Y (1997) Glutamate transporter GLAST is expressed in the radial glia-astrocyte lineage of developing mouse spinal cord. J Neurosci 17:9212-9219.

Shimazaki T, Shingo T, Weiss S (2001) The ciliary neurotrophic factor/leukemia inhibitory factor/gp130 receptor complex operates in the maintenance of mammalian forebrain neural stem cells. J Neurosci 21:7642-7653.

Smith CM, Luskin MB (1998) Cell cycle length of olfactory bulb neuronal progenitors in the rostral migratory stream. Dev Dyn 213:220-227.

Steindler DA, Pincus DW (2002) Stem cells and neuropoiesis in the adult human brain. Lancet 359:1047-1054.

Stemple DL, Anderson DJ (1992) Isolation of a stem cell for neurons and glia from the mammalian neural crest. Cell 71:973-985.

Sternberger NH, Itoyama Y, Kies MW, Webster HD (1978) Myelin basic protein demonstrated immunocytochemically in oligodendroglia prior to myelin sheath formation. Proc Natl Acad Sci U S A 75: 2521-2524.

Stoffel W, Hillen H, Giersiefen H (1984) Structure and molecular arrangement of proteolipid protein of central nervous system myelin. Proc Natl Acad Sci U S A 81:5012-5016.

Stump G, Durrer A, Klein AL, Lutolf S, Suter U, Taylor V (2002) Notch1 and its ligands Delta-like and Jagged are expressed and active in distinct cell populations in the postnatal mouse brain. Mech Dev 114:153-159.

Sultana S, Sernett SW, Bellin RM, Robson RM, Skalli O (2000) Intermediate filament protein synemin is transiently expressed in a subset of astrocytes during development. Glia 30:143-153.

Sun XZ, Takahashi S, Cui C, Zhang R, Sakata-Haga H, Sawada K, Fukui Y (2002) Normal and abnormal neuronal migration in the developing cerebral cortex. J Med Invest 49:97-110.

Takahashi M, Osumi N (2002) Pax6 regulates specification of ventral neurone subtypes in the hindbrain by establishing progenitor domains. Development 129:1327-1338.

Tanigaki K, Nogaki F, Takahashi J, Tashiro K, Kurooka H, Honjo T (2001) Notch1 and Notch3 instructively restrict bFGF-responsive multipotent neural progenitor cells to an astroglial fate. Neuron 29:45-55.

Temple S, Raff MC (1985) Differentiation of a bipotential glial progenitor cell in a single cell microculture. Nature 313:223-225.

Terskikh AV, Easterday MC, Li L, Hood L, Kornblum HI, Geschwind DH, Weissman IL (2001) From hematopoiesis to neuropoiesis: evidence of overlapping genetic programs. Proc Natl Acad Sci U S A 98:7934-7939.

Tropepe V, Sibilia M, Ciruna BG, Rossant J, Wagner EF, van der Kooy D (1999) Distinct neural stem cells proliferate in response to EGF and FGF in the developing mouse telencephalon. Dev Biol 208:166-188.

Tropepe V, Hitoshi S, Sirard C, Mak TW, Rossant J, van der Kooy D (2001) Direct neural fate specification from embryonic stem cells: a primitive mammalian neural stem cell stage acquired through a default mechanism. Neuron 30:65-78.

Trotter J, Schachner M (1989) Cells positive for the O4 surface antigen isolated by cell sorting are able to differentiate into astrocytes or oligodendrocytes. Brain Res Dev Brain Res 46:115-122.

Tsai RY, McKay RD (2002) A nucleolar mechanism controlling cell proliferation in stem cells and cancer cells. Genes Dev 16:2991-3003.

Uchida N, Buck DW, He D, Reitsma MJ, Masek M, Phan TV, Tsukamoto AS, Gage FH, Weissman IL (2000) Direct isolation of human central nervous system stem cells. Proc Natl Acad Sci U S A 97:14720-14725.

Ueda S, Aikawa M, Kawata M, Naruse I, Whitaker-Azmitia PM, Azmitia EC (1996) Neuro-glial neurotrophic interaction in the S-100 beta retarded mutant mouse (Polydactyly Nagoya). III. Transplantation study. Brain Res 738:15-23.

van Praag H, Schinder AF, Christie BR, Toni N, Palmer TD, Gage FH (2002) Functional neurogenesis in the adult hippocampus. Nature 415:1030-1034.

Vescovi AL, Reynolds BA, Fraser DD, Weiss S (1993) bFGF regulates the proliferative fate of unipotent (neuronal) and bipotent (neuronal/astroglial) EGF-generated CNS progenitor cells. Neuron 11:951-966.

Weiss S, Dunne C, Hewson J, Wohl C, Wheatley M, Peterson AC, Reynolds BA (1996) Multipotent CNS stem cells are present in the adult mammalian spinal cord and ventricular neuroaxis. J Neurosci 16:7599-7609.

Wen PH, Friedrich VL, Jr., Shioi J, Robakis NK, Elder GA (2002) Presenilin-1 is expressed in neural progenitor cells in the hippocampus of adult mice. Neurosci Lett 318:53-56.

Yang HY, Lieska N, Shao D, Kriho V, Pappas GD (1994) Proteins of the intermediate filament cytoskeleton as markers for astrocytes and human astrocytomas. Mol Chem Neuropathol 21:155-176.

Yang Q, Hamberger A, Wang S, Haglid KG (1996) Appearance of neuronal S-100 beta during development of the rat brain. Brain Res Dev Brain Res 91:181-189.

Ying QL, Stavridis M, Griffiths D, Li M, Smith A (2003) Conversion of embryonic stem cells into neuroectodermal precursors in adherent monoculture. Nat Biotechnol 21:183-186.

Zappone MV, Galli R, Catena R, Meani N, De Biasi S, Mattei E, Tiveron C, Vescovi AL, Lovell-Badge R, Ottolenghi S, Nicolis SK (2000) Sox2 regulatory sequences direct expression of a (beta)-geo transgene to telencephalic neural stem cells and precursors of the mouse embryo, revealing regionalization of gene expression in CNS stem cells. Development 127:2367-2382.

Zheng T, Steindler DA, Laywell ED (2002) Transplantation of an indigenous neural stem cell population leading to hyperplasia and atypical integration. Cloning Stem Cells 4:3-8.

Zhou Q, Anderson DJ (2002) The bHLH transcription factors OLIG2 and OLIG1 couple neuronal and glial subtype specification. Cell 109:61-73.

Zhu G, Mehler MF, Zhao J, Yu Yung S, Kessler JA (1999) Sonic hedgehog and BMP2 exert opposing actions on proliferation and differentiation of embryonic neural progenitor cells. Dev Biol 215:118-129.

Zimmerman L, Parr B, Lendahl U, Cunningham M, McKay R, Gavin B, Mann J, Vassileva G, McMahon A (1994) Independent regulatory elements in the nestin gene direct transgene expression to neural stem cells or muscle precursors. Neuron 12:11-24.

Chapter 4

Isolation of Stem Cells from Multiple Sites in the CNS

Mahendra S. Rao and Larysa Pevny

INTRODUCTION

The adult nervous system is composed of a large diversity of cell types that arise from a sheet of morphologically indistinguishable epithelial cells termed the 'neural plate'. Once induced, the stem cells of the neural plate undergo rapid expansion and a combination of epigenetic and genetic mechanisms act to specify regional fate along the prospective anteroposterior and mediolateral axes of the plate. Coupled with these patterning mechanisms is the differentiation of neuroepithelial progenitors into the three major postmitotic cell types that constitute the mature CNS: neurons, oligodendrocytes and astrocytes. It is now becoming apparent that multipotent neural stem cells (cells defined by the ability to self-renew and differentiate into neurons, astrocytes and oligodendrocytes *in vitro*) are present throughout the development of the nervous system, initially in the cells of the neural plate and then in the ventricular zone (VZ) of the neural tube, and persist into adulthood in certain locations (reviewed in Barres, 1999; Momma et al., 2000; Temple and Alvarez-Buylla, 1999). The relationship between "stem cell" populations at different stages of ontogeny and different rostrocaudal and dorsoventral locations remains unclear. Stem cells isolated at different developmental stages do share expression of some universal molecular characteristics, maintain the ability to self-renew and to give rise to neurons and glia *in vitro*. However, these stem cells are also regionally specified, express unique molecular markers and respond differently to growth factors.

In this chapter we summarize the data on types of stem cells present at different stages and regions of the developing and adult CNS. Specifically, we consider the cellular and molecular characteristics that are shared and/or unique amongst neural stem cells at defined stages of ontogeny. Considering this evidence, we propose that it may be possible to distinguish between neural stem cell populations based on particular combination of markers to prospectively identify stem cells *in vivo*.

From: *Neural Stem Cells: Development and Transplantation*
Editor: Jane E. Bottenstein © 2003 Kluwer Academic Publishers, Norwell, MA

Neural stem cells of the embryonic central nervous system

Vertebrate neural development begins with the allocation of a group of ectodermal cells toward a neural fate; these cells are termed neuroepithelial precursor (NEP) stem cells and comprise the early neural plate (Kalyani et al., 1997). NEP cells arise through a series of inductive interactions, first described by Spemann and Mangold in 1924, and ultimately give rise to the entire nervous system. Studies in amphibian embryos have proposed that early gastrula ectoderm differentiates into neural ectoderm by a "default" molecular mechanism. Such that, neural inducing signals, including Noggin, Chordin and Follistatin, induce neural fate in early ectodermal cells by antagonizing the epidermalizing activity of Bone Morphogenetic Protein (BMP) (reviewed in Harland, 2000). However, the conservation of attenuation of BMP signals as the mechanism to initiate neurogenesis in amniotes has yet to be clearly demonstrated. For example, experiments from chick suggest that Chordin is not sufficient to elicit neural induction nor do BMP4 and BMP7 inhibit neural induction by the organizer (Streit et al., 1998). Similar observations have been made in mouse, where mutations in genes encoding candidate neural inducers within the organizer region (the mouse node) still exhibit neural differentiation (Bachiller et al., 2000; McMahon et al., 1998; Tam and Behringer, 1997). Moreover, elimination of the entire node through genetic mutations fails to block neural differentiation (Ang and Rossant, 1994; Klingensmith et al., 1999; Weinstein et al., 1994). Thus, neural tissue may be initiated by signals derived from other cell types. Members of other families of signaling molecules, most notably members of the fibroblast growth factors (FGFs) recently have been proposed as early-acting factors that imitate neural induction (Streit et al., 2000; Wilson et al., 2000; Wilson and Rubenstein, 2000). Furthermore, mouse embryonic stem cells have been shown to default to a neural fate (Tropepe et al., 2001) however; this appears to require autocrine fibroblast growth factor signaling ((Ying et al., 2003) reviewed in (Stavridis and Smith, 2003)).

The newly induced cells of the early neural plate (NEP cells) are largely morphogenetically, cellularly and molecularly homogenous (See Figure 1). *In vitro*, dissociated NEP cells undergo self-renewal, and single NEP cells can differentiate into neurons, astrocytes and oligodendrocytes (Kalyani et al., 1997; Mujtaba et al., 1999). They require basic fibroblast growth factor (bFGF) for their proliferation both *in vivo* and *in vitro*. Studies in mouse and rat have demonstrated that during the period of neural induction about 90% of the cells that comprise the neural plate have stem cell properties (Cai et al., 2002). Thus, NEP cells represent one of the earliest identifiable neural stem cell populations *in vivo*. These morphogenetic and cellular properties also correlate with the expression of general molecular markers. Neural induction results in the activa-

tion of regulatory genes that are characterized by early and broad expression
domains in the neural plate and serve
to define NEP cells. These include,
among others (Table 1), proteins such
as nestin, E-NCAM, Musashi, Notch1,
HES1 and SOXB1 factors (Collignon
et al., 1996; Frederiksen and McKay,
1998; Hockfield and McKay, 1985;
Lendahl et al., 1990; Pevny et al., 1998;
Sakakibara et al., 1996; Sakakibara and
Okano, 1997; Sasai, 2001; Weinstein
et al., 1994; Wood and Episkopou,
1999). As described below the expression of a number of these markers is
maintained in proliferating neural progenitors throughout ontogeny and thus
may serve to universally identify neural stem cells.

After neural induction the neural
plate undergoes a series of morpho-
genic movements to form a tube consisting of prominent vesicles anteriorly,
which represent the anlage of the forebrain, midbrain and hindbrain, and a
thin portion posteriorly, which develops into the spinal cord. The initially
homogenous population of dividing
cells in the neural tube is patterned over
several days to generate neurons, oligodendrocytes and astrocytes in a characteristic spatial and temporal profile
with proliferating neural cells restricted
to the inner ventricular zone (Altman

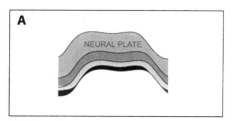

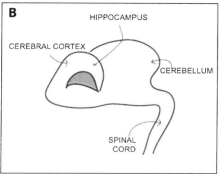

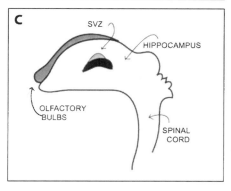

**Figure 1: Potential neural stem cells during
CNS ontogeny**. Areas of the embryonic and
adult CNS from which neural stem cells have
been isolated. **A.** Neural plate **B.** Embryonic
neural tube **C.** Adult neuroaxis.

and Bayer, 1984; Bayer and Altman, 1991; Lillien, 1998; McConnell, 1995;
Rakic, 1988; Wentworth, 1984). Lineage tracing studies *in vivo* and clonal culture experiments *in vitro* have demonstrated that stem cells at this stage of development are located within the proliferative ventricular zone. Similar to the
cells of the neural plate, stem cells found in the ventricular zone of the early
neural tube require bFGF signaling for their survival and proliferation *in vitro*.
Consistent with this requirement *in vitro*, FGFR and FGF null mice in addition

Table 1. Potential markers for neural stem populations. Markers that are known to be relatively specific for stem cell populations are listed. Current data does not allow us to determine if any of these markers have the requisite specificity and sensitivity to be used as a single marker for all potential stem cell populations in the adult brain. Some combination of markers however may uniquely specify all neural stem cell populations.

Potential universal stem cell markers	Comments
Nestin	Probably expressed in all dividing neural populations
Musashi	Is also expressed by progenitor populations
Hu	Is also expressed by progenitor populations
Neuralstemnin	Recently described factor that may be relatively specific to dividing populations
Sox1	Appears relatively specific to neural stem cell populations though persists (transiently) in some progenitor cell populations
Sox2	All neurosphere forming cells appear to express Sox2 and labeling appears localized to regions enriched in stem cells
Response to acetylcholine	Has been used to identify stem cell populations
Telomerase activity/TERT expression	Present in all stem cell populations tested but also shown to be present in nonstem cell populations
Low Hoechst/Rhodamine staining	Appears to be specific for quiescent stem cell populations but does not identify rapidly dividing stem cell populations
ABCG2 expression	Appears relatively specific *in vivo* and *in vitro*
Aldefluor labeling	May be a nonspecific method of identifying stem cell populations
Absence of differentiation markers	Has been used successfully to enrich for stem cell populations at multiple states of development

to the effect on neuronal and glial populations, show a diminished ventricular domain (Ortega et al., 1998; Raballo et al., 2000; Vaccarino et al., 1999).

Around mid-embryogenesis the ventricular zone is much reduced in size and additional zones of mitotically active progenitors can be identified (Figure 1B). Mitotically active cells derived from the ventricular zone that accumulate adjacent to the VZ have been termed subventricular (SVZ) cells. This SVZ is later called the subependymal zone as the ventricular zone diminishes in size to a single layer of ependymal cells. The SVZ is prominent in the forebrain and can be identified as far back as the fourth ventricle. No SVZ can be detected in more caudal regions of the brain and if it exists it is likely a very small population of cells. An additional germinal matrix that is derived from the rhombic lip of the fourth ventricle, called the external granule layer generates the granule cells of the cerebellum.

Around the time the SVZ can be clearly demarcated an additional stem cell population can be isolated and propagated in culture. This second stem cell population has been termed the epidermal growth factor (EGF) dependent stem cells (Reynolds et al., 1992; Reynolds and Weiss, 1992; Reynolds and Weiss, 1996). In the presence of EGF, dissociated neural cells proliferate and form floating multicellular structures called neurospheres (Nakamura et al., 2000; Reynolds and Weiss, 1996). Most of the cells in the neurosphere are clonally derived from a single CNS stem cell/progenitor and are thought to possess the characteristics of CNS stem cells i.e., they have self-renewing activity and are multipotent, able to differentiate into either neurons or glia (Nakamura et al., 2000; Reynolds and Weiss, 1996). Sequential clonal analysis suggests that the stem cell population constitutes a fraction of the cells present in any sphere which undergo self renewal and differentiate into neurons, astrocytes and oligodendrocytes *in vitro*. EGF-dependent stem cells can be isolated from the entire rostrocaudal axis from E14 onwards and, as described below, EGF-dependent stem cells have been isolated from adult tissue (Ciccolini, 2001; Laywell et al., 2000; Reynolds and Weiss, 1996).

Several lines of evidence suggest that in the developing embryo the neurosphere forming stem cell population likely resides in the subventricular zone in regions where a defined SVZ exists, and it likely resides outside the ventricular zone in more caudal brain regions including the spinal cord. Retroviral labeling experiments indicate that subpopulations of cells within the SVZ are multipotent. High EGFR and EGF expression are seen in the SVZ but not in the early VZ (Burrows et al., 2000; Kalyani et al., 1997). EGF knockouts do not alter the size or prominence of the ventricular zone that contains FGF-dependent neuroepithelial stem cells but affects later neuronal and glial survival (Kornblum et al., 1998; Thomson et al., 1998). In addition, it has been difficult to isolate neurospheres with EGF alone from regions of the brain where an SVZ cannot be morphologically identified, and EGF-dependent neurosphere forming stem cells cannot be isolated at stages prior to the formation of the SVZ. Further, microdissection experiments have shown that the region that contains the largest neurosphere forming ability includes the SVZ and its immediate environs (reviewed in Morshead et al., 1998). Thus a neurosphere forming stem cell population that is dependent on EGF is present during late embryonic development, and it is likely to be localized to the subventricular zone/region throughout the rostralcaudal axis. Large numbers of EGF-dependent neurosphere forming stem cells are present in cranial regions where the size of the SVZ is large and smaller numbers are present in more caudal regions.

It has recently been proposed that radial glial cells in the embryonic nervous system have stem cell characteristics. Radial glia have their cell bodies in the VZ and extend a long radial process to the pial surface (Schmechel and Rakic, 1979). These cells have traditionally been thought to provide a migratory scaffold along which newly generated neurons migrate from the VZ to postmitotic areas (Rakic, 1988). Radial glia persist till late perinatal ages and transform into astrocytes as a normal process of development (Levitt et al., 1981). The results of recent *in vivo* studies suggest that radial glial cells comprise the majority of progenitor cells that give rise to neurons of the cortical germinal zone and may also function as a self renewing multipotent population (Gray and Sanes, 1992; Hartfuss et al., 2001; Malatesta et al., 2000; Noctor et al., 2001; Noctor et al., 2002). It is important to note however that markers characteristic of radial glial cells such as RC1 and RC2 and the expression of GFAP, vimentin, GLAST, and other markers are not seen immediately, after neural tube closure when dividing stem cells can be readily identified and expression of radial glial markers when present is always in a subset of the proliferating cells. In addition, while radial glial cells divide, the rate of cell division is not consistent with their being the predominant dividing stem cell population *in vivo*, at least at early stages in development. Consistent with these variations, an elegant fate mapping study of radial glial cells has recently demonstrated that the neurogenic potential of radial glia is region specific, such that radial glia generate the majority of cortical projection neurons but not interneurons originating in the ventral telencephalon (Malatesta et al., 2003). Nevertheless the ability of radial glia to generate neurons, and astrocytes in culture and *in vivo* suggest that they may represent one type of multipotent cell present in early development. Furthermore, as discussed below, it has recently been hypothesized that adult SVZ cells might be derived from embryonic radial glial cells that retain neuroepithelial stem cell characteristics into adulthood (Alvarez-Buylla et al., 2001).

Thus, during early neurogenesis a single population of stem cells is present which is localized to the ventricular zone. At somewhat later developmental stages at least two additional populations of stem cells can be isolated–the predominant proliferating populations becoming localized to the subventricular zone, and a smaller population of cells (radial glia and neuropithelial cells) being localized to the diminishing ventricular zone. Cells in the SVZ are morphologically distinct and can be distinguished from VZ cells based on factor dependence and the cell types that they generate *in vivo*. Differences between VZ derived NEP cells and SVZ derived neurosphere forming stem cells are summarized in Table 2. The predominant difference between these populations appears to be their growth factor dependence, positional marker expression, and subtypes of neurons that are generated.

Table 2. Differences between VZ and SVZ derived stem cells. The differences between the two major populations of neural stem cells identified in the developing brain are listed. Note that both populations of cells share several similarities including expression of nestin, Sox1 and Sox2, and absence of most lineage-specific markers.

Stem Cell Populations	VZ/NSC Cells	SVZ/NSC Cells
EGFR expression	Low or undetectable	High and essential for isolation
Cytokine dependency	FGF required for proliferation	EGF is necessary and sufficient
Presence of cilia	Present in vivo and in vitro	Absent
Interkinetic nuclear movement	Feature of normal development	Not observed in vivo
Neurotransmitter response	Glutamate causes proliferation	Glutamate inhibits proliferation
GFAP expression	Not observed	May define the stem cell population in vivo
Types of neurons generated	Projection neurons	Primarily GABAergic interneurons
Radial glial differentiation	Normal aspect of development	Ability to generate radial glia unknown
Neural Crest differentiation	Demonstrated in vitro and in vivo	Unknown
Positional markers	Differ between VZ and SVZ	Differ between VZ and SVZ

Developmental regionalization, molecular and cellular heterogeneity of embryonic CNS stem cells.

In addition to changes in proliferation rate, growth factor dependence, and differentiation ability, stem/progenitor cells in the CNS are regionalized by patterning molecules. Patterning in the proliferating neuroepithelium is initiated at the time of neural induction and occurs along the rostrocaudal and dorsoventral axis. For example, in the developing spinal cord, inductive signals emanating from the underlying axial mesoderm and overlying ectoderm act to regionalize progenitors along the dorsoventral (DV) axis of the ventricular zone (Jessell, 2000). Differential expression of several transcription factors of the Pax and homeodomain families define distinct progenitor domains along the DV axis of the neural tube that subsequently correspond to specific neuronal fates (Figure 2). Moreover, the fate of progenitors in one domain, whether they differentiate into motor neurons or interneurons, can be altered by misexepressing in it homeodomain factors characteristic of another subdomain (Briscoe et al., 2000).

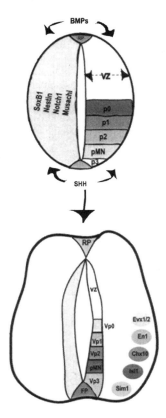

Figure 2: Regionalization of the vertebrate spinal cord. A. Schematic illustrating the patterning mechanisms that generate different neuronal subtypes along the dorsoventral axis of the spinal cord (see also Anderson et al., 2001; Briscoe et al., 2000). The proliferating cells of the early neural tube uniformly express a set of pan-neural markers, including SoxB1, Nestin, Notch and Musashi (left-yellow). Coincident with the expression of these universal progenitor markers and in response to opposing diffusion gradients of Sonic Hedgehog (SHH) and BMPs the ventricular zone cells are subdivided into discrete zone of homeodomain transcription factor gene expression (right vP3-vP1), **B.** Neurons (circles) in the developing spinal cord can be uniquely identified by their expression of a characteristic set of transcription factors, examples of which are given in the circles. The different types of interneuron and motor neurons derive from corresponding dorsoventral progenitor domains in the ventricular layer. FP, floor plate; RP, roof plate

Similar domains of transcription factor expression have been shown to regionalize the ventricular zone along the rostrocaudal axis (reviewed in Lumsden and Krumlauf, 1996; Wilson and Rubenstein, 2000).

Consistent with this regional restriction of cell fate *in vivo*, a number of studies have demonstrated the importance of cell autonomous mechanisms in maintaining identity of neuroepithelial cells *in vitro*. It has now been clearly shown that positional markers that define the rostrocaudal and dorsoventral identity of stem cells persist over multiple generations *in vitro*. (Hitoshi et al., 2002; Nakagawa et al., 1996; Zappone et al., 2000). For example, neural stem cell colonies derived from cortex and spinal cord of embryonic day (E14.5) differentially express regional marker genes along the anteroposterior axis (Zappone et al., 2000) and this expression persists for at least forty generations.

In addition, direct transplantation of uncultured neural progenitors has suggested that they are more restricted in the subtypes of cells they generate. For example, progenitor cells from the cortical ventricular zone of middle-staged ferret embryos can generate neurons in stage-appropriate layers when transplanted to older but not younger hosts (Desai and McConnell, 2000). Also, when

SVZ progenitors that normally generate only interneurons of the olfactory bulb, are transplanted to the embryonic nervous system, they do not give rise to long projection neurons, which are normally generated from endogenous progenitor cells (Lim et al., 1997).

What remains unclear is whether progenitor cells expressing particular regional transcription factors are committed in their fate (see below). For example, such cells may remain plastic until they have withdrawn from the cell cycle and left the ventricular zone (McConnell and Kaznowski, 1991; reviewed by Anderson, 2001; Edlund and Jessell, 1999). Moreover, transplantation of pieces of ventricular zone tissue to different locations along the rostrocaudal axis has resulted in respecification of transcription factor gene expression (Guthrie, 1996). Similarly, it has recently been demonstrated that embryonic progenitors maintain expression of markers of regional identity *in vitro* but can be respecified when grafted to heterologous sites *in vivo* (Hitoshi et al., 2002).

While patterning influences may be overcome in some conditions it is important to note that overcoming such inherent influences is difficult and generally never complete. It therefore raises the possibility that effective transplant strategies may require selection not only of stem cells but also stem cells appropriate for that particular region of the CNS. Thus, the importance of patterning clearly is more significant in the nervous system than in the hematopoietic system where mobilized stem cells from any region of the rostrocaudal axis appear effective at repopulating the bone marrow.

Stem cells in the adult CNS

Although the vast majority of cells in the mammalian nervous system are born during the embryonic and early postnatal period, new neurons are continuously added in certain regions of the adult brain (Altman and Das, 1965). Early ^{3}H-thymidine labeling studies (Altman, 1962) and recent bromdeoxyuridine (BrdU) and retroviral labeling experiments have demonstrated two major sites of ongoing neurogenesis in the adult brain: the SVZ of the lateral ventricle and the hippocampus. Proliferating progenitors residing in the SVZ migrate to the olfactory bulb and differentiate into local interneurons (Lois and Alvarez-Buylla, 1993; Luskin et al., 1993), whereas in the hippocampus, progenitors in the subgranular zone differentiate to granule cells in the dentate gyrus (Kuhn et al., 1996). These new neurons are thought to be derived from a population of neural stem cells. Consistent with this, it has been shown that dividing SVZ-derived neuroblasts ablated with antimitotic agents *in vivo* can be regenerated from a pool of slowly dividing stem cells (Doetsch et al., 1999). Moreover, mutations in Querkopf generate greatly reduced numbers of olfactory neurons in the adult mouse due to a loss of the SVZ stem cells (Rietze et al., 2001). Taking cells from the adult brain and propagating them *in vitro* has demonstrated the presence of adult neural stem cells, that have the capacity for self-renewal and are able to generate the three major CNS cells types: neurons, astrocytes and oligo-dendrocytes (Gage et al., 1995; Gritti et al., 1996; Lois and Alvarez-Buylla, 1993; Morshead et al., 1994; Palmer et al., 1997; Reynolds and Weiss, 1992).

Recent studies demonstrate there may be two potential sources of adult neural stem cells the ependymal layer and the subventricular zone of the lateral ventricle of the brain (Chiasson et al., 1999; Doetsch et al., 1999; Johansson et al., 1999a; Johansson et al., 1999b; reviewed by Barres, 1999). Ependymal cells are the remnants of the proliferating ventricular zone and are therefore a logical candidate for adult multipotent stem cells. Ependymal cells are relatively quiescent *in vivo* but retain the ability enter the cell cycle, and may respond to injury by proliferation. Infusion of FGF and EGF can cause a proliferation of ependymal cells and retroviral lineage analysis has suggested that individual cells can generate astrocytes and neurons in at least some regions of the brain (Johansson et al., 1999b; Josephson et al., 1998). Ependymal tumors express both neuronal and glial markers, thus it is reasonable to assume that ependymal cells are multipotent. Whether they possess sufficient self-renewal ability however has been questioned. Van der Kooy and colleagues have pointed out that most neurospheres do not consist of ciliated cells (Chiasson et al., 1999). Neurospheres derived from ciliated ependymal cells do not undergo significant self-renewal, and neurospheres that undergo self-renewal can be isolated from other regions of the brain. Thus, if ependymal cells represent an adult stem cell population they can at best represent a minority population of the neurosphere generating cells.

The SVZ is the location of a second population of stem cells present in early development (see above), and undifferentiated cells can be identified in the SVZ even at late adult stages. Multipotent cells with the ability to self-renew and form neurospheres in culture can be isolated from the cortical SVZ. Retroviral labeling of SVZ cells has suggested regional heterogeneity and has shown that the SVZ consists of a mixture of stem and progenitor cells (Levison and Goldman, 1997; Luskin et al., 1993; Price et al., 1987; Williams, 1995; reviewed in Alvarez-Buylla and GarciaVerdugo, 2002). Tritiated thymidine injections, to kill actively dividing cells at later stages of subventricular zone development, have suggested that approximately 1% of the cells are slowly dividing stem cells (Morshead et al., 1994) that can regenerate the remaining cells in the SVZ.

It is not clear however which cell in this heterogeneous population equates to the neurosphere forming stem cell. Different groups have suggested different locations and different properties. In the adult, neurosphere forming stem cells may be localized to the type B GFAP+ astrocytic cells in the SVZ or the type C

GFAP⁻ cells in the SVZ (reviewed in Alvarez-Buylla et al., 2000; Garcia-Verdugo et al., 1998) or may represent a distinct population that has not been clearly defined as yet (discussed in Capela and Temple, 2002). Recent compelling experiments demonstrate that the majority of EGF-responsive cells in the adult SVZ that generate neurospheres are actually derived from the rapidly dividing transit-amplifying C cells (Doetsch et al., 2002). Thus, at least one population of multipotent stem cells exists in the adult SVZ; this population is relatively quiescent but can enter the cell cycle and participate in repair and regeneration in the olfactory bulb and the hippocampus.

Multipotential cells can also be isolated from non-neurogenic regions of the adult mammalian CNS such as the spinal cord. Recent studies have suggested that the ependymal cells lining the central canal of the postnatal spinal cord possess properties of neural stem cells (Palmer et al., 1995; Shihabuddin et al., 2000; Weiss et al., 1996). These cells, like the cells of the SVZ and hippocampus, undergo self-renewal and mutipotent differentiation *in vitro* and may represent SVZ-like cells.

Others have argued that stem cell populations exist in the parenchyma of the adult brain throughout the rostrocaudal axis. These cortical stem cells appear to be characterized by PSA-NCAM immunoreactivity, grow as neurospheres and are multipotent as assessed by differentiation into neurons, astrocytes and oligodendrocytes. Indeed their differentiation potential may be wider than that of most neural stem cells: they may also be capable of differentiating into peripheral nervous system derivatives such as Schwann cells (Marmur et al., 1998). It is unclear however, how many such cells are present *in vivo* and whether these cells participate in repair and regeneration. Retroviral lineage analysis does not show the presence of multipotent stem cells in the parenchyma and there is little evidence of neurogenesis in response to parenchymal damage to cortical precursors. Given the clear demonstration of their multipotentiality in culture and the evidence that these neurosphere forming cells arise from the parenchyma the operating assumption is that these PSA-NCAM cells represent either a quiescent population of stem cells that fail to respond to environmental signals, due to the presence of inhibitory signals or represent dedifferentiated cells that have reentered the cell cycle. Both possibilities are consistent with the available data and indirect evidence for the ability of postmitotic cells to reenter the cell cycle.

In an intriguing set of experiments Macklis and colleagues (see Chapter 12 in this volumne) have shown that when subsets of neurons are specifically ablated without large scale localized tissue damage, then these neurons are replaced by adjacent cells which appear to differentiate and integrate into appropriate neuronal circuits. These neurons even send axons across the corpus collosum or project to the thalamus. Their data suggests the presence of a cortical progenitor or possibly a stem cell population that does not participate in repair, as its response is inhibited in most conditions (Magavi et al., 2000). Brewer (1999) has suggested that postmitotic cells, specifically neurons, can reenter the cell cycle and can differentiate into neurons and astrocytes. Specifically, he retrogradely labeled projection neurons and showed that in dissociated culture these cells initiate cell division and single cells differentiate into neurons and astrocytes. While neither data demonstrates that the new neurogenesis that was observed was of biological significance or that the cells that proliferated were truly multipotent, it nevertheless provides importance evidence that quiescent populations exist that can be induced to reenter the cell cycle and possibly contribute to neurosphere formation *in vitro*.

Experiments described by Kondo and Raff (2000) showed that glial progenitor cells could be dedifferentiated and then induced to differentiate into neurons. Along the same line, Doestch et al. (2000), provide compelling evidence that progenitor cells in the adult brain retain stem cell properties. Specifically, that after exposure to high concentrations of EGF, type C neuroblasts of the SVZ function as stem cells *in vitro*. Further, it has been have suggested that astrocytes at some stage in development are multipotent and competent to differentiate into neurons, astrocytes and oligodendrocytes (Laywell et al., 2000; Laywell and Steindler, 2002).

It is critical to determine whether one is working with endogenous stem cells that have been maintained in a similar stem cell state in culture or whether one is working with a non stem cell population that has acquired stem cell characteristics after being maintained in culture. This is an important and perplexing problem that needs to be resolved and is perhaps the source of much of the controversy in the field in so far as markers characteristic of stem cell populations is concerned. If one uses the criteria of forming neurospheres as evidence for the presence of stem cells, then one cannot distinguish between endogenous, i.e., true stem cells that participate in normal development, quiescent stem populations that can be reactivated in culture; or other cells that have somehow acquired stem-like characteristics (or lost differentiated cell markers). Indeed, recent results have suggested that the ability to form a sphere and grow in non-adherent cell culture conditions is not a property that is unique to stem cells. Ependymal cells, astrocytes, oligodendrocyte precursors and neuronal progenitor cells can all readily form neurosphere-like aggregates which can be passaged at least for a limited time period. Overall the data suggest that a need exists to be able to localize the origin of the neurosphere forming cell, to distinguish one population of neurosphere forming cell from another, and to identify those neurosphere forming stem cells that are truly stem cell in character.

Shared cellular and molecular properties of embryonic and adult neural stem cells

Neural stem cells isolated from embryonic and adult CNS are defined by common properties. First, cells isolated from the embryonic ventricular and subventricular zone, those surrounding the adult left ventricles and subgranular zone (SGZ) of the hippocampus, and those from the central canal of the adult spinal cord all share the ability to form neurospheres, the ability to self-renew, and the ability of single cells to differentiate into neurons, astrocytes and oligodendrocytes *in vitro* (Gritti et al., 1996; Johe et al., 1996; Shihabuddin et al., 1997). Second, both embryonic and adult neural stem cells of the CNS can differentiate appropriately in a new host region after transplantation (Brustle et al., 1995; Campbell et al., 1995; Fishell, 1995; Vicario-Abejon et al., 1995). For example, adult hippocampal stem cells can give rise to specific and region appropriate cell types not only in the hippocampus but also when transplanted to the olfactory bulb, cerebellum and retina (Gage et al., 1995; Suhonen et al., 1996; Takahashi et al., 1998). Stem cells derived from the human embryonic nervous system and expanded *in vitro* by oncogenic immortalization exhibit a similarly broad developmental potential when transplanted *in vivo* (Flax et al., 1998). Such transplanted human NSCs can migrate over long distances to colonize different sites of differentiation, especially after transplantation into the neonatal brain (Brustle et al., 1998). These data indicate that CNS stem cells have broader potential than the cell types they normally generate *in vivo* and may share several common markers

Universal markers of cells with stem cell potential in the CNS include a number of transcription factors, such as members of the Sox, Pax, HES and BFAP gene families, members of the Notch signaling pathway, the RNA binding protein musashi1, the intermediate filament protein nestin, and others (Table 2). These expression profiles support the likelihood of common/generic molecular mechanisms shared by neural stem cells throughout their ontogeny. These conserved signaling pathways may serve to maintain generic cellular properties that define the stem cell state, such as the ability to self-renew and multi-lineage differentiation. For example, one of the better characterized molecular pathways conserved in neural stem cells throughout their ontogeny is the Notch signaling pathway. This pathway appears to play an essential role in the maintenance of a stem/progenitor cells pool as well as play a role in regulating asymmetric vs. symmetric division. Both during embryogenesis (Chambers et al., 2001; Gaiano et al., 2000; Ishibashi et al., 1994) and in adulthood expression of notch1 or one of its downstream regulators, such as HES-1, inhibits neuronal differentiation and results in the maintenance of a progenitor state. The exact mechanism by which notch signaling regulates cell fate is not completely determined. Recently, for example, numerous studies in vertebrates have suggested that rather than simply inhibiting neuronal differentiation and maintaining a neural progenitor state, notch may in some contexts promote the acquisition of glial identity (Furukawa et al., 2000; Gaiano et al., 2000; Hojo et al., 2000; Morrison et al., 2000; Scheer et al., 2001). This is consistent with the possibility that, as discussed above, certain glial cell types (radial glia, astrocytes) may be multipotent progenitors.

Recent experiments have also raised the possibility that molecules involved in the consolidation of neural fate during primary neural induction also play a role in adult neurogenesis. For example, Noggin is expressed in ependymal cells, suggesting that it may function to promote neurogenesis. In support of this hypothesis, overexpression of BMP in ependymal cells leads to the reduction of SVZ proliferation and abolishes neuroblast regeneration in the SVZ (Lim et al., 2000). Further, direct comparison of the function of these conserved molecular pathways *in vivo*, in the embryo as well as in the adult, will begin to elucidate similarities and/or differences in the molecular mechanisms during neural stem cell differentiation.

Moreover, several recent studies have presented evidence to challenge the stem cell dogma that stem cells which persist after embryogenesis are restricted in potential to forming only the cell types characteristic of the tissue from which the were isolated. For example, genetically labeled neural stem cells transplanted into an irradiated host give rise to mature blood cell types (Bjornson et al., 1999) and skeletal muscle (Galli et al., 2000). It is important, however, to consider whether the cells that form neurospheres *in vitro* represent "transformed" cells that *in vivo* do not possess stem cells characteristics. It may be possible that the transformation of differentiated cells can result in dedifferentiation or transdifferentiation to a stem cell state.

These studies also raise the possibility that stem cells from different tissues may be more closely related than previously assumed and may share common molecular regulators. Indeed several investigators have argued for the concept of "stemness" or a molecular signature that may be universal to stem cell populations irrespective of the tissue source from which they are identified. Indeed, several investigators have profiled gene expression in different stem cell populations (Ivanova et al., 2002; Ramalho-Santos et al., 2002; Terskikh et al., 2001) and have noted that embryonic, hematopoietic and neural stem cells share many similarities at the transcriptional level. These investigators have proposed that these shared transcripts that are selectively and commonly expressed in two or more types of stem cells define a functionally conserved group of genes evolved to participate in basic stem cell functions, including stem cell self-renewal and that this overlapping set of gene products represents a molecular signature of stem cells.

While the concept of stemness is reasonable, it has been difficult to find truly universal markers that are shared by all stem cell populations irrespective of their tissue origin. Two possible candidates include Tert expression and by extension telomerase activity (Cai et al., 2002) and expression of ABCG2 (Zhou et al., 2001). Expression of both of these molecules has been described in hematopoietic stem cells, ES cells and neural stem cells. A novel seven transmembrane receptor that is expressed in hematopoietic stem cells (HSCs) and is present in proliferating zones of the nervous system (Terskikh et al., 2001) represents a candidate universal stem cell marker.

These results raise the possibility that it may be possible to identify markers that are shared by the multiple types of stem cells present in the nervous system as well as markers that may be shared among stem cells isolated from different tissues.

Identification of neural stem cells types using a combination of universal markers and stem cell subtype markers.

To date it remains unclear whether there exists a generic neural stem cell, as found in the hematopoietic system. It appears that the CNS consists of heterogenic stem cells, all retaining the ability to self-renew, differentiate into neurons and glia, and express a set of universal markers, but restricted in their potency. To understand exactly what characteristics define a neural stem cell it is first necessary to elucidate the lineage relationship between the various types of stem cells and how they contribute to the formation and maintenance of the central nervous system. To do this certain methodologies need to be developed for the isolation of neural stem cells from selected regions of the CNS during defined developmental stages. A number of groups have suggested methods by which stem cell populations can be isolated from mixed cultures of cells. One such method includes a negative selection criterion that takes advantage of the observation that stem cells do not express markers characteristic of differentiated cells. Rao and colleagues have used the absence of expression of neuronal, astrocytic and oligodendroglial markers to enrich for stem cells from late fetal stages (Rao, 1999; Cai et al., 2002). Rietze (2001) in similar experiments have suggested two potential markers that can be used to enrich for neural stem cells in adults. They showed that low PNA (peanut agglutinin) and HSA (heat stable antigen) staining combined with size selection can be used to select for stem cell populations from neurosphere cultures. Using a similar negative selection strategy Maric et al. (2003) used surface ganglioside epitopes emerging on differentiating CNS cells to isolate neural progenitors from E13 rat telencephalon by FACS.

Based on the above results it is possible to identify and isolate stem cells from a mixed population. However, it is difficult to use the absence of expression of markers to localize stem cells *in vivo* given the multiplicity of markers required. As an alternative, parallel approaches have identified positive selection markers that may be used to identify neural stem cells. It has been suggested that AC133 may be an additional neural stem cell marker (Uchida et al., 2000). Within a neurosphere derived from adult tissue the population of cells that are Hoechst low and Rhodhamine 123 low are enriched for stem cells (Hulspas and Quesenberry, 2000; Quesenberry et al., 1999). The efflux is likely mediated by ABCG2 expression that is present on neural stem cells during development and is downregulated in differentiated cells (Cai et al., 2002; Goodell et al., 1996). FACS sorting based on changes in calcium response to neurotransmitter application or growth factor delivery can also be used to identify stem and progenitor cell populations (Mandler et al., 1988; St. John et al., 1986).

An alternative approach to positively select for neural stem cells is through the generation of mouse lines in which the expression of a drug-selection marker or a live cell marker such as green fluorescent protein (EGFP) is driven by the regulatory domains of a universal neural stem cell marker. For example, transgenic lines have been generated carrying EGFP under the control of the neural specific enhancer for the nestin gene (Roy et al., 2000; Sawamoto et al., 2001). FACS analysis showed that nestin-EGFP expression directly correlates with multipotency and density of neurosphere initiating cells, thereby permitting the high enrichment of neural cells isolated from the embryonic cerebral cortex. Similarly, to develop an *in vivo* system for analyzing neurogenesis. Transgenic mice have been generated that express EGFP under the control of the regulatory regions of the SOX2 gene, a universal neural stem cell marker. In this mouse line, EGFP expression is confined to progenitor cell populations during early development and persists in selected populations in the adult (Figure 3; unpublished data). Moreover, clonal analysis of SOX2-EGFP-positive cells demonstrates that multipotential stem cells isolated from both the embryonic CNS and the adult CNS all express SOX2-EGFP.

Thus, positive and negative selection criteria can be used to define populations of stem cells at any stage of development. These markers, either singly or in concert, may help localize stem cells *in vivo* and their expression in neurospheres may help define whether a particular neurosphere contains a multipotent stem cell.

Universal stem cell markers provide a means to identify cells which fulfill the basic criteria of a stem cell, self-renewal and multipotent differentiation, and thus define shared features when removed from their normal milieu. However, it has been previously shown that positional markers that define the rostrocaudal identity of stem cells persist over multiple generations *in vitro*

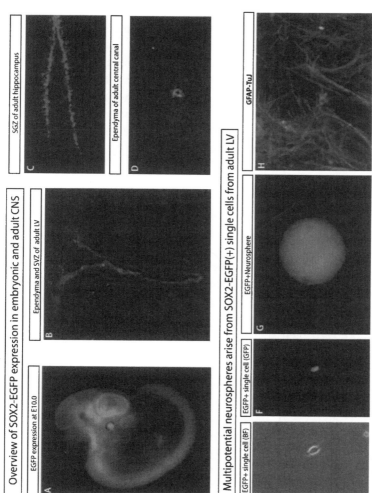

Figure 3: Sox-2 expression identifies stem cell populations of the embryonic and adult nervous system. A. Enhanced green fluorescent protein (EGFP) expression in an 10 day old embryo (E10) from a SOX2-EGFP mouse line. EGFP is expressed throughout the neuroepithelium. **B-C.** Coronal section through adult forebrain showing the lateral ventricle (LV) labeled with EGFP (B) and EGFP-expressing cells confined to the subgranular zone of the hippocampus. **E-H.** Time lapse of a multipotent neurosphere arising from a single EGFP+ cell isolated from the adult SVZ.

(Hitoshi et al., 2002; Nakagawa et al., 1996; Zappone et al., 2000). It is therefore reasonable to assume that markers characteristic of the cell that generates a neurosphere *in vitro* will persist at least over the initial passages. Thus, if markers exist that distinguish between the potential stem cell candidates, it may be possible to determine if all neurospheres are derived from the same population or are a heterogeneous population whose properties depend on the particular stem cell population that generated them.

A potential problem with this approach is the possibility that neurospheres may not contain only a stem cell population. Stem cells differentiate in response to environmental signals and it is difficult to ensure homogeneity of the microenvironment in a neurosphere culture. Cells at the margin are exposed to different conditions as compared to the cells within a neurosphere and thus even at early stages of its formation, a neurosphere may be a heterogeneous population. It is therefore critical that apart from determining the expression of markers, one also test the properties of the cells rigorously by passaging and cloning.

Combining universal stem cell markers in double labeling or cell isolation experiments may allow one to bypass this problem. For example, if a GFAP+ cell that formed a neurosphere co-expressed TERT, had high telomerase activity, expressed Sox1 and Sox2, and had high ABCG2 expression and activity, then it would be reasonable to assume in subsequent experiments that this represented a stem cell population. Equally important would be demonstrating the absence of markers that define dividing populations of cells that are not multipotent but more restricted such as A2B5, CD26, CD44, and others (see above). Experiments along similar lines in the hematopoietic system have helped define the stem cell population *in vivo* that has the highest self-renewing capacity and has allowed investigators to begin to probe fundamental aspects of stem cell biology.

Two important points worth emphasizing are that no single marker is likely to work and even combinations of markers that work are likely to be useful only in specific situations. For example AC133 may turn out to be a useful marker for stem cells in the nervous system but given that it is also expressed by hematopoietic stem cells (D'Arena et al., 2002; Majka et al., 2000) and may be expressed by ES cells it may be difficult to use AC133 as a sole selection marker. Likewise generalized methods of isolation such as forward versus side scatter, Hoechst labeling or aldefluor selection (Cai et al., 2002; Murayama et al., 2002) cannot be used to select between neural and non-neural stem cell populations.

Similarly negative selection criteria will need to be modified depending on the mixture of cells that need to be removed to enrich the stem cell population. For example, in the hippocampus A2B5 expression is quite limited at early stages of differentiation, while PSA-NCAM and CD24 expression is abundant.

A2B5 may not, therefore, be a critical component of any stem cell selection cocktail for hippocampal stem cells. In contrast, in the adult spinal cord the largest population of dividing cells are glial progenitors including A2B5$^+$ cells and CD44$^+$ astrocyte precursors (our unpublished results). In this case separating glial progenitors from stem cells will provide maximal enrichment and these markers will be critical. Whatever strategy is selected, it will have to be tested to ensure that it is appropriate for particular stages in development and the region of brain that is being tested. Universal markers will be an important component of any generalized strategy that will have to be supplemented with markers appropriate for the mixture of cells from which stem cells are to be enriched.

A final point worth discussing is the discrimination between sorting from tissue or acutely harvested cells grown in culture for short time periods (1-7 days). Properties of cells, their ability to differentiate, growth factor receptor expression, and particularly cell surface markers used for selection, may be altered and stem cell markers may not display the same specificity as documented during *in vivo* development. RC1 is a good example of a relatively specific radial glial marker whose expression is seen in a large proportion of cultured cells that are clearly not radial glia. CD44 expression can be induced in a variety of cells in response to growth factor signaling, and nestin expression is seen in multiple differentiated cell types maintained in culture. The problem of sorting from cultured and passaged cells is further exacerbated by data suggesting that cells may transdifferentiate in culture. Transdifferentiation does not recapitulate the developmental process precisely and transdifferentiated cells are likely to express a mixture of stem cell and other differentiated cell markers. One cannot therefore simply rely on published data on the specificity of antibodies demonstrated to be specific *in vivo* or in acutely dissociated cell cultures. Rather, a rigorous analysis of acceptable markers appropriate for that particular system will have to be undertaken. The currently described markers represent a good starting point but should be used with caution. We would suggest that methods that rely on generalized properties of stem-like cells might be the first set of selection strategies to attempt. Tert expression, aldefluor uptake, ABCG2-dependent efflux, positive selection with markers known to be expressed by stem-like cells such as Sox1, Sox2, or AC133 may represent good candidate selection molecules.

CONCLUSION

Selecting neural stem cell populations is intrinsically different from selecting hematopoietic stem cells due to the variable sources of cells, the variety of stem cells described, and our ability to maintain neural stem cells in culture for prolonged time periods. Success is likely given the number of markers that have now become available. To ensure success, however, one will have to tailor the selection strategy to the source of stem cells and the type of stem cell one wishes to isolate. A rigorous evaluation of all selection strategies in independent laboratories will permit the development of a series of selection protocols that may then allow identification of stem cells from any source.

Acknowledgements

The authors gratefully acknowledge the input and suggestions of members of our laboratories and colleagues in the stem cell field. The NIH supported LP. MSR was supported by the NIA, CNS foundation, NINDS and the ALS Center.

REFERENCES

Altman J (1962) Are new neurons formed in the brains of adult mammals? Science 135:1127-1128.
Altman J, Das GD (1965) Autoradiographic and histological evidence of postnatal neurogenesis in rats. J Comp Neurol 124:319-335.
Altman J, Bayer SA (1984) *The Development of the Rat Spinal Cord.* Berlin, Heidelberg, New York, Tokyo: Springer.
Alvarez-Buylla A, Herrara DG, Wichterle H (2000) The subventricular zone: source of neuronal precursors for brain repair. Progress in Brain Research 127:1-11.
Alvarez-Buylla A, Garcia-Verdugo JM, Tramontin AD (2001) A unified hypothesis on the lineage of neural stem cells. Nature Review of Neuroscience 2:287 293.
AlvarezBuylla A, GarciaVerdugo JM (2002) Neurogenesis in the adult subventricular zone. Journal of Neuroscience 22:629 634.
Anderson D (2001) Stem cells and pattern formation in the nervous system: the possible versus the actual. Neuron 30:19-35.
Ang S-L, Rossant J (1994) *HNF-3 beta* is essential for node and notochord formation in mouse development. Cell 78:561-574.
Bachiller D, Klingensmith J, Kemp C, Belo JA, Anderson RM, May SR, McMahon JA, McMahon A, Harland R, Rossant J, DeRobertis EM (2000) The organiser factors Chordin and Noggin are required for mouse forebrain development. Nature 404:658-661.
Barres BA (1999) A New Role for Glia; Generation of Neurons. Cell 97:667-670.
Bayer SA, Altman J (1991) Neocortical development. New York.
Bjornson CRR, Rietze RL, Reynolds BA, Magli C, Vescovi AL (1999) Turning Brain into Blood: A Hematopoietic Fate Adopted by Adult Neural Stem Cells In Vivo. Science 283:534-537.
Brewer GJ (1999) Regeneration and proliferation of embryonic and adult rat hippocampal neurons in culture. Exp Neurol 159:237-247.
Briscoe J, Pierani A, Jessell TM, Ericson J (2000) A homeodomain protein code specifies progenitor cell identity and neuronal fate in the ventral neural tube. Cell 101.

Brustle O, Maskos U, McKay RDG (1995) Host-guided migration allows targeted introduction of neurons into the embryonic brain. Neuron 15:1275-1285.

Brustle O, Choudhary K, Karram K, Huttner A, Murray K, Dubois-Dalcq M, McKay R (1998) Chimeric brains generated by intraventricular transplantation of fetal human brain cells into embryonic rats. Nat Biotechnol 16:1041-1044.

Burrows RC, Lillien L, Levitt P (2000) Mechanisms of progenitor maturation are conserved in the striatum and cortex. Dev Neurosci 22:7-15.

Cai J, Wu Y, Mirua T, Pierce JL, Lucero MT, Albertine KH, Spangrude GJ, Rao MS (2002) Properties of a fetal multipotent neural stem cell (NEP cell). Dev Biol 251:221-240.

Campbell K, Olsson M, Bjorklund A (1995) Regional incorporation and site-specific differentiation of striatal precursors transplanted to the embryonic forebrain ventricle. Neuron 15:1259-1273.

Capela A, Temple S (2002) LeX/SSEA-1 is expressed by adult mouse CNS stem cells, identifying them as non-ependymal. Neuron 35:865-875.

Chambers CB, Peng Y, Nguyen H, Gaiano N, Fishell G, Nye JS (2001) Spatiotemporal selectivity of response to Notch1 signals in mammalian forebrain precursors. Development 128:689-702.

Chiasson BJ, Tropepe V, Morshead CM, van der Kooy D (1999) Adult mammalian forebrain ependymal and subependymal cells demonstrate proliferative potential, but only subependymal cells have neural stem cell characteristics. Journal of Neuroscience 19:4462-4471.

Ciccolini F (2001) Identification of two distinct types of multipotent neural precursors that appear sequentially during CNS development. Mol Cell Neurosci 17:895-907.

Collignon J, Sockanathan S, Hacker A, Cohen-Tannoudji M, Norris D, Rastan S, Stevanovic M, Goodfellow PN, Lovell-Badge R (1996) A comparison of the properties of Sox-3 and Sry, and two related genes Sox-1 and Sox-2. Development 122:509-522.

D'Arena G, Savino L, Nunziata G, Cascavilla N, Matera R, Pistolese G, Carella AM (2002) Immonophenotypic profile of AC133-positive cells in bone marrow, mobilized peripheral blood and umbilical cord blood. Leuk Lymphoma 43:869-873.

Desai AR, McConnell SK (2000) Progressive restriction in fate potential by neural progenitors during cerebral cortical development. Development 127:2863-2872.

Doetsch F, Caille I, Lim DA, Garcia-Verdugo JM, Alvarez-Buylla A (1999) Subventricular Zone Astrocytes Are Neural Stem Cells in the Adult Mammalian Brain. Cell 97:703-716.

Doetsch F, Petreanu L, Caille I, Garcia-Verdugo JM, Alvarez-Buylla A (2002) EGF converts transit-amplifying neurogenic precursors in the adult brain into multipotent stem cells. Neuron 36:1021-1034.

Edlund T, Jessell TM (1999) Progression from extrinsic to intrinsic signaling in cell fate specification: a view from the nervous system. Cell 96:211-224.

Fishell G (1995) Striatal precursors adopt cortical identities in response to local cues. Development 121:803-812.

Flax JD, Aurora S, Yang C, Simonin RL, Wills AM, Billinghurst LL, Jendoubi M, Sidman RL, Wolfe JH, Kim SU, Snyder EY (1998) Engraftable human neural stem cells respond to developmental cues, replace neurons, and express foreign genes. Nat Biotechnol 16:1033-1039.

Frederiksen K, McKay RD (1998) Proliferation and differentiation of rat neuroepithelial precursor cells *in vivo*. Journal of Neuroscience 9:1144-1151.

Furukawa T, Mukherjee A, Bao ZZ, Morrow EM, Cepko C (2000) rax, Hes1, and Notch1 promote the formation of Muller glia by postnatal retinal progenitor cells. Neuron 26:383-394.

Gage FH, Ray J, Fisher LJ (1995) Isolation, characterisation and use of stem cells from the CNS. Ann Rev Neurosci 18:159-192.

Gaiano N, Nye JS, Fishell G (2000) Radial glial identity is promoted by Notch1 signaling in the murine forebrain. Neuron 26:395-404.

Galli R, Borrelo U, Gritti A, Minasi MG, Bjornson C, Coletta M, Mora M, De Angelis MS, Fiocco R, Cossu G, Vescovi AL (2000) Skeletal myogenic potential of human and mouse neural stem cells. Nat Neurosci 3:986-991.

Garcia-Verdugo JM, Doetsch F, Wichterle H, Lim DA, Alvarez-Buylla A (1998) Architecture and cell types of the adult subventricular zone: in search of stem cells. Journal of Neurobiology 36:234-248.

Goodell MA, Brose K, Paradis G, Conner AS, Mulligan RC (1996) Isolation and functional properties of murine hematopoietic stem cells that are replicating in vivo. J Exp Med 183:1797.

Gray GE, Sanes JR (1992) Lineage of radial glia in the chicken optic tectum. Development 114:271-283.

Gritti A, Parati EA, Cova L, Frolichsthal P, Galli R, Wanke E, Faravelli L, Morasuttin DJ, Roisen F, Nickel DD, Vescovi AL (1996) Multipotential stem cells from the adult mouse brain proliferate and self-renew in response to basic fibroblast growth factor. J Neurosci 16:1091-1100.

Guthrie S (1996) Patterning the hindbrain. Curr Opin Neurobiol 6:41-48.

Harland R (2000) Neural induction. Current Opinion in Genetics and Development 10:357-362.

Hartfuss E, Galli R, Heins N, Gotz M (2001) Characterization of CNS precursor subtypes and radial glia. Dev Biol 229:15-30.

Hitoshi S, Tropepe V, Ekker M, van der Kooy D (2002) Neural stem cell lineages are regionally specified, but not committed, within distinct compartments of the developing brain. Development 129:233-244.

Hockfield S, McKay RD (1985) Identification of major cell classes in the developing mammalian nervous system. Journal of Neuroscience 5:3310-3328.

Hojo M, Ohtsuka T, Hashimoto C, Gradwohl G, Guillemot F, Kageyama R (2000) Glial cell fate specification modulated by the bHLH gene Hes5 in mouse retina. Development 127:2515-2522.

Hulspas R, Quesenberry PJ (2000) Characterization of neurosphere cell phenotypes by flow cytometry. Cytometry 40:245-250.

Ishibashi M, Moriyoshi K, Sasai Y, Shiota K, Nakanishi S, Kageyama R (1994) Persistent expression of helix-loop-helix factor HES-1 prevents mammalian neural differentiation in the central nervous system. EMBO 13:1799-1805.

Ivanova NB, Dimos JT, Schaniel C, Hackney JA, Moore KA, Lemischka IR (2002) A stem cell molecular signature. Science 298:601-604.

Jessell TM (2000) Neuronal specification in the spinal cord: inductive signals and transcriptional codes. Nat Rev Genet 1:20-29.

Johansson CB, Svensson M, Wallstedt L, Janson AM, Frisen J (1999a) Neural stem cells in the adult human brain. Exp Cell Res 253:733-736.

Johansson CB, Momma S, Clarke DL, Risling M, Lendahl U, Frisen J (1999b) Identification of a neural stem cell in the adult mammalian central nervous system. Cell 96:25-34.

Johe KK, Hazel TG, Muller T, Dugich-Djordjevic MM, McKay RDG (1996) Single factors direct the differentiation of stem cells from the fetal and adult central nervous system. Genes Dev 10:3129-3140.

Josephson R, Muller T, Pickel J, Okabe S, Reynolds K, Turner PA, Zimmer A, McKay R (1998) POU transcription factors control expression of CNS stem cell-specific genes. Development 125:3087-3100.

Kalyani A, Hobson K, Rao MS (1997) Neuroepithelial stem cells from the embryonic spinal cord: isolation, characterisation, and clonal analysis. Dev Biol 186:202-223.

Klingensmith J, Ang S-L, Bachiller D, Rossant J (1999) Neural Induction and Patterning in the Mouse in the Absence of the Node and Its Derivatives.Developmental Biology. Developmental Biology 216:535-549.

Kondo T, Raff MC (2000) Oligodendrocyte precursor cells reprogrammed to become multipotential CNS stem cells. Science 289:1754-1757.

Kornblum HI, Hussain R, Weisen J, Miettinen P, Zurcher SD, Chow K, Dernyck R, Werb Z (1998) Abnormal astrocyte development and neuronal death in mice lacking the epidermal growth factor receptor. J Neurosci Res 53:697-717.

Kuhn HG, Dickinson-Anson H, Gage FH (1996) Neurogenesis in the dentate gyrus of the adult rat: age related decrease of neuronal progenitor proliferation. J Neurosci 16:2027-2033.

Laywell ED, Steindler DA (2002) Glial stem-like cells: implications for ontogeny, phylogeny and CNS regeneration. Prog Brain Res 138:435-450.

Laywell ED, Rakic P, Kukekov VG, Holland EZ, Steindler DA (2000) Identification of a multipotent astrocytic stem cell in the immature and adult mouse brain. Proc Nat Acad Sci USA 97.

Lendahl U, Zimmerman LB, McKay RDG (1990) CNS stem cells express a new class of intermediate filament protein. Cell 60:585-595.

Levison SW, Goldman JE (1997) Multipotential and lineage restricted precursors coexist in the mammalian perinatal subventricular zone. J Neurosci Res 48:83-94.

Levitt P, Cooper ML, Rakic P (1981) Coexistance of neuronal and glial precursor cells in the cerebral ventricular zone of the fetal monkey: an ultrastuctural immunoperoxidase anaysis. J Neurosci 1:27-39.

Lillien L (1998) Neural progenitors and stem cells: mechanisms of progenitor heterogeniety. Current Opinion in Neurobiology 8:37-44.

Lim DA, Fishell G, Alvarez-Buylla A (1997) Postnatal mouse subventricular zone neuronal precursors can migrate and differentiate within multiple levels of the developing neuraxisw. Proc Nat Acad Sci USA 94:14832-14836.

Lim DA, Tramontin AD, Trevejo JM, Herrera DG, Garcia-Verdugo JM, Alvarez-Buylla A (2000) Noggin Antagonizes BMP Signaling to Create a Niche for Adult Neurogenesis. Neuron 28:713-726.

Lois C, Alvarez-Buylla A (1993) Proliferating subventricular zone cells in the adult mammalian forebrain can differentiate into neurons and glia. Proc Natl Acad Sci USA 90:2074-2077.

Lumsden A, Krumlauf R (1996) Patterning the vertebrate neuraxis. Science 274:1009-1115.

Luskin MB, Parnavelas JG, Barfield JA (1993) Neurons, astrocytes and oligodendrocytes of the rat cerebral cortex originate from separate progenitor cells: an ultrastructural analysis of clonally related cells. J Neuroscience 13:1730-1750.

Magavi SS, Leavitt BR, Macklis JD (2000) Induction of neurogenesis in the neocortex of adult mice. Nature 405:951-955.

Majka M, Ratajczak J, Machalinski B, Carter A, Pizzini D, Wasik MA, Gewirtz AM, Ratajczak MZ (2000) Expression, regulation and function of AC133, a putative cell surface marker of primitive human haematopoietic cells. Folia Histochem Cytobiol 38:53-63.

Malatesta P, Hartfuss E, Gotz M (2000) Isolation of radial glial cells by fluorescent-activated cell sorting reveals a neuronal lineage. Development 127:5253-5263.

Malatesta P, Hack MA, Hartfuss E, Kettenmann H, Klinkert W, Kirchhoff F, Gotz M (2003) Neuronal or Glial Progeny: Regional Differences in Radial Glial Fate. Neuron 37:751-764.

Mandler RN, Schaffner AE, Novotny EA, Lange GD, Barker JL (1988) Flow cytometric analysis of membrane potential in embryonic rat spinal cord cells. J Neuosci Methods 22.

Maric D, Maric I, Chang YH, Barker JL (2003) Prospective cell sorting of embryonic rat neural stem cells and neuronal and glial progenitors reveals effects of basic firbroblast growth factor and epidermal growte factor on self-renewal and differentiation. J Neurosci 23:240-251.

Marmur R, Mabie PC, Gokhan S, Song Q, Kessler JA, Mehler MF (1998) Isolation and developmental characterization of cerebral cortical multipotent progenitors. Dev Biol 204:577-591.

McConnell SK (1995) Constructing the cerebral cortex: neurogenesis and fate determination. Neuron 15:791-803.

McConnell SK, Kaznowski CE (1991) Cell cycle depndence of laminar determination in developing neocortex. Science 254.

McMahon J, Takada S, Zimmerman LB, Fan CM, Harland R, McMahon AP (1998) Noggin mediated antagonism of BMP signaling is required for growth and patterning of the neural tube and somite. Genes and Development 12:1438-1452.

Momma S, Johansson CB, Frisen J (2000) Get to know your stem cells. Current Opinion in Neurobiology 10:45-49.

Morrison S, Perez SE, Qiao Z, Verdi JM, Hicks C, Anderson D (2000) Transient Notch activation initiates an irreversible switch from neurogenesis to gliogenesis by neural crest stem cells. Cell 499-510.

Morshead CM, Craig CG, van der Kooy D (1998) In vivo clonal analyses reveal the properties of endogenous neural stem cell proliferation in the adult mammalian forebrain. Development 125:2251-2261.

Morshead CM, Reynolds BA, Craig CG, McBurney MW, Staines WA, Morasutti DJ, Weiss S, van der Kooy D (1994) Neural stem cells in the adult mammalian forebrain: a relatively quiescent subpopulation of subependymal cells. Neuron 13:1071-1082.

Mujtaba T, Piper DR, Kalyani A, Groves AK, Lucero MT, Rao MS (1999) Lineage-restricted neural precursors can be isolated from both the mouse neural tube and cultured ES cells. Dev Biol 214:113-127.

Murayama A, Matsuzaki Y, Kawaguchi A, Shimazaki R, Okano H (2002) Flow cytometric analysis of neural stem cells in the developing and adult mouse brain. J Neurosci Res 69:837-847.

Nakagawa Y, Kaneko T, Ogura T, Suzuki T, Torii M, Kaibuchi K, Arai K, Nakamura S, Nakafuku M (1996) Roles of cell-autonomous mechanisms for differential expression of region-specific transcription factors in neuroepithelial cells. Development 122:2449-2464.

Nakamura N, Mitamura T, Takahashi T, Kobayashi T, Mekada E (2000) Importance of the major extracellular domain of CD9 and the epidermal growth factor (EGF)-like domain of heparin-binding EGF-like growth factor for up-regulation of binding and activity. J Biol Chem 275:18284-18290.

Noctor SC, Flint AC, Weissman TA, Dammerman RS, Kreigstein AR (2001) Neurons derived from radial glial cells establish radial units in neocortex. Nature 409:714-720.

Noctor SC, Flint AC, Weissman TA, Wong WS, Clinton BK, Kriegstein AR (2002) Dividing precursor cells of the embryonic cortical ventricular zone have morphological and molecular characteristics of radial glia. J Neurosci 22:3161-3173.

Ortega S, Ittmann M, Tsang SH, Ehrlich M, Basilico C (1998) Neuronal defects and delayed wound healing in mice lacking fibroblast growth factor 2. Proc Natl Acad Sci U S A 95:5672-5677.

Palmer TD, Ray J, Gage FH (1995) FGF-2 responsive neuronal progenitors reside in proliferative and quiescent regions of the adult rodent brain. Molecular and Cellular Neuroscience 6:474-486.

Palmer TD, Takahashi J, Gage FH (1997) The adult rat hippocampus contains primordial neural stem cells. Molecular and Cellular Neuroscience 8:389-404.

Pevny LH, Sockanathan S, Placzek M, Lovell-Badge R (1998) A role for SOX1 in neural determination. Development 125:1967-1978.

Price J, Turner D, Cepko C (1987) Lineage analysis in the vertebrate nervous system by retrovirus-mediated gene transfer. Dev Biol 84:156-160.

Quesenberry PJ, Hulspas R, Joly C, Benoit B, Engstrom C, Rielly J, Savarese T, Pang L, Recht L, Ross A, Stein G, Stewart M (1999) Correlates between hematopoiesis and neuropoiesis: neural stem cells. J neurotrauma 16:661-666.

Raballo R, Rhee J, Lyn-Cook R, Leckman JF, Schwartz ML, Vaccarino FM (2000) Basic fibroblast growth factor (Fgf2) is necessary for cell proliferation and neurogenesis in the developing cerebral cortex. J Neurosci 20:12-23.

Rakic P (1988) Specification of cerebral cortical areas. Science 241:170-176.

Ramalho-Santos M, Yoon S, Matsuzaki Y, Mulligan RC, Melton DA (2002) "Stemness": transcriptional profiling of embryonic and adult stem cells. Science 298:597-600.

Rao M (1999) Multipotent and Restricted Precursors in the Central Nervous System. The Anatomical Record 257:137-148.

Reynolds BA, Weiss S (1992) Generation of neurons and astrocytes from isolated cells of the adult mammalian central nervous system. Science 255:1707-1710.

Reynolds BA, Weiss S (1996) Clonal and population analyses demonstrate that an EGF-responsive mammalian embryonic CNS precursor is a stem cell. Dev Biol 175:1-13.

Reynolds BA, Tetzlaff W, Weiss S (1992) A multipotent EGF-responsive striatal embryonic progenitor cell produces neurons and astrocytes. J Neurosci 12.

Rietze RL, Valcanis H, Brooker GF, Thomas T, Voss AK, Bartlett PF (2001) Purification of a pluripotent neural stem cell from the adult mouse brain. Nature 412:736-739.

Roy NS, Benraiss A, Wang S, Fraser RA, Goodman R, Couldwell WT, Nedergaard M, Kawaguchi A, Okano H, Goldman SA (2000) Promoter-targeted selection and isolation of neural progenitor cells from the adult ventricular zone. J Neurosci Res 59:321-331.

Sakakibara S, Okano H (1997) Expression of neural RNA-binding proteins in the postnatal CNS: implications of their roles in neuronal and glial cell development. Journal of Neuroscience 17:8300-8312.

Sakakibara S, Imai T, Hamaguchi K, Okabe M, Aruga J, Nakajima K, Yasutomi D, Nagata T, Kurihara Y, Uesugi S, Miyata T, Ogawa M, Mikohiba K, Okano H (1996) Mouse Musashi-1, a neural RNA-1 binding protein highly enriched in the mammalian CNS stem cell. Developmental Biology 176:230-242.

Sasai Y (2001) Roles of Sox factors in neural determination: conserved signaling in evolution? International Journal of Developmental Biology 45:351-326.

Sawamoto K, Nakao N, Kakishita K, Ogawa Y, Toyama Y, Yamamoto A, Yamaguchi M, Mori K, Goldman SA, Itakura T, Okano H (2001) Generation of dopaminergic neurons in the adult brain from mesencephalic precursor cells labeled with nestin-GFP transgene. J Neurosci 21:3895-3903.

Scheer N, Groth A, Hans S, Campos-Ortega JA (2001) An instructive function for Notch in promoting gliogenesis in the zebrafish retina. Development 128:1099-1107.

Schmechel DE, Rakic P (1979) Arrested proliferation of radial glial cells during midgestation in rhesus monkey. Nature 277:303-305.

Shihabuddin LS, Ray J, Gage FH (1997) FGF-2 is sufficient to isolate progenitors found in the adult mammalian spinal cord. Exp Neurol 148:577-586.

Shihabuddin LS, Horner PJ, Ray J, Gage FH (2000) Adult spinal cord stem cells generate neurons after transplantation in the adult dentate gyrus. J Neurosci 20:8727-8735.

St. John PA, Kell WM, Mazzetta JS, Lange GD, Barker JL (1986) Analysis and isolation of embryonic mammalian neurons by flourescence-activated cell sorting. J Neurosci 6:1492-1512.

Stavridis M, Smith A (2003) Neural differentiation of embryonic stem cells. Biochem Soc Trans 31:45-49.

Streit A, Berliner AJ, Papanayoton c, Serulnik A, Stern C (2000) Initiation of neural induction by FGF signalling before gastrulation. Nature 406:74-78.

Streit A, Lee KJ, Woo I, Roberts C, Jessell TM, Stern CD (1998) Chordin regulates primitive streak development and the stability of induced neural cells but is not sufficient for neural induction in the chick embryo. Development 125:507-519.

Suhonen JA, Peterson DA, Ray J, Gage FH (1996) Differentiation of adult hippocampus-derived progenitors into olfactory neurons in vivo. Nature 383:624-627.

Takahashi M, Palmer TD, Takahashi J, Gage FH (1998) Widespread intergration and survival of adult-derived neural progenitor cells in the developing optic retina. Mol Cell Neurosci 12:340-348.

Tam PPL, Behringer RR (1997) Mouse gastrulation: the formation of a mammalian body plan. Mech Dev 68:3-25.

Temple S, Buylla-Alvarez A (1999) Stem cells in the adult mammalian central nervous system. Current Opinion in Neurobiology 9:135-141.

Terskikh AV, Earsterday MC, Li L, Hood L, Kornblum HI, Gerschwind DH, Weissman IL (2001) From hematopoiesis to neuropoiesis: evidence of overlapping genetic programs. Proc Nat Acad Sci USA 98:7934-7939.

Thomson JS, Marshall WS, Trojanowski JQ (1998) Neural differentiation of rhesus embryonic stem cells. APMIS 106.

Tropepe V, Hitoshi S, Sirand C, Mak TW, Rossant J, van der Kooy D (2001) Direct neural fate specification from embryonic stem cells: a primitive mammalian neural stem cell stage acquired through a default mechanism. Neuron 30:65-78.

Ushida N, Buck DW, D. H, Reitsma MJ, Masek M, Phan TV, Tsukamoto AS, Gage FH, Weissman IL (2000) Direct isolation of human central nervous system stem cells. Proc Natl Acad Sci U S A 97:14720-14725.

Vaccarino FM, Schwartz ML, Raballo R, Nilsen J, Rhee J, Zhou M, Doetschman T, Coffin JD, Wyland JJ, Hung YT (1999) Changes in cerebral cortex size are governed by fibroblast growth factor during embryogenesis. Nat Neurosci 2:246-253.

Vicario-Abejon C, Cunningham MG, McKay RDG (1995) Cerebellar precursors transplanted to the neonatal dentate gyrus express features characteristic of hippocampal neurons. J Neurosci 15:6351-6363.

Weinstein DC, Ruiz I Altaba A, Chen WS, Hoodless PA, Jessell TM, Darnell JEJ (1994) The winged-helix transcription factor *HNF-3 beta* is required for notochord development in the mouse embryo. Cell 78:575-588.

Weiss S, Dunne C, Hewson J, Wohl C, Wheatley M, Peterson AC, Reynolds BR (1996) Multipotent CNS stem cells are present in the adult mammalian spinal cord and ventricular neuroaxis. J Neuroscience 16.

Wentworth LE (1984) The development of the cervical spinal cord of the mouse embryo. II. A Golgi analysis of sensory, commissural, and association cell differentiation. J Comp Neurol 222:96-115.

Williams B (1995) Precursor cell types in the germinal zone of the cerebral cortex. Bioessays 17:391-393.

Wilson SI, Rubenstein JL (2000) Induction and dorsoventral patterning of the telencephalon. Neuron 28:641-651.

Wilson SI, Graziano E, Harland R, Jessell TM, Edlund T (2000) An early requirement for FGF signaling in the acquisition of neural cell fate in chick embryos. Curr Biol 10:421-429.

Wood H, Episkoppu V (1999) Comparative expression of the mouse *Sox1*, *Sox2* and *Sox3* genes from pre-gastrulation to early somite stages. Mech Dev 86:197-201.

Ying Q-L, Stavridis M, Griffiths D, Li M, Smith A (2003) Conversion of embryonic stem cells into neuroectodermal precursors in adherent monoculture. Nature Biotechnology 21:183-186.

Zappone MS, Galli R, Catena R, Meani N, De Biasi S, Mattei E, Tiveron C, Vescovi AL, Lovell-Badge R, Ottolenghi S, Nicolia SK (2000) Sox2 regulatory sequences direct expression of a (beta)-geo transgene to telencephalic neural stem cells and precursors of the mouse embryo, revealing regionalization of gene expression in CNS stem cells. Development 127:2367-2382.

Zhou S, Schuetz JD, Bunting KD, Colapietro AM, Sampath J, Morris JJ, Lagutina I, Grosveld GC, Osawa M, Nakauchi H, Sorrentino BP (2001) The ABC transporter Bcrp1/ABCG2 is expressed in a wide variety of stem cells and is a molecular determinant of the side-population phenotype. Nat Med 7:1028-1034.

Chapter 5

Neural Cells Derived From Embryonic Stem Cells

Mark J. Tomishima and Lorenz Studer

INTRODUCTION

Embryonic stem (ES) cells are renewable cells capable of differentiating into all cell types of an organism. ES cells have the unique ability to divide in culture indefinitely without transformation or loss of differentiation potential (Suda, 1987). Gain and loss-of-function studies using ES cell technology for creating genetically modified mice revolutionized molecular biology and our understanding of mammalian nervous system development. The recent demonstration of homologous recombination in human ES cells sets the stage for a new revolution in our understanding of human nervous system development and disease (Zwaka and Thomson, 2003). The isolation of human ES cells, and the ability to induce neural differentiation *in vitro*, has renewed interest in developing strategies for cell therapy and brain repair. In this chapter, we will define ES cells and related pluripotent cell types, review protocols for directing ES cell differentiation into specific neural fates and provide an outlook for ES cell research in basic studies and in neural repair.

ES and ES-like cells

ES cell lines are derived from a transient population of cells present in the early embryo. The blastocyst is a developmental stage before the embryo has implanted, and within this structure are a small population of cells called the inner cell mass (ICM). These cells are progenitors of all the cells that form the developing embryo, and they can be harnessed at this pluripotent state if removed from the blastocyst and cultured appropriately. Stable cell lines derived from the ICM that retain the full differentiation potential are called ES cells. The gold standard for demonstrating ES cell function is the chimeric contribution to all organs including the germ-line after injecting cultured ES cells into a developing blastocyst.

From: *Neural Stem Cells: Development and Transplantation*
Edited by: J.E. Bottenstein © 2003 Kluwer Academic Publishers

Much of our understanding of ES cells comes from the study of two related cell types called embryonal carcinoma (EC) and embryonic germ (EG) cells. A set of remarkable experiments in the early 1970's demonstrated a direct relationship between ES and EC cells. When normal early mouse embryos are placed ectopically into the kidney or testis of an adult mouse, large tumors develop (Solter et al., 1970, Stevens, 1970). These tumors, called teratomas or teratocarcinomas, contain cell types from all three germ layers often with patches of organized tissues (such as hair or teeth) as well as undifferentiated ES-like cells. The ES-like cells within tumors can be isolated and propagated *in vitro* and are named EC cells (Evans, 1972). Teratocarcinomas also form spontaneously from germ cells (Stevens and Hummel 1957), and numerous EC cell lines have been isolated from such tumors. Each EC cell line behaves differently (Kleinsmith and Pierce 1964, Martin and Evans 1975), and in general they do not contribute equally to all tissues when reintroduced into an early mouse embryo (Martin 1980). While some EC lines differentiate into a wide variety of cell types, others are called "nullipotent" because they cannot be induced to differentiate (Bernstine et al. 1973). Almost all EC cell lines are aneuploid and likely contain other genetic alterations that contribute to their ability to divide rapidly in tumors and in culture (Smith 2001).

A third pluripotent cell type called embryonic germ cells (EG) cells are derived from the primordial germ cells in an embryo, the progenitors of the male and female germ cells. Their differentiation potential is similar to that of ES cells (Rohwedel et al. 1996, Ohtaka et al. 1999), with one notable difference: EG cells are unique in the expression of imprinted genes (Surani et al. 1998, Howell et al. 2001). Some chimeras created with EG cells develop normally, and can contribute EG cells to the germline (Labosky et al. 1994, Stewart et al. 1994). In other cases, the embryo develops abnormally with phenotypes characteristic of an imprinting defect (Tada et al. 1998). EC, EG and ES cells are morphologically similar, express many of the same genes, require similar growth conditions, and are capable of differentiating into all three germ layers. Further, all three cell types can clonally give rise to teratomas or teratocarcinomas. One interesting question is whether undifferentiated cells isolated from an ES cell-derived teratocarcinoma are genetically similar to the input ES cells or whether tumor growth selects for cells that efficiently proliferate but do not differentiate (Smith 2001).

Advantages of using ES cells

The most striking advantages of using ES and ES-like cells to study the nervous system are: 1) Unlimited growth in culture, 2) full differentiation potential into any adult (neural) cell type, 3) access to the earliest stages of neural development and 4) ease of introducing stable genetic modifications. ES cells can also be used to perform screens of gene function *in vitro*, which provides several advantages over *in vivo* screens. In such genetic screens, there is a reduced need for producing and maintaining large colonies of mice to study a genotype of interest. Mutations are usually produced as heterozygous lesions in the genome, but can be converted to homozygosity *in vitro* (Mortensen et al. 1992). Once an interesting phenotype has been identified *in vitro*, ES-derived mutant mice can be made rapidly to verify that the phenotype occurs *in vivo*. Moreover, many mutants are embryonic lethal, making it impossible to study gene function at later developmental stages. Embryonic lethality can often be bypassed *in vitro*, since embryonic lethality does not necessarily equal cell lethality (Tsai et al. 2000, Ge et al. 2002). ES cells offer an extremely powerful approach to systematically study the genetics of human nervous system development.

Limitations of ES cells

One limitation of ES cells for neural repair strategies is illustrated by the close relationship between ES, EG and EC cells. As noted above, ES cells are capable of forming teratomas and teratocarcinomas when grafted in their undifferentiated state. Therefore, neural transplantation strategies should be designed with this caveat in mind (Freed, 2002, Brustle et al. 1997, Bjorklund et al. 2002). Genetic stability is often thought of as an advantage of using ES cells, but little work has been done to prove this point. Recently, it was shown that mouse ES cells are less susceptible to point mutagenesis *in vitro* than most other cell types, yet are more prone to lose entire chromosomes than their somatic counterparts (Cervantes et al. 2002). Therefore, it will be important to examine the genetic stability of ES cells in culture, particularly for the primate lines of potential therapeutic use. ES cell lines have only been isolated from a limited set of species including mouse, monkey and human ES cells (see Challenges for the Future section for discussion), so only these species can be studied through ES cells. The use of human ES cells has raised many ethical questions that are hotly debated in the scientific and general public. Currently, there is a debate among scientists about the ethics of creating a mouse-human hybrid embryo (DeWitt, 2002). A chick-human hybrid embryo has already been described after transplanting undifferentiated human ES cells at the stage of organogenesis (Goldstein et al. 2002). One final limitation (that is a fascinating biological problem) is that we do not clearly understand the molecular codes required for controlling and directing the differentiation of ES cells into all of the hundreds of individual specific cell types that form the brain. Understanding this language will provide unique insights into brain development and provide the basis for translational studies aiming at cell replacement in the diseased brain. In the absence of this knowledge, ES-derived cultures inevitably will contain contaminating cell types that interfere with the interpretation of both translational and basic studies.

Producing ES cell lines

ES cell lines are derived by coculturing isolated blastocysts with fibroblast feeder layers (Figure 1). The inner cell mass of the blastocyst-stage embryo contains the progenitor cells that give rise to ES cells, though the precise lineage relationship is not yet known. Typically, preimplantation blastocysts are washed out of the uterus, and are then placed directly on a fibroblast feeder layer. In some protocols, the inner cell mass is isolated and separated from other cell types prior to coculture with the feeder layers. Colonies with various morphologies grow on the feeder layer, and ES cell colonies are identified based on morphology and gene expression (Figure 1). It is not known how fibroblasts maintain mammalian ES cells in an undifferentiated state. Mouse embryonic fibroblasts (MEFs) are normally used to derive ES cell lines from all species, but human ES cells have been maintained on human embryonic or adult fibroblasts. This is a welcome step toward clinical applications of ES cells, since coculturing human ES cells with mouse feeders might allow for pathogen transfer across species (Richards et al. 2002). Leukemia inhibitory factor (LIF) is produced by fibroblast feeders and inhibits the spontaneous differentiation of mouse ES cells. However, primate ES cells also depend on factors produced by fibroblast feeders (Thomson et al. 1995, Thomson et al. 1998, Xu et al. 2001) but do not respond to LIF (Thomson et al. 1998). Identification of the elusive factors produced by fibroblasts will be essential for our understanding of human ES cell biology. The addition of basic fibroblast growth factor (bFGF or FGF2) in conjunction with a serum-free medium can substitute for serum-containing media (Xu et al. 2001). After isolation and propagation, ES cells can be passaged without the feeder layer. Whereas mouse ES cells are commonly cultured on tissue culture plastic coated with gelatin in defined medium supplemented with LIF, primate ES cells can be maintained on matrigel or laminin in fibroblast-conditioned medium (Xu et al. 2001).

Methods for *in vitro* neural differentiation

Embryoid body formation

The classic method for inducing neural differentiation from ES cells involves the formation of embryoid bodies (EB). EBs are formed by forcing undifferentiated ES cells to grow as aggregates in suspension. As the EB grows,

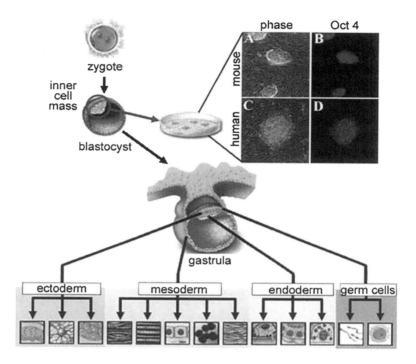

Figure 1. Deriving ES cell lines. After fertilization, the single-celled zygote develops into the preimplantation blastocyst within 3-10 days depending on the species. Within the blastocyst is a collection of cells called the inner cell mass (ICM). The ICM later develops into the gastrula, ultimately becoming all tissue types in the adult organism. If the blastocyst is cultured at the appropriate developmental stage, ES cell lines can be derived from the ICM by co-culture with fibroblast feeder cells. Both mouse (**A**) and human (**C**) ES cell colonies can be identified based on morphology. Putative ES cell lines can be confirmed by examining gene expresssion. For example, mouse (**B**) and human (**D**) ES cells express Oct 4, one of the genes expressed exclusively by ES cells.

cell-to-cell interactions and the release of soluble signaling molecules in the aggregate cause ES cells to differentiate. Derivatives of all three germ layers are found in EBs (Doetschman et al. 1985, for review, see Weiss and Orkin, 1996), and EB culture serves as a screening tool to demonstrate pluripotency of putative ES cell lines. Neural cells are not efficiently produced under standard EB conditions, leading to a number of variations on this protocol that enhance neural induction, or that select for and expand neural precursors in EBs.

Retinoic acid

One method for increasing the neural population from EBs is exposure to retinoic acid (RA) (Bain et al. 1995; for an excellent review and protocol, see

Bain et al. 1998). RA is a vitamin A-derivative released by the mesoderm during development, and has a strong neural inducing and neural patterning effect (for review, see Maden 2002). Similar strategies were proposed by a number of groups (Franchiard et al. 1995, Strubing et al. 1995). In the original 4-/4+ protocol (Bain et al. 1995), EBs are grown for 4 days in standard medium before being placed in medium containing 0.5 µM RA for an additional 4 days in suspension culture (Figure 2). RA-treated EBs are mainly composed of neural progenitors that express nestin and sox1 among other neural markers (Liu et al. 2000, Wichterle et al. 2002). After 8 days in suspension culture, EBs containing neural progenitors are allowed to attach to tissue culture plastic. Over the next few days, aggregates flatten out and form a monolayer of flat non-neuronal cells and cells that have a neuronal appearance (~40%). Most of the glial cells produced are astrocytes (Strubing et al. 1995, Fraichard et al. 1995, Angelov et al. 1998), although oligodendrocytes and microglia are present in lower numbers (Fraichard et al. 1995, Angelov et al. 1998, Liu et al. 2000). All three protocols primarily yield glutamatergic and GABAergic neurons, as determined by immunohistochemistry and electrophysiology (Bain et al. 1995, Fraichard et al. 1995, Strubing et al. 1995, Finley et al. 1996). The Bain protocol yields neurons that are roughly 70% glutamatergic, 25% GABAergic and 5% glycineric (Gottlieb, 2002).

The presence of mostly glutamate and GABA neurons is somewhat paradoxical, since RA caudalizes the nervous system by inducing the Hox gene cascade that leads to formation of the hindbrain and more posterior cell fates (Maden, 2002). Glutamate and GABA neurons are typically found in the most anterior regions of the brain. Two different studies using slightly different protocols have shown that RA-exposed EBs can produce motor neurons, a cell type produced in the posterior nervous system. Renoncourt et al. (1998) described the derivation of ES-derived motor neurons *in vitro* using a 2 µM RA (2-/7+) protocol. They also provided a more detailed molecular characterization including the expression of region-specific transcription factors demonstrating the presence of caudal and ventral CNS cell types, including hindbrain and spinal cord motor neurons and interneurons. A large proportion of motor neurons were thought to correspond to cranial regions since they expressed Phox2b. Islet1-negative presumptive interneurons were also identified. In contrast, Wichterle et al. (2002) reported that exposing EBs to 2 µM RA (2-/3+) leads mainly to rostral spinal cord motor neurons (Phox2b-negative) and interneurons that lack dorsal/ventral specification. EB-derived neural precursors could be converted to ventral cell types by adding SHH to cultures. Future studies are necessary to define the neuronal subtypes produced by the various EB-RA protocols. However, it is interesting to note that RA-treated EBs appear to produce either rostral spinal cord motor neurons (Wichterle et al. 2002), or the rostral

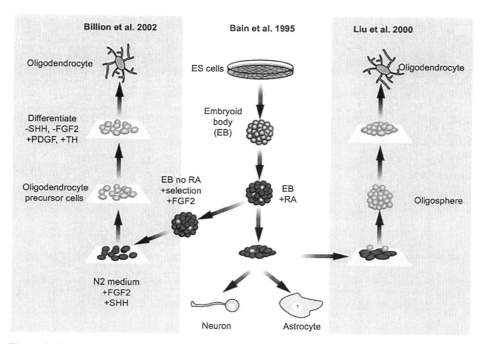

Figure 2. Retinoic acid-enhanced neural cell production from embryoid bodies (EBs). Center.
In the Bain et al. protocol (1995), ES cells are removed from adherent culture and grown as embryoid bodies (EBs) before exposure to retinoic acid (RA). Exposure to RA increases the number of neural precursors present in the EB. After RA exposure, EBs are placed into adherent culture conditions and neural precursors emerge from EBs before forming primarily neurons and astrocytes. Two protocols have been developed to enhance oligodendrocyte production from this basic protocol. **Left side.** Billion et al. 2002 used genetic selection to produce a purified population of neural precursors after removing the RA in suspension culture. Purified neural precursors were subsequently grown in adherent culture in the presence of FGF2 and SHH, allowing oligodendrocyte precursor cells (OPCs) to emerge. Extended culture in PDGF/thyroid hormone without SHH/FGF2 caused OPCs to mature into oligodendrocytes. **Right side.** Liu et al. 2000 collected the loosely-adherent cells produced in the Bain protocol, and propagated them as oligospheres. These oligospheres differentiated into oligodendrocytes after conversion back to adherent culture. N2 medium is described in Bottenstein and Sato (1979).

hindbrain cranial motor neurons (Renoncourt et al. 1998). These results suggest that additional signals might be required to derive cell types specific to more caudal spinal cord regions. The caudalizing effects of RA are not limited to EB-derived neural cells: ES cells differentiated into neurons by coculture with stromal cells are also caudalized into rostral spinal cord motor neurons after exposure to RA (Wichterle et al. 2002; see Stromal-derived inducing activity section below).

Little is known about the effects of RA on EBs derived from human ES cells. Schuldiner et al. (2001) reported that exposing human EBs to either nerve growth factor (NGF) or 1 μM RA increased the number of presumptive neurons compared with untreated control EBs (39% and 52% respectively versus 21%). RT-PCR analysis demonstrated gene expression patterns expected of serotonergic and dopaminergic cells, but no further characterization was performed. In contrast, data by Carpenter et al. (2001) indicate that the standard Bain protocol is not effective for human cells and proposed a modified protocol, where EBs are formed for 4 days in the presence of 10 μM RA prior to plating on fibronectin in a complex, defined medium. Dividing neural precursors were detected 3 days after plating EBs, and these precursors were capable of differentiating into mature neurons and astrocytes after extended culture. Neuronal subtype analysis revealed that 3% of these neurons were positive for tyrosine hydroxlyase, the rate limiting enzyme in the synthesis of dopamine, suggesting the derivation of catecholaminergic neurons.

With minor modifications, the Bain protocol has also been adapted to derive purified populations of glial cells from mouse ES cells (Figure 2). Liu et al. (2000) altered the 4-/4+ protocol to produce a high number of oligodendrocytes. The initial stage of the protocol (4-/4+) is identical to the Bain protocol. However, in the next step the loosely-adherent cells were selectively collected and propagated as free-floating spheres. These structures give rise to high percentages of oligodendrocytes or oligodendrocyte precursor cells and are termed oligospheres (Liu et al. 2000), a name first defined for similar structures derived from primary neural precursors (Avellana-Adalid, 1996, Zhang et al. 1998).

Martin Raff and colleagues also used RA-exposed EBs to produce mouse oligodendrocytes (Figure 2; Billion et al. 2002). They harnessed ES cell genetics to select for purified neuroepithelial precursors through positive and negative selection. These genetic tricks allowed synchronous differentiation to FGF2-dependent neural precursors. Exposing these precursors to SHH increased the percentage of neural cells that express oligodendrocyte precursor markers from 12% to 40–85%. Extended culture in platelet-derived growth factor (PDGF)-AA and thyroid hormone led to a further maturation of oligodendrocytes in roughly half of the cells.

HepG2 conditioned medium

A second method to increase the production of neural precursors from EBs is to expose them to the conditioned media of a hepatocarcinoma cell line (HepG2) (Rathjen et al. 2002). When EBs are formed in HepG2 conditioned medium (the modified EBs are called EBMs), cells in EBMs adopt a neuroectodermal fate (Figure 3). Similar to RA-treated EBs, EBM aggregates do not express genes found in mesoderm and endoderm and primarily express neural markers such as sox 1, 2, and nestin. Dissociated EBs and EBMs were subjected to flow cytometry using antibodies against NCAM, a cell-surface adhesion protein widely expressed in the nervous system (Rutishauser et al. 1992). Almost all of the cells in EBMs express NCAM on the cell surface, whereas less than half of the cells in control EBs express NCAM. After extended suspension culture, ~80% of the EBM aggregates contained neurofilament-positive cells that have a neuronal morphology, compared to ~20% in the control EBs. In addition to neuronal differentiation, EBM-derived cells can produce neural crest or glial fates under appropriate culture conditions. Characterization of the expression of region specific genes in EBMs was inconclusive, yet was most similar to unspecified anterior neuroectoderm. It will be essential to further characterize EBM-derived neuronal subtypes. However, the preliminary data suggest that HepG2 conditioned medium, unlike RA treatment, might promote neural induction without exerting an obvious patterning effect. The factors present in HepG2

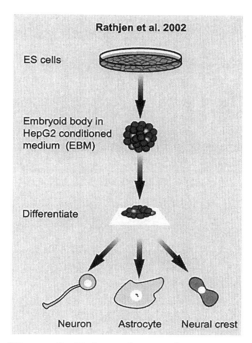

Figure 3. Enhanced neural precursor production from embryoid bodies (EBs) in the presence of HepG2 conditioned media. Rathjen et al. (2002) converted EBs grown in the presence of conditioned media from a hepatocarcinoma cell line (HepG2) to aggregates highly enriched in neural precursors (EBMs). EBMs primarily form neurons after adherent culture, but can be coaxed into neural crest derivatives or astrocytes through the manipulation of culture conditions.

conditioned medium have yet to be purified. Some evidence indicates that at least two separable components are responsible for this activity, and one of these components is apparently a known extracellular matrix molecule (Rathjen et al. 2002).

Multistep EB protocols

A third strategy to harness EBs for neural cell production, called the five-step protocol, selects for and propagates mouse CNS precursors from EBs without requiring the addition of exogenous factors such as conditioned media or retinoic acid (Figure 4; Okabe et al. 1996). Undifferentiated mouse ES cells (stage 1) are first grown in suspension allowing EB formation for 4 days (stage

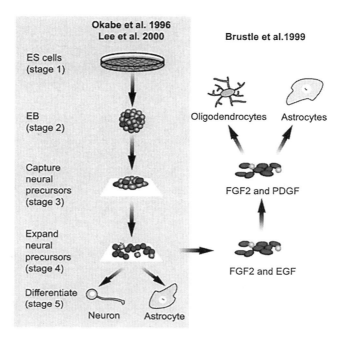

Figure 4. Selective isolation and propagation of ES-derived neural precursor cells.
Left side. Okabe et al. (1996) and Lee et al. (2000) described the 5-step protocol, a method
that does not require adding neural-inducing factors to developing embryoid bodies (EBs).
ES cells (stage 1) are first grown as EBs (stage 2), before plating under conditions that
favor isolation of neural precursors (stage 3). CNS precursors are expanded with FGF2
(stage 4) before differentiation into astrocytes and neurons by FGF2 withdrawal (stage 5).
Right side. Brustle et al. (1999) modified the culture conditions of the 5-stage protocol to
enhance oligodendrocyte production. Sequential culture in FGF2, FGF2/PDGF, and FGF2/
thyroid hormone enhanced astrocyte and oligodendrocyte production.

2). CNS precursors are selected (stage 3) for 8 days on tissue culture plastic in
serum-free medium and in the presence of high concentrations of fibronectin.
The precursors are subsequently expanded (stage 4) for 6 days in medium containing FGF2 and laminin before differentiation (stage 5) into neurons induced
by FGF2 withdrawal.

Extensive characterization of gene expression in this protocol (Lee et al.
2000, Kim et al. 2002) revealed a close relationship in the temporal sequence of
gene expression between *in vitro* and *in vivo* development. For example, Otx1
and 2 are expressed in the developing neuroectoderm and are essential for normal forebrain and midbrain development. *In vivo*, Otx2 is expressed very early
in the epiblast before being restricted to anterior neuroectoderm, whereas Otx1
is expressed in neuroectoderm in the dorsal telencephalon. Similar to the developing nervous system, Otx2 is present in undifferentiated ES cells and is ex-

pressed at lower levels in stages 2 and 3, whereas Otx1 is not expressed until stage 3. Several genes known to control midbrain and hindbrain development were detected during stages 3 and 4. At this stage, the cell population is relatively uniform and most of the cells express nestin protein.

The efficiency of neuronal differentiation varies between 50% (Westmoreland et al. 2001) to 72% (Lee et al. 2000) as assessed by the expression of β-Tubulin III (TuJ1). Among ES-derived neurons, ~7% express tyrosine hydroxlyase, the rate limiting enzyme in the production of dopamine (Lee et al. 2000, Kim et al. 2002), while most of the other neurons are GABAergic (Westmoreland et al. 2001) or glutamatergic (Okabe et al. 1996). Interestingly, unlike the retinoic acid protocols, no cholinergic neurons have been reported with this protocol (Okabe et al. 1996) and the effects of RA exposure on the various stages have not yet been defined.

The proportion of dopaminergic and serotonergic neurons produced in this multistep EB protocol can be increased by the application of SHH and FGF8 at the neural precursor stage (stage 4), or by exposure to ascorbic acid (AA) during differentiation (stage 5; Lee et al. 2000). The effects of SHH and FGF8 on early ES-derived neural precursors are thought to mirror developmental patterning events first defined in explant cultures (Ye et al. 1998) – FGF8 inducing isthmic organizer (midbrain/hindbrain) fates and sonic hedgehog providing the ventralizing signal. The mechanism of AA action on dopamine neuron differentiation is still unknown. However, it is interesting to note that its effects are shared among ES-derived and primary midbrain precursor cells (Yan et al. 2001). The dopaminergic nature of the TH-positive neurons was confirmed by the lack of DBH and GABA expression, detection of synaptic dopamine release by HPLC and by extensive electrophysiological recordings in TH-positive cells (Lee et al. 2000, Kim et al. 2002). A further demonstration of how developmental studies can be applied to ES-cell culture is the application of FGF4 instead of FGF2 at stage 4. Similar to the work in explants (Ye et al. 1998), FGF4 exposure at stage 4 dramatically increases the ratio of serotonergic versus dopaminergic neurons (Kim et al. 2002).

Directed *in vitro* differentiation via epigenetic factors can be complemented with the expression of neuron subtype specific transcription factors. This strategy was utilized to further increase DA neuron production using the multistep EB protocol by generating mouse ES cells stably overexpressing Nurr1, a transcription factor that plays a role in generation and maturation of midbrain dopaminergic neurons during mouse development (Zetterstrom et al. 1997). In the absence of any extrinsic factors, Nurr1 overexpression in mouse ES cells increased the percentage of TH-positive neurons obtained at stage 5 from 5 to 50%. In combination with patterning factors such as SHH and FGF8, Nurr1-ES cells generate up to 78% TH-positive neurons. (Kim et al. 2002). These Nurr1-

ES derived TH-positive neurons appear to selectively express markers of midbrain dopaminergic neurons, and transplantation of these neurons restored function in Parkinsonian rats. In addition, the authors provided electrophysiological evidence of mature dopamine neuron types *in vivo*. These studies are an impressive demonstration how developmental signals can be harnessed to direct the differentiation of ES cells into appropriate neuronal subtypes and how such technology may impact therapeutic strategies.

Glial cells were selectively produced using a modified multistep EB protocol (Brustle et al. 1999). EB-derived neural precursors were expanded and passaged in medium containing FGF2, FGF2 with EGF, and finally FGF2 and PDGF. These conditions selected for ES-derived glial precursors that have a round or bipolar morphology and express the A2B5 antigen. After growth factor withdrawal, glial precursors differentiated into astrocytes (~36% GFAP-positive cells) and oligodendrocytes (~38% O4-positive cells). In addition to the expression of cell type specific markers and morphologies, grafted ES-derived oligodendrocytes were capable of remyelination *in vivo* (see *In vivo* section below).

Zhang et al. (2001) were the first to adapt the multistep EB protocol to the distinct developmental timing of human ES cells. In this protocol ES cells were aggregated to form short-term EBs (4 days) prior to plating on tissue culture treated plastic. EB-derived neural precursors were expanded in defined medium with FGF2 for 8-10 days forming neural rosettes, structures of layered neuroepithelial cells arranged in a concentric fashion and first described in mouse embryonal carcinoma cell cultures maintained in N2 medium without FGF2 (Darmon et al., 1981). These neural rosettes were enzymatically isolated and grown in suspension for extended periods. Similar to neurospheres derived from primary neural precursor cells, such human ES-derived neural spheres could be expanded and propagated in serum-free medium supplemented with FGF2. However, unlike neurospheres, EGF appears to be ineffective for propagating human ES-derived neural precursors. Terminal differentiation of these precursor cells yielded glutamatergic and GABAergic neurons as well as astrocytes.

Multistep EB differentiation protocols have also been adapted for the neural differentiation of autologous ES cells derived via nuclear transfer (ntES cells; Wakayama et al. 2001). These studies demonstrated the efficient derivation of midbrain dopamine neurons from ntES cells derived from adult somatic nuclei including cumulus cells or cells derived from tail tip biopsies. However, this study also revealed significant variability in the efficiency of dopamine neuron generation among various ntES and ES cells using the EB multistep protocol. This apparent lack of robustness could provide a significant hurdle for the application of the multistep EB protocol to a wider range of ES and ntES cell types (e.g., for screening libraries of mutant ES cells).

Default induction

Work in the frog Xenopus laevis in the early 1990's led to the formulation of the neural default hypothesis. According to this model, the induction of neural fate from ectoderm occurs in the absence of instructive signaling. The default hypothesis ascribes the neural "inducing" activity of the organizer, as defined by the pioneering experiments of Spemann and Mangold nearly 80 years ago (Spemann and Mangold, 1924), to the expression of BMP antagonists such as follistatin (Hemmati-Brivanlou et al. 1994), chordin (Sasai et al. 1995) and noggin (Smith and Harland, 1992). This model predicts that in the absence of BMP signaling, early ectodermal cells will spontaneously differentiate into neural progeny (see review by Munoz-Sanjuan and Brivanlou, 2002). While aspects of this hypothesis remain controversial for mammalian nervous system development, there is considerable evidence that blocking the BMP pathway in mouse ES cells is required for neural differentiation. Transfecting undifferentiated ES cells with the BMP antagonists chordin or noggin causes neural differentiation (Gratsch and O'Shea, 2002). An alternative method for blocking paracrine BMP signaling is to culture cells in suspension at very low densities (Tropepe et al. 2001). Under these conditions ES cells differentiate into neurosphere-like aggregates when grown in defined medium in the presence of LIF and in the absence of any cell-cell interactions or any factors derived from other germ layers. Whereas ~70% of the ES cells do not survive the transition from adherent culture to low-density, suspension culture, most of the remaining cells express nestin. A small fraction of these nestin-expressing cells (0.2%) acquire neural stem cell characteristics. These ES-derived neural precursors can self renew or differentiate into neurons and glia. Dissociating primary colonies to form secondary colonies requires FGF2 in addition to LIF. Gene expression analysis of such spheres confirmed that epidermal, mesodermal, and late endodermal genes are downregulated. However, unexpectedly, cells expressed the early endodermal marker GATA4. While expression of some forebrain and hindbrain markers was reported, no data was provided on neuronal subtype composition.

Recently, another direct route from ES cells to neural precursors has been described (Ying et al. 2003). Using a cell line that expresses green fluorescent protein (GFP) upon neural induction, they found that plating mouse ES cells onto gelatin-coated, tissue culture plastic in a defined medium without LIF converts ES cells to neural progenitors (see Darmon et al., 1981 for the first description of this using mouse embryonic carcinoma cells). One day after plating, cells still have characteristics of ES cells, but begin conversion to neural precursors on the second day after induction. Cells with the morphology and gene expression of neurons are found 4 days after induction, and under optimal conditions ~60% of the cells become neurons. Under standard conditions, the generation of large numbers of GABAergic neurons and a few TH-positive presumptive dopamine neurons were detected. The proportion of dopamine neurons was increased by exposing cultures to SHH and FGF8 indicating that ES-derived neural progenitors generated under these conditions are receptive to patterning by exogenous factors. Minor modifications to the protocol allowed for the derivation of both astrocytes and oligodendrocytes. Whereas these data appear to confirm the default hypothesis, the work further demonstrated that inhibition of FGF signaling via expression of a dominant negative FGF receptor abrogates the neural differentiation response (Ying et al. 2003).

Direct neural induction from primate ES cells has not been reported. However, direct induction of parthenogenetic primate stem cells without EB formation has been used in combination with a multistep patterning protocol to generate specific neural cell types (Cibelli et al. 2002). Pluripotent primate stem cells were created via parthenogenetic activation of an unfertilized egg cell, and have all the major characteristics of embryonic stem cells. They express high levels of telomerase, can be propagated indefinitely *in vitro*, and give rise to derivatives of all the three germ layers *in vitro* and *in vivo*. Such pluripotent parthenogenetic cells were differentiated into neural fates by plating on gelatin in a defined medium with FGF2. Neuronal and astrocytic differentiation was obtained and dopaminergic neurons could be generated at high efficiencies (Cibelli et al. 2002). Future studies will have to address whether these findings apply only to parthenogenetic cells or are generally applicable to primate ES cells.

Stromal-derived inducing activity (SDIA)-positive factors for neural induction

In contrast to the default model, there is evidence that positive signals are required for neural induction in addition to BMP inhibition (for review, see Stern, 2002). It has long been hypothesized that mesodermal factors acting on the overlying ectoderm are essential for neural induction. *In vitro*, ES cells are neurally-induced when co-cultured with mesodermal bone marrow stromal cells. The stromal cell line (PA6), traditionally used to maintain hematopoietic stem cells in long-term culture, neuralizes mouse and monkey ES cells with high efficiencies (Kawasaki et al., 2000; Kawasaki et al., 2002). The factor(s) responsible for neural induction/differentiation remain to be identified, but based on preliminary studies the inducing activity (named Stromal-Derived Inducing Activity, or SDIA) probably consists of a transmembrane protein or a soluble protein tightly bound to the cell surface. A number of stromal cell types may express SDIA, although only PA6 has so far been reported to exhibit this activity under normal conditions. Several additional cell types that do not generally cause neural induction can be converted to neural-inducing cell types after paraformaldehyde fixation. This result led to the hypothesis that at least two mechanisms guide neural induction: a positive and a negative factor. Whereas PA6 would express only the positive factor, other cell lines would express both positive and negative factors. The negative factor could be eliminated by paraformaldehyde fixation. These results also support the notion that SDIA is a cell-surface factor, since a fixed cell should not produce a soluble factor. Intriguingly, this method produces a large number of midbrain dopamine neurons without the addition of SHH and FGF8. It is estimated that about 40-50% of the ES cells become TuJ1-positive neurons (Morizane et al. 2002, Kawasaki et al. 2000). Among TuJ1-positive cells, ~20-30% are positive for TH (Morizane et al. 2002, Kawasaki et al. 2000). These TH-positive cells are composed of midbrain dopamine neurons since they lack DBH, express Nurr1 and Ptx3 (Kawasaki et al. 2000), and exhibit potassium-evoked dopamine release. Other neuron types are 18% GABAergic, 9% cholinergic, and 2% serotonergic (Kawasaki et al. 2000). Cholinergic motor neuron production can be further enhanced in SDIA-induced mouse neurons by the addition of RA (Wicherele et al. 2002). PA6 cells do not express the BMP antagonists, making it unlikely that stromal cells cause neural induction by simply blocking BMP signaling. Furthermore, blocking BMP receptors was not sufficient to induce neural differentiation (Kawasaki et al. 2000). Identifying the molecular nature of SDIA is an important step for our understanding of neural induction in ES cells. Further studies are also needed to address whether SDIA provides positional information that could explain the high propensity of midbrain dopamine neuron generation.

In vitro neural cell function

Proving that an ES-derived neural cell *in vitro* is equivalent to a primary neuron, astrocyte, or oligodendrocyte is a difficult task and many approaches have been used to this end (see review by Perrier and Studer, 2003, on *in vitro* characterization of midbrain dopamine neurons). Cellular morphology at the light and electron microscopic level can be used to identify specific neural cell types. Synapse formation in neurons and maturation of oligodendrocytes, particularly the formation of myelin, are morphological events that can be readily observed using transmission electron microscopy and provide very strong evidence for obtaining proper phenotypes. The presence of enzymes necessary for the biosynthesis or uptake of specific neurotransmitters is another essential criterion for the identification of ES-derived neuronal subtypes. However, such biochemical characterization is not sufficient for a full description of neuronal subtypes. For example, dopaminergic neurons exist in 12 different regions in the brain and the properties of the various dopaminergic neuron subtypes are distinct. Therefore, identifying the presence or absence of region-specific transcription factors is required to define subclasses of neurons expressing a given neurotransmitter. Functional tests for the *in vitro* characterization of ES-derived neurons include measuring evoked neurotransmitter release using HPLC or electrophysiology to demonstrate tetrodotoxin-sensitive action potentials and firing patterns and pharmacological definition of receptor subtype composition.

The demonstration of astrocyte function is challenging as their *in vivo* function remains a hotly debated area of research (Matthias et al. 2003, for review see Fields and Stevens-Graham, 2002). It is also not clear how much regional specification impacts on astrocyte function. Different phenotypes have been reported for astrocytes derived from different regions of the nervous system, but the underlying changes in gene expression have not been well documented. Conversely, oligodendrocyte function is well known and electron microscopic evidence of myelin deposition around host axons is good evidence of complete differentiation. These morphological studies can be complemented by functional measurements of the action potential velocity in axons.

In vivo neural cell function

The finding that ES cells are capable of developing into neural cells *in vitro* outside the environment of a developing embryo is remarkable. While the *in vitro* evidence suggests that ES-derived neural cells are virtually identical to their *in vivo* counterparts, the extent to which they are capable of participating in and responding to a normal or injured brain remains to be addressed. Transplantation studies challenge the properties of ES-derived neurons for survival, migration, differentiation and synaptic integration into the host circuitry. Grafted oligodendrocytes must identify exposed axons and produce myelin to permit saltatory conduction and astrocytes need to integrate into existing glial networks. An important variable in all transplantation studies is the developmental stage of the host brain. Whereas the developing brain contains signals that will guide both endogenous and grafted precursors cells into appropriate fates and functions the adult brain lacks most of these cues, particularly those important for neurogenesis and axon guidance.

One obvious experiment to test the effects of the environment on ES cell development is the transplantation of undifferentiated ES cells. However, as noted earlier, undifferentiated ES cells introduced into adult animals tend to form teratomas, particularly when introduced in large numbers. Smaller numbers of ES cells grafted into the brain can differentiate into neural cells that are capable of leading to functional recovery in a Parkinson's disease model (Bjorkland et al. 2002). These findings can be interpreted that ES cells spontaneously differentiate into neural cells at low enough concentrations similar to the model proposed for ES cells *in vitro* (Tropepe et al. 2001). However, even grafts with small cell numbers produced tumors in a significant percentage of all animals and ES differentiation appeared to occur independent of environmental cues. In fact, ES cells grafted into the kidney capsule differentiated into the same neural cell types as those grafted into the CNS (Deacon et al. 1998).

Transplantation studies revealed that the ES-derived primitive neural stem cells (Tropepe et al. 2001) may have a more broad developmental potential than nervous system-derived neural stem cells. Chimeric mice could be created with the ES-derived stem cells, but not with neural stem cells isolated from the nervous system. These results in addition to their *in vitro* findings (see above) prompted the authors to hypothesize that the ES-derived neural stem cell is a more primitive type of stem cell.

In contrast to these studies with more primitive cell types, most grafting to the CNS has been performed with ES-derived cells at the neural precursor cell stage and implanted into the developing CNS. ES-derived mouse neural precursors differentiate into all three major lineages when transplanted in utero into the ventricles of an embryonic rat (Brustle et al. 1997). Similar results have been reported for human ES-derived neural cells introduced into the brain of neonatal rats (Zhang et al. 2001, Reubinoff et al. 2001). A remarkable finding of most of these studies is that ES-derived cells were often indistinguishable from the surrounding host tissue. Only the presence of graft-specific makers allowed proper identification of donor versus host cells. Axons from ES-derived neurons projected long distances, following normal routes used by host neurons. Oligodendrocytes also participated normally in development, migrating to white matter tracts where they myelinated passing axons. Interestingly, the ES-derived progeny were produced in the same temporal sequence found during development, with neurons appearing prior to glial cells (Brustle et al. 1997). Neuroepithelial precursors formed small rosette-like structures near the ventricles, and continued to proliferate at least two weeks after transplantation.

In the adult CNS, glial cells continue to be born throughout life in most regions of the brain and primary neural precursors can readily adopt glial fates *in vivo*. Animal models of demyelination have therefore obtained particular consideration for testing the potential of ES-derived neural precursors. For example, ES-derived glial precursors implanted into a rat model of Pelizaeus-Merzbacher syndrome exhibit widespread differentiation into oligodendrocytes with electron microscopic evidence of remyelination (Brustle et al. 1999). Similar results were obtained with ES-derived precursors grafted into the spinal cord after chemical demyelination or in myelin-deficient shiverer mice (Liu et al. 2000). However, other studies indicated that mechanical destruction of a brain area might allow for the differentiation of grafted of ES-derived precursors into neurons as well as glial cells (McDonald et al. 1999).

While limited neuronal differentiation is observed after grafting ES-derived neural precursors, efficient neuronal cell replacement might require more complex *in vitro* differentiation steps prior to implantation. Wichterle et al. (2002) demonstrated that mouse ES-derived motor neurons and interneurons could participate in the development of the chick spinal cord. Motor neurons transplanted into a developing chick spinal cord migrated to the ventral spinal cord, while mouse-derived interneurons were distributed along the dorsal/ventral axis of the chick spinal cord. These predifferentiated neurons grew axons into nerve roots following the normal pathways and appeared to contact target muscles. Correctly aligned pre- and post-synaptic specializations were detected between motor neurons and the muscle target, but no functional data was provided to conclusively demonstrate the contribution of ES-derived axons to muscle function. Motor neuron replacement in the adult CNS will be a vast challenge that needs to address cell migration, differentiation, axonal outgrowth and pathfinding to appropriate targets in addition to the derivation of proper phenotypes *in vitro*. A more realistic goal for neuronal cell replacement is the treatment of Parkinson's disease. Extensive experience with fetal tissue grafts has identified many of the parameters required for successful outcomes (Lindvall et al. 2000) and obtaining a proper source of dopamine neurons has remained a limiting factor in developing this therapy (Dunnett, 1999).

Kim et al. (2002) predifferentiated ES-derived neural precursors to adopt a midbrain dopaminergic cell fate before implanting them into an animal model of Parkinson's disease. Transplanting these ES-derived neurons led to the survival and integration of dopaminergic neurons into the brain. The behavior of the Parkinsonian animals improved, and electrophysiological experiments showed that transplanted dopamine neurons retained midbrain-specific synaptic properties. This study has provided conclusive evidence that ES-derived neural progeny can functionally repair deficits in an animal model of a neural disease. Despite these advances, more work is required to assure the safety and long-term function of such grafts and to identify diseases amenable to such approaches.

Challenges for the future

Embryonic stem cell technology has played a crucial role in the development of modern mouse genetics. The availability of complete maps of the human and mouse genome offer enormous possibilities for the next revolution in ES cell technology. The application of genomic technologies will allow a much more precise characterization of the molecular events that guide neural induction, regional subtype specification, terminal neuronal and glial differentiation as well as many maturation events such as axonal pathfinding. Initial *in vitro* ES cell differentiation studies heavily relied on developmental biology to design rational directed neural differentiation protocols (Lee et al. 2000, Kim et al. 2002, Wichterle et al. 2002). The striking similarities in the molecular signals that control normal CNS development *in vivo* and differentiation of ES cells *in vitro* suggest that future efforts may take advantage of *in vitro* ES cell differentiation systems. These genetic screens could identify important genes and predict the *in vivo* gene function. Unlike *in vivo* studies, ES cells can be easily adapted for high-throughput assays. Such ES cell based genome-wide approaches could revolutionize our understanding of mammalian brain development. The technology to perform such genome-wide functional studies is still being developed for ES cells. However, early attempts using genome-wide RNAi interference in *C. elegans* (Ashrafi et al., 2003) to screen for body weight regulation have provided glimpses of the power of such an approach. Another unique potential of ES cell technology is the presence of thousand of specific heterozygous mutant ES cells already established. These lines can be readily converted to homozygosity *in vitro* for performing functional screens. Moreover gene trapping efforts in ES cells (Gossler et al. 1989, Zambrowicz et al. 1998) have been scaled up to a level where ~ 200,000 individual ES cell lines are available with unique insertions in specific genes.

Despite these exciting perspectives, significant additional work is required to refine *in vitro* differentiation protocols. Whereas the derivation of certain neuronal and glial subtypes from mouse ES cells is becoming routine, the derivation of other cell types has still not been reported. Furthermore the protocols for directed neural differentiation of human ES cells are still in their infancy and not all the conditions developed for mouse ES cells can be readily translated for the human ES cell applications. In particular, the derivation of oligodendrocytes has proven to be unexpectedly difficult in human cells (Studer, 2001), whereas oligodendrocytes can be readily obtained from mouse ES cells. The generation of pure populations of differentiated cells types is essential not only for the development of precise *in vitro* assays but also for the potential use of ES-derived neurons and glia for cell therapy. Purification of ES-derived neural progeny via promoter-driven selection is an additional strategy that needs to be further developed in our attempts to obtain homogenous populations of cells.

Efforts are also being undertaken to develop ES cell lines for all model organisms. Findings in developmental biology often differ between various model organisms, and it is not always clear whether these results reflect differences among species or are due to the different assays used in these species. ES cells might provide a common assay for answering these questions. There is evidence for ES-like cells derived from horse (Saito et al. 2002), chick (Pain et al. 1996, but see also Soodeen-Karamath and Gibbins, 2001), mink (Sukoyan et al., 1993), pig (Chen et al. 1999), zebrafish (Ghosh and Collodi, 1994, Sun et al. 1995), and a relative of zebrafish called medakafish (Hong et al. 1996; Hong et al. 1998). Despite intensive efforts, ES cells have not yet been obtained from rats (Buehr et al., 2003) as well as many other species. The lack of rat ES cells is unfortunate in view of the important role this animal has played in neurobiology. In the case of human studies, ES cells are the only system that offers systematic experimental access to the earliest stages of nervous system development.

Human ES cells are also an important tool for the study and treatment of neural disease. The derivation of specialized cells such as midbrain dopamine neurons for cell transplantation in Parkinson's disease is a powerful approach that has received a lot of public attention. However, ES cells can also be used to model neural diseases. Such an approach could be accomplished not only from animals but ultimately from humans using early mutant embryos left over from *in vitro* fertilization attempts of couples with a high risk for genetic disease. In such cases, embryos are screened for the genetic defect and only the normal embryos are implanted while the mutant embryos are typically discarded. ES cell lines derived from such mutant embryos could shed light on various aspects of these diseases and provide unlimited resources of neural tissue for basic studies. An even bolder step towards this end is the use of nuclear transfer ("somatic cell nuclear transfer" or "therapeutic cloning"). In this approach somatic nuclei from an adult cell are transferred into the enucleated egg cell obtained from unrelated donors. Blastocysts developed from such a reconstructed egg cell are used as a source for deriving nuclear transfer ES (ntES) cells (Wakayama et al., 2001; Munsie et al., 2000). The most obvious goal of therapeutic cloning is the generation of immunocompatible tissues for transplantation therapy (Wakayama et al. 2001, Rideout et al. 2002). However, this technology could also be used to generate ntES cells from individuals suffering from unknown (genetic) diseases or from cells that underwent somatic mutations such as cancer cells. Libraries of human ES cell lines relating to specific disease conditions could be invaluable for many diseases, particularly those where access to sufficient tissue is critical.

The work on cloning also provides a unique glimpse into the amazing properties of the egg cytoplasm that within hours can reprogram an adult specialized cell nucleus derived from a simple fibroblast into a totipotent cell capable of forming all tissues of an adult organism. Research that unravels the molecular nature of such reprogramming events will revolutionize our understanding on how cell fate decisions are made and maintained and will likely impact on stem cell as well as neurobiology. However, it has to be stressed that standard nuclear transfer techniques have not yet been successfully applied to humans and early attempts have caused a lot of controversy due to the need for creating and subsequently destroying a nuclear transfer blastocyst, and due to fears that this technology could allow reproductive cloning. Furthermore cloning has proven to be inefficient. To create Dolly the cloned sheep, 277 nuclear transfer attempts were required (Wilmut et al., 1997). Currently, the generation of a stable mouse ES line from a given mouse requires about 20 nuclear transfer attempts (Wakayama et al., 2001). Whereas this number is closer to practicality, future improvements will likely come from a better basic understanding of the reprogramming process. The questions and fears raised by both human ES cell and cloning technology are considerable and need to be carefully addressed to realize the full potential of these approaches. However, for the first time it is now conceivable that every neuron type at any given developmental stage can be produced in a culture dish. Using ntES cells this could be achieved in unlimited numbers, autologous to a given host, and endowed with targeted genetic modification required for both basic and therapeutic applications. Thus, the applications for ES cells in both basic and applied studies are vast and will contribute significantly to our understanding of brain function.

Acknowledgements

Our own work cited was supported by the Michael J. Fox Foundation for Parkinson's disease and by the National Institutes of Health. Design of Figure 1 was adapted from the NIH Report: "Stem Cells: Scientific Progress and Future Research Directions."

REFERENCES

Angelov DN, Arnhold S, Andressen C, Grabsch H, Puschmann M, Hescheler J, Addicks K (1998) Temporospatial relationships between macroglia and microglia during *in vitro* differentiation of murine stem cells. Dev Neurosci 20:42-51.

Ashrafi K, Chang FY, Watts JL, Fraser AG, Kamath RS, Ahringer J, Ruvkun G (2003) Genome-wide RNAi analysis of Caenorhabditis elegnas fat regulatory genes. Nature 421:268-272.

Avellana-Adalid V, Nait-Oumesmar B, Lachapelle F, Baron-Van Evercooren A (1996) Expansion of rat oligodendrocyte progenitors into proliferative "oligospheres" that retain differentiation potential. J Neurosci Res 45:558-570.

Bain G, Kitchens D, Yao M, Huettner JE, Gottlieb DI (1995) Embryonic stem cells express neuronal properties *in vitro*. Dev Biol 168:342-357.

Bain G, Yao M, Huettner JE, Finley MFA, Gottlieb DI (1998) Neuronlike cells derived in culture from P19 embryonal carcinoma and embryonic stem cells. In: Culturing nerve cells, 2nd Ed. (Gary Banker and Kimberly Goslin, eds.), pp189-211. Cambridge, MA: The MIT Press.

Bernsteine EG, Hooper ML, Grandchamp S, Ephrussi B (1973) Alkaline phosphatase activity in mouse teratoma. Proc Natl Acad Sci USA 70:3899-3903.

Billon N, Jolicoeur C, Ying QL, Smith A, and Raff M (2002) Normal timing of oligodendrocytes development from genetically engineered, lineage-selectable mouse ES cells. J Cell Sci 115:3657-3665.

Bjorklund LM, Sanchez-Pernaute R, Chung S, Andersson T, Chen IY, McNaught KS, Brownell AL, Jenkins BG, Wahlestedt C, Kim KS, Isacson O (2002) Embryonic stem cells develop into functional dopaminergic neurons after transplantation in a Parkinson's rat model. Proc Natl Acad Sci USA 99:2344-2349.

Bottenstein JE, Sato GH (1979) Growth of a rat neuroblastoma cell line in serum-free supplemented medium. Proc Natl Acad Sci USA 76: 514-517.

Brustle O, Jones KN, Learish RD, Karram K, Choudhary K, Wiestler OD, Duncan ID, McKay RDG (1999) Embryonic stem cell-derived glial precursors: a source of myelinating transplants. Science 285:754-756.

Brustle O, Spiro AC, Karram K, Choudhary K, Okabe S, McKay RD (1997) *In vitro*-generated neural precursors participate in mammalian brain development. Proc Natl Acad Sci USA 94:14809-14814.

Buehr M, Nichols J, Stenhouse F, Mountford P, Greenhalgh CJ, Kantachuvesiri S, Brooker G, Mullins J, and Smith AG (2003) Rapid loss of oct-4 and pluripotency in cultured rodent blastocysts and derivative cell lines. Biol Reprod 68:222-229.

Carpenter MK, Inokuma MS, Denham J, Mujtaba T, Chiu C-P, Rao MS (2001) Enrichment of neurons and neural precursors from human embryonic stem cells. Exp Neurol 172:383-397.

Cervantes RB, Stringer JR, Shao C, Tischfield JA, Stambrook PJ (2002) Embryonic stem cells and somatic cells differ in mutation frequency and type. Proc Natl Acad Sci USA 99:3586-3590.

Chen LR, Shiue YL, Bertolini L, Medrano JF, BonDurant RH, Anderson GB (1999) Establishment of pluripotent cell lines from porcine preimplantation embryos. Theriogenology 52:195-212.

Cibelli JB, Grant KA, Chapman KB, Cunniff K, Worst T, Green HL, Walker SJ, Gutin PH, Vilner L, Tabar V, Dominko T, Kane J, Wettstein PJ, Lanza RP, Studer L, Vrana KE, West MD (2002) Parthenogenetic stem cells in nonhuman primates. Science 295:819.

Darmon M, Bottenstein J, Sato G (1981) Neural differentiation following culture of embryonal carcinoma cells in a serum-free defined medium. Dev Biol 85: 463-473.

Deacon T, Dinsmore J, Costantini L, Ratliff J, Isacson O (1998) Blastula-stage stem cells can differentiate into dopaminergic and serotonergic neurons after transplantation. Exp Neurol 149:28-41.

DeWitt N (2002) Biologists divided over proposal to create human-mouse embryos. (News) Nature 420:255.

Doetschman TC, Eistetter H, Katz M, Schmidt W, Kemler R (1985) The *in vitro* development of blastocyst-derived embryonic stem cell lines: Formation of visceral yolk sac, blood islands and myocardium. J Embryol Exp Morphol 87:27-45.

Dunnett SB (1999) Repair of the damaged brain. The Alfred Meyer Memorial Lecture 1998. Neuopathol Appl Neurobiol 25:351-362.

Evans MJ (1972) The isolation and properties of a clonal tissue culture strain of pluripotent mouse teratoma cells. J Embryol Exp Morphol 28:163-176.

Fields RD, Stevens-Graham B (2002) New insights into neuron-glia communication. Science 298:556-562.

Finley MF, Kulkarni N, Huettner JE (1996) Synapse formation and establishment of neuronal polarity by P19 embryonic carcinoma cells and embryonic stem cells. J Neurosci 16:1056-1065.

Fraichard A, Chassande O, Bilbaut G, Dehay C, Savatier P, Samarut, J (1995) *In vitro* differentiation of embryonic stem cells into glial cells and functional neurons. J Cell Sci 108:3181-3188.

Freed C (2002) Will embryonic stem cells be a useful source of dopamine neurons for transplant into patients with Parkinson's disease? Proc Natl Acad Sci USA 99:1755-1757.

Ge W, Marinowich K, Wu X, He F, Miyamoto A, Fan G, Weinmaster G, Sun YE (2002) Notch signaling promotes astrogliogenesis via direct CSL-mediated glial gene activation. J Neurosci Res 69:848-860.

Ghosh C, Collodi P (1994) Culture of cells from zebrafish (Brachydanio rerio) blastula-stage embryos. Cytotechnology 14:21-26.

Goldstein RS, Drukker M, Reubinoff BE, Benvenisty N (2002) Integration and differentiation of human embryonic stem cells transplanted to the chick embryo. Dev Dyn 225:80-86.

Gossler A, Joyner AL, Rossant J, Skarnes WC (1989) Mouse embryonic stem cells and reporter constructs to detect developmentally regulated genes. Science 244:463-465.

Gottlieb DI (2002) Large-scale sources of neural stem cells. Annu Rev Neurosci 25:381-407.

Gratsch TE, O'Shea KS (2002) Noggin and chordin have distinct activities in promoting lineage commitment of mouse embryonic stem (ES) cells. Dev Biol 245:83-94.

Hemmati-Brivanlou A, Kelly OG, Melton DA (1994) Follistatin, an antagonist of activin, is expressed in the Spemann organizer and displays neuralizing activity. Cell 77:283-295.

Hong Y, Winkler C, Schartl M (1996) Pluripotency and differentiation of embryonic stem cell lines from the medakafish (Oryzias latipes). Mech Dev 60:33-44.

Hong Y, Winkler C, Schartl M (1998) Production of medakafish chimeras from a stable embryonic stem cell line. Proc Natl Acad Sci USA 95:3679-3694.

Howell CY, Bestor TH, Ding F, Latham KE, Mertineit C, Trasler JM, Chaillet JR (2001) Genomic imprinting disrupted by a maternal effect mutation in the Dnmt1 gene. Cell 104:829-838.

Kawasaki H, Mizuseki K, Nishikawa S, Kaneko S, Kuwana Y, Nakanishi S, Nishikawa S-I, Sasai Y (2000) Induction of midbrain dopaminergic neurons from ES cells by stromal cell-derived inducing activity. Neuron 28:31-40.

Kawasaki H, Suemori H, Mizuseki K, Watanabe K, Urano F, Ichinose H, Haruta M, Takahashi M, Yoshikawa K, Nishikawa S, Nakatsuji N, Sasai Y (2002) Generation of dopaminergic neurons and pigmented epithelia from primate ES cells by stromal cell-derived inducing activity. Proc Natl Acad Sci USA 99:1580-1585.

Kim JH, Auerbach JM, Rodriguez-Gomez JA, Velasco I, Gavin D, Lumelsky N, Lee SH, Nguyen J, Sanchez-Pernaute R, Bankiewicz K, McKay R (2002) Dopamine neurons derived from embryonic stem cells function in an animal model of Parkinson's disease. Nature 418:50-56.

Kleinsmith LJ, Pierce GB (1964) Multipotentiality of single embryonal carcinoma cells. Cancer Res 24:1544-1552.

Laborsky PA, Barlow DP, Hogan BL (1994) Mouse embryonic germ (EG) cell lines: transmission through the germline and differences in the methylation imprint of insulin-like growth factor 2 receptor (Igf2r) gene compared with embryonic stem (ES) stem cell lines. Development 120:3197-3204.

Lee S-H, Lumelsky N, Studer L, Auerbach JM, McKay RDG (2000) Efficient generation of midbrain and hindbrain neurons from mouse embryonic stem cells. Nat Biotechnol 18:675-679.

Lindvall O, Hagell P (2000) Clinical observations after neural transplantation in Parkinson's disease. Prog Brain Res 127:299-320.

Liu S, Qu Y, Stewart TJ, Howard MJ, Chakrabortty S, Holekamp TF, McDonald JW (2000) Embryonic stem cells differentiate into oligodendrocytes and myelinate in culture and after spinal cord transplantation. Proc Natl Acad Sci USA 97:6126-6131.

Maden M (2002) Retinoid signaling in the development of the central nervous system. Nat Rev Neurosci 3:843-853.

Martin GR (1980) Teratocarcinomas and mammalian embryogenesis. Science 209:768-776.

Martin GR, Evans MJ (1975) The formation of embryoid bodies *in vitro* by homogeneous embryonal carcinoma cell cultures derived from isolated single cell. In: Teratomas and differentiation (Sherman MI and Solter D, eds.), pp.169-187. New York, NY: Academic Press.

Matthias K, Kirchhoff F, Seifert G, Huttmann K, Matyash M, Kettenmann H, Steinhauser C (2003) Segregated expression of AMPA-type glutamate receptors and glutamate transporters defines distinct astrocyte populations in the mouse hippocampus. J Neurosci 23:1750-1758.

McDonald JW, Liu X-Z, Qu Y, Liu S, Turetsky D, Mickey SK, Gottlieb DI, Choi DW (1999) Transplanted embryonic stem cells survive, differentiate, and promote recovery in injured rat spinal cord. Nat Med 5:1410-1412.

Morizane A, Takahashi J, Takagi Y, Sasai Y, Hashimoto N (2002) Optimal conditions for *in vivo* induction of dopaminergic neurons from embryonic stem cells through stromal cell-derived inducing activity. J Neurosci Res 69:934-939.

Mortensen RM, Conner DA, Chao S, Geisterfer-Lowrance AA, Seidman JG (1992) Production of homozygous mutant ES cells with a single targeting construct. Mol Cell Biol 12:2391-2395.

Muncie MJ, Michalska AE, O'Brien CM, Trounson AO, Pera MF, Mountford PS (2000) Isolation of pluripotent embryonic stem cells from reprogrammed adult mouse somatic cell nuclei. Curr Biol 10:989-992.

Munoz-Sanjuan I, Brivanlou AH (2002) Neural induction, the default model and embryonic stem cells. Nat Rev Neurosci 3:271-280.

Ohtaka T, Matsui Y, Obinata M (1999) Hematopoietic development of primordial germ cell-derived mouse embryonic germ cells in culture. Biochem Biophysical Res Comm 260:475-482.

Okabe S, Forsberg-Nilsson K, Spiro AC, Segal M, McKay RD (1996) Development of neuronal precursor cells and functional postmitotic neurons from embryonic stem cells *in vitro*. Mech Dev 59:89-102.

Pain B, Clark ME, Shen M, Nakazawa H, Sakurai M, Samarut J, Etches RJ (1996) Long-term *in vitro* culture and characterisation of avian embryo stem cells with multiple morphogenetic potentialities. Development 122:2339-2348.

Perrier AL, Studer L (2003) Making and repairing the mammalian brain - *in vitro* production of dopaminergic neurons. Annu Rev Cell Dev Biol. In press.

Rathjen J, Haines BP, Hudson KM, Nesci A, Dunn S, Rathjen PD (2002) Directed differentiation of pluripotent cells to neural lineages: homogeneous formation and differentiation of a neuroectoderm population. Development 129:2649-2661.

Renoncourt Y, Carroll P, Filippi P, Arce V, Alonso S (1998) Neurons derived from ES cells express homeoproteins characteristic of motoneurons an interneurons. Mech Dev 79:185-197.

Reubinoff BE, Itsykson P, Turetsky T, Pera MF, Reinhartz E, Itzik A, Ben-Hur T (2001) Neural progenitors from human embryonic stem cells. Nat Biotechnol 19:1134-1140.

Richards M, Fong CY, Chan WK, Wong PC, Bongso A (2002) Human feeders support prolonged undifferentiated growth of human inner cell masses and embryonic stem cells. Nat Biotechnol 20:933-936.

Rideout 3rd WE, Hochedlinger K, Kyba M, Daley GQ, Jaenisch R (2002) Correction of a genetic defect by nuclear transplantation and combined cell and gene therapy. Cell 109:17-27.

Rohwedel J, Sehlmeyer U, Shan J, Meister A, Wobus A (1996) Primordial germ cell-derived mouse embryonic germ EG cells *in vitro* resemble undifferentiated stem cells with respect to differentiation capacity and cell cycle distribution. Cell Biol Int 20:579-587.

Rutishauser U (1992) NCAM and its polysialic acid moiety: a mechanism for pull/push regulation of cell interactions during development? Dev Suppl 99-104.

Saito S, Ugai H, Sawai K, Yamamoto Y, Minamihasi A, Kurosaka K, Kobayashi Y, Murata T, Obata Y, Yokoyama K (2002) Isolation of embryonic stem-like cells from equine blastocysts and their differentiation *in vitro*. FEBS Lett 531:389-396.

Sasai Y, Lu B, Steinbeisser H, Geissert D, Gont LK, De Robertis EM (1994) Xenopus chordin: a novel dorsalizing factor activated by organizer-specific homeobox genes. Cell 79:779-790.

Schuldiner M, Eiges R, Eden A, Yanuka O, Iskovitz-Eldor J, Goldstein RS, Benvenisty N (2001) Induced neuronal differentiation of human embryonic stem cells. Brain Res 913:201-205.

Smith A (2001) Embryo-derived stem cells: of mice and men. Annu Rev Cell Dev Biol 17:435-462.

Smith WC, Harland RM (1992) Expression cloning of noggin, a new dorsalizing factor localized to the Spemann organizer in Xenopus embryos. Cell 70:829-840.

Solter D, Skreb N, Damjanov I (1970) Extrauterine growth of mouse egg cylinders results in malignant teratoma. Nature 227:503-504.

Soodeen-Karamath S, Gibbins AM (2001) Apparent absence of oct 3/4 from the chicken genome. Mol Reprod Dev 58:137-148.

Spemann H, Mangold H (1924) Uber Induktion von Embryonalanlagen durch Implantation artfremder Organisatoren. Wilh Roux' Arch Entw Mech Organ 100:599-638. English translation in: Foundations of experimental embryology (Willier BH, Oppenheimer JM, eds.), pp.144-184. New York, NY: Hafner Press.

Stern CD (2002) Induction and initial patterning of the nervous system - the chick embryo enters the scene. Curr Opin Genet Dev 12:447-451.

Stevens LC (1970) The development of transplantable teratocarcinomas from intratesticular grafts of pre- and postimplantation mouse embryos. Dev Biol 21:364-382.

Stevens LC, Hummel KP (1957) A description of spontaneous congenital testicular teratomas in strain 129 mice. J Natl Cancer Inst 18:719-747.

Stewart CL, Gadi I, Bhatt H (1994) Stem cells from primordial germ cells can reenter the germ line. Dev Biol 161:626-628.

Strubing C, Ahnert-Hilger G, Shan J, Wiedenmann B, Hescheler J, Wobus AM (1995) Differentiation of pluripotent embryonic stem cells into the neuronal lineage *in vitro* gives rise to mature inhibitory and excitatory neurons. Mech Development 53:275-287.

Studer L (2001) Stem cells with brainpower. Nat Biotechnol 19:105-109.

Suda Y, Suzuki M, Ikawa Y, Aizawa S (1987) Mouse embryonic stem cells exhibit indefinite proliferative potential. J Cell Physiol 133:197-201.

Sukoyan MA, Yatolin SY, Golubitsa AN, Zhelezova AI, Semenova LA, Serov OL (1993) Embryonic stem cells derived from morulae, inner cell mass, and blastocysts of mink: comparisons of their pluripotencies. Mol Reprod Dev 36:148-158.

Sun L, Bradford CS, Ghosh C, Collodi P, Barnes DW (1995) ES-like cell cultures derived from early zebrafish embryos. Mol Mar Biotl Biotechnol 4:193-199.

Surani MA (1998) Imprinting and the initiation of gene silencing in the germ line. Cell 93:309-312.

Tada T, Tada M, Hilton K, Barton SC, Sado T, Takagi N, Surani MA (1998) Epigenotype switching of imprintable loci in embryonic germ cells. Dev Genes Evol 207:551-561.

Thomson JA, Itskovitz-Eldor J, Shappiro SS, Waknitz MA, Swiergiel JJ, Marshall VS, Jones JM (1998) Embryonic stem cell lines derived from human blastocysts. Science 282:1145-1147.

Thomson JA, Kalishman J, Golos TG, Durning M, Harris CP, Becker RA, Hearn JP (1995) Isolation of a primate embryonic stem cell line. Proc Natl Acad Sci USA 92:7844-7848.

Tropepe V, Hitoshi S, Sirard C, Mak TW, Rossant J, van der Kooy D (2001) Direct neural fate specification from embryonic stem cells: a primitive mammalian neural stem cell stage acquired through a default mechanism. Neuron 30:65-78.

Tsai M, Wedemeyer J, Ganiatsas S, Tam S-Y, Zon LI, Galli SJ (2000) *In vivo* immunological function of mast cells derived from embryonic stem cells: an approach for the rapid analysis of even embryonic lethal mutations in adult mice *in vivo*. Proc Natl Acad Sci USA 97:9186-9190.

Wakayama T, Tabar V, Rodriguez I, Perry AC, Studer L, Mombaerts P (2001) Differentiation of embryonic stem cell lines generated from adult somatic cells by nuclear transfer. Science 292:740-743.

Weiss MJ, Orkin SH (1996) *In vitro* differentiation of murine embryonic stem cells. J Clin Invest 97:591-595.

Westmoreland JJ, Hancock CR, Condie BG (2001) Neuronal development of embryonic stem cells: a model of GABAergic neuron differentiation. Biochem Biophy Res Commun 284:674-680.

Wichterle H, Lieberam I, Porter JA, Jessell TA (2002) Directed differentiation of embryonic stem cells into motor neurons. Cell 110:385-397.

Wilmut I, Schnieke AE, McWhir J, Kind AJ, Campbell KH (1997) Viable offspring derived from fetal and adult mammalian cells. Nature 385:810-813.

Xu C, Inokuma MS, Denham J, Golds K, Kundu P, Gold JD, Carpenter MK (2001) Feeder-free growth of undifferentiated human embryonic stem cells. Nat Biotechnol 19:971-974.

Yan J, Studer L, McKay RD (2001) Ascorbic acid increases the yield of dopaminergic neurons derived from basic fibroblast growth factor expanded mesencephalic precursors. J Neurochem 76:307-311.

Ye W, Shimamura K, Rubenstein JL, Hynes MA, Rosenthal A. (1998) FGF and SHH signals control dopaminergic and serotonergic cell fate in the anterior neural plate. Cell 93:755-766.

Ying QL, Stavridis M, Griffiths D, Li M, Smith A (2003) Conversion of embryonic stem cells into neuroectodermal precursors in adherent monoculture. Nat Biotechnol 21:183-186.

Zambrowicz BP, Friedrich GA, Buxton EC, Lilleberg SL, Person C, Sands AT (1998) Disruption and sequence identification of 2,000 genes in mouse embryonic stem cells. Nature 392:608-611.

Zetterstrom RH, Solomin L, Jansson L, Hoffer BJ, Olson L, Perlmann T (1997) Dopamine neuron agenesis in Nurr1-deficient mice. Science 276:248-250.

Zhang SC, Lundberg C, Lipsitz D, O'Connor LT, Duncan ID (1998) Generation of oligodendroglial progenitors from neural stem cells. J Neurocytol 27:475-489.

Zhang SC, Wernig M, Duncan ID, Brustle O, Thomson JA (2001) *In vitro* differentiation of transplantable neural precursors from human embryonic stem cells. Nat Biotechnol 19:1129-1133.

Zwaka TP, Thomson JA (2003) Homologous recombination in human embryonic stem cells. Nat Biotechnol 21:319-321.

Chapter 6

On the Origin of Newly Made Neural Cells in the Adult Organism: Does Transdifferentiation Occur?

Éva Mezey

INTRODUCTION

When we break a bone or cut our skin the injured tissues heal. Initially, the skin is painful and red, then a thin new layer of cells grows beneath a protective scab. Finally, the scab falls off and the new skin is fully functional. We cannot watch broken bones being repaired, but we all accept the fact that this happens. When the brain or the spinal cord is injured, on the other hand, function is not always restored. What makes the nervous system so different from other organs? How is repair accomplished in other parts of the body? How are old cells renewed or replaced? These questions have kept generations of biologists and physicians busy, but it seems we now have partial answers to them.

In 1894 the first classification of tissues based on their renewal potential was published by Giulio Bizzozero, a professor at the University of Pavia in Italy between 1846 and 1901 (Bizzozero, 1894). After reviewing the available literature and his own data, Bizzozero concluded that all tissues belong to one of three categories: 'labile', 'stable' or 'everlasting'. Cells of labile tissues show signs of continuous reproduction by mitosis throughout life, with new cells replacing the lost elements. To this category he assigned the testis, ovary, lymph nodes, sebaceous glands, stomach, gut, bone marrow, and the spleen. Stable tissues are made up of cells, which stop dividing at birth or some time later, but which can also undergo regeneration in response to pathological conditions during postnatal life. Bizzozero placed liver, bones, and smooth muscle among the stable tissues. As an example of regeneration in pathological conditions, Bizzozero cited the experiments of Ponfick, who removed a large portion of the liver in experimental animals and found subsequent regeneration leading to the restoration of its initial mass. Everlasting tissues include the brain and striated muscle, which are made up of postmitotic cells. According to Bizzozero, their reservoir of germinal cells is exhausted early in embryonic development, and he concluded that these tissues lack the potential to reproduce and regenerate even in pathological conditions. The giants and founding fathers of neurobiol-

From: *Neural Stem Cells: Development and Transplantation*
Edited by: J.E. Bottenstein © 2003 Kluwer Academic Publishers

ogy all seemed to agree that no new neurons are formed in humans after birth. Confirming Bizzozero's classification, the early works of Ramon y Cajal (Cajal and May, 1959) also suggested that no neurons in the central nervous system (CNS) of higher vertebrates show mitotic characteristics. Thus the dogma of "no adult neurogenesis" became widely accepted.

NEURAL STEM CELLS IN THE CNS

As with most dogmas, the suggestion that adult neurogenesis never occurs was challenged when new techniques were applied to the problem. In 1960, ^{3}H-thymidine incorporation began to be used to study cell proliferation (Messier and Leblond, 1960). Thymidine is one of the four nucleic acids from which DNA is synthesized when cells divide and their genomes are replicated. Since tritium has a long halflife, cells labeled with tritiated thymidine can be followed over long periods of time.

Subventricular zone (SVZ)

Using this technique Smart (1961) was able to show that there are a "collection of undifferentiated, mitotically active cells, which appears during embryonic development, plays an important part in the production of cells for the cerebral cortex and persists into adult life retaining, at least in rat and mice, its ability to form new cells" in the subependymal zone of the lateral ventricle. Subsequently, Altman (1962) gave rats tritiated thymidine, made electrolytic lesions in the lateral geniculate nucleus and then found labeled stellate and small pyramidal neurons there. He concluded that cells labeled before the lesions gave rise to the neurons he saw.

Hippocampus

Almost 15 years passed before Kaplan and Hinds (1997) performed another seminal experiment. They injected three-month-old rats with [^{3}H] thymidine and 30 days later discovered labeled cells in the granular cell layers of the dentate gyrus in the hippocampus and in the olfactory bulb that had all of the morphological features of neurons. More than a decade after Kaplan and Hinds's paper appeared, Lois and Alvarez-Buylla (1993) used thymidine incorporation to show that the subventricular zone of adult mammals contains cells that proliferate spontaneously *in vivo*. The same authors demonstrated that SVZ cells differentiate directly into neurons and glia in explant cultures. Labeling with ^{3}H-thymidine showed that 98% of the new neurons were derived from precursor cells that underwent their last division *in vivo*. It was soon accepted that there are progenitor cells in two regions of the brain – the SVZ and the hippocampal subgranular layer (SGL) – responsible for adult neurogenesis (Cameron and Gould, 1994; Eriksson et al., 1998; Gould et al., 1999; Kornack and Rakic, 1999).

Other Sources

In a careful analysis of the SVZ cells, Doetch and coworkers concluded that the SVZ contains four cell types: immature precursors, astrocytes, migrating neuroblasts, and ependymal cells. They have demonstrated that the SVZ astrocytes act as stem cells and can give rise to neurons (Doetsch et al., 1999). For a cell to be considered a neural stem cell it has to satisfy the following criteria: it can generate cell type(s) of the nervous system, it has a capacity for self renewal, and it can go through asymmetric cell division to generate new cells as well as replenish itself. Neural stem cells were thought to give rise to either neuronal or glial cells as terminally differentiated cell types. Kondo and Raff (2000), however, have shown that cell fates are rather complex; e.g., oligodendrocyte precursor cells have an even greater developmental potential than previously thought. In response to extracellular signals they can revert to multipotent neural stem cells, which can then give rise to neurons, astrocytes, and oligodendrocytes. Such changes in cell fate would be called "transdifferentiation" by developmental biologists. In fact, one of the most common events of this sort is the conversion of radial glia to astrocytes. The role of radial glial cells as guides for migrating neurons has been long known (Parmar et al., 2002). Recent *in vitro* and *in vivo* evidence suggest that a large subset of radial glia generates neurons and the progeny of radial glial cells does not differ from the progeny of precursors labeled from the ventricular surface. In addition to embryonic development, radial glia have been shown to be neurogenic also in adulthood suggesting that they are a subset of CNS precursor cells (Hartfuss et al., 2001; Gotz et al., 2002). In fact Chanas-Sacre et al. (2000) suggested that a significant portion of neuronal stem cells in the CNS might derive from radial glia .

In addition to the SVZ and the hippocampal SGL, additional brain regions appear to contain neural progenitors, among them the rodent septum and striatum, which contain cells that give rise to neurons *in vitro* in response to fibroblast growth factor (FGF2) stimulation (Palmer et al., 1995) and the cerebral cortex which contains cells that can be induced to generate layer- and region specific cortical neurons which establish connections with appropriate target sites (Magavi et al., 2000; Magavi and Macklis, 2002). The cerebral cortex also contains specific oligodendrocyte progenitors (Reynolds and Hardy, 1997). In addition, the combination of epidermal growth factor (EGF) and basic fibroblast growth factor (FGF2) induced cells isolated from the spinal cord of the mouse to differentiate into neurons and macroglia, adding spinal cord to the growing list of regions that contain CNS progenitor cells (Weiss et al., 1996). A variety of factors have been suggested to play a role in the differentiation of all the above-mentioned progenitor cells. These include FGF2 (Weiss et al., 1996; Qian et al., 1997), EGF (Weiss et al., 1996), retinoic acid (Wohl and Weiss, 1998), brain-derived neurotropic factor (BDGF), glial-derived neurotrophic factor (GDNF), and neurotrophin 3 (NT3).

NEURAL STEM CELLS OUTSIDE THE CNS

Olfactory epithelium

Among the basal cells of the olfactory epithelium, a stem cell has been identified which divides and differentiates into new sensory neurons throughout adult life (see Calof et al., 1998). Sicard and coworkers dissected rat olfactory epithelium from the nasal septum and grew the dissociated cells in medium containing epidermal growth factor for 5 days. Only supporting cells and keratin-positive horizontal basal cells survived at the end of this period. Then they stressed the cells (by either passaging or mechanical stress) to induce neuronal differentiation, and based on their results suggested that olfactory sensory neurons can arise from a non-neuronal precursor, probably the horizontal basal cell (Sicard et al., 1998). Later the same group showed that neurons identified with specific immunological reagents, could form in the absence of growth factors, complex media, explants or feeder layers of glia or other nonepithelial cells. The neurons were bipolar in form as olfactory sensory neurons are (Feron et al., 1999).In the epithelium there are also cells that give rise to glial (ensheathing) cells that are unique in the CNS because they resemble both Schwann cells and astrocytes (Au and Roskams, 2003). It is not yet clear, however, whether the glial and neuronal cells derive from the same or different precursors.

Neural Crest Stem Cells

The cells that make up the peripheral nervous system arise in the neural crest. During formation of the neural tube from ectoderm, a group of cells separate from the neural tube to form a mass along its dorsolateral margin. These cells give rise to the neural crests. Cells of the neural crest migrate to different locations in the body to form the paravertebral and prevertebral ganglionic chains as well as chromaffin tissue that later forms the adrenal medulla. The neural crest generates the dorsal root ganglia, autonomic ganglia, cranial nerve ganglia, enteric ganglia, Schwann cells, and satellite cells, as well as some nonneural tissues (LaBonne and Bronner-Fraser, 1998). Recently, neural crest stem cells

(NCSCs) were identified in the adult rat gut. These cells could renew themselves in culture, but to a lesser degree than fetal gut NCSCs. Postnatal gut NCSCs could produce neurons with a variety of neurotransmitters, but their plasticity seemed restricted; they were unable to make certain types of neurons that are generated during fetal development (Kruger et al., 2002). Some NCSCs may originate in the spinal cord neuroepithelium, which contains nestin-positive cells that are similar to previously characterized NCSCs. These cells differentiate into peripheral neurons, smooth muscle, and Schwann cells in both mass and clonal culture. Clonal analysis of neuroepithelial cells demonstrates that a common progenitor cell can generate both CNS and PNS cells. The differentiation into NCSCs seems to be regulated by bone morphogenetic protein BMP-2/4 (Mujtaba et al., 1998).

NEURAL CELLS FROM OTHER TISSUE STEM CELLS

In the last decade many studies by a variety of groups addressed the possibility that peripheral stem cells might enter the CNS and differentiate into neural cells.

Blood

Total Bone Marrow (BM)

Basic studies. Krivit et al. (1987) were the first to call attention to the possible entry of circulating blood cells into the human brain as a means of improving neurological problems. In their study, two siblings had the same neurodegenerative autosomal recessive storage disease (metachromatic leukodystrophy) that results in a progressive deterioration of the nervous system before inevitably death occurs due to the low levels of an enzyme. One of the siblings showed the characteristic progression of the disease and died. The other sibling received a bone marrow transplant from a nonaffected compatible sibling and has not developed the intellectual and neurologic impairment that the first one did. The authors concluded that the transplanted bone marrow somehow contributed to the CNS and arrested the progression of the disease (Krivit et al., 1987). A year later Hickey and Kimura showed that a subset of CNS cells, namely the perivascular microglia, is bone marrow derived (Hickey and Kimura, 1988). These findings were later confirmed in human brain parenchyma which normally seems to contain mononuclear leukocytes as well as microglia (Unger et al., 1993). Subsequently, retroviral tagging of bone marrow cells as well as gender mismatched transplants were used to demonstrate that in addition to perivascular microglia, parenchymal microglia as well as a small portion of the astrocytes in the brain could potentially come from bone marrow in rodents (Eglitis and Mezey, 1997). Based on incorporation of carbon, Kaur et al. (2001) confirmed the monocytic origin of brain microglia and Priller et al. (2001b) used bone marrow taken from enhanced green fluorescent protein (EGFP) labeled mice to show that 4 months after transplant 20% of brain microglia exhibited the green fluorescence suggesting that they came from the donor bone marrow.

The observation that microglia are likely to originate in bone marrow proved that exogenous cells can indeed populate the CNS in the adult. This was not too surprising, since microglia are closely related to monocytes and macrophages, and there was not much resistance to the suggestion in the scientific community. In fact, experiments began to be designed to determine whether microglia could be used as a vector to introduce certain agents or replace missing proteins (enzymes) in the brain.

It was more difficult to convince biologists to accept the experimental results of Kabos et al. (2002) who described the isolation and successful propagation of neural progenitor cells from adult rat bone marrow. Unfractionated bone marrow was cultured *in vitro* with EGF and FGF2 and gave rise to cellular spheres containing elements that labeled with neuronal and glial markers. This suggested that some cells in the bone marrow have the capacity to generate not only microglial cells but also neuronal and gial cells in the rat.

In spite of the emerging *in vitro* data on bone marrow stem cells, it was not until *in vivo* data started to surface that neuroscientists worried that a century-old dogma might be on the verge of disappearing. Two independent laboratories, using different techniques, have demonstrated that transplanted bone marrow can generate neurons in adult rodents. Brazelton et al. (2000) used irradiated recepient mice and transplanted them with bone marrow from mice that express EGFP in all cells. One to six months later they found many cells with neuronal morphology which contained neuron-specific markers (NeuN, 200-kilodalton neurofilament, and class III beta- tubulin) in the olfactory bulb. Mezey et al. (2000) transplanted transgenic female mice (McKercher et al., 1996) that are born without white blood cells with wild type littermate male donor bone marrow, and used the Y chromosome as the donor-specific marker. The advantage of this technique is that there is no need to irradiate the mice and that all cells that are bone marrow-derived will be marked (i.e. there is no "own" bone marrow to compete with). They found many Y chromosome labeled cells in a variety of brain areas and some of these cells expressed a neuron-specific marker, NeuN (Mullen et al., 1992). The cells described in these two studies clearly came from the donor bone marrow, they expressed neuron-specific markers, and (in the case of the GFP-tagged cells) they had a neuronal morphology. It is not clear though, that they actually function as neurons. In a similar study to evaluate the potential of bone marrow to generate mature neurons in adult mice, Priller et al. (Priller et al., 2001a) transferred the enhanced green fluorescent protein gene into BM cells using a retroviral vector. They achieved stable, high-level long-term EGFP expression and found EGFP-expressing cells in the brain of transplanted mice. Fifteen months after transplantation, fully developed GFP-positive Purkinje neurons were found in all the mice studied. Based on the very characteristic morphology of the cells and their expression of glutamic acid decarboxylase, which is known to be present in these GABAergic neurons, the newly generated Purkinje cells appeared to be functional. A similar finding was reported by Weimann et al. (2003) who found donor-derived Purkinje cells in the brains of gender-mismatched transplant patients.

A recently published study from our laboratory confirmed the above findings in human brains. We examined brain samples from females who had received bone marrow transplants from male donors one to nine months before they died of their underlying diseases. A combination of immunocytochemistry (to identify neurons using NeuN and Kv2.1 as specific markers) and fluorescent in situ hybridization histochemistry (FISH) was used to search for Y chromosome-positive neurons. We found cells containing Y-chromosomes in several brain regions in all four patients. Most of these cells were nonneuronal, but in addition to endothelial cells and cells in the white matter, neurons in the hippocampus and in the cerebral cortex were labelled. Since the Kv2.1 antibody labels the plasma membrane we could identify small cortical pyramidal cells based on their morphology. The youngest patient (a three year old child) who also lived the longest time following transplantation had the greatest number of donor-derived neurons (7 donor derived neurons among 10,000 cells). The distribution of the labeled neurons in the sections examined was not homogeneous since clusters of Y chromosome-positive cells were present in areas that were otherwise negative suggesting that single progenitor cells underwent clonal expansion and differentiation (Mezey et al., 2003a). Based on the above studies, we conclude that there are elements in the bone marrow that are capable of entering the brain and generating neural cells there. In fact, it may be possible to increase the number of bone marrow-derived cells that enter the CNS. Corti et al. (2002b) have shown that inducing the proliferation of bone marrow cells with granulocyte colony stimulating factor (G-CSF) and stem cell factor (SCF) results in a significant increase in the number of BM derived neurons in the treated animals in all brain regions examined.

The exact nature of the bone marrow cells that enter the brain is not yet known. According to our present knowledge there are two distinct populations of stem cells in the bone marrow: hematopoietic stem cells (HSC) which give rise to blood cells, and the mesenchymal (also called stromal) stem cells (MSC), which renew the bone, cartilage and connective tissue.

Functional studies: Several groups of workers have tried to determine whether bone marrow stem cell are used by the body for brain repair or whether they could be induced to contribute to the process.

Li and coworkers transplanted bromdeoxyuridine (BrdU)-tagged adult bone marrow nonhematopoietic cells into mouse striatum following middle cerebral artery occlusion (MCAO). Four weeks later the implanted cells had migrated more than 2 mm away from the injection site toward the ischemic brain areas. One percent of the cells expressed a neuron-specific marker (NeuN) and 8 % expressed GFAP, an astrocyte-specific protein. Interestingly, they observed a significant improvement of function in the transplanted mice compared to non-transplanted controls (Li et al., 2000).

Hess et al. (2002) transplanted male EGFP-labeled bone marrow into female mice and used the Y chromosome and the fluorescent protein to track the bone marrow-derived cells. The recipient underwent middle cerebral artery occlusion (MCAO), and 3-14 days later bone marrow-derived cells had contributed to the vasculature endothelium in the ischemic zone. Some bone marrow-derived cells also expressed the neuronal marker NeuN. In a similar experiment Mahmood et al. (2001a) injected male bone marrow cells into the ischemic border zone of traumatic brain injuries (TBI) in female rats. They observed significant functional improvement in the rotarod tests of strength 14 and 28 days postoperatively compared to rats that did not receive bone marrow injections. Histological analysis of the brains suggested that the bone marrow cells survived, proliferated and migrated toward the injury site in the recipient's brains. Immunohistochemistry showed the presence of astrocytic and neuronal markers in cells of bone marrow origin.

Sasaki asked whether bone marrow-derived cells might be able to remyelinate axons in the rat spinal cord. Irradiation was used to destroy the myelin in the dorsal funiculus. Acutely isolated bone marrow cells from LacZ transgenic mice were transplanted into the demyelinated dorsal column lesions of immunosuppressed rats. The cells injected were a mixture of hematopoietic and nonhematopoietic stem and precursor cells and lymphocytes. An intense blue beta-galactosidase reaction was observed in the transplantation zone and the bone marrow cells remyelinated the spinal cord. Interestingly, the differentiated elements resembled Schwann cells except that each one wrapped more than a single axon (Sasaki et al., 2001).

There has been a longstanding interest in possible regeneration of retinal neurons due to the fact that a large percentage of blindness can be attributed to the loss of these cells in humans. Tomita et al. (2002) injected stem cell-enriched, EGFP marked bone marrow into the vitreous body of rats following mechanical injury of the retina. Two weeks later they found that the bone marrow cells integrated into the retina and gave rise to retinal neurons that expressed specific markers. Interestingly, the new cells were not restricted to the area of the lesion, but were found in non-injured areas of the retina as well.

Bone Marrow Mesenchymal stem cells (BMSC)

Basic studies: It was a big surprise to neurobiologists when in the summer of 2000 two groups of researchers showed that rat and human BMSCs *in vitro* can adopt a neuronal phenotype, expressing neuron specific proteins. With an optimal protocol, almost 80% of the cells expressed neuronal markers in addition to extending long processes that terminated in typical growth cones and filopodia. Woodbury et al. (2000) found that clonal cell lines, established from single cells, yielded both undifferentiated and neuronal cells. Sanchez-Ramos et al. (2000) reached a similar conclusion; they demonstrated that human and mouse BMSCs can be induced to differentiate into neural cells *in vitro* by EGF or BDNF. Cells in these cultures also expressed glial fibrillary acidic protein (GFAP) and neuron-specific nuclear protein (NeuN). When labeled human or mouse BMSCs were cultured with rat fetal mesencephalic or striatal cells, a small fraction of BMSC-derived cells differentiated into neuron-like cells expressing NeuN and glial cells expressing GFAP. Hung et al. (2002) isolated size-sieved stem (SS) cells (a subpopulation of MSC) from human bone marrow and propagated them *in vitro* to demonstrate that SS cells could be induced to differentiate into neural cells under experimental cell culture conditions. Five hours after exposure to antioxidant agents ({beta}-mercaptoethanol {±} retinoic acid) in serum-free conditions, SS cells expressed the protein for nestin, neuron-specific enolase (NSE), neuron-specific nuclear protein (NeuN), and neuron-specific tubulin-1 (TuJ-1), and the mRNA for NSE and Tau. Immunofluorescence showed that almost all the cells (>98%) expressed NeuN and TuJ-1. Deng et al. (2001) found that by treating human BMSCs *in vitro* with agents that increase the intracellular level of cAMP, about 25% of the hBMSCs differentiated into cells with a typical neuronal morphology and expressed increased levels of neuron-specific proteins such as NSE along with vimentin. Kohyama et al, (2001) used specific inducers, and coating to generate neurons from marrow stroma. These cells formed neurites, and expressed neuron-specific markers and genes. In addition to the "morphological" markers these cells were also shown to respond to depolarizing stimuli just as functional mature neurons do. Even mature osteoblasts isolated from the stromal cell population could be efficiently converted into functional neurons. This "transdifferentiation or metadifferentiation" was enhanced by Noggin, an inhibitor of bone morphogenetic protein, which has long been suspected to play a key role in cell fate determination.

Functional studies: Azizi et al. (1998) injected human BMSCs into the rat striatum. Five to 72 days later they found that about 20% of the infused cells had engrafted without any evidence of an inflammatory response. They also showed that the cells migrated from the injection site along known migratory pathways of neural stem cells and stopped making collagen and fibronectin suggesting a change in character. Since no further characterization of the implanted cells was performed, we do not know if they were neural in nature. The authors suggested that BMSCs may be useful vehicles for autotransplantation in both cell and gene therapy. Zhao et al. (2002a) also implanted human BMSCs into rat brains one week after inducing cortical brain ischemia. Two and six weeks later the animals were assessed for sensorimotor function and showed significantly better performance compared to controls. Immunohistological analyses showed that the transplanted human cells expressed markers for astrocytes, oligodendrocytes and neurons. None of the cells derived from the graft resembled mature neurons, however; they were small and had few processes, all of which were short. Thus, the authors concluded that the functional improvement must be due to factors secreted by the implanted BMSCs.

Kopen et al. injected mouse MSCs into the lateral ventricle of newborn mice. They found several regions containing donor derived astrocytes and also found neurofilament positive donor-derived cells in the brainstem, suggesting that the injected MSC cells underwent neuronal differentiation in the brain (Kopen et al., 1999).

Chopp and colleagues (Lu et al., 2001a; Mahmood et al., 2001b) performed a number of studies to determine whether BMSCs can be useful in brain repair and to learn whether one route of delivery might be better than others. They found a significant improvement in motor function tests in rats that were injected with BMSC intravenously following traumatic brain injury (TBI). The outcome was similar when they injected MSCs into the carotid artery on the side of the lesion. In addition to the functional improvement, they also determined that there is a significant engraftment of BMSCs in these brains, 19% when the cells are pretreated with NGF and BDNF and 14.4% without pretreatment (Lu et al., 2001b). They also studied a stroke model, middle cerebral artery occlusion. Following ischemia and reperfusion the rats received a local injection of MSC into the ischemic boundary zone of the stroke. Significant improvement was observed in somatosensory functions and, if the BMSCs cells were pretreated with NGF, motor function (Chen et al., 2001a). Finally, they looked at a model of Parkinson's disease in which mice were treated with MPTP, an agent that kills dopaminergic neurons. Intrastriatal injections of cultured BMSCs resulted in a statistically significant improvement in rotarod test results vs. controls. Histological examination confirmed that many of the implanted cells survived and that some of them expressed neuronal markers as well as tyrosine hydroxylase, an enzyme required for dopamine synthesis (Li et al., 2001).

Akiyama et al. (2002) isolated collagen type I, fibronectin and CD44 positive stromal cells from green fluorescent protein (GFP)-expressing mice. They transplanted these cells by direct microinjection into demyelinated spinal cords of immunosuppressed rats. There appeared to be both central and peripheral remyelination as demonstrated by electron microscopy. GFP-positive cells and myelin profiles observed in the remyelinated spinal cord indicated that the donor-isolated stromal cells generated the new myelin. The GFP-positive cells were colocalized with myelin basic protein. Furthermore the conduction velocities of axons traveling through the injured region were also improved.

In another recent study Corti et al. (2002) examined the spinal cord, and dorsal root ganglia (DRG) of adult mice after they were given GFP-labeled bone marrow. The authors found few GFP-positive cells coexpressing neuronal markers (TuJ1, NF, and NeuN) in either the spinal cord or the sensory ganglia of uninjured mice. These cells were small and had short cytoplasmic processes. Cells with both GFP and GFAP were found only in the spinal cord. Large numbers of GFP-positive cells contained F4/80, a microglial marker. These cells had a typical microglial appearance (Corti et al., 2002a).

Hematopoietic stem cells (HSCs)

The most important job of the hematopoietic stem cells (HSCs) is to replace aging blood cells and keep the white cell population within the normal physiological range at all times. In the last decade data emerged in the literature hinting that HSCs might have more plasticity than previously thought. It was almost a decade ago that Lin et al. (1995) showed that the human cell surface molecule CD34, a HSC marker that may regulate early events in blood cell migration, is expressed in several cell types, including cells in the neural tube of the mouse embryo and neurons in the adult mouse. The first indication that the HSC and the brain stem cells might have common potencies came from Bjornson et al. (1999) who showed that labeled neural stem cells produced a variety of blood cells, including myeloid and lymphoid cells, and early hematopoietic cells in irradiated host animals. Krause et al. (2001) performed a detailed study of the distribution of HSC-derived somatic cells after they injected single isolated HSCs into irradiated host mice, and found that the cells differentiated into epithelial cells of the liver, lung, GI tract, and skin suggesting their "tremendous differentiative capacity." When GFP-expressing HSC were administered intravenously, the cells entered the CNS and differentiated into microglia there. Four months after transplantation about one fourth of the brain microglial population was of donor origin (Priller et al., 2001b). In addition to microglia, Bonilla et al.

(2002) demonstrated that adult HSC can generate oligodendrocytes after they are transplanted into brains of normal neonatal mouse. While a study using cDNA microarray found a large set of common expressing genes between HSC and mouse neurospheres (a population greatly enriched for neural progenitor cells; Terskikh et al., 2001), it is still not clear what the differences are between these two populations.

The idea that HSCs are multipotent, let alone pluripotent, is far from being generally accepted, and studies contradicting those above have been published. In a very thorough study Wagers et al. (2002) transplanted a single green fluorescent protein (GFP)-marked HSC into lethally irradiated recipients. Although the single HSCs reconstituted peripheral blood leukocytes in the animals, they did not contribute appreciably to tissues other than blood, including brain, kidney, gut, liver, and muscle. Interestingly, though, even in this "negative" study the researchers reported seeing a fully developed Purkinje cell of donor origin.

Human Cord Blood (HCB)

The blood remaining in the umbilical cord after birth contains hematopoietic precursors and has become an important source of hematopoietic stem cells (Broxmeyer et al., 1989). Several groups studied the differentiation potential of human cord blood in order to determine if it could be used for tissue regeneration.

Basic studies: The stromal cell population in bone marrow has been the focus of much attention because this cell population can be expanded and differentiated into cells with many different phenotypes. In addition to hematopoetic stem cells, there are rare cells in HCB that are negative for CD34, the HSC marker. These have been called multipotent stem cells (UC-MC) and they may be equivalent to the BMSCs from the bone marrow (Erices et al., 2000; Rosada et al., 2002). Exposure of UC-MCs to basic fibroblast growth factor (bFGF) and human epidermal growth factor (hEGF) in culture induces expression of neural and glial markers in these cells (Goodwin et al., 2001; Bicknese et al., 2002). Ha et al. (2001) also showed that cultured human cord blood monocytes (from newborn umbilical blood) express neural markers and resemble neuronal morphology. Sanchez-Ramoz et al. (2001) used a combination of retinoic acid and nerve growth factor to treat HCB cells and noticed a change in their appearence. The cultured cells started to express neuronal and glial mRNAs and proteins. The astrocytic marker GFAP was present in both treated and untreated cord blood cells suggesting an unexpected plasticity in these cells.

Functional studies: Chen et al. (2001b) and Lu et al. (2002) infused HCB cells intravenously after stroke (MCAO) and traumatic brain injury (TBI) in rats. In both cases they observed a significant functional improvement in the treated animals. Histology confirmed that some HCB cells were reactive for the GFAP protein and the neuronal markers NeuN and microtubule-associated protein 2 suggesting HCB derived neurogenesis in the brains. Also, brain tissue from ischemic brains attracts HCB cells *in vitro*, suggesting that damaged brain areas might release factors that attract stem cells into the site of the injury (Chen et al., 2001b).

Recently, Zigova et al. (2002) began to characterize mononuclear cells from human umbilical cord blood *in vitro* and *in vivo*. The cryopreserved human cells are available in unlimited quantities and it is believed that they may represent a source of cells with possible therapeutic and practical value. Their previous molecular and immunocytochemical studies on cultured HCB cells revealed their ability to respond to nerve growth factor (NGF) by increased expression of markers typical of neural stem cells. In addition, the DNA microarray detected downregulation of several genes associated with development of blood cell lines. To further explore the survival and phenotypic properties of transplanted HCB cells [cultured with DMEM and fetal bovine serum or exposed to retinoic acid (RA) and nerve growth factor (NGF)] into the developing rat brain, cells were injected into the anterior part of subventricular zone of 1-day-old pups and the brains were studied a month later. The results showed that about 20% of the injected cells survived and expressed neuronal and glial markers (Zigova et al., 2002).

Retinal pigment epithelium (PE)

When one looks up the word transdifferentiation in a biology dictionary the one example that is given refers to the eye: "Change of a cell or tissue from one differentiated state to another. Rare, and has mainly been observed with cultured cells. In newts the pigmented cells of the iris transdifferentiate to form lens cells if the existing lens is removed "(Lackie and Dow, 1999).

It has long been known that in amphibians and chicks the neural retina is capable of regeneration (see Reh and Pittack, 1995). This regeneration occurs through the transdifferentiation of retinal pigment epithelial (PE) cells into neural cells. Several groups studied this phenomenon and suggested the involvement of regulatory factors in the process. In Xenopus, where retinal regeneration spontaneously occurs, Sakaguchi et al. (1997) found that FGF2 promotes the transdifferentiation of PE cells into neurons *in vitro*. In the quail it was confirmed that the FGF signaling pathway is involved in this process and a critical role for the gene Mitf was suggested (Mochii et al., 1998; Araki et al., 2002). Further studies in chick retina have indicated that FGF2 and insulin together promote the transdifferentiation (Fischer and Reh, 2001). In addition to FGF2, neuronal cell adhesion molecules also play a role, and retinal neurons can transdifferentiate into PE cells suggesting a bidirectional plasticity between retinal neurons and PE cells (Opas et al., 2001). Tropepe et al. (2000) were the first to report that mammalian retina also has a capacity to regenerate. They studied single pigmented cells and single neural retinal cells from E14 and adult mouse eyes and found that in the presence of EGF and FGF2 these cells can form small spherical colonies that can further proliferate even without exogenous growth factors and eventually form photoreceptor cells, bipolar neurons and Müller glia. Studies in mice also demonstrated the important role of the Mitf gene and have confirmed the key role of FGF2 in the transdifferentiation process (Galy et al., 2002). In the rat, neural progenitors from the ciliary body were shown to respond to FGF2 stimulation (Ahmad et al., 2000). Retinal progenitor cells were also isolated from E17 rats and when FGF2 and neurotrophin 3 were added *in vitro*, these cells differentiated into neurons and astrocytes, but never oligodendrocytes (Yang et al., 2002a). Similarly, the human retina between the 10th-13th weeks of gestation, contains progenitor cells with similar differentiation capabilities (Yang et al., 2002b). Finally, another region in the adult rat eye, the corneal limbal epithelium, was also shown to contain progenitor cells that can expand and differentiate into neurons and glia. The limbal epithelium, which, like the neuroepithelium, is ectodermally derived, participates in the regeneration of the cornea throughout life. When limbal epithelial cells are cultured in the presence of mitogens, they begin to express nestin, a neural progenitor marker. Bone morphogenetic protein (BMP) seems to be involved in the process (Zhao et al., 2002b).

Skeletal Muscle

Almost a decade ago Tajbakhsh et al. (1994) reported that some mouse neural tube cells can differentiate into skeletal muscle cells *in vitro* and *in vivo*. Later, in addition to confirming the mouse data, Galli et al. (2000) showed that acutely isolated and clonally derived neural stem cells from both mice and humans differentiate into skeletal myotubes *in vitro* and *in vivo* following transplantation. Recently epidermal growth factor together with basic fibroblast growth factor (bFGF) were found to induce newborn and adult muscle tissue to differentiate into a neural stem cell-like elements. Primary muscle cells and secondary expanded clones formed spherical aggreagates containing neuron-, astrocyte-, and oligodendrocyte-like cells. When transplanted into brains, the muscle-derived neural stem cells developed a neuronal phenotype (Torrente et al., 2002).

Adipose Tissue

Like bone marrow, human adipose tissue is a mesodermal derivative and a progenitor cell population can be isolated from it. These cells can easily be maintained in culture and proliferate well. They are known as processed lipoaspirate (PLA) cells, since they are an incidental byproduct of liposuction procedures. The cells express many markers in common with bone marrow MSCs, and a few distinct proteins. In addition to giving rise to mesodermal cell types, the PLA cells can be induced to make neuronal (NeuN), astrocytic (GFAP) and oligodendrocytic (galactocerebroside) markers (Zuk et al., 2002).

Skin

In 1984 Nurse et al. reported that after sensory nerve injury there is a significant increase in the Merkel-cell population of the skin in the rat. Merkel cells are specialized nerve cells, and the observation was surprising since it suggested that nerve cells are renewed after traumatic injury. Toma studied juvenile and adult rodent skin to explore the origin of these regenerating nerve cells. He isolated neural precursors from dissociated skin after culturing the cells in the presence of EGF and FGF2. These precursors formed spheres, 60% of which started to express nestin after three passages even without the addition of growth factors. When kept in culture for long periods, the cells expressed additional neuronal markers, such as beta-tubulin, neuron-specific enolase and glutamic acid decarboxyglase (GAD) as well as the oligodendrocyte marker CNPase. These authors found a similar population of multipotent cells in the adult human scalp (Toma et al., 2001), and interestingly, Lako et al. (2002) reported that skin progenitor cells from hair follicles are capable of fully restoring the heamatopoietic system. Since Toma et al. (2001) used hairy skin (rodent skin and human scalp) in both of his studies, hair follicle progenitors might in fact be responsible for the effects he observed.

Dental Pulp

In the human dental pulp there are stem cells that form dentin and connective tissue cells. When these progenitor cells are cultured and then transplanted into mice they form a dental pulp-like tissue. Colonies of cells derived from the dental pulp do not seem to be identical. While most generate abundant amounts of dentin, some make very little, but are capable of differentiating into adipose and neural-like cells (Gronthos et al., 2002).

Gastrointestinal (GI) tract

Zulewski et al, (2001) isolated pancreatic islets from E16 and adult rat pancreas slices and cultured the cells in the presence of FGF2 and EGF. They observed cells that separated from the islets, grew processes, and started to express nestin, the neural progenitor marker. They named these cells nestin-positive islet derived progenitors (NIPs). Later, they demonstrated that the human pancreas has similar progenitor cells and that the human NIPs contain a subpopulation of side population cells (similar to the side population cells found in bone marrow) that express nestin (Lechner et al., 2002). In the human gastrointestinal tract nestin-positive cells have also been described in the stomach, and small and large intestinal enteric plexi, but their role has yet to be determined (Vanderwinden et al., 2002).

CONCLUSIONS AND FUTURE DIRECTIONS

In all developmental biology textbooks it is stated that gastrulation marks the onset of changes in cellular behavior that result in formation of the organism. As a consequence of gastrulation, the embryo becomes a trilaminar entity, with an outer layer of ectoderm, an inner layer of endoderm and an intermediate layer of mesoderm. Cells of these layers generate specific organs and tissues that characterize the adult. For the last 50 years we were taught that strict rules govern development. For example, we were told that once a cell made a commitment to a certain dermal lineage, that commitment was irrevocable. While some noncommitted or partially committed stem cells were acknowledged to choose their fates later in life, they were thought to have tissue specificity; that is a stem cell residing in a particular tissue could only differentiate into cells characteristic of that tissue. Thus a hematopoetic stem cell would give rise to new blood cells; a liver stem cell would make new liver cells etc. And there was one more fundamental rule: the nervous system was unique in its inability to renew itself in adulthood.

It was not until neural stem cells were discovered (and suggested to be responsible for adult neurogenesis in restricted areas) that the foundation of textbook dogmas began to shake a bit. Soon a large number of studies emerged from workers in different fields suggesting the unthinkable - that stem cells isolated from a variety of organs seemed to ignore cell (and dermal) lineage boundaries and exhibited more plasticity in their fate choices than previously thought (Figure 1). Numerous studies reported that stem cells can adapt to the microenvironment they find themselves in, and forget or ignore the germ layer origin from which they arose (Bjornson et al., 1999; Clarke and Frisen, 2001;

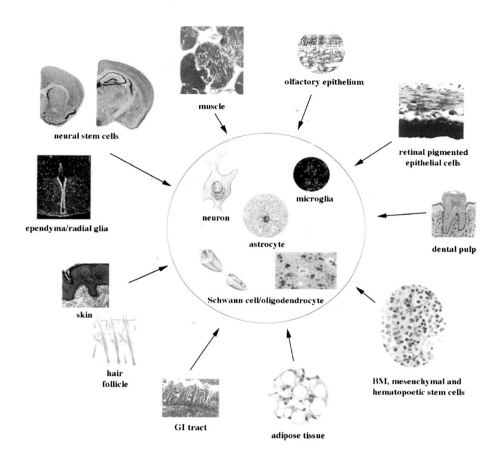

Figure 1. Summary of all tissues able to generate neural cells *in vitro* and/or *in vivo*.

Krause et al., 2001; Poulsom et al., 2002; Theise and Krause, 2002; Greco and Recht, 2003).

Not everyone agreed with this, however. Two major ideas were floated by the "non-believers". They suggested that marked stem cells might occasionally fuse with differentiated cells and that the fusion product would be marker positive and have the phenotype of the differentiated partner (Terada et al., 2002; Ying et al., 2002). Studies supporting this idea were performed *in vitro* though using embryonic stem cells and rather artificial conditions. In a recent study, Spees et al. (2003) claimed that fusion might be common. They cultured human mesenchymal stem cells with small airway epithelial cells which were heat shocked for 30 minutes at 47°C before the stem cells were added. In the absence of stem cells many of the epithelial cells became multinucleated. Heat shock induced fusion has been reported in the past (Ahkong et al., 1973; Ohno-Shosaku and Okada, 1984; Antonov, 1990; Gasser and Most, 1999; Bateman et al., 2000) and this seems to be another example of the same phenomenon. It is not surprising that added BMSCs were caught up in the process. Two recently published studies demonstrated fusion of bone marrow cells with liver cells *in vivo*(Vassilopoulos et al., 2003; Wang et al., 2003). Both groups used the same recipient mouse that lacks a vital enzyme for liver cells to survive and showed that these defective liver cells fuse with healthy transplanted bone marrow cells and are thus capable of producing the enzyme and survive. These studies are very interesting and elegant, but we have to disagree with the conclusion that suggests that fusion must be responsible for all observed and suggested "transdifferentiation" phenomena. While polyploidy is a well known occurrence in plants there are no data to suggest that with the exception of liver and myocytes it is common in mature mammals. In fact, in healthy livers the polyploidy is due to modified cell division cycles suggesting that the polyploid genome may provide protection against loss of tumor suppressors. This may be very important for an organ that is responsible for detoxification (Guidotti et al., 2003). Guidotti et al observed that over 40% of hepatocytes are polyploid by 5-6 weeks of age in healthy rats. In their study using video imaging, they did not comment on cell fusion, but concluded that the polyploidy in the healthy liver is due to nuclear division that is not followed by cell division. Thus, although now cell fusion has been demonstrated *in vivo* in a transgenic (needy) liver, we still have no reason to dismiss data showing transdifferentiation in different organs of the adult body (see Brownstein and Mezey, 2003).

A second criticism has been aimed at studies based on gender-mismatched bone marrow transplants in which the Y chromosome is used as a marker to track donor-derived cells. The presence of Y^+ cells might be explained by microchimerism; male cells might have persisted in organs of adult women who were once pregnant with male fetuses. It has been suggested that women might sometimes deny such pregnancies or even be unaware of them. In a thorough study, Tran et al. (2003) analyzed buccal epithelial cells of female patients who had received bone marrow transplants from a male relative years before. Buccal cell spreads contain the whole cell (as opposed to tissue sections) and as such are ideal for X and Y chromosomal fluorescence in situ hybridization (FISH). After performing FISH on several thousand cells, the authors concluded that many differentiated epithelial cells were Y chromosome-positive, but fusion could not explain this. Furthermore, they studied one patient whose donor and son were both available for genotyping. Using a panel of Y chromosomal microsatellite markers, they showed that the Y chromosomes present in the cheek cells of the patient were identical to the donor's but not the son's, showing in this instance that microchimerism could not have occurred.

In still another study Wagers et al. (2002) published a note indicating that they found "little evidence for developmental plasticity of adult HSCs." They gave GFP-marked bone marrow to irradiated recipients and also used parabiotic animals with a shared circulation to show that no green HSC-derived cells were present in the brains of the recipient mice. Actually, they found one fully developed donor-derived Purkinje cell in one of the brains examined. They dismissed the phenomenon as a rare event, and concluded that "transdifferentiation" of circulating HSCs and/or their progeny is extremely uncommon, if it occurs at all. Their experiment was quite different from other studies in which HSCs were shown to be multipotent. The differences are clearly described in a recent paper (Theise et al., 2003) suggesting that the Wagers' et al. (2002) study was too narrowly focused to refute earlier claims. Their study may also illustrate another general problem in the stem cell field: problems with tracking lineage. Expression of fluorescent proteins has grown very popular for following cell fates because they are so easy to detect. At this point the majority of the data published are based on cell tracking using GFP or EGFP (or LacZ) tags. Not much is known and even less has been published about the reliability of these gene expression systems. In our experience there is a serious discrepancy between the presence of donor-derived cells vs. fluorescent cells in specific tissues. When we use gender mismatched fluorescent bone marrow we see many more Y chromosome-positive than green cells, suggesting that gene silencing occurs. We have also noticed marked variability, even among littermates, in the expression of EGFP (Mezey et al., 2003b). A similar phenomenon has been reported in LacZ Rosa mice (Theise et al., 2003) indicating that the problem is not unique. It appears that negative results obtained with reporter-tagged cells have to be interpreted with caution. At the very least, multiple cells and recipients should be used in such studies.

In summary, based on the numerous studies that have appeared in the last 3-5 years it seems safe to say that stem cells throughout the body have much greater potential and a much less restricted choice of fate than we previously thought. More studies are needed so that we can understand the differences among stem cell pools, how cell fates are chosen, what kind of factors affect differentiation and dedifferentiation of progenitors and even differentiated cells, and whether cells ever completely lose their capacity to change. If it turns out that many stem cells in the adult are capable of site or signal specific alterations of fate and are able to cross lineage boundaries that were previously thought impenetrable, we shall have to rethink some of our present dogmas and modify our textbooks. If we come to accept the notion that most if not all, stem cells retain a good deal of plasticity, the word "transdifferentiation" may have little utility.

Acknowledgement

The author would like to thank Michael J. Brownstein for his help in editing the manuscript as well as for all the stimulating discussions on the project.

REFERENCES

Ahkong QF, Cramp FC, Fisher D, Howell JI, Tampion W, Verrinder M, Lucy JA (1973) Chemically-induced and thermally-induced cell fusion: lipid-lipid interactions. Nat New Biol 242:215-217.

Ahmad I, Tang L, Pham H (2000) Identification of neural progenitors in the adult mammalian eye. Biochem Biophys Res Commun 270:517-521.

Akiyama Y, Radtke C, Kocsis JD (2002) Remyelination of the rat spinal cord by transplantation of identified bone marrow stromal cells. J Neurosci 22:6623-6630.

Altman J (1962) Are new neurons formed in the brains of adult mammals? Science 135:1127-1128.

Antonov PA (1990) Thermofusion of cells. Biochim Biophys Acta 1051:279-281.

Araki M, Takano T, Uemonsa T, Nakane Y, Tsudzuki M, Kaneko T (2002) Epithelia-mesenchyme interaction plays an essential role in transdifferentiation of retinal pigment epithelium of silver mutant quail: localization of FGF and related molecules and aberrant migration pattern of neural crest cells during eye rudiment formation. Dev Biol 244:358-371.

Au E, Roskams AJ (2003) Olfactory ensheathing cells of the lamina propria *in vivo* and *in vitro*. Glia 41:224-236.

Azizi SA, Stokes D, Augelli BJ, DiGirolamo C, Prockop DJ (1998) Engraftment and migration of human bone marrow stromal cells implanted in the brains of albino rats—similarities to astrocyte grafts. Proc Natl Acad Sci U S A 95:3908-3913.

Bateman A, Bullough F, Murphy S, Emiliusen L, Lavillette D, Cosset FL, Cattaneo R, Russell SJ, Vile RG (2000) Fusogenic membrane glycoproteins as a novel class of genes for the local and immune-mediated control of tumor growth. Cancer Res 60:1492-1497.

Bicknese AR, Goodwin HS, Quinn CO, Henderson VC, Chien SN, Wall DA (2002) Human umbilical cord blood cells can be induced to express markers for neurons and glia. Cell Transplant 11:261-264.

Bizzozero G (1894) An address on the growth and regeneration of the organism. Br Med J 38:728-732.

Bjornson CR, Rietze RL, Reynolds BA, Magli MC, Vescovi AL (1999) Turning brain into blood: a hematopoietic fate adopted by adult neural stem cells *in vivo* [see comments]. Science 283:534-537.

Bonilla S, Alarcon P, Villaverde R, Aparicio P, Silva A, Martinez S (2002) Haematopoietic progenitor cells from adult bone marrow differentiate into cells that express oligodendroglial antigens in the neonatal mouse brain. Eur J Neurosci 15:575-582.

Brazelton TR, Rossi FM, Keshet GI, Blau HM (2000) From marrow to brain: expression of neuronal phenotypes in adult mice. Science 290:1775-1779.

Brownstein MJ, Mezey É (2003) Children of the Rib: How Potent Are Adult Stem Cells? PreClinica 2:50-52.

Broxmeyer HE, Douglas GW, Hangoc G, Cooper S, Bard J, English D, Arny M, Thomas L, Boyse EA (1989) Human umbilical cord blood as a potential source of transplantable hematopoietic stem/progenitor cells. Proc Natl Acad Sci USA 86:3828-3832.

Cajal S, May R (1959) Degeneration and regeneration of the nervous system. In, p 750. New York: Hafner.

Calof AL, Mumm JS, Rim PC, Shou J (1998) The neuronal stem cell of the olfactory epithelium. J Neurobiol 36:190-205.

Cameron HA, Gould E (1994) Adult neurogenesis is regulated by adrenal steroids in the dentate gyrus. Neuroscience 61:203-209.

Chanas-Sacre G, Rogister B, Moonen G, Leprince P (2000) Radial glia phenotype: origin, regulation, and transdifferentiation. J Neurosci Res 61:357-363.

Chen J, Li Y, Wang L, Lu M, Zhang X, Chopp M (2001a) Therapeutic benefit of intracerebral transplantation of bone marrow stromal cells after cerebral ischemia in rats. Journal of the Neurological Sciences 189:49-57.

Chen J, Sanberg PR, Li Y, Wang L, Lu M, Willing AE, Sanchez-Ramos J, Chopp M (2001b) Intravenous administration of human umbilical cord blood reduces behavioral deficits after stroke in rats. Stroke 32:2682-2688.

Clarke D, Frisen J (2001) Differentiation potential of adult stem cells. Current Opinion in Genetics & Development 11:575-580.

Corti S, Locatelli F, Donadoni C, Strazzer S, Salani S, Del Bo R, Caccialanza M, Bresolin N, Scarlato G, Comi GP (2002a) Neuroectodermal and microglial differentiation of bone marrow cells in the mouse spinal cord and sensory ganglia. J Neurosci Res 70:721-733.

Corti S, Locatelli F, Strazzer S, Salani S, Del Bo R, Soligo D, Bossolasco P, Bresolin N, Scarlato G, Comi GP (2002b) Modulated generation of neuronal cells from bone marrow by expansion and mobilization of circulating stem cells with *in vivo* cytokine treatment. Experimental Neurology 177:443-452.

Deng W, Obrocka M, Fischer I, Prockop DJ (2001) *In vitro* differentiation of human marrow stromal cells into early progenitors of neural cells by conditions that increase intracellular cyclic AMP. Biochem Biophys Res Commun 282:148-152.

Doetsch F, Caille I, Lim DA, Garcia-Verdugo JM, Alvarez-Buylla A (1999) Subventricular zone astrocytes are neural stem cells in the adult mammalian brain. Cell 97:703-716.

Eglitis MA, Mezey E (1997) Hematopoietic cells differentiate into both microglia and macroglia in the brains of adult mice. Proc Natl Acad Sci U S A 94:4080-4085.

Erices A, Conget P, Minguell JJ (2000) Mesenchymal progenitor cells in human umbilical cord blood. Br J Haematol 109:235-242.

Eriksson PS, Perfilieva E, Bjork-Eriksson T, Alborn AM, Nordborg C, Peterson DA, Gage FH (1998) Neurogenesis in the adult human hippocampus. Nat Med 4:1313-1317.

Feron F, Mackay-Sim A, Andrieu JL, Matthaei KI, Holley A, Sicard G (1999) Stress induces neurogenesis in non-neuronal cell cultures of adult olfactory epithelium. Neuroscience 88:571-583.

Fischer AJ, Reh TA (2001) Transdifferentiation of pigmented epithelial cells: a source of retinal stem cells? Dev Neurosci 23:268-276.

Galli R, Borello U, Gritti A, Minasi MG, Bjornson C, Coletta M, Mora M, De Angelis MG, Fiocco R, Cossu G, Vescovi AL (2000) Skeletal myogenic potential of human and mouse neural stem cells. Nat Neurosci 3:986-991.

Galy A, Neron B, Planque N, Saule S, Eychene A (2002) Activated MAPK/ERK kinase (MEK-1) induces transdifferentiation of pigmented epithelium into neural retina. Dev Biol 248:251-264.

Gasser A, Most J (1999) Generation of multinucleated giant cells *in vitro* by culture of human monocytes with Mycobacterium bovis BCG in combination with cytokine-containing supernatants. Infect Immun 67:395-402.

Goodwin HS, Bicknese AR, Chien SN, Bogucki BD, Oliver DA, Quinn CO, Wall DA (2001) Multilineage differentiation activity by cells isolated from umbilical cord blood: Expression of bone, fat, and neural markers. Biology of Blood and Marrow Transplantation 7:581-588.

Gotz M, Hartfuss E, Malatesta P (2002) Radial glial cells as neuronal precursors: a new perspective on the correlation of morphology and lineage restriction in the developing cerebral cortex of mice. Brain Res Bull 57:777-788.

Gould E, Reeves AJ, Graziano MS, Gross CG (1999) Neurogenesis in the neocortex of adult primates. Science 286:548-552.

Greco B, Recht L (2003) Somatic plasticity of neural stem cells: Fact or fancy? J Cell Biochem 88:51-56.

Gronthos S, Brahim J, Li W, Fisher LW, Cherman N, Boyde A, DenBesten P, Robey PG, Shi S (2002) Stem cell properties of human dental pulp stem cells. J Dent Res 81:531-535.

Guidotti JE, Bregerie O, Robert A, Debey P, Brechot C, Desdouets C (2003) Liver cell polyploidization: A pivotal role for binuclear hepatocytes. J Biol Chem.

Ha Y, Choi JU, Yoon DH, Yeon DS, Lee JJ, Kim HO, Cho YE (2001) Neural phenotype expression of cultured human cord blood cells *in vitro*. Neuroreport 12:3523-3527.

Hartfuss E, Galli R, Heins N, Gotz M (2001) Characterization of CNS precursor subtypes and radial glia. Dev Biol 229:15-30.

Hess DC, Hill WD, Martin-Studdard A, Carroll J, Brailer J, Carothers J (2002) Bone Marrow as a Source of Endothelial Cells and NeuN-Expressing Cells After Stroke. Stroke 33:1362-1368.

Hickey WF, Kimura H (1988) Perivascular microglial cells of the CNS are bone marrow-derived and present antigen *in vivo*. Science 239:290-292.

Hung S-C, Cheng H, Pan C-Y, Tsai MJ, Kao L-S, Ma H-L (2002) *In vitro* Differentiation of Size-Sieved Stem Cells into Electrically Active Neural Cells. Stem Cells 20:522-529.

Kabos P, Ehtesham M, Kabosova A, Black KL, Yu JS (2002) Generation of neural progenitor cells from whole adult bone marrow. Exp Neurol 178:288-293.

Kaplan MS, Hinds JW (1977) Neurogenesis in the adult rat: electron microscopic analysis of light radioautographs. Science 197:1092-1094.

Kaur C, Hao AJ, Wu CH, Ling EA (2001) Origin of microglia. Microsc Res Tech 54:2-9.

Kohyama J, Abe H, Shimazaki T, Koizumi A, Nakashima K, Gojo S, Taga T, Okano H, Hata J, Umezawa A (2001) Brain from bone: Efficient "meta-differentiation" of marrow stroma-derived mature osteoblasts to neurons with Noggin or a demethylating agent. Differentiation 68:235-244.

Kondo T, Raff M (2000) Oligodendrocyte precursor cells reprogrammed to become multipotential CNS stem cells. Science 289:1754-1757.

Kopen GC, Prockop DJ, Phinney DG (1999) Marrow stromal cells migrate throughout forebrain and cerebellum, and they differentiate into astrocytes after injection into neonatal mouse brains. Proc Natl Acad Sci U S A 96:10711-10716.

Kornack DR, Rakic P (1999) Continuation of neurogenesis in the hippocampus of the adult macaque monkey. Proc Natl Acad Sci U S A 96:5768-5773.

Krause DS, Theise ND, Collector MI, Henegariu O, Hwang S, Gardner R, Neutzel S, Sharkis SJ (2001) Multi-organ, multi-lineage engraftment by a single bone marrow-derived stem cell. Cell 105:369-377.

Krivit W, Lipton ME, Lockman LA, Tsai M, Dyck PJ, Smith S, Ramsay NK, Kersey J (1987) Prevention of deterioration in metachromatic leukodystrophy by bone marrow transplantation. Am J Med Sci 294:80-85.

Kruger GM, Mosher JT, Bixby S, Joseph N, Iwashita T, Morrison SJ (2002) Neural crest stem cells persist in the adult gut but undergo changes in self-renewal, neuronal subtype potential, and factor responsiveness. Neuron 35:657-669.

LaBonne C, Bronner-Fraser M (1998) Induction and patterning of the neural crest, a stem cell-like precursor population. J Neurobiol 36:175-189.

Lackie JM, Dow JAT (1999) The dictionary of cell and molecular biology, Third edition Edition. London: Academic Press.

Lako M, Armstrong L, Cairns PM, Harris S, Hole N, Jahoda CAB (2002) Hair follicle dermal cells repopulate the mouse haematopoietic system. Journal of Cell Science 115:3967-3974.

Lechner A, Leech CA, Abraham EJ, Nolan AL, Habener JF (2002) Nestin-positive progenitor cells derived from adult human pancreatic islets of Langerhans contain side population (SP) cells defined by expression of the ABCG2 (BCRP1) ATP-binding cassette transporter. Biochem Biophys Res Commun 293:670-674.

Li Y, Chen J, Wang L, Zhang L, Lu M, Chopp M (2001) Intracerebral transplantation of bone marrow stromal cells in a 1-methyl-4-phenyl-1,2,3,6-tetrahydropyridine mouse model of Parkinson's disease. Neurosci Lett 316:67-70.

Li Y, Chopp M, Chen J, Wang L, Gautam SC, Xu YX, Zhang Z (2000) Intrastriatal transplantation of bone marrow nonhematopoietic cells improves functional recovery after stroke in adult mice [In Process Citation]. J Cereb Blood Flow Metab 20:1311-1319.

Lin G, Finger E, Gutierrez-Ramos JC (1995) Expression of CD34 in endothelial cells, hematopoietic progenitors and nervous cells in fetal and adult mouse tissues. Eur J Immunol 25:1508-1516.

Lois C, Alvarez-Buylla A (1993) Proliferating subventricular zone cells in the adult mammalian forebrain can differentiate into neurons and glia. Proc Natl Acad Sci U S A 90:2074-2077.

Lu D, Mahmood A, Wang L, Li Y, Lu M, Chopp M (2001a) Adult bone marrow stromal cells administered intravenously to rats after traumatic brain injury migrate into brain and improve neurological outcome. Neuroreport 12:559-563.

Lu D, Li Y, Wang L, Chen J, Mahmood A, Chopp M (2001b) Intraarterial administration of marrow stromal cells in a rat model of traumatic brain injury. J Neurotrauma 18:813-819.

Lu D, Sanberg PR, Mahmood A, Li Y, Wang L, Sanchez-Ramos J, Chopp M (2002) Intravenous administration of human umbilical cord blood reduces neurological deficit in the rat after traumatic brain injury. Cell Transplant 11:275-281.

Magavi SS, Macklis JD (2002) Induction of neuronal type-specific neurogenesis in the cerebral cortex of adult mice: manipulation of neural precursors in situ. Brain Res Dev Brain Res 134:57-76.

Magavi SS, Leavitt BR, Macklis JD (2000) Induction of neurogenesis in the neocortex of adult mice. Nature 405:951-955.

Mahmood A, Lu D, Yi L, Chen JL, Chopp M (2001a) Intracranial bone marrow transplantation after traumatic brain injury improving functional outcome in adult rats. J Neurosurg 94:589-595.

Mahmood A, Lu D, Wang L, Li Y, Lu M, Chopp M (2001b) Treatment of traumatic brain injury in female rats with intravenous administration of bone marrow stromal cells. Neurosurgery 49:1196-1203; discussion 1203-1194.

McKercher SR, Torbett BE, Anderson KL, Henkel GW, Vestal DJ, Baribault H, Klemsz M, Feeney AJ, Wu GE, Paige CJ, Maki RA (1996) Targeted disruption of the PU.1 gene results in multiple hematopoietic abnormalities. Embo J 15:5647-5658.

Messier B, Leblond C (1960) Cell proliferation and migration as revealed by radioautography after injection of thymidine-H3 into male rats and mice. Am J Anat 106:247-285.

Mezey E, Chandross KJ, Harta G, Maki RA, McKercher SR (2000) Turning blood into brain: cells bearing neuronal antigens generated *in vivo* from bone marrow. Science 290:1779-1782.

Mezey E, Key S, Vogelsang G, Szalayova I, Lange GD, Crain B (2003a) Transplanted bone marrow generates new neurons in human brains. Proc Natl Acad Sci U S A 100:1364-1369.

Mezey E, Nagy A, Szalayova I, Key S, Bratincsak A, Baffi J, Shahar T (2003b) Comment on "Failure of bone marrow cells to transdifferentiate into neural cells *in vivo*". Science 299:1184; author reply 1184.

Mochii M, Ono T, Matsubara Y, Eguchi G (1998) Spontaneous transdifferentiation of quail pigmented epithelial cell is accompanied by a mutation in the Mitf gene. Dev Biol 196:145-159.

Mujtaba T, Mayer-Proschel M, Rao MS (1998) A common neural progenitor for the CNS and PNS. Dev Biol 200:1-15.

Mullen RJ, Buck CR, Smith AM (1992) NeuN, a neuronal specific nuclear protein in vertebrates. Development 116:201-211.

Nurse CA, Macintyre L, Diamond J (1984) Reinnervation of the rat touch dome restores the Merkel cell population reduced after denervation. Neuroscience 13:563-571.

Ohno-Shosaku T, Okada Y (1984) Facilitation of electrofusion of mouse lymphoma cells by the proteolytic action of proteases. Biochem Biophys Res Commun 120:138-143.

Opas M, Davies JR, Zhou Y, Dziak E (2001) Formation of retinal pigment epithelium *in vitro* by transdifferentiation of neural retina cells. Int J Dev Biol 45:633-642.

Palmer TD, Ray J, Gage FH (1995) FGF-2-responsive neuronal progenitors reside in proliferative and quiescent regions of the adult rodent brain. Mol Cell Neurosci 6:474-486.

Parmar M, Skogh C, Bjorklund A, Campbell K (2002) Regional specification of neurosphere cultures derived from subregions of the embryonic telencephalon. Mol Cell Neurosci 21:645-656.

Poulsom R, Alison MR, Forbes SJ, Wright NA (2002) Adult stem cell plasticity. J Pathol 197:441-456.

Priller J, Persons DA, Klett FF, Kempermann G, Kreutzberg GW, Dirnagl U (2001a) Neogenesis of cerebellar Purkinje neurons from gene-marked bone marrow cells *in vivo*. Journal of Cell Biology 155:733-738.

Priller J, Flugel A, Wehner T, Boentert M, Haas CA, Prinz M, Fernandez-Klett F, Prass K, Bechmann I, de Boer BA, Frotscher M, Kreutzberg GW, Persons DA, Dirnagl U (2001b) Targeting gene-modified hematopoietic cells to the central nervous system: Use of green fluorescent protein uncovers microglial engraftment. Nature Medicine 7:1356-1361.

Qian X, Davis AA, Goderie SK, Temple S (1997) FGF2 concentration regulates the generation of neurons and glia from multipotent cortical stem cells. Neuron 18:81-93.

Reh TA, Pittack C (1995) Transdifferentiation and retinal regeneration. Semin Cell Biol 6:137-142.

Reynolds R, Hardy R (1997) Oligodendroglial progenitors labeled with the O4 antibody persist in the adult rat cerebral cortex *in vivo*. J Neurosci Res 47:455-470.

Rosada C, Justesen J, Melsvik D, Ebbesen P, Kassem M (2002) The Human Umbilical Cord Blood: A Potential Source for Osteoblast Progenitor Cells. Calcif Tissue Int 4:4.

Sakaguchi DS, Janick LM, Reh TA (1997) Basic fibroblast growth factor (FGF-2) induced transdifferentiation of retinal pigment epithelium: generation of retinal neurons and glia. Dev Dyn 209:387-398.

Sanchez-Ramos J, Song S, Cardozo-Pelaez F, Hazzi C, Stedeford T, Willing A, Freeman TB, Saporta S, Janssen W, Patel N, Cooper DR, Sanberg PR (2000) Adult bone marrow stromal cells differentiate into neural cells *in vitro*. Exp Neurol 164:247-256.

Sanchez-Ramos JR, Song SJ, Kamath SG, Zigova T, Willing A, Cardozo-Pelaez F, Stedeford T, Chopp M, Sanberg PR (2001) Expression of neural markers in human umbilical cord blood. Experimental Neurology 171:109-115.

Sasaki M, Honmou O, Akiyama Y, Uede T, Hashi K, Kocsis JD (2001) Transplantation of an acutely isolated bone marrow fraction repairs demyelinated adult rat spinal cord axons. Glia 35:26-34.

Shetty AK, Turner DA (1998) *In vitro* survival and differentiation of neurons derived from epidermal growth factor-responsive postnatal hippocampal stem cells: inducing effects of brain-derived neurotrophic factor. J Neurobiol 35:395-425.

Sicard G, Feron F, Andrieu JL, Holley A, Mackay-Sim A (1998) Generation of neurons from a nonneuronal precursor in adult olfactory epithelium *in vitro*. Ann N Y Acad Sci 855:223-225.

Smart I (1961) The subependymal layer of the mouse brain and its cell production as shown by autoradiography after thymidine-3H injection. J Comp Neurol 116:325-348.

Spees JL, Olson SD, Ylostalo J, Lynch PJ, Smith J, Perry A, Peister A, Wang MY, Prockop DJ (2003) Differentiation, cell fusion, and nuclear fusion during ex vivo repair of epithelium by human adult stem cells from bone marrow stroma. Proc Natl Acad Sci U S A.

Tajbakhsh S, Vivarelli E, Cusella-De Angelis G, Rocancourt D, Buckingham M, Cossu G (1994) A population of myogenic cells derived from the mouse neural tube. Neuron 13:813-821.

Terada N, Hamazaki T, Oka M, Hoki M, Mastalerz DM, Nakano Y, Meyer EM, Morel L, Petersen BE, Scott EW (2002) Bone marrow cells adopt the phenotype of other cells by spontaneous cell fusion. Nature 416:542-545.

Terskikh AV, Easterday MC, Li L, Hood L, Kornblum HI, Geschwind DH, Weissman IL (2001) From hematopoiesis to neuropoiesis: evidence of overlapping genetic programs. Proc Natl Acad Sci U S A 98:7934-7939.

Theise ND, Krause DS (2002) Toward a new paradigm of cell plasticity. Leukemia 16:542-548.

Theise ND, Krause DS, Sharkis S (2003) Comment on "Little evidence for developmental plasticity of adult hematopoietic stem cells". Science 299:1317; author reply 1317.

Toma JG, Akhavan M, Fernandes KJL, Barnabe-Heider F, Sadikot A, Kaplan DR, Miller FD (2001) Isolation of multipotent adult stem cells from the dermis of mammalian skin. Nature Cell Biology 3:778-784.

Tomita M, Adachi Y, Yamada H, Takahashi K, Kiuchi K, Oyaizu H, Ikebukuro K, Kaneda H, Matsumura M, Ikehara S (2002) Bone marrow-derived stem cells can differentiate into retinal cells in injured rat retina. Stem Cells 20:279-283.

Torrente Y, Belicchi M, Pisati F, Pagano SF, Fortunato F, Sironi M, D'Angelo MG, Parati EA, Scarlato G, Bresolin N (2002) Alternative sources of neurons and glia from somatic stem cells. Cell Transplant 11:25-34.

Tran S, Pillemer SR, Dutra A, Barrett J, Brownstein MJ, Key S, Pak E, Leakan RA, Yamada KM, Baum BJ, Mezey E (2003) Human bone marrow-derived cells differentiate into buccal epithelial cells *in vivo* without fusion. Lancet.

Tropepe V, Coles BL, Chiasson BJ, Horsford DJ, Elia AJ, McInnes RR, van der Kooy D (2000) Retinal stem cells in the adult mammalian eye. Science 287:2032-2036.

Unger ER, Sung JH, Manivel JC, Chenggis ML, Blazar BR, Krivit W (1993) Male donor-derived cells in the brains of female sex-mismatched bone marrow transplant recipients: a Y-chromosome specific in situ hybridization study. J Neuropathol Exp Neurol 52:460-470.

Vanderwinden JM, Gillard K, De Laet MH, Messam CA, Schiffmann SN (2002) Distribution of the intermediate filament nestin in the muscularis propria of the human gastrointestinal tract. Cell Tissue Res 309:261-268.

Vassilopoulos G, Wang PR, Russell DW (2003) Transplanted bone marrow regenerates liver by cell fusion. Nature 422:901-904.

Wagers AJ, Sherwood RI, Christensen JL, Weissman IL (2002) Little evidence for developmental plasticity of adult hematopoietic stem cells. Science 297:2256-2259.

Wang X, Willenbring H, Akkari Y, Torimaru Y, Foster M, Al-Dhalimy M, Lagasse E, Finegold M, Olson S, Grompe M (2003) Cell fusion is the principal source of bone-marrow-derived hepatocytes. Nature 422:897-901.

Weimann JM, Charlton CA, Brazelton TR, Hackman RC, Blau HM (2003) Contribution of transplanted bone marrow cells to Purkinje neurons in human adult brains. Proc Natl Acad Sci U S A.

Weiss S, Dunne C, Hewson J, Wohl C, Wheatley M, Peterson AC, Reynolds BA (1996) Multipotent CNS stem cells are present in the adult mammalian spinal cord and ventricular neuroaxis. J Neurosci 16:7599-7609.

Wohl CA, Weiss S (1998) Retinoic acid enhances neuronal proliferation and astroglial differentiation in cultures of CNS stem cell-derived precursors. J Neurobiol 37:281-290.

Woodbury D, Schwarz EJ, Prockop DJ, Black IB (2000) Adult rat and human bone marrow stromal cells differentiate into neurons [In Process Citation]. J Neurosci Res 61:364-370.

Yang P, Seiler MJ, Aramant RB, Whittemore SR (2002a) Differential lineage restriction of rat retinal progenitor cells *in vitro* and *in vivo*. J Neurosci Res 69:466-476.

Yang P, Seiler MJ, Aramant RB, Whittemore SR (2002b) *In vitro* isolation and expansion of human retinal progenitor cells. Exp Neurol 177:326-331.

Ying QL, Nichols J, Evans EP, Smith AG (2002) Changing potency by spontaneous fusion. Nature 416:545-548.

Zhao LR, Duan WM, Reyes M, Keene CD, Verfaillie CM, Low WC (2002a) Human bone marrow stem cells exhibit neural phenotypes and ameliorate neurological deficits after grafting into the ischemic brain of rats. Exp Neurol 174:11-20.

Zhao X, Das AV, Thoreson WB, James J, Wattnem TE, Rodriguez-Sierra J, Ahmad I (2002b) Adult Corneal Limbal Epithelium: A Model for Studying Neural Potential of Non-Neural Stem Cells/Progenitors. Developmental Biology 250:317-331.

Zigova T, Song S, Willing AE, Hudson JE, Newman MB, Saporta S, Sanchez-Ramos J, Sanberg PR (2002) Human umbilical cord blood cells express neural antigens after transplantation into the developing rat brain. Cell Transplant 11:265-274.

Zuk PA, Zhu M, Ashjian P, De Ugarte DA, Huang JI, Mizuno H, Alfonso ZC, Fraser JK, Benhaim P, Hedrick MH (2002) Human adipose tissue is a source of multipotent stem cells. Mol Biol Cell 13:4279-4295.

Zulewski H, Abraham EJ, Gerlach MJ, Daniel PB, Moritz W, Muller B, Vallejo M, Thomas MK, Habener JF (2001) Multipotential Nestin-Positive Stem Cells Isolated From Adult Pancreatic Islets Differentiate Ex Vivo Into Pancreatic Endocrine, Exocrine, and Hepatic Phenotypes. Diabetes 50:521-533.

Chapter 7

Neural Stem Cell Purification and Clonal Analysis

Alexandra Capela, Stanley Tamaki and Nobuko Uchida

INTRODUCTION

During embryonic development, cell proliferation occurs at an incredibly fast pace. The immense proliferative and differentiative potential of stem cells is used to fuel the need for growth and diversification of a developing organism. Stem cells are generally defined as cells that are capable of self-renewal (creation of more stem cells) and multilineage differentiation (generation of an array of differentiated cells characteristic of a particular organ or system (reviewed by Potten and Loeffler, 1990; Morrison et al., 1997; Fuchs and Segre, 2000). Tissue-specific somatic stem cells in the adult animal also contribute to the normal cycle of regeneration of high turnover tissues such as the skin, hair, lining of the small intestine and blood cells (Fuchs and Segre, 2000).

Much of our understanding of stem cell-based systems comes from studies of hematopoiesis. The hematopoietic stem cell (HSC) is the most well characterized stem cell, and a great deal is known about its molecular characteristics and regulation. Given this, the HSC has become a standard in the study of stem cell biology, guiding experiments on other, less well understood stem cell types. The hematopoietic and nervous systems appear to represent extreme opposites with respect to regeneration potential and tissue organization. Hematopoietic stem cells (HSC) exist throughout the developmental period and into adulthood, generating large numbers of progeny throughout life. Enormous numbers of blood cells (calculated to be 10^{10}) are generated daily in humans. This incredible proliferative capacity has been utilized in reconstitution experiments of lethally ablated animals leading to unequivocal identification of the HSC. HSCs reside in the bone marrow, associated with stromal cells and are amenable to single cell separation and purification. In addition, adult HSCs can be purified from peripheral blood upon cytokine stimulation (mobilized HSCs). In contrast, the nervous system, with its wealth of diverse and intricately connected cells, was initially thought to be built during the developmental period from precursors with limited plasticity. Moreover, the irreversibility of tissue loss after injury to the adult brain or spinal cord led to the dogma that the mammalian

From: *Neural Stem Cells: Development and Transplantation*
Edited by: Jane E. Bottenstein © 2003 Kluwer Academic Publishers, Norwell, MA

nervous system had little regenerative potential. Yet over the course of the last decade it has become clear that the nervous system is built using multipotent, environmentally-responsive stem cells, and that neural stem cells (NSC) are present and active in the adult nervous system where they are involved in continued neuron generation. Along with the description of NSCs came an outpouring of excitement for the possibility of neural regeneration, which spilled from science research centers into the public domain. Thus despite fundamental differences in the structure of blood and neural tissue, both are generated from stem cells present throughout life.

The relative quiescence of NSCs in the adult and the structural rigidity of the brain parenchyma has made the nervous system recalcitrant to classical ablation/reconstitution studies used to identify HSCs. Thus the NSC remains more enigmatic, with identification relying mostly on retrospective methods – in which the presence of a stem cell is deduced by the formation of a stem-like clone. Despite this, the NSC offers an important advantage over the HSC as a model for studying aspects of stem cell biology. A significant limitation of HSCs is their resistance to extensive expansion (self-renewal) in culture. In contrast, NSCs can be more easily expanded *in vitro* and this feature has enabled their retrospective identification (Reynolds and Weiss, 1992). Culturing of heterogeneous primary single cell brain preparation can result in rapid elimination of committed progenitors and differentiated cells through cell death (Galli et al., 2003). The undifferentiated NSCs are forced into active proliferation by mitogen stimulation leading to their positive selection.

NSCs capable of multilineage differentiation and self-renewal have been identified by *in vitro* selection of cells derived from the embryonic murine septum, cortex, striatum, spinal cord and human fetal forebrain (Temple, 1989; Davis and Temple, 1994; Reynolds and Weiss, 1996; Mayer-Proschel et al., 1997; Svendsen et al., 1998; Carpenter et al., 1999; Qian et al., 2000) . NSCs have also been isolated from adult regions using this method. Neurogenesis continues throughout life in a variety of species including birds, rodents, non-human primates and humans (Goldman and Nottebohm, 1983; Lois and Alvarez-Buylla, 1993; Kempermann et al., 1997; Eriksson et al., 1998; Gould et al., 1999; Kornack and Rakic, 1999). In contrast to the widespread neurogenesis of the embryonic brain, neurogenesis in the adult brain occurs in very restricted places, the subventricular zone (SVZ) surrounding the lateral ventricles which continually feeds new neurons into the olfactory bulbs through the rostral migratory stream (RMS) and the dentate gyrus (DG) of the hippocampus (reviewed by Kempermann et al., 1997; Alvarez-Buylla and Garcia-Verdugo, 2002). Recently, it was reported that hippocampal pyramidal neurons can be regenerated following hypoxia upon growth factor-induced mobilization of endogenous progenitor cells located in the periventricular area and/or hippocampal parenchyma (Nakatomi et al., 2002; see Chapter 12 in this volume).

The extensive proliferative capacity of NSCs has allowed the establishment of NSC lines derived from adult rodent and human SVZ and hippocampus through culture in the presence of the growth factors FGF2 and EGF (Reynolds and Weiss, 1992; Morshead et al., 1994; Gritti et al., 1996; Palmer et al., 1997; Kukekov et al., 1999; Palmer et al., 2000; Roy et al., 2000b). These cells grow as adherent monolayers or as free-floating neurospheres and possess the properties of self-renewal and multilineage differentiation expected of NSCs.

In this chapter we review the methodologies used for NSC purification and clonal analysis and discuss how NSC purification is critical for our understanding of the biology of these rare and exciting cells.

WHY IS IT IMPORTANT TO IDENTIFY MARKERS FOR STEM CELLS?

Until very recently, NSCs were identified largely retrospectively, based on their selective growth advantage *in vitro*. Neural tissue is dissociated to single cells and plated in a culture dish in the presence of growth factors. While most neural cells have limited proliferative potential, NSCs can grow successfully in this environment and quickly generate large multicell clones. These clones can be sub-cloned to generate more stem cells, demonstrating self-renewal. Thus, a single cell grown in a culture dish with appropriate growth factors that proliferates extensively and generates both neuronal and glial cells is operationally defined as a NSC. However, using this method to define stem cells is cumbersome. Most importantly, by the time we recognize it as such, the original NSC is lost as it has already divided and generated progeny, and is thus unavailable for study. This significantly limits our ability to investigate the characteristics of NSCs as they exist *in vivo*, rather we can only study the culture-expanded NSCs emanating from these clones. The description of unique cell surface markers has been critical for the identification of other somatic stem cell types, such as HSCs, allowing them to be isolated (Morrison et al., 1995). Hence much research effort has been placed on finding markers for NSCs, which would allow their enrichment for *in vitro* studies and, most importantly, would enable their identification in situ, revealing their endogenous niches and behavior *in vivo*. Identification of NSC markers is thus a critical step towards making progress in understanding NSC biology (see Chapter 3 in this volume).

METHODS FOR ISOLATING NEURAL STEM CELLS

Methods used for isolating/enriching NSCs include the generation of viable single cell suspensions, which are used to screen collections of monoclonal antibodies against cell surface antigens. This critical cell dissociation step has been developed by researchers interested in studying neural tissue in culture, for which large numbers of highly viable cells have to be obtained. Other techniques for NSC purification take advantage of their physical and physiological properties such as size, buoyancy and dye binding/influx/efflux kinetics.

Immunopanning and immunomagnetic sorting

Immunopanning and immunomagnetic sorting are two simple and widely used techniques for cell enrichment. Both methods employ the use of solid matrices to which antibodies directed against specific cell surface antigens are bound. Cell suspensions are placed in contact with the antibody-bound matrix and the cells expressing specific surface antigens form a complex with the matrix. Cell populations are either positively selected or removed by negative selection. Both methods are inexpensive and yield highly enriched populations.

Immunopanning

For immunopanning, tissue culture dishes are incubated with antibodies diluted in protein-free buffer at basic pH in order to electrostatically attach the antibody to the plastic. The plates are washed to remove unbound antibody and blocked with protein to prevent non-specific cell attachment. Single cell suspensions are incubated in the plates. Cells are bound via antibodies directed against a cell surface epitope or by the use of secondary antibodies to cells preincubated with a primary antibody. After a 1-2 hr incubation period, the unbound cells are removed by gentle washing and collected in the case of negative depletion. For positive selection, attached cells are recovered by forcefully pipeting plates with media, using cell scrapers, or mild enzyme treatment.

Sequential immunopanning procedures have been used to isolate highly purified astrocyte precursor cells (APCs) from the developing rat optic nerve. Dishes coated with Thy1.1 and A2B5 antibodies are first used to deplete the cell suspension of meningeal, microglial (Thy1.1), and oligodendrocyte progenitor (A2B5) cells, respectively. APCs are then isolated from the negatively selected cells using a positive selection dish coated with the C5 antibody (Mi and Barres, 1999). Using a similar approach, neuronal precursors are isolated from the spinal cords of rat embryos by negative selection in dishes coated with A2B5 antibody followed by positive selection in E-NCAM coated dishes (Mayer-Proschel et al., 1997). Oligodendrocyte progenitors of the O4$^+$/O1$^-$ phenotype are also isolated from neonatal cortices using negative selection first (O1) followed by positive selection (O4; Ingraham et al., 1999). Immunopanning techniques typically result in high viability and purity of the selected cell population (>95%).

Immunomagnetic sorting

Immunomagnetic selection employs the use of a powerful magnet to capture cells indirectly labeled with iron beads. Two types of beads are commercially available: "nanobeads" approximately 50 nm in diameter (Miltenyi Biotec) and "microbeads" around 5 μm in diameter (Dynal). Single cell suspensions are incubated with primary antibodies that recognize cell surface epitopes on target cells. Secondary antibodies conjugated to beads are added to bind to the primary antibody-cell complex. A strong magnet is used to immobilize the cells associated with the beads; bead-free cells are removed by washing.

The beads are biocompatible and do not seem to affect the viability or proliferative potential of rodent brain cells. Mouse embryonic cortical cells have been separated into distinct subpopulations using both types of beads without any deleterious effect on their survival or growth at clonal densities (Capela and Temple, unpublished observations). Others have reported success in purifying rat glial cells (Wright et al., 1997) as well as Notch$^+$ ependymal cells (Johansson et al., 1999) from adult mice. Although both types of beads perform equally well in separating highly pure cell populations, the use of nanobeads for positive selection is preferred because their small size does not interfere with clonal analysis assays. In addition, beads are not toxic to recipient animals transplanted with bead-bound cells (Uchida et al., 1998).

A variety of primary antibodies conjugated with beads as well as custom antibody-bead conjugation kits are now commercially available making immunomagnetic sorting rapid and simple. In general, immunomagnetic sorting is faster and more reproducible than immunopanning; this is because beads can be used in solution, making the attachment of target cells more efficient. Overall, both methods are inexpensive, reliable, and yield relatively pure viable populations of target cells.

Fluorescence-activated cell sorting

Immunopanning as well as immunomagnetic sorting have been mostly used to select restricted progenitors or differentiated cell populations for which specific antibody markers are known and not to actively purify NSCs. One reason for this is that until very recently, no specific NSC surface markers had been identified. On the other hand, the recent advances in the NSC field are occurring at a time point when powerful cell separation techniques like fluorescence-activated cell sorting (FACS) are becoming more accessible to a wider scientific community.

FACS involves sophisticated instrumentation that combines fluidic, optical, and electronic systems to achieve the physical separation of discrete subpopulations of cells. Typically, unique cell populations can be resolved based on light scatter properties, immunofluorescent labels and dye binding characteristics. Briefly, cells are moved in a fluid medium through the path of a focused laser excitation source, while fluorescence and light scatter measurements are collected on a per cell basis. The cells are then segregated into individual microdroplets whereas droplets containing cells of interest can be electrostatically charged, deflected, and collected in individual containers, including tissue culture plates. Purified subpopulations of cells can then be subjected to further analyses.

In addition to immunofluorescent labels and dye binding characteristics, fluorescent reporter gene products such as green fluorescent protein (GFP) can be employed to identify specific cell populations.

The use of GFP as a reporter gene for the expression of candidate intracellular neural stem cell markers.

Nestin, Musashi1 and Sox2 are molecules that are expressed very early during neural development and thus are candidates for NSC markers. Nestin is an intermediate filament protein present in early neural progenitors (Hockfield and McKay, 1985; Lendahl et al., 1990) whose expression declines in more differentiated progeny. Musashi1 is an RNA binding protein whose expression pattern in the early CNS is similar to that of Nestin: it is expressed in neural precursor cells that can generate neurons and glia (Sakakibara et al., 1996). Sox2 is a transcription factor involved in neuroepithelial determination (Li et al., 1998; Zappone et al., 2000; Avilion et al., 2003). Regulatory sequences in the sox2 gene have been shown to direct expression of a reporter gene to NSCs in the mouse embryonic telencephalon (Zappone et al., 2000). In addition, Doestch and colleagues (Doetsch et al., 1999b; Doetsch et al., 1999a) showed that adult mouse subventricular zone GFAP[+] cells divide, are neurogenic *in vivo* and generate self-renewing multipotent neurospheres *in vitro*, and are therefore NSCs. Although the protein is only expressed around birth in mice, GFAP mRNA is present earlier in the developing neuroepithelium, in cells that have neurogenic potential (Malatesta et al., 2000). Unfortunately, all of the important markers referred to so far are intracellular, limiting the usefulness of antibodies raised against them to isolate live NSCs. The generation of transgenic mice or primary cells carrying reporter genes such as GFP, under the control of developmentally regulated promoters and/or enhancers of NSC candidate markers provided the much needed solution. Using this approach, putative NSCs can be purified/enriched by GFP-based FACS analysis.

Transgenic mice. A team lead by Dr. Okano generated a transgenic mouse that expresses enhanced green fluorescent protein (EGFP) under the control of the neural-specific enhancer of the nestin gene (Kawaguchi et al., 2001; Murayama et al., 2002). Cell suspensions derived from the forebrain of transgenic mouse embryos were then subjected to FACS analysis, and three populations were defined based on EGFP fluorescence intensity. All three cell subpopulations were cultured under conditions that favor the generation of multipotent, self-renewing neurospheres (*in vitro* diagnosis of NSCs) to reveal a correlation between the intensity of EGFP fluorescence and the capacity to generate neurospheres: cells with the highest fluorescence (F^{++}) generate the majority of neurospheres. Cells that do not fluoresce generate negligible numbers of neurospheres, as do F^+ cells (medium fluorescence). In agreement with this result, virtually all F^{++} cells express musashi1 and are negative for neuronal markers. Interestingly, an estimated 20% of F^{++} cells generate neurospheres, meaning that not all F^{++} are true NSCs.

Generation of neurospheres by cells derived from the periventricular area of the adult forebrain known to harbor NSCs revealed similar results. However, the frequency of neurosphere formation was significantly higher in the F^+ subpopulation than in the F^{++}, probably reflecting NSC heterogeneity. Independently of the developmental stage analyzed, Nestin expression correlated well with an immature phenotype.

Transfection or infection methods. Another type of promoter-based separation of cell populations that uses GFP as reporter gene is the one pioneered by Goldman and colleagues, initially employed to purify neuronal and oligodendroglial populations (Wang et al., 1998; Roy et al., 1999; Roy et al., 2000a; Roy et al., 2000b; Wang et al., 2000; Nunes et al., 2003) from rodent and human brains. This methodology was then used to enrich for early progenitors and NSCs from fetal and adult human CNS cells. Human fetal and adult forebrain ventricular zone (VZ) cells were either transfected with a plasmid vector containing GFP under the nestin enhancer (E/nestin:EGFP; Roy et al., 2000a; Roy et al., 2000b) or infected with two adenoviruses containing GFP under the control of the musashi1 promoter or the nestin enhancer (Keyoung et al., 2001). After infection, GFP$^+$ cells were selected by FACS and tested for NSC characteristic growth and differentiation *in vitro* and *in vivo*. Both musashi1$^+$ and Nestin$^+$ cells gave rise to neurospheres that produced neuronal and glial progeny (multipotency) and more neurospheres (self-renewal). Upon xenotransplantation into the ventricular space of E17 rat embryos, both human-derived musashi1$^+$ and nestin$^+$ cells generated neuronal progeny that migrated to appropriate locations in the rat cortex. Upon transplantation into P0-2 rats, human cells generated mainly glia and few neurons located mainly in the RMS and olfactory bulb.

Similarly to the transgenic methodology described before, selection of nestin[+] and musashi1[+] cells results in a substantial enrichment in neurosphere generating, multilineage engrafting NSCs, although these sorted populations also include a high percentage of non-NSCs. One drawback of these techniques (infection or transfection followed by FACS) is the requirement to culture cells for a certain period of time. Therefore, fresh isolation of NSCs is not possible using this method. On the other hand, it offers the unique opportunity to enrich for specific cells (including nontransgenic cells such as those derived from human tissues) based on intracellular markers.

Neural stem cell surface markers

A breakthrough study by Morrison and colleagues (Morrison et al., 1999) showed that mouse neural crest stem cells (NCSCs) which generate all the cells in the peripheral nervous system (PNS) can be isolated from the E14.5 sciatic nerve by FACS using antibodies against the surface marker p75 (the low affinity neurotrophin receptor) and P_0 (a peripheral myelin protein). The p75[+]P_0^- fraction contains cells that behave like NCSCs *in vitro*: generate multipotent clones containing neurons, Schwann cells and smooth muscle-like myofibroblasts. Upon transplantation into host chick embryos, freshly isolated p75[+]P_0^- cells generate neurons and glia. This way, it was demonstrated that neurogenic potential is endogenous to freshly isolated p75[+]P_0^- cells and not a culture-acquired property. The capacity to directly isolate NCSCs by flow cytometry provided, for the first time, the unique opportunity to test the potential of neural crest stem cells *in vivo* upon transplantation, bypassing culture periods that might irreversibly change the properties of such cells.

A parallel approach was taken by Uchida et al., (2000) for the isolation of human neural stem cells (hNSCs) from the fetal forebrain. They demonstrated that the prospective hNSC has the CD133[+]CD34[-]CD45[-] phenotype, and generates multilineage self-renewing neurospheres *in vitro* (Figure 1). Interestingly, the antigen recognized by the CD133 is a transmembrane protein initially described as a human hematopoietic stem cell (HSC) marker (Miraglia et al., 1997; Yin et al., 1997). The frequency of neurosphere generation is around 1 in 31 cells. This subpopulation is expandable *in vitro* for more than 15 passages. Moreover, upon transplantation into the ventricles of neonatal immunodeficient NOD-Scid mice, both *in vitro* expanded or freshly isolated CD133[+]CD34[-]CD45[-] cells generate neurons and glia in the SVZ, olfactory bulb and hippocampal dentate gyrus. Purified CD133[+]CD34[-]CD45[-] cells can be stably transduced with a lentivirus containing GFP as a reporter gene (Tamaki et al., 2002) for the ready identification and morphological visualization of human cells engrafted into the mouse brain.

Another study used sequential sorts based on cell size, differential binding of peanut agglutinin lectin (PNA) and expression of the heat stable antigen (HSA or CD24) to separate a highly enriched population (80%) of adult mouse SVZ NSCs (Rietze et al., 2001). The NSC has a diameter larger than 12 μm and the PNA^{lo}HSA^{lo} phenotype. When single PNA^{lo}HSA^{lo} cells were cultured, around 80% generated multilineage self-renewing neurospheres, which is, by far, the best NSC frequency attained for any isolated neural stem cell (both NCSCs or hNSCs). However, given that only 63% of the total SVZ stem cells were selected, stem cells with a different phenotype do exist. In addition, contrary to the previous two studies, stem cell purification was done by negative selection, or lack of marker expression, which does not permit their localization *in vivo*. More recently, Capela and Temple (Capela and Temple, 2002) demonstrated that the adult mouse SVZ NSC is non-ependymal and expresses the LewisX trisaccharide epitope 3-fucosyl-N-acetyllactosamine (also defined as SSEA-1, FAL or LeX). Similarly to NCSC and hNSC, the discovery of this antigen makes it possible to positively purify NSCs and to locate them *in vivo*, in contrast to the method developed by Rietze et al. (2001). LeX$^-$ cells generate a negligible amount of neurospheres; hence, the use of antibodies against LeX also permits the recovery of all NSCs in the LeX$^+$ subpopulation. Not all LeX$^+$ cells are NSCs; only 17% of the LeX$^+$ cells generate multipotent self-renewing neurospheres. Other surface markers could be used to refine this identification method, for example the markers used by Reitze and colleagues: HSA and PNA. However, there is no overlap between LeX expression and mCD24 (HSA); because mCD24 is present on ependymal cells and neuroblasts, this showed that NSCs are not ependymal cells, as corroborated by others (Chiasson et al., 1999; Doetsch et al., 1999b; Laywell et al., 2000). Other markers, such as PNA, should be explored as a means to aid the LeX-based purification of the adult mouse NSC population further.

Dye binding properties of neural stem cells

Various studies in the mouse hematopoietic field have shown that a subpopulation (the so-called "side population" or SP) of bone marrow cells with specific dye binding properties include HSCs. The characteristic feature of these cells is their low red and blue Hoechst 33342 dye fluorescence as measured by dual wavelength analysis (Goodell et al., 1997) due to the high activity of proteins of the ABC transporter superfamily (Zhou et al., 2001). An additional explanation for the low Hoechst fluorescence in HSCs is that accessibility of

Figure 1. Schematic illustration of isolation of human CNS-NSC and limiting dilution analysis. Human NSC-derived from fetal brain can be isolated based on expression of the cell surface marker CD133. These CD133[+] cells can be plated at limiting cell numbers (from 1 cell per well to 100 cells per well) into a 96-well plate with an automated cell deposition unit. Culture conditions were adapted from the work of Carpenter et al. (1999) and are identical for all cell densities used (limited dilution assay as well as bulk cultures). Briefly, sorted cells were incubated in Ex-vivo 15 medium containing 0.2 mg/ml heparin, 60 µg/ml N-acetyl-cysteine (NAC), supplemented with N2 (Bottenstein and Sato, 1979), 20 ng/ml FGF2, 20 ng/ml EGF and 10 ng/ml LIF (expansion medium) and were maintained in a 37°C humidified incubator (Uchida et al., 2000; Tamaki et al., 2002). After 6 and 8 weeks of culture the wells are examined and the number of neurosphere negative wells are counted at each cell dilution to calculate frequency (see text). Single sorted CD133[+] cells can generate clonal neurospheres. To test for self-renewal, the clonogenic neurospheres are dissociated into single cells by enzymatic treatment with (0.5 mg/ml in PBS containing 0.1% human serum albumin) and recultured in expansion medium (Tamaki et al., 2002). To test for multilineage differentiation capability, dissociated cells are plated in poly-ornithine-coated dishes and incubated in Ex-vivo 15 medium containing 0.2 mg/ml heparin, 60 µg/ml NAC and supplemented with N2, neurotrophic factors (10 ng/ml BDNF, 10 ng/ml GDNF) and 1 µg/ml laminin. After 1-2 weeks, cultures were fixed and stained with cell type-specific antibodies (Uchida et al., 2000). Figures adapted from Uchida et al. (2000) PNAS 97:14720-14725 (Figure 3a.b).

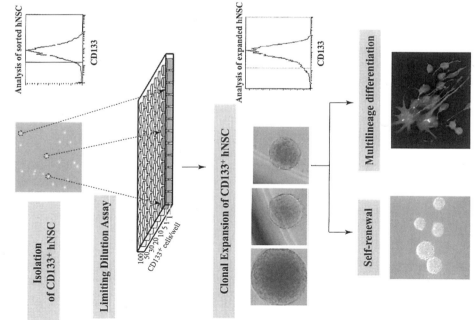

the dye to DNA might be reduced in these cells probably due to chromatin structure.

Recently, a similar analysis for the presence of an "SP" on neural cells was carried out. Hulspas and Quesenberry (2000) demonstrated the existence of an SP in freshly isolated (uncultured) cells and neurospheres derived from the brains of mouse embryos. The percentage of cells with the Hoechst^{low} phenotype was much higher in neurospheres when compared to freshly isolated cells. Most interestingly, when cultured separately, the neurosphere derived SP sorted population generates neurospheres and has the same Hoechst binding characteristics as the parent neurospheres indicating the existence of self renewing cells in the SP. Another study has also determined the existence of an SP population in freshly isolated mouse striatum cells from embryo to adult (Murayama et al., 2002). Although the neurosphere generation capacity of sorted SP cells was not tested, the study shows that SP cells have other characteristics attributed to NSCs such as expression of nestin and Notch1. Nevertheless, the SP is not phenotypically homogeneous; for instance, some SP cells are Nestin$^-$ and only a small percentage of adult periventricular SP cells are CD24^{low}/PNA^{low}. In addition, neurosphere generation is not exclusive of SP cells; cells that are Hoechst^{high} also generate neurospheres (Hulspas and Quesenberry, 2000). In conclusion, the presence of a highly efficient dye efflux mechanism and/or poor DNA accessibility is a characteristic of a subpopulation of NSCs and thus can be used in enrichment protocols.

Other selection methods

Another method for NSC enrichment is based on the buoyant density of cells: fractionation of live cells can be accomplished by density gradient centrifugation. For instance, discontinuous Percoll gradients have been used to fractionate crude CNS cell suspensions into enriched populations of oligodendrocytes, neurons and progenitor cells (Lisak et al., 1981; Maric et al., 1997; Silverman et al., 1999; Dori et al., 2000). Using carefully optimized Percoll gradients, Palmer et al., (1999) fractionated freshly isolated adult rat hippocampal and cortical cells into various subpopulations. Culture of the fractionated subpopulations showed that the most immature cells have surprisingly high densities, accumulating in the bottom of the gradient. Around 20% of these low-buoyancy cells derived from adult hippocampus and cortex generate clones containing neurons and glia and are therefore NSCs. Although only one-fifth of the NSCs can be recovered using this method, it provides a simple and reliable way to enrich a population in NSCs for further studies.

CLONAL LINEAGE ANALYSES

Understanding the types of progeny that stem cells generate – their lineage relationships – is fundamental for understanding a stem cell system (see Chapter 2 in this volume). For the HSC, this was accomplished by clonal studies (reviewed in Weissman, 2000; Weissman et al., 2001). In this method, single progenitor cells are allowed to develop into clones. The numbers and types of progeny in each clone are analyzed to give information about the progenitor cell types that are present in the population. Lineage relationships can then be deduced from the clone compositions. A more direct method of lineage analysis is to actually watch the cell divisions of a stem or progenitor cell. This method was pioneered in early developmental studies, for example in the full description of the nematode C. elegans development from the egg to the final worm (Sulston and Horvitz, 1977; Sulston et al., 1983). Direct observation provides more concrete information about the lineage relationships of progenitor cells, as well as the spatial and temporal events that unfold as the lineage progresses, such as the cell cycle length, cell movements, and the occurrence of cell death. For larger animals developing in utero, direct observation of lineages cannot be accomplished; hence *in vitro* methods have been established for following clone growth.

Defining the *in vitro* lineage potential of a given cell, either a stem cell or a progenitor cell, relies on recording the divisions these cells undergo while growing in a culture dish. At a defined time point, the group of cells (clone) generated by the single founder cell is fixed and stained with cell type specific antibodies so that clone composition can be identified. This apparently straightforward cell culture assay is far from being a simple one. For instance, cells need to be plated apart enough so that clone superimposition does not occur; in addition, at such low densities, cell growth may not be optimal and therefore, supplements, growth factors, or cellular extracts need to be added to the culture medium in order to attain optimal clone development. An alternative method to highly supplemented media is to grow cells at clonal density on a feeder layer consisting of neural cells from a different animal species. In that situation, the feeder cells can be distinguished from the test cells using antibodies (Malatestra et al., 2000).

Single cell and low density studies

Pioneering work described by Sally Temple (1989) set a high standard for clonal analysis: single neuroepithelial cells derived from the septum of rat embryos were plated into microwells of Terasaki plates by micromanipulation under microscopy guidance. Each microwell was then carefully inspected for the presence of only one cell to ensure clonality. From this study, it was determined that there are three basic types of cells in the embryonic septal region. Two undergo a limited number of divisions and generate pure clones (containing only neurons or only glia) or mixed clones of neurons and glia. The other generates large mixed clones containing neurons, glia and cells with characteristics reminiscent of the original founder cell. The author suggested for the first time that these cells might be CNS NSCs which generate multiple types of progeny and can self-renew. Using the same clonal analysis technique, Davis and Temple (1994) demonstrated that cells with similar *in vitro* growth characteristics can be isolated from the embryonic rat cortex and their prevalence is around 10-20% of the total cells. Both studies relied on the presence of feeder cells growing on the sides of the culture wells or on glial cell extracts in order to provide the right conditions for the growth of the septal or cortical single cells; however, similar results were obtained when early cortical progenitors were cultured in a defined medium containing FGF2 and the supplements N2 (Bottenstein and Sato, 1979) and B27 (Qian et al., 1997, 1998, 2000).

The introduction of time-lapse video microscopy (TLVM) further this group to determine the exact lineage relationship between stem cells and their progenitors (Qian et al., 1998, 2000). With TLVM it is possible to follow every cell division, from the founder cell until the time at which the clone is fixed and processed for cell type analysis. This way, faithful reconstitutions of the lineage trees for all founder cells in the video screen can be accomplished. Also, clonality no longer needs to be ensured by single cell plating once it is possible to videotape the behavior of more than one cell, as long as developing clones stay in frame. Finally, by matching immunocytochemical data and lineage tree information acquired through repeated analysis of the videotapes, it is possible to accurately determine the lineage relations between the founder cell and its progeny. Using TLVM, it was found that generation of neurons and glia by cortical NSCs isolated from E10.5 mouse embryos is temporally regulated: neurons are generated first, usually via asymmetric divisions, and glia arise later in the lineage, via symmetric divisions (Qian et al., 2000). These clonal analyses results reflect a recapitulation of the *in vivo* sequence of cell genesis.

Independently of the clonal analysis technique chosen, the total number of clones obtained is recorded; the types of clones generated (pure neuronal, pure glial and pure stem) are quantified as a percentage of total clones. This way, any attempt to enrich the initial population in NSCs (or neuronal or glial progenitors) would lead to an increase in the incidence of that particular type of clone. Hence, the study of clonal cortical NSC development provides an excellent model to address how neuronal and glial cell types are generated and also how neuronal diversity is accomplished.

Another type of clonal analysis relies not on the development of adherent clones but on the generation of multipotent, self-renewing neurospheres by cells derived from adult mouse SVZ plated on untreated culture dishes (Reynolds and Weiss, 1992; Vescovi et al., 1993; Gritti et al., 1996; Reynolds and Weiss, 1996; Gritti et al., 1999). Single neurospheres are then collected, plated on a poly-L-lysine substratum to encourage differentiation, and finally fixed and stained in order to determine if all cell types (neurons, oligodendrocytes and astrocytes) are present; single neurospheres are also dissociated and plated at low density or as single cells to demonstrate self-renewal by generating more neurospheres. The percentage of neurosphere generators, hence NSCs, is calculated as the percentage of neurospheres vs total plated cells. Neurosphere generation is hence a widely accepted diagnostic assay for NSCs in the rodent and human brain, and clonal analysis indicates that around 1% of the adult mouse SVZ cells are NSCs. Interestingly, a very recent study indicates that the "transit-amplifying progenitors" of the adult mouse SVZ NSC is also capable of generating neurospheres that are both multipotent and self-renewable (Doetsch et al., 2002), suggesting that neurosphere generation capacity is not exclusive of the slowly dividing NSCs but also of their immediate progenitors.

The common observation that neurospheres can fuse together (identical to adherent clone superimposition), thus underestimating the number of NSCs/ "transit-amplifying progenitors" in the plated cell population, Gritti and et al. (1999) used a method for clonal analysis in which migration and fusion of neurospheres is greatly diminished by plating the cells in a methylcellulose-based semi-solid matrix. However, migration and fusion can still occur on methylcellulose (Hulspas et al., 1997) making the low density (1-5 cells per well) or single cell plating approach a much better neurosphere generation assay (Rietze et al., 2001; Capela and Temple, 2002; Engstrom et al., 2002).

Limiting-dilution studies

Each stem cell is a clonogenic progenitor for multiple differentiated cell lineages, and also possesses the capacity for self-renewal. Clonal analysis as described above has been used to characterize self-renewal and differentiation of NSCs *in vitro*. For HSC studies, clonogenic assays have been established for each of the hematolymphoid and myeloid lineages. These assays include the spleen colony assay for erythrocyte, megakaryocyte and myeloid lineage, the thymus colony forming assay for T lymphocyte lineage, and a B lymphocyte lineage clonogenic assay (reviewed in Spangrude et al., 1991; Morrison et al., 1995; Weissman, 2000; Weissman et al., 2001). Bone marrow stroma-based assays were developed for long-term culture, including long-term colony initiating cells (LTC-IC) (Sutherland et al., 1990) and cobblestone area-forming cells (CAFC; Muller-Sieburg et al., 1986; Ploemacher et al., 1991; Weilbaecher et al., 1991). In these assays, limiting dilution analysis has been utilized extensively to measure the frequency of single-hit events (Lefkovits and Waldmann, 1999).

Uchida et al. (2000) have adapted this approach to determine the frequency of neurosphere-initiating cells (NS-IC) in human fetal brain. In this case, sorted "test" cell populations derived from fetal brains were plated in a series of limiting cell numbers into 96-well plates. Cultures were incubated for 6-8 weeks and the wells that did not contain neurospheres were scored as negative. The proportion of wells that are negative for neurosphere formation is plotted against the number of sorted cells plated per well. Traditionally, limiting dilution analysis is presented as a plot of the logarithm of the fraction of negative wells on the y-axis versus a linear scale of the number of input cells on the x-axis. Linear correlation of log percentage negative wells and number of cells plated indicate that a neurosphere was initiated from a single-hit event. Based on the Poisson distribution, interpolation of the frequency can be determined at the cell concentration at which 37% of the wells are negative for growth. This linear regression analysis was used to determine the frequency of NS-IC in a population of $CD133^+CD34^-CD45^-$ sorted fetal brain cells. The frequency of NS-IC derived from unsorted total fetal brain cells is 1/880. Moreover, the $CD133^-$ (which represent 96 % of fetal brain cells), and $CD133^+CD34^-CD45^-$ subpopulations have NS-IC frequencies of 1/4860 and 1/31, respectively. Therefore, the $CD133^+CD34^-CD45^-$ subpopulation is 24-fold enriched for NS-IC relatively to total fetal brain cells, which parallels the frequency of $CD133^+CD34^-CD45^-$ cells in the total enzyme-dissociated fetal brain cell suspension. Finally, it was concluded that the $CD133^+CD34^-CD45^-$ subpopulation, purified from human fetal brains of 16 to 20 gestational weeks, contained virtually all the NS-IC activity and therefore all the hNSCs.

THE IMPORTANCE OF NEURAL STEM CELL MARKERS FOR PLASTICITY, TRANSPLANTATION AND LINEAGE RELATIONS

Recent studies indicate that cultured mouse and human NSCs derived from adult tissue have unexpected plasticity: they can generate many non-neuronal tissues. For instance, ex-vivo expanded NSCs (neurospheres) derived from adult mouse brain can generate blood cells when injected into an irradiated mouse (Bjornson et al., 1999), skeletal muscle when transplanted into regenerating muscle sites (Galli et al., 2000) or even derivatives of the three germ layers upon injection into mouse and chick blastocysts (Clarke et al., 2000). One criticism of these experiments is that they describe very rare events that have proven difficult to repeat. The fact that cells make these fate transitions rarely could indicate that the expanded NSCs were in fact abnormal. We know that extensive growth of cells in culture can lead to genetic changes related to cell transformation, which increases the ability of a cell to take on diverse fates (Gao and Hatten, 1994). Another reason why these experiments have to be viewed with some skepticism, is because of the recently described novel type of rare cell fusion that can occur between implanted cells and blastocyst cells (Terada et al., 2002; Ying et al., 2002). Thus, it is difficult to be sure that the results of the embryonic transplantation experiments are truly a reflection of the capacity of the implanted cells, rather than a hybrid generated through cell fusion. Moreover, the difficulty in reproducing some of these results by other laboratories has cast a shadow of uncertainty as to whether plasticity is a real phenomenon (Morshead et al., 2002; Wagers et al., 2002).

A further criticism of most of the studies on NSC plasticity to date is the fact that expanded rather than freshly isolated NSCs were used. Again, it is impossible to determine if such plasticity is only the result of long-term culture of NSCs in the presence of exogenous of growth factors or if it does occur *in vivo*. Moreover, although neurospheres allow expansion of the rare NSC, they do contain restricted progenitors and ultimately, one has to question if lineage jumping is a NSC feature at all, although conceptually, it is more easily acceptable that a NSC does so than a neuron suddenly making a muscle cell or a myeloid cell.

Another important point to bear in mind is the assessment of NSC potential upon transplantation. For instance, although cultured adult rodent NSCs seem to have a relatively broad potential when engrafted into different areas of the brain, freshly isolated cells show a much more restricted potential (Herrera et al., 1999) even if transplanted into an embryonic environment where endogenous NSCs are actively engaged in the generation of various types of neurons (Lim et al., 1997). Transplantation of freshly dissociated embryo-derived NSCs back into the embryo reveals high plasticity: diverse neuronal progeny, including projection neurons, are generated (Brustle et al., 1995; Fishell 1995; Olsson et al., 1997). However, NSCs derived from newborn mice lose this plasticity and no longer can generate projection neurons (Lim et al., 1997). Is it because NSCs change their properties over time, or is it the result of feedback signaling from committed progenitors that are co-transplanted with NSCs? This is another question that transplantation of freshly purified NSCs can help address.

Identification and purification of NSCs provides yet another important possibility: that of studying the relationship between NSCs at different developmental ages and even between NSCs and other tissue-specific stem cells. This has not been adequately addressed due to the paucity of specific NSC markers. Is there a universal stem cell gene expression profile? Is there a gene expression blueprint that distinguishes NSCs from HSCs or other stem cells? Is this blueprint developmentally regulated? These are complex and fascinating questions that are now being explored (see Chapter 3 in this volume).

Geschwind et al. (2001) set the stage for unraveling the gene expression pattern of CNS progenitors using powerful molecular biology techniques to identify genes involved in NSC/progenitor cell proliferation and function (Geschwind et al., 2001). To achieve this goal, cDNA isolated from neurospheres derived from neonatal mouse cortex was "subtracted" from that of sister neurospheres subjected to 24 h of differentiation conditions. This approach lead to the discovery of novel genes specifically expressed in germinal areas of the embryonic and adult mouse brain. Building on this work, Terskikh et al. (2001) showed that some of these genes are common to HSCs, suggesting that NSCs and HSCs share genetic programs and signaling strategies, results later corroborated by Ivanova et al. (2002) and Ramalho-Santos et al. (2002). Again, these studies use cultured neurospheres as the source of NSCs and thus are subject to the criticisms previously discussed. The use of purified, freshly isolated, NSCs as a starting material is the next step, and one which can now be undertaken.

In conclusion, significant progress has been made towards establishing methods for the identification and purification of NSC populations. These initial steps have already provided much information about NSCs, and strongly indicate that 'neural stem cell' is a broader term that encompasses diverse cell sub-populations with different characteristics. As more phenotypic characteristics are revealed, molecules that can serve as markers to unequivocally identify and purify NSC populations should emerge. This will bring us to a new level of understanding of these important cells, which offer enormous promise for regeneration and repair of the diseased and damaged nervous system.

Acknowledgements

We would like to thank Sally Temple, Michael Reitsma, Yakop Jacobs, Ruud Hulspas and Nelson Jumbe for their suggestions and comments.

REFERENCES

Alvarez-Buylla A, Garcia-Verdugo JM (2002) Neurogenesis in adult subventricular zone. J Neurosci 22:629-634.

Avilion AA, Nicolis SK, Pevny LH, Perez L, Vivian N, Lovell-Badge R (2003) Multipotent cell lineages in early mouse development depend on SOX2 function. Genes Dev 17:126-140.

Bjornson CR, Rietze RL, Reynolds BA, Magli MC, Vescovi AL (1999) Turning brain into blood: a hematopoietic fate adopted by adult neural stem cells *in vivo*. Science 283:534-537.

Bottenstein JE, Sato GH (1979) Growth of a rat neuroblastoma cell line in serum-free supplemented medium. Proc Natl Acad Sci USA 76: 514-517.

Brustle O, Maskos U, McKay RD (1995) Host-guided migration allows targeted introduction of neurons into the embryonic brain. Neuron 15:1275-1285.

Capela A, Temple S (2002) LeX/ssea-1 is expressed by adult mouse CNS stem cells, identifying them as nonependymal. Neuron 35:865-875.

Carpenter MK, Cui X, Hu ZY, Jackson J, Sherman S, Seiger A, Wahlberg LU (1999) *In vitro* expansion of a multipotent population of human neural progenitor cells. Exp Neurol 158:265-278.

Chiasson BJ, Tropepe V, Morshead CM, van der Kooy D (1999) Adult mammalian forebrain ependymal and subependymal cells demonstrate proliferative potential, but only subependymal cells have neural stem cell characteristics. J Neurosci 19:4462-4471.

Clarke DL, Johansson CB, Wilbertz J, Veress B, Nilsson E, Karlstrom H, Lendahl U, Frisen J (2000) Generalized potential of adult neural stem cells. Science 288:1660-1663.

Davis AA, Temple S (1994) A self-renewing multipotential stem cell in embryonic rat cerebral cortex. Nature 372:263-266.

Doetsch F, Garcia-Verdugo JM, Alvarez-Buylla A (1999a) Regeneration of a germinal layer in the adult mammalian brain. Proc Natl Acad Sci U S A 96:11619-11624.

Doetsch F, Caille I, Lim DA, Garcia-Verdugo JM, Alvarez-Buylla A (1999b) Subventricular zone astrocytes are neural stem cells in the adult mammalian brain. Cell 97:703-716.

Doetsch F, Petreanu L, Caille I, Garcia-Verdugo JM, Alvarez-Buylla A (2002) EGF converts transit-amplifying neurogenic precursors in the adult brain into multipotent stem cells. Neuron 36:1021-1034.

Dori A, Maric D, Maric I, Masalha R, Barker JL, Silverman WF (2000) Striatal matrix neurons of the rat differentiate in culture from dissociated fetal progenitor cells isolated by buoyant density centrifugation. Neurosci Lett 282:77-80.

Engstrom CM, Demers D, Dooner M, McAuliffe C, Benoit BO, Stencel K, Joly M, Hulspas R, Reilly JL, Savarese T, Recht LD, Ross AH, Quesenberry PJ (2002) A method for clonal analysis of epidermal growth factor-responsive neural progenitors. J Neurosci Methods 117:111-121.

Eriksson PS, Perfilieva E, Bjork-Eriksson T, Alborn AM, Nordborg C, Peterson DA, Gage FH (1998) Neurogenesis in the adult human hippocampus. Nat Med 4:1313-1317.

Fishell G (1995) Striatal precursors adopt cortical identities in response to local cues. Development 121:803-812.

Fuchs E, Segre JA (2000) Stem cells: a new lease on life. Cell 100:143-155.

Galli R, Gritti A, Bonfanti L, Vescovi AL (2003) Neural stem cells: an overview. Circ Res 92:598-608.

Galli R, Borello U, Gritti A, Minasi MG, Bjornson C, Coletta M, Mora M, De Angelis MG, Fiocco R, Cossu G, Vescovi AL (2000) Skeletal myogenic potential of human and mouse neural stem cells. Nat Neurosci 3:986-991.

Gao WQ, Hatten ME (1994) Immortalizing oncogenes subvert the establishment of granule cell identity in developing cerebellum. Development 120:1059-1070.

Geschwind DH, Ou J, Easterday MC, Dougherty JD, Jackson RL, Chen Z, Antoine H, Terskikh A, Weissman IL, Nelson SF, Kornblum HI (2001) A genetic analysis of neural progenitor differentiation. Neuron 29:325-339.

Goldman SA, Nottebohm F (1983) Neuronal production, migration, and differentiation in a vocal control nucleus of the adult female canary brain. Proc Natl Acad Sci U S A 80:2390-2394.

Goodell MA, Rosenzweig M, Kim H, Marks DF, DeMaria M, Paradis G, Grupp SA, Sieff CA, Mulligan RC, Johnson RP (1997) Dye efflux studies suggest that hematopoietic stem cells expressing low or undetectable levels of CD34 antigen exist in multiple species. Nat Med 3:1337-1345.

Gould R, Freund C, Palmer F, Knapp PE, Huang J, Morrison H, Feinstein DL (1999) Messenger RNAs for kinesins and dynein are located in neural processes. Biol Bull 197:259-260.

Gritti A, Frolichsthal-Schoeller P, Galli R, Parati EA, Cova L, Pagano SF, Bjornson CR, Vescovi AL (1999) Epidermal and fibroblast growth factors behave as mitogenic regulators for a single multipotent stem cell-like population from the subventricular region of the adult mouse forebrain. J Neurosci 19:3287-3297.

Gritti A, Parati EA, Cova L, Frolichsthal P, Galli R, Wanke E, Faravelli L, Morassutti DJ, Roisen F, Nickel DD, Vescovi AL (1996) Multipotential stem cells from the adult mouse brain proliferate and self-renew in response to basic fibroblast growth factor. J Neurosci 16:1091-1100.

Herrera DG, Garcia-Verdugo JM, Alvarez-Buylla A (1999) Adult-derived neural precursors transplanted into multiple regions in the adult brain. Ann Neurol 46:867-877.

Hockfield S, McKay RD (1985) Identification of major cell classes in the developing mammalian nervous system. J Neurosci 5:3310-3328.

Hulspas R, Quesenberry PJ (2000) Characterization of neurosphere cell phenotypes by flow cytometry. Cytometry 40:245-250.

Hulspas R, Tiarks C, Reilly J, Hsieh CC, Recht L, Quesenberry PJ (1997) *In vitro* cell density-dependent clonal growth of EGF-responsive murine neural progenitor cells under serum-free conditions. Exp Neurol 148:147-156.

Ingraham CA, Rising LJ, Morihisa JM (1999) Development of O4+/O1- immunopanned pro-oligodendroglia *in vitro*. Brain Res Dev Brain Res 112:79-87.

Ivanova NB, Dimos JT, Schaniel C, Hackney JA, Moore KA, Lemischka IR (2002) A stem cell molecular signature. Science 298:601-604.

Johansson CB, Momma S, Clarke DL, Risling M, Lendahl U, Frisen J (1999) Identification of a neural stem cell in the adult mammalian central nervous system. Cell 96:25-34.

Kawaguchi A, Miyata T, Sawamoto K, Takashita N, Murayama A, Akamatsu W, Ogawa M, Okabe M, Tano Y, Goldman SA, Okano H (2001) Nestin-EGFP transgenic mice: visualization of the self-renewal and multipotency of CNS stem cells. Mol Cell Neurosci 17:259-273.

Kempermann G, Kuhn HG, Gage FH (1997) More hippocampal neurons in adult mice living in an enriched environment. Nature 386:493-495.

Keyoung HM, Roy NS, Benraiss A, Louissaint A, Jr., Suzuki A, Hashimoto M, Rashbaum WK, Okano H, Goldman SA (2001) High-yield selection and extraction of two promoter-defined phenotypes of neural stem cells from the fetal human brain. Nat Biotechnol 19:843-850.

Kornack DR, Rakic P (1999) Continuation of neurogenesis in the hippocampus of the adult macaque monkey. Proc Natl Acad Sci U S A 96:5768-5773.

Kukekov VG, Laywell ED, Suslov O, Davies K, Scheffler B, Thomas LB, O'Brien TF, Kusakabe M, Steindler DA (1999) Multipotent stem/progenitor cells with similar properties arise from two neurogenic regions of adult human brain. Exp Neurol 156:333-344.

Laywell ED, Rakic P, Kukekov VG, Holland EC, Steindler DA (2000) Identification of a multipotent astrocytic stem cell in the immature and adult mouse brain. Proc Natl Acad Sci U S A 97:13883-13888.

Lefkovits I, Waldmann H (1999) Limiting dilution analysis of cells in the immune system. Oxford: Oxford University Press.

Lendahl U, Zimmerman LB, McKay RD (1990) CNS stem cells express a new class of intermediate filament protein. Cell 60:585-595.

Li M, Pevny L, Lovell-Badge R, Smith A (1998) Generation of purified neural precursors from embryonic stem cells by lineage selection. Curr Biol 8:971-974.

Lim DA, Fishell GJ, Alvarez-Buylla A (1997) Postnatal mouse subventricular zone neuronal precursors can migrate and differentiate within multiple levels of the developing neuraxis. Proc Natl Acad Sci U S A 94:14832-14836.

Lisak RP, Pleasure DE, Silberberg DH, Manning MC, Saida T (1981) Long term culture of bovine oligodendroglia isolated with a Percoll gradient. Brain Res 223:107-122.

Lois C, Alvarez-Buylla A (1993) Proliferating subventricular zone cells in the adult mammalian forebrain can differentiate into neurons and glia. Proc Natl Acad Sci U S A 90:2074-2077.

Malatesta P, Hartfuss E, Gotz M (2000) Isolation of radial glial cells by fluorescent-activated cell sorting reveals a neuronal lineage. Development 127:5253-5263.

Maric D, Maric I, Ma W, Lahojuji F, Somogyi R, Wen X, Sieghart W, Fritschy JM, Barker JL (1997) Anatomical gradients in proliferation and differentiation of embryonic rat CNS accessed by buoyant density fractionation: alpha 3, beta 3 and gamma 2 GABAA receptor subunit co-expression by post-mitotic neocortical neurons correlates directly with cell buoyancy. Eur J Neurosci 9:507-522.

Mayer-Proschel M, Kalyani AJ, Mujtaba T, Rao MS (1997) Isolation of lineage-restricted neuronal precursors from multipotent neuroepithelial stem cells. Neuron 19:773-785.

Mi H, Barres BA (1999) Purification and characterization of astrocyte precursor cells in the developing rat optic nerve. J Neurosci 19:1049-1061.

Miraglia S, Godfrey W, Yin AH, Atkins K, Warnke R, Holden JT, Bray RA, Waller EK, Buck DW (1997) A novel five-transmembrane hematopoietic stem cell antigen: isolation, characterization, and molecular cloning. Blood 90:5013-5021.

Morrison SJ, Uchida N, Weissman IL (1995) The biology of hematopoietic stem cells. Annu Rev Cell Dev Biol 11:35-71.

Morrison SJ, Shah NM, Anderson DJ (1997) Regulatory mechanisms in stem cell biology. Cell 88:287-298.

Morrison SJ, White PM, Zock C, Anderson DJ (1999) Prospective identification, isolation by flow cytometry, and *in vivo* self-renewal of multipotent mammalian neural crest stem cells. Cell 96:737-749.

Morshead CM, Benveniste P, Iscove NN, van der Kooy D (2002) Hematopoietic competence is a rare property of neural stem cells that may depend on genetic and epigenetic alterations. Nat Med 8:268-273.

Morshead CM, Reynolds BA, Craig CG, McBurney MW, Staines WA, Morassutti D, Weiss S, van der Kooy D (1994) Neural stem cells in the adult mammalian forebrain: a relatively quiescent subpopulation of subependymal cells. Neuron 13:1071-1082.

Muller-Sieburg CE, Whitlock CA, Weissman IL (1986) Isolation of two early B lymphocyte progenitors from mouse marrow: a committed pre-pre-B cell and a clonogenic Thy-1-lo hematopoietic stem cell. Cell 44:653-662.

Murayama A, Matsuzaki Y, Kawaguchi A, Shimazaki T, Okano H (2002) Flow cytometric analysis of neural stem cells in the developing and adult mouse brain. J Neurosci Res 69:837-847.

Nakatomi H, Kuriu T, Okabe S, Yamamoto S, Hatano O, Kawahara N, Tamura A, Kirino T, Nakafuku M (2002) Regeneration of hippocampal pyramidal neurons after ischemic brain injury by recruitment of endogenous neural progenitors. Cell 110:429-441.

Nunes MC, Roy NS, Keyoung HM, Goodman RR, McKhann G, Jiang L, Kang J, Nedergaard M, Goldman SA (2003) Identification and isolation of multipotential neural progenitor cells from the subcortical white matter of the adult human brain. Nat Med 9:439-447.

Olsson M, Campbell K, Turnbull DH (1997) Specification of mouse telencephalic and mid-hindbrain progenitors following heterotopic ultrasound-guided embryonic transplantation. Neuron 19:761-772.

Palmer TD, Takahashi J, Gage FH (1997) The adult rat hippocampus contains primordial neural stem cells. Mol Cell Neurosci 8:389-404.

Palmer TD, Willhoite AR, Gage FH (2000) Vascular niche for adult hippocampal neurogenesis. J Comp Neurol 425:479-494.

Palmer TD, Markakis EA, Willhoite AR, Safar F, Gage FH (1999) Fibroblast growth factor-2 activates a latent neurogenic program in neural stem cells from diverse regions of the adult CNS. J Neurosci 19:8487-8497.

Ploemacher RE, van der Sluijs JP, van Beurden CA, Baert MR, Chan PL (1991) Use of limiting-dilution type long-term marrow cultures in frequency analysis of marrow-repopulating and spleen colony-forming hematopoietic stem cells in the mouse. Blood 78:2527-2533.

Potten CS, Loeffler M (1990) Stem cells: attributes, cycles, spirals, pitfalls and uncertainties. Lessons for and from the crypt. Development 110:1001-1020.

Qian X, Davis AA, Goderie SK, Temple S (1997) FGF2 concentration regulates the generation of neurons and glia from multipotent cortical stem cells. Neuron 18:81-93.

Qian X, Goderie SK, Shen Q, Stern JH, Temple S (1998) Intrinsic programs of patterned cell lineages in isolated vertebrate CNS ventricular zone cells. Development 125:3143-3152.

Qian X, Shen Q, Goderie SK, He W, Capela A, Davis AA, Temple S (2000) Timing of CNS cell generation: a programmed sequence of neuron and glial cell production from isolated murine cortical stem cells. Neuron 28:69-80.

Ramalho-Santos M, Yoon S, Matsuzaki Y, Mulligan RC, Melton DA (2002) "Stemness": transcriptional profiling of embryonic and adult stem cells. Science 298:597-600.

Reynolds BA, Weiss S (1992) Generation of neurons and astrocytes from isolated cells of the adult mammalian central nervous system. Science 255:1707-1710.

Reynolds BA, Weiss S (1996) Clonal and population analyses demonstrate that an EGF-responsive mammalian embryonic CNS precursor is a stem cell. Dev Biol 175:1-13.

Rietze RL, Valcanis H, Brooker GF, Thomas T, Voss AK, Bartlett PF (2001) Purification of a pluripotent neural stem cell from the adult mouse brain. Nature 412:736-739.

Roy NS, Wang S, Harrison-Restelli C, Benraiss A, Fraser RA, Gravel M, Braun PE, Goldman SA (1999) Identification, isolation, and promoter-defined separation of mitotic oligodendrocyte progenitor cells from the adult human subcortical white matter. J Neurosci 19:9986-9995.

Roy NS, Benraiss A, Wang S, Fraser RA, Goodman R, Couldwell WT, Nedergaard M, Kawaguchi A, Okano H, Goldman SA (2000a) Promoter-targeted selection and isolation of neural progenitor cells from the adult human ventricular zone. J Neurosci Res 59:321-331.

Roy NS, Wang S, Jiang L, Kang J, Benraiss A, Harrison-Restelli C, Fraser RA, Couldwell WT, Kawaguchi A, Okano H, Nedergaard M, Goldman SA (2000b) *In vitro* neurogenesis by progenitor cells isolated from the adult human hippocampus. Nat Med 6:271-277.

Sakakibara S, Imai T, Hamaguchi K, Okabe M, Aruga J, Nakajima K, Yasutomi D, Nagata T, Kurihara Y, Uesugi S, Miyata T, Ogawa M, Mikoshiba K, Okano H (1996) Mouse-Musashi-1, a neural RNA-binding protein highly enriched in the mammalian CNS stem cell. Dev Biol 176:230-242.

Silverman WF, Alfahel-Kakunda A, Dori A, Barker JL (1999) Separation of dorsal and ventral dopaminergic neurons from embryonic rat mesencephalon by buoyant density fractionation: disassembling pattern in the ventral midbrain. J Neurosci Methods 89:1-8.

Spangrude GJ, Smith L, Uchida N, Ikuta K, Heimfeld S, Friedman J, Weissman IL (1991) Mouse hematopoietic stem cells. Blood 78:1395-1402.

Sulston JE, Horvitz HR (1977) Post-embryonic cell lineages of the nematode, Caenorhabditis elegans. Dev Biol 56:110-156.

Sulston JE, Schierenberg E, White JG, Thomson JN (1983) The embryonic cell lineage of the nematode Caenorhabditis elegans. Dev Biol 100:64-119.

Sutherland HJ, Lansdorp PM, Henkelman DH, Eaves AC, Eaves CJ (1990) Functional characterization of individual human hematopoietic stem cells cultured at limiting dilution on supportive marrow stromal layers. Proc Natl Acad Sci U S A 87:3584-3588.

Svendsen CN, ter Borg MG, Armstrong RJ, Rosser AE, Chandran S, Ostenfeld T, Caldwell MA (1998) A new method for the rapid and long term growth of human neural precursor cells. J Neurosci Methods 85:141-152.

Tamaki S, Eckert K, He D, Sutton R, Doshe M, Jain G, Tushinski R, Reitsma M, Harris B, Tsukamoto A, Gage F, Weissman I, Uchida N (2002) Engraftment of sorted/expanded human central nervous system stem cells from fetal brain. J Neurosci Res 69:976-986.

Temple S (1989) Division and differentiation of isolated CNS blast cells in microculture. Nature 340:471-473.

Temple S (2001) The development of neural stem cells. Nature 414:112-117.

Terada N, Hamazaki T, Oka M, Hoki M, Mastalerz DM, Nakano Y, Meyer EM, Morel L, Petersen BE, Scott EW (2002) Bone marrow cells adopt the phenotype of other cells by spontaneous cell fusion. Nature 416:542-545.

Terskikh AV, Easterday MC, Li L, Hood L, Kornblum HI, Geschwind DH, Weissman IL (2001) From hematopoiesis to neuropoiesis: evidence of overlapping genetic programs. Proc Natl Acad Sci U S A 98:7934-7939.

Uchida N, Tsukamoto A, He D, Friera AM, Scollay R, Weissman IL (1998) High doses of purified stem cells cause early hematopoietic recovery in syngeneic and allogeneic hosts. J Clin Invest 101:961-966.

Uchida N, Buck DW, He D, Reitsma MJ, Masek M, Phan TV, Tsukamoto AS, Gage FH, Weissman IL (2000) Direct isolation of human central nervous system stem cells. Proc Natl Acad Sci U S A 97:14720-14725.

Vescovi AL, Reynolds BA, Fraser DD, Weiss S (1993) bFGF regulates the proliferative fate of unipotent (neuronal) and bipotent (neuronal/astroglial) EGF-generated CNS progenitor cells. Neuron 11:951-966.

Wagers AJ, Sherwood RI, Christensen JL, Weissman IL (2002) Little evidence for developmental plasticity of adult hematopoietic stem cells. Science 297:2256-2259.

Wang S, Roy NS, Benraiss A, Goldman SA (2000) Promoter-based isolation and fluorescence-activated sorting of mitotic neuronal progenitor cells from the adult mammalian ependymal/subependymal zone. Dev Neurosci 22:167-176.

Wang S, Wu H, Jiang J, Delohery TM, Isdell F, Goldman SA (1998) Isolation of neuronal precursors by sorting embryonic forebrain transfected with GFP regulated by the T alpha 1 tubulin promoter. Nat Biotechnol 16:196-201.

Weilbaecher K, Weissman I, Blume K, Heimfeld S (1991) Culture of phenotypically defined hematopoietic stem cells and other progenitors at limiting dilution on Dexter monolayers. Blood 78:945-952.

Weissman IL (2000) Stem cells: units of development, units of regeneration, and units in evolution. Cell 100:157-168.

Weissman IL, Anderson DJ, Gage F (2001) Stem and progenitor cells: origins, phenotypes, lineage commitments, and transdifferentiations. Annu Rev Cell Dev Biol 17:387-403.

Wright AP, Fitzgerald JJ, Colello RJ (1997) Rapid purification of glial cells using immunomagnetic separation. J Neurosci Methods 74:37-44.

Yin AH, Miraglia S, Zanjani ED, Almeida-Porada G, Ogawa M, Leary AG, Olweus J, Kearney J, Buck DW (1997) AC133, a novel marker for human hematopoietic stem and progenitor cells. Blood 90:5002-5012.

Ying QL, Nichols J, Evans EP, Smith AG (2002) Changing potency by spontaneous fusion. Nature 416:545-548.

Zappone MV, Galli R, Catena R, Meani N, De Biasi S, Mattei E, Tiveron C, Vescovi AL, Lovell-Badge R, Ottolenghi S, Nicolis SK (2000) Sox2 regulatory sequences direct expression of a (beta)-geo transgene to telencephalic neural stem cells and precursors of the mouse embryo, revealing regionalization of gene expression in CNS stem cells. Development 127:2367-2382.

Zhou S, Schuetz JD, Bunting KD, Colapietro AM, Sampath J, Morris JJ, Lagutina I, Grosveld GC, Osawa M, Nakauchi H, Sorrentino BP (2001) The ABC transporter Bcrp1/ABCG2 is expressed in a wide variety of stem cells and is a molecular determinant of the side-population phenotype. Nat Med 7:1028-1034.

Chapter 8

Embryonic and Neural Stem Cell Lines

Ryan M. Fryer, Mahesh Lachyankar, Steven R. Gullans & Evan Y. Snyder

INTRODUCTION

Recognition of the existence of cells with stem-like qualities in the nervous system (Snyder et al., 1992) had a profound impact on both developmental neuroscience and translational neuroscience because it, at once, offered insight into previously unsuspected plasticity within the nervous system and provided a tangible means for potentially exploiting it for therapeutic purposes. The finding that stem cells can generate neural tissue even beyond the typical window of brain organogenesis (McKay, 1997; Snyder, 1998; Gage, 2000) has obvious implications for the repair of the nervous system in patients that have debilitating or fatal neurodegenerative disorders such as Parkinson's Disease; amyotrophic lateral sclerosis (ALS); inherited neurogenetic diseases, e.g., leukodystrophies, gangliosidoses, and lysosomal storage diseases; or acquired maladies such as spinal cord injury (SCI), stroke, or traumatic brain injury (TBI). The range of potential therapeutic targets towards which stem cell biology may be applied is still very much in the exploration stage; similarly, the limitations of such an approach are also being delineated.

Terminology in the stem cell field is, at present, controversial, confusing, inconsistent, and imprecise. We will next, therefore, indicate how we will use various terms in this chapter. A neural "stem" cell, i.e., the most primordial of cells in the nervous system, in our view, must have the potential to differentiate into all of the rich diversity of cells within all three fundamental neural lineages throughout all regions of the nervous system including the myriad types of neurons, the variety of astrocytes, and the various oligodendrocytes. Each stem cell must also self-renew sufficiently to yield the vast number of cells necessary to construct the functional mammalian brain during embryogenesis. At some phases in the stem cell life cycle, this need for adequate numbers will demand symmetrical cell divisions, i.e., a stem cell dividing to yield 2 immature, stem-like daughter cells; at other times, asymmetrical divisions will be required, i.e., a cell dividing to yield a committed, postmitotic neural cell as well as another proliferative, immature stem cell. Neural "progenitor" cells, in our lexicon, are

From: *Neural Stem Cells: Development and Transplantation*
Edited by: Jane E. Bottenstein © 2003 Kluwer Academic Publishers, Norwell, MA

cells with a more restricted potential, perhaps limited to a range of cell types or to cell types of a particular region of the nervous system. The potential can still be quite broad, however. More restricted still are neural "precursor" cells. These cells, while still immature and appearing earlier in a developmental pathway than a given terminally-differentiated, mature neural cell, have a limited number of developmental options, i.e., it may even be unipotent or bipotent (McKay, 1997).

Neural stem cells (NSCs) have been found throughout the embryonic and fetal CNS including the cerebral cortex, striatum, hippocampus, cerebellum, basal forebrain, and spinal cord, principally in germinal zones, e.g., the ventricular zone. They have also been found in the neural crest, a germinal structure that ultimately gives rise to the peripheral, autonomic, and enteric nervous systems. Presumptive NSCs have also been isolated from the adult CNS. At that "developmental" stage, they have principally been found in two "secondary germinal zones", the subgranular zone of the hippocampus and the subventricular zone of the forebrain, the only two regions with persistent neurogenic potential. Neural stem-like cells have also been isolated from adult "non-neurogenic regions" such cerebral parenchyma (including cortex) and spinal cord (Temple, 2001). While the presence of NSCs is not unexpected in the nervous system during organogenesis, their existence in the "formed" "post-developmental" nervous system remains somewhat mysterious and even controversial. Are such cells really present in nature or simply "created" by the act of investigators pulling cells from their *in vivo* context and manipulating them *in vitro* by exposing them to fate-altering mitogens or genes? If they are, in fact, present in the older nervous system, do they serve a purpose or are they simply vestiges of the developmental process with no precise role?

Classification of Stem Cells

Embryonic stem cells (ESCs) reside in the inner cell mass (ICM) of the blastocyst, a stage prior to the generation of the 3 fundamental germ layers that ultimately give rise to all of the body's organs and tissues and probably the stem cells that invest them. NSCs represent a class of stem cells known as "tissue-resident" or "tissue-derived" stem cells, i.e., stem cells that have been allocated to reside within a particular organ or structure. Tissue-resident stem cells might be generated *in vitro* from ESCs by recapitulating the process pursued *in vivo* during development. For example, it appears that NSCs can be found in cultures of ESCs, although only approximately 0.2% of ESCs in most ESC lines spontaneously become neural cells with stem-like qualities (Tropepe et al., 2001). As the signals directing ESCs towards a neural lineage become better understood, this efficiency is likely to increase. Relatively efficient procedures have been derived for doing this, typically involving the application of retinoic acid to embryoid bodies. Whether these procedures in any way emulate what happens naturally *in vivo* is unclear. Also, most procedures for generating NSCs from ESCs never do so to the total exclusion of nonneural cell types or, unfortunately, residual ESC-like cells that can still yield teratomas.

The yield of NSCs isolated directly from embryonic or fetal neuroectodermal-derived tissue is much higher. While there seems to be an abundance of NSCs in early embryonic or fetal neural tissue, their frequency appears to decline rapidly with age. In some instances this may simply reflect the fact that their representation among the cell types of the brain is diluted by the generation of more restricted progenitors and a greater number of differentiated cells as development proceeds (Temple, 2001). It is also possible that the absolute number of stem cells decreases with age and/or that the number of self-renewing divisions within a given stem cell is limited and becomes exhausted with age, i.e., senescence (see Chapter 9 in this volume).

An intriguing feature of NSCs , while in their most robust state, is their ability to model the developmental physiology of the intact animal. These cells undergo repeated asymmetric cell divisions, first producing neurons and later glia, reproducing spontaneously the normal neuron-glial order found in embryonic development (Qian et al., 2000). Harnessing the capacity of NSCs to yield a variety of neural cell types may provide treatment strategies based not only the derivation of a single desired cell type, but multiple cell types that might be required to reconstruct the neural fabric of a given injured CNS structure, e.g., not just the neurons but also glial cells, interneurons, and others.

Whether other types of tissue-resident stem cells (e.g. hematopoietic stem cells, bone marrow mesenchymal stem cells, etc.) can transdifferentiate into neural cells is currently unclear and quite controversial (Bjornson et al., 1999; Castro et al., 2002; Clarke et al., 2000; D'Amour and Gage, 2002; Morshead et al., 2002; see Chapter 2 in this volume). Since there have been conflicting reports. Furthermore, the fact that neural and hematopoietic progenitor cells seem to share overlapping gene expression profiles simply adds to the confusion and uncertainty (Geschwind et al., 2001). It remains unclear whether stem cells from non-neural organs can or should be transplanted effectively and safely for neural therapies.

Equally unclear and controversial is the degree to which endogenous stem cells in the nervous system can affect repair without recourse to the transplantation of exogenous stem cells. Can a sufficient number of endogenous cells from the CNS, or even from other non-neural organs, migrate to appropriate areas of the CNS during degeneration and stress to yield functional neural cells? While the ability of the brain to repair itself via its endogenous progenitor pool is an interesting concept (see Chapter 12 in this volume), to understand its potential and limitations, one first needs to gain a clearer understanding of why recovery in the most devastating of injuries does not already occur spontaneously. For example, what triggers and inhibits NSC functional neurogenesis and differentiation (Kruger and Morrison, 2002)? Is the inability of endogenous progenitors to promote significant recovery in many circumstances simply a matter of an inadequate supply of cells and/or growth factors, both of which might be supplemented from exogenous stores, or is there something more fundamental afoot?

The isolation and availability of NSC lines of human origin will also be reviewed in this chapter. Indeed, many human stem cell lines are starting to be used for transplantation into normal and diseased animal models. However, greater insight into some of their characteristics, particularly the degree to which they differ from rodent cells, will be important prior to their use in humans.

ADVANTAGES OF USING STEM CELL LINES

There are obvious advantages to using NSC lines for both basic and applied scientific investigations including within the pharmaceutical and biotechnology industries. Most NSCs used in almost all publications to date, unless they were freshly abstracted from the brain, placed immediately in a culture dish and never passaged, should actually be regarded as "cell lines" (see Chapter 9 for discussion of cell strains vs. cell lines). The need for expansion and passaging, however, is typically required not only because of the small absolute number of stem cells in the brain, but also because observation of the progeny of a single cell over time and in different situations over time is usually the only way by which the presence of stem-like properties can be affirmed. Reliably sensitive and specific markers for neural stem cells do not yet exist (see discussion in Chapter 3 in this volume); hence an operational definition of a stem cell is still de rigueur. Only a small number of studies have, in fact, used primary neural stem cells. However, it is possible that primary cells could be obtained through fluorescence activated cell sorting (FACS) for unambiguously defining surface markers, as in hematopoietic cell isolation, remains controversial and probably elusive, but, when achieved, would aid in procuring populations enriched for NSCs for culture and subsequent use (see Chapter 7 in this volume).

Clonal NSC lines can be used to dissect biological processes *in vitro* and, following transplantation can mirror what signals are prevalent in a specific CNS region *in vivo* at a given time in development under various normal and abnormal conditions. The use of a clonal stem cell line is actually the only way to insure consistency from experiment-to-experiment. When used judiciously, stem cell lines can yield insights into basic properties inherent to endogenous stem cells as well as their behavior during development and in disease states. These observations have suggested strategies for the use of exogenous NSCs in therapeutic transplantation paradigms, e.g., their ability to cross-correct basic genetic defects (Flax et al., 1998) or to replace missing neural cells (Rosario et al., 1997; Snyder et al., 1997; Yandava et al., 1999). Importantly, data gathered from rodent NSC lines have been applicable to their human, albeit more finicky, counterparts (McKay, 1997). The multipotency of NSC lines in culture, e.g., their differentiation into multiple cell types with diverse morphologies and immunopositivity for distinct markers, has predicted their multipotency *in vivo* following transplantation into intact and lesioned animals, including subhuman primate models. Indeed, Ourednik et al. (2001) have recently affirmed the capacity of human NSCs for extensive migration within the developing monkey brain, participating in cerebrogenesis and the formation of neurons and glia throughout the primate brain (Figure 1). Such findings may suggest the potential of human NSCs for transplantation-based therapies in genuine clinical practice.

RODENT NEURAL CELL LINES

A prototypical NSC murine clone that we and others have found to be very instructive and useful is called clone C17.2 and has now become very well-characterized *in vivo* and *in vitro* (Ryder et al., 1990; Snyder et al., 1992). A brief review of the isolation of these cells is provided as well as their properties. NSCs isolated by different techniques appear to evince the same properties.

Rodent NSC Isolation

The protocols described in this section were developed for NSC isolation from mouse; however, most of these approaches can be applied to the rat. NSCs can be effectively propagated *in vitro* through both epigenetic and genetic means. The key is to keep NSCs in the cell cycle until differentiation is desired. Multipotency is preserved, differentiation is held in abeyance, and engraftment is optimal when the cells are cycling. *In vivo*, the best engraftment is attained if the cells exit the cell cycle within the brain rather than in the culture dish *ex vivo*.

The epigenetic approach typically entails placing defined mitogens into serum-free culture medium. The mitogens most commonly used have been epidermal growth factor (EGF) and/or basic fibroblast growth factor (FGF2, also called FGF2). Sonic hedgehog (SHH) has recently also been described as a mitogen for neural progenitors, though it has not been used for routine passaging of cells. The genetic approach has typically involved transducing isolated stem/progenitor cells with propagating-enhancing genes such as v-myc or large

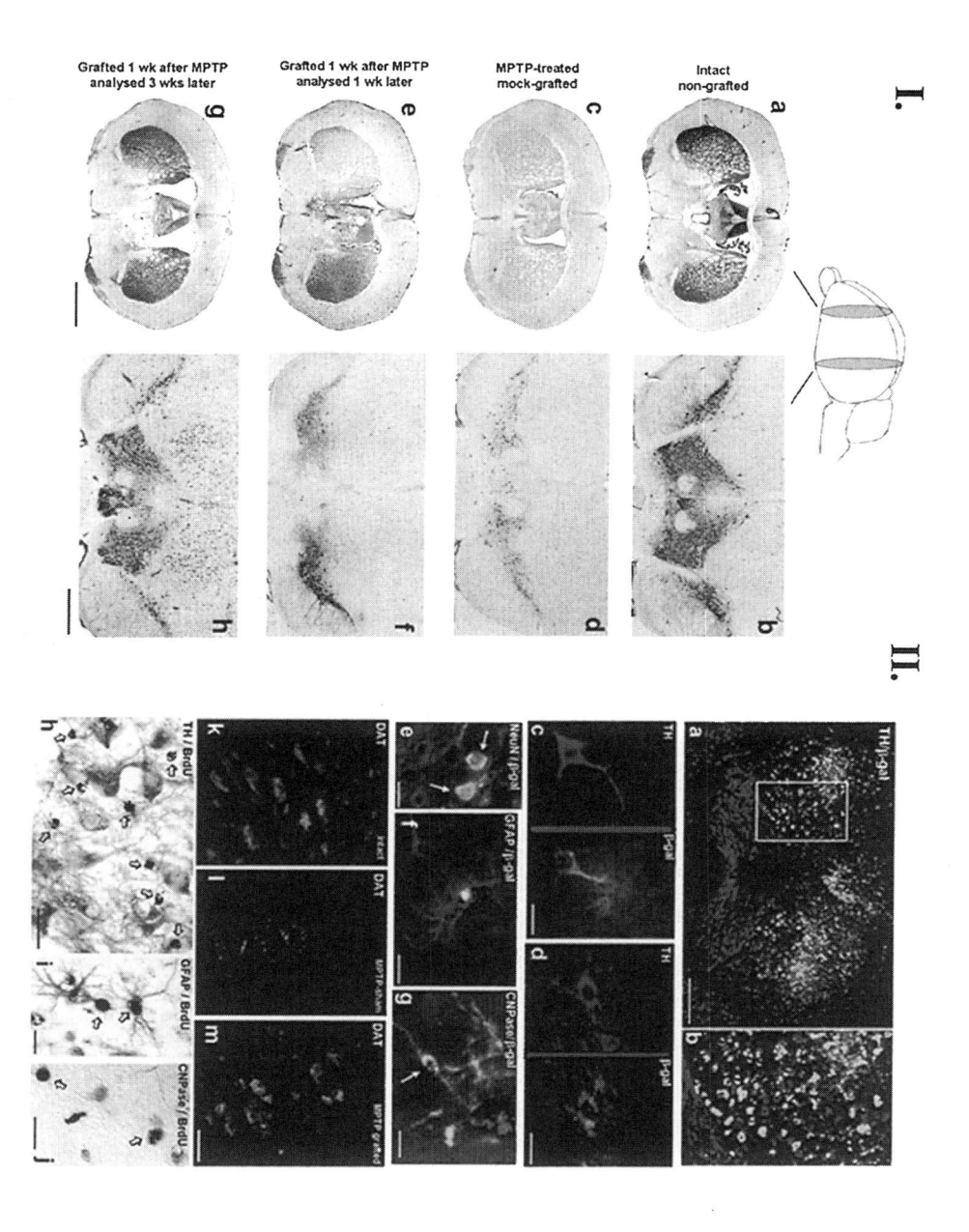

I.

Grafted 1 wk after MPTP analysed 3 wks later — g, h

Grafted 1 wk after MPTP analysed 1 wk later — e, f

MPTP-treated mock-grafted — c, d

Intact non-grafted — a, b

II.

TH/β-gal — a, b; TH — c, d; NeuN/β-gal — e; GFAP/β-gal — f; CNPase/β-gal — g; β-gal — h; TH/BrdU — h; GFAP/BrdU — i; CNPase/BrdU — j; DAT — k, l, m

Figure 1 I: Tyrosine hydroxylase (TH) expression in mesencephalon and striatum of aged mice following MPTP lesioning and unilateral NSC engraftment into the substantia nigra/ventral tegmental area (SN/VTA). Schematic on top indicates the levels of the analyzed transverse sections along the rostrocaudal axis of the mouse brain. Representative coronal sections through the striatum are presented in the left column (**A, C, E, G**) and through the SN/VTA area in the right column (**B, D, F, H**). (**A, B**) Immunodetection of TH (black cells) shows the normal distribution of dopamine (DA)-producing TH$^+$ neurons in coronal sections in the intact SN/VTA (**B**) and their projections to the striatum (**A**). (**C, D**) Within one week, 1-methyl-4-phenyl1-1,2,3,6-tetrahydropyridine (MPTP) treatment caused extensive and permanent bilateral loss of TH immunoreactivity in both the striatum (**C**) and the mesostriatal nuclei (**D**), which lasted for at least seven weeks. Shown in this example, and matching the time point in (**G, H**), is the situation in a mock-grafted animal four weeks after grafting, with substantial recovery of TH synthesis within the ipsilateral DA nuclei (**F**) and their ipsilateral striatal projections (**E**). (**G, H**). By three weeks post-transplant, however, the asymmetric distribution of TH expression disappeared, giving rise to TH immunoreactivity in the midbrain (**H**) and striatum (**G**) of both hemispheres that approached that of intact controls (**A, B**) and gave the appearance of mesostriatal restoration. Similar observations were made when NSCs were injected four weeks after MPTP treatment (not shown). Bars: 2 mm (left), 1 mm (right). **II: Immunohistochemical analyses of TH$^+$, dopamine transporter (DAT$^+$, and BrdU$^+$ cells in MPTP treated and grafted mouse brains.** Aged mice exposed to MPTP, transplanted one week later with NSCs, and killed after three weeks. The following combinations of markers were evaluated: TH (red) with β-gal (green) (**A–D**); NeuN (red) with β-gal (green) (**E**); GFAP (red) with β-gal (green) (**F**); CNPase (green) with β-gal (red) (**G**); as well as TH (brown) and BrdU (black) (**K**); GFAP (brown) with BrdU (black) (**L**); and CNPase (brown) with BrdU (black) (**M**). Anti-DAT-stained areas are revealed in green in the SN of intact brains (**H**), and in mock-grafted (**I**) and NSC-grafted (**J**) MPTP-treated brains. (**A, B**) Low-power overview of the SN/VTA of both hemispheres. The majority of TH$^+$ cells (red) in (**A**) within the SN are of host origin (~90%), with a much smaller proportion being of donor derivation (green, ~10%). A representative close-up of such a donor-derived TH$^+$ cell in the SN can be seen in (**D**). Although a substantial proportion of NSCs differentiated into TH$^+$ neurons, many of these resided ectopically, dorsal to the SN (boxed area in (**A**), enlarged in (**B**); high-power view of an ectopic donor-derived (green) cell that was also TH$^+$ (red) in (**C**)), where ~90% of the cells were donor-derived compared with ~10% host-derived. Note the almost complete absence of a green β-gal-specific signal in the SN+VTA while, many of the ectopic TH$^+$ cells were double-labeled and thus NSC-derived, appearing yellow-orange in higher power under a red/green filter in (**B**). (**E-G**) NSC-derived non-TH neurons [NeuN$^+$; (**E**), arrow], astrocytes (GFAP$^+$; (**F**)), and oligodendrocytes [CNPase$^+$; (**G**), arrow] were also seen, within the mesencephalic nuclei and dorsal to them. (**H-J**) The green DAT-specific signal in (**J**) suggests that the reconstituted mesencephalic nuclei in the NSC-grafted mice were functional DA neurons comparable to those seen in intact nuclei (**H**) but not in MPTP-lesioned, sham-engrafted controls (**I**). This additionally suggests that the TH$^+$ mesostriatal DA neurons affected by MPTP are, indeed, functionally impaired. Note that sham-grafted animals (**I**) contain only punctate residual DAT staining within their dysfunctional fibers, while DAT staining in control (**H**) and engrafted (**J**) animals was normally and robustly distributed within processes and throughout their cell bodies. (**K-M**) Proliferative BrdU$^+$ cells after MPTP treatment and/or grafting (arrows) were confined to glial cells, while TH$^+$ neurons (**K**) were BrdU$^-$. This finding suggested that the reappearance of TH$^+$ host cells was not the result of neurogenesis but rather the recovery of host TH$^+$ neurons. Bars: 90 μm (**A**); 20 μm (**C-E**); 30 μm (**F**); 10 μm (**G**); 20 μm (**H-J**); 25 μm (**K**); 10 μm (**L**); 20 μm (**M**). From *Nature Biotechnology* (2002) 20:1103-1110.

T-antigen. Myc appears to be a gene related to cell cycle regulation and to some fundamental properties of stem cell self-renewal. Despite the differences in isolation and subsequent propagation techniques, the similarities between neural stem cell lines are striking.

The C17.2 clone, used extensively in the literature, was originally derived from a 4-day old neonatal mouse cerebellum (Ryder et al., 1990) and is readily propagated as monolayer in fetal calf serum-containing medium following retroviral-mediated transduction with v-myc. It was one of over 40 lines similarly established and grouped into clonally-related families based on the location of unique viral insertion sites. All lines were negative for tumorigenicity when injected into the brains of newborn mice. Some clones were then infected with a second retroviral vector encoding lacZ, the gene for E.Coli β-galactosidase, allowing these stem cells and all of their progeny to be recognized *in vitro* and *in vivo* as blue cells following processing with Xgal histochemistry or as immunopositive cells following reaction with an anti-βgal antibody (Snyder et al., 1992). This unique second retroviral insertion site also allows clonality to be affirmed.

Kilpatrick and Bartlett (1993) and others (Drago et al., 1991) have reported similar protocols for isolating neuroepithelial cells from the telencephalon and mesencephalon of embryonic day 10 (E10) mice. Progenitor cells from the rat mesencephalon have been isolated and reported to differentiate into dopaminergic neurons in the presence of interleukin1 (Ling et al., 1998), further enhanced by the addition of interleukin11, LIF, and GDNF. Similar results have been obtained with human mesencephalic precursor cells (Storch et al., 2001).

In addition to multipotent neuroepithelial cell lines, the generation of neural progenitor lines with more lineage restriction has been reported, e.g., neuronal-restricted or glial-restricted. It remains controversial, however, whether such cells are genuinely restricted to one lineage or simply have not been placed in environments with signals sufficient for unveiling or inducing a broader potential (Mayer-Proschel et al., 1997; Mujtaba et al., 1999). Indeed, one cell line RN33b (Onifer et al., 1993; Whittemore and White, 1993), was initially reported to be restricted solely to a neuronal fate, yet in the hands of other investigators (Lundberg et al., 2002) was found to yield glia as well. Attempts at isolating lineage-restricted lines have been made by immunoselection for cell surface markers, e.g., highly polysialated neural cell adhesion molecules (PSA-NCAM) for neuron-restricted precursors and A2B5 for glial-restricted precursors. These markers themselves, however, may not be sufficiently lineage-specific for such a task.

Of late, there has been a growing awareness of the possibility that the properties of a cell might change if passaged in culture for extensive periods, particularly following chronic exogenous mitogen exposure (EGF and/or FGF2). Interestingly, it appears that transducing the cells with genes that function within the normal cell cycle and mitotic regulatory mechanisms [e.g., myc (Sah et al., 1997; Flax et al., 1998; Villa et al., 2000) or TERT (Roy et al., 2003) which regulates telomerase] not only maintain the cells in a stem-like state *in vitro* but do not seem to subvert their normal differentiation potential. Perhaps this is due to the fact that the cells regulate exogenous copies of these genes in the same manner in which they regulate their own endogenous copies. Empirically, these clones appear to be more resistant to passage- and time-dependent alterations in fate and function than their mitogen-propagated counterparts. They also seem to be less prone to senescence and age-related changes.

Cells, even within the same clone, that acquire a more rapid doubling time may overtake the rest of the cellular population in the dish and confer a character to the culture that does not reflect its original properties. Neoplastic transformation may also occur and the potential of the cells may change such that they no longer accurately reflect the normal biological potential of the cells they were originally derived from. This may explain surprisingly poor and unexpected differentiation potential by NSCs in certain circumstances. It may also account for certain controversial observations of transdifferentiation (Bjornson et al., 1999; Castro et al., 2002; Clarke et al., 2000; D'Amour and Gage, 2002; Morshead et al., 2002; see Chapter 6 in this volume). Because of the risks of such unknown mutations, potentially introduced during extended passage, most NSC transplant biologists use NSCs that have been passaged only 4-8 times (Morshead et al., 2002). To be useful NSC lines must also be able to be cryopreserved and thawed without altering their fundamental properties. To date, that does not seem to be a problem, but one must always be vigilant for such changes.

Rodent NSC Properties

Maintaining a NSC in a proliferative state in culture does not appear to subvert its ability to respond to normal developmental cues *in vivo* following transplantation or to undergo integration into host circuitry. Upon introduction into the *in vivo* environment, NSCs withdraw from the cell cycle and interact with host cells and host-derived cues (including vasculature, extracellular matrix, diffusible factors, etc.), which trigger differentiation of the NSCs into site-appropriate and developmental stage-appropriate neurons and/or glia. Following implantation into primary or secondary germinal zones, murine NSCs migrate and differentiate in a temporally- and regionally-appropriate direction. It was this type of behavior that suggested the potential of NSCs for cell replacement and was affirmed in two different paradigms. First, NSCs were used to replace pyramidal neurons in a circumscribed region of the neocortex of adult mice whose own pyramidal neurons in that area were induced to die by apoptosis (Snyder et al., 1997). In a second example (Rosario et al., 1997), NSCs transplanted into the external germinal layer of a mouse mutant (meander tail) whose cerebellar granule neurons failed to develop, shifted their differentiation fate to yield more of that deficient neuronal cell type, repopulating the granule cell-deficient anterior lobe. When transplanted into a mouse mutant shiverer whose oligodendrocytes rather than neurons were dysfunctional, the NSC shifted its differentiation pattern to yield a greater proportion of that cell type with symptomatic relief in some animals (Yandava et al., 1999).

Interestingly, in many NSC transplants, a significant subpopulation of NSCs will remain as quiescent, undifferentiated cells intermixed among more differentiated host and donor cells. This subpopulation, too, probably represents the normal segregation of NSC progeny during development and, as we note below, may play an unanticipated important role in therapeutic contexts.

A true stem cell, as opposed to a progenitor, should be able to participate in normal development along the entire neuraxis, independent of the region from which it was originally isolated. For example, a true stem cell isolated from the hindbrain should be able to integrate not only into the developing cerebellum but also throughout the remainder of the immature and adult central and peripheral nervous systems in a cytoarchitecturally-appropriate manner. How these properties emerge is a pivotal question in developmental neurobiology. Clones of NSCs that can be tracked and isolated can now help in the exploration of this important question.

That such engrafted NSCs can also express foreign genes *in vivo* suggested that they might also be engineered *ex vivo* to import specific bioactive, often therapeutic factors in a "Trojan Horse" fashion into the host brain, where they might release these gene products in a site-specific manner to help promote survival, regeneration, repair, or rescue. This technique has been used not only to replace missing essential gene products in inherited neurodegenerative disorders, e.g., in lysosomal storage diseases (Lacorazza et al., 1996; Snyder et al., 1995), but also to provide extra proteins, such as neurotrophic factors, that might promote regeneration, differentiation, connectivity, and neuroprotection of host cells. Liu et al. (1999) used NSCs to deliver neurotrophin-3 (NT-3) to the spinal cord and Akerud et al. (2001) used NSCs to deliver the neuroprotective factor, GDNF, to the striatum. In this latter experiment, the integrated NSCs not only gave rise to neurons, astrocytes, and oligodendrocytes, but also maintained high levels of GDNF *in vivo* for at least 4 months. In a mouse lesioned with 6-hydroxydopamine, a model of Parkinson's disease, the intrastriatal implantation of these GDNF-overexpressing NSCs prevented the degeneration of dopaminergic neurons in the substantia nigra and reduced behavioral impairment.

That NSCs exhibited a tropism for intracranial pathology was first demonstrated by Aboody et al. (2000) during their examination of experimental brain tumors. They observed that murine NSCs (as well as human NSCs), when implanted into intracranial gliomas *in vivo* distributed themselves throughout the tumor bed and migrated in juxtaposition to tumor cells while expressing foreign genes. Furthermore, they demonstrated that the delivery of cytosine deaminase, which converts 5-fluorocytosine to the oncolytic 5-fluorouracil, by transplanted NSCs injected caused a reduction in tumor mass, suggesting that these inherently migratory NSCs could be used therapeutically. In fact, even when implanted to the contralateral hemisphere, the NSCs could cross the midline to home in specifically to the area of pathology. It would be particularily beneficial to use NSCs not only first to destroy brain tumors but then subsequently use them to repair the damaged tissue through differentiation into lost cellular phenotypes (Noble, 2000). Such a "home run", however, has not yet been accomplished.

Although cell replacement and gene therapy are considered to be the typical therapeutic applications of NSCs, an additional role has recently been suggested (Ourednik et al., 2002; Park et al., 2002b; Steindler, 2002). Ourednik et al. (2002) demonstrated that NSCs appear to possess the inherent ability to rescue dysfunctional neurons. The model used was the induction of dopamine dysfunction in aged mice by the toxin MPTP, which causes permanent dysfunction of mesostriatal dopamine neurons by oxidative stress with eventual fiber degradation and cell death. Implantation of NSCs unilaterally into the midbrains of such dopamine-impaired mice was associated with reconstitution bilaterally of the mesostriatal system anatomically and functionally. Although a subpopulation of the donor NSCs spontaneously differentiated into new dopaminergic neurons, the majority of the dopaminergic neurons in this reconstituted mesostriatal system was actually of host origin. They had been "rescued" from death and their function had been "restored" by association with donor NSCs that migrate and distributed themselves bilaterally, particularly those in an undifferentiated or glial-differentiated state. In other words, donor-derived NSCs that had not differentiated into replacement neurons nevertheless had a very powerful therapeutic impact, perhaps more important than those that had followed a neuronal lineage. Without having been genetically engineered, these non-neuronal cells appeared spontaneously and constitutively to express neuroprotective substances that accounted for the "rescue" of host cells (Figure2).

This complex interaction between a degenerating or injured host and transplanted NSCs was highlighted in another recent study (Park et al., 2002b). NSCs were placed into the large infarction cavity of animals subjected to severe hypoxic-ischemic injury. The NSCs were initially supported by a three-dimen-

I.

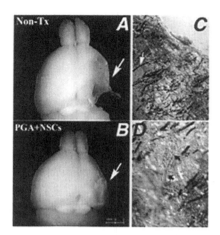

II.

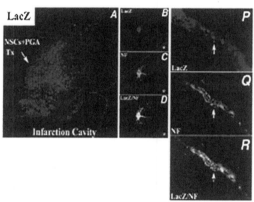

III.

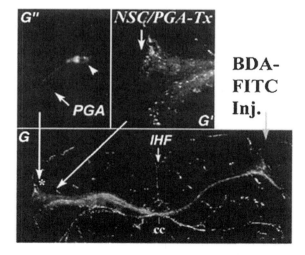

Figure 2. I. Implantation of NSC–polyglycolic acid (PGA) scaffold complexes following extensive hypoxic ischemic brain injury and necrosis. (A) Brain of an untransplanted mouse subjected to right HI injury with extensive infarction and cavitation of the ipsilateral right cortex, striatum, thalamus, and hippocampus (arrow). (B) Brain of a similarly injured mouse implanted with an NSC–Scaffold complex, generated *in vitro* into the infarction cavity 7 days after the induction of HI (arrow; n = 60). At maturity [age-matched to the animal pictured in (A)], the NSC–scaffold complex appears in this whole mount to have filled the cavity (arrow) and become incorporated into the infarcted cerebrum. (C, D) Higher magnification of representative coronal sections in which the parenchyma appears to have filled in spaces between the dissolving black polymer fibers [white arrow in (C)] and supports neovascularization by host tissues as seen in (D). A blood vessel is indicated by the closed black arrow in (D); open arrow in (D) points to a degrading black polymer fiber. Scale bars: (C, D) 100 μm. **II. Characterization *in vivo* of the neural composition of NSC–Scaffold complexes within the HI-injured brain.** (A) Two weeks following transplantation of an NSC–Scaffold complex into the infarction cavity (arrow), donor-derived cells immunostained for β-gal, showed robust engraftment within the injured region (n = 15). (B-D) Donor-derived cells, identified by the β-gal antibody in (B), coimmunostained in (C) for neural cell type–specific antigens. The neuronal marker neurofilament (NF) (n = 15) and the oligodendroglial marker CNPase (n = 15) are merged in (D), indicating the two neural cell types most damaged by HI, are coexpressed by the donor-derived cells. Note: LacZ-expressing, donor-derived neurons like those in **B-D** stained similarly for neuronal markers NeuN and MAP2. Interestingly, not only were donor-derived neural cells present, but also host-derived cells seemed to have entered the NSC–scaffold complex, migrating along and becoming adherent to the scaffold matrix. An intermingling of host with donor cells was seen for neuronal elements; see **P-R**. Donor-derived (lacZ⁺) neurons (NF⁺) within the complex appeared to send processes along the PGA fibers out of the scaffold into host parenchyma as in **P-R** (n = 15). (P-R) By six weeks following engraftment, (P) donor-derived lacZ⁺ cells (arrow) appeared to extend (Q) many exceedingly long, complex NF⁺ processes along the length of the disappearing scaffold and into host parenchyma, [merged image in (R)], where arrow indicates the same cell. **III. Long-distance neuronal connections extend from the transplanted NSC–Scaffold complexes in the HI-injured brain toward presumptive target regions in the intact contralateral hemisphere.** To confirm the suggestion in II. (P-R) that long-distance processes project from the injured cortex into host parenchyma, a series of anterograde and retrograde tract tracings are illustrated on the donor-derived β-gal⁺ neurons from II. (P). (G) Biotinylated dextran amine (BDA)-FITC was injected into the contralateral intact cortex and external capsule (green arrow) 8 weeks after implantation of the NSC–scaffold complex into the infarction cavity (n = 10). Axonal projections (green) are visualized by the retrograde transport of BDA, leading back across the interhemispheric fissure (IHF) via the corpus callosum (cc) and emanating from cells in the NSC–PGA complex within the damaged ipsilateral cortex and penumbra. These projections are seen at progressively higher magnification in G' (region indicated by arrow to **G**) and G" (region indicated by arrow and asterisk in (**G**). In G", the retrogradely BDA-labeled perikaryon of a representative neuron adherent to a dissolving PGA fiber is well visualized. Reprinted from *Nature Biotechnology* (2002) 16:1111-1117.

sional biodegradable synthetic scaffold that transiently provided a platform to help bridge large cystic lesions and support the NSCs in space before "dissolving" after 2-to-4 weeks. The scaffold of sustained intimate exposure of donor and host cells to each other. It also trapped molecules emanating from the injured brain as well as from the NSCs so that their contact with target receptors could be maximized. New cerebral parenchyma began to form and fill in the gaps, including the reformation of cortical tissue and the likely reestablishment of neural connections. The new neural networks that formed were derived from the donor cells as well as from the host brain. Donor-derived neurons, which were now numerous, were capable of directed, target-appropriate neurite outgrowth (and extension into the opposite hemisphere) without the need for specific external induction with growth factors or genetic manipulation of the cells or host (Figure 2). The new parenchyma became vascularized by the host, suggesting that the NSCs were also expressing angiogenic factors. Of additional interest was the observation that inflammation and glial scarring seemed to be reduced, not enhanced, by the provision of exogenous NSCs in this situation, probably contributing to the reconstitution of cerebral tissue. Similar phenomena were observed in experimental models of spinal cord injury (Teng et al., 2002).

HUMAN NEURAL STEM CELLS

The *in vitro* and *in vivo* properties and therapeutic potential of human neural stem and progenitor cells have recently been reviewed (Martinez-Serrano et al., 2001). Despite some important differences, principally related to a much longer cell cycle, human NSC (hNSCs) appear to retain many of the appealing properties of their murine counterparts regarding their potential as therapeutic agents . Therefore, insights gleaned from rodent NSCs have been broadly applicable to human systems.

Isolation and Generation of Human NSC Clones

The isolation, propagation, characterization, cloning, and transplantation of NSCs from the human CNS have largely followed a blueprint established by prior experience with murine NSCs (including clone C17.2 as well as growth factor-expanded lines). Some of the techniques used to isolate neural stem/progenitor cells from human tissue will be reviewed.

From Freshly Dissociated Tissue

Cells have been isolated from numerous CNS regions including the telencephalon, cerebellum, striatal eminence, ventral mesencephalon, hippocampus, cortex, diencephalon, and forebrain. In some protocols the cells are maintained as adherent monolayer cultures and in others as suspension cultures using a variety of substrata and growth factors. Although a number of cell culture systems have been developed for maintaining human neural stem cells in culture, and many of these have provided successful graft material, the optimal technique for inexhaustible passaging without senescence, that does not involve genetic manipulation, probably still needs to be devised (Buc-Caron, 1995; Carpenter et al., 1999; Flax et al., 1998; Rubio et al., 2000; Svendsen et al., 1996; Svendsen et al., 1997; Svendsen et al., 1998; Vescovi et al., 1999; Villa et al., 2000). Fortunately, genetic manipulations to date have been safe and effective.

In rodent studies, it has been observed that immature uncommitted NSCs have a dual responsiveness to both EGF and FGF2 (Kilpatrick and Bartlett, 1993; Kitchens et al., 1994; Kornblum et al., 1990). In the absence of highly specific surface markers for stem cells, this dual responsiveness was chosen many years ago in our laboratory for both the screening and enhancement of a starting population of dissociated primary human neural tissue for "stem-like" cells. Neural cells were dissociated from the telencephalon, principally the ventricular zone thought to contain the major population of NSCs (Gage, 2000), from a human fetal cadaver as soon after fetal demise as possible. The earlier in gestation the fetus, the more abundant and useful the cells abstracted. Cells expressing both EGF and FGF2 receptors were selected by initially growing the cells as a polyclonal population in serum free medium supplemented with both EGF or FGF2 and then sequentially transferring the cells from medium containing first one and then the other mitogen. Maintenance in any particular medium for 2-3 weeks prior to the next transfer was sufficient for selection and, although large numbers of cells failed to survive passage at each step, the remaining culture consisted of passageable, immature, proliferative cells that appeared to expressed both the EGF and FGF2 receptors. Some of these cells were maintained in FGF2-containing media for genetic manipulation via retroviral-mediated transduction by v-myc and subsequent cloning. To assess the role of NSC *in vivo* following transplantation, some populations were infected with an amphotropic replication-incompetent retroviral lacZ vector. Retroviral insertion sites also helped with affirming that certain populations were monoclonal.

Post Mortem Human Tissue

Recently, progenitor cells have been isolated from human brains that have been post mortem for as long as several hours. Palmer et al., (2001 provide a full description of their methods for the *in vitro* culture of these cells. Although they showed differences between young and old specimens, cells from adult tissues were capable of expansion for >30 population doublings before senescence. On the other hand, cells from an 11 week old neonate grew logarithmically for >70 population doublings before showing signs of a significant reduction in growth rate. Neurons were spontaneously generated at all stages in these cultures and complete differentiation occurred following withdrawal of growth factors. This effect was enhanced by forskolin, an activator of adenylate cyclase, and retinoic acid, previously shown to enhance differentiation in other cellular models (Thompson et al., 1984). Despite clear differences in time to senescence, both neonatal and adult progenitors did yield similar proportions of neurons and astrocytes with little oligodendrocyte formation. Consistent with a reported reduction in progenitor cell activity in adulthood, tissues from young individuals were found to yield significantly more cells-per-gram and these cells had a higher proliferative capacity. The use of these cells circumvents the social and ethical issues raised by the derivation of human embryonic stem cells or the isolation of neural stem cells from fetal tissue. Their utility as graft material in transplant studies still needs to be established, however.

Fluorescence Activated Cell Sorting (FACS)

Neural crest stem cells have been isolated from mammalian peripheral nerve (Morrison et al., 1999). These cells were selected by positive labeling for the low affinity neurotrophin receptor, $p75^{NTR}$, and negative labeling for a peripheral myelin marker protein, P0. They demonstrated the formation of colonies of multipotent cells that could differentiate into multiple phenotypes self-renew *in vivo* as suggested by BrdU incorporation, and can be used for transplantation.

Uchida et al., (2000; see Chapter 7 in this volume) directly isolated human CNS stem cells from fetal brain tissue using antibodies to cell surface markers and subsequent FACS analysis. They demonstrated that these human NSCs bore the following surface marker phenotypes: $CD133^+$, $5E12^+$, $CD34^-$, $CD45^-$, and $CD24^{-/lo}$. Additionally, cells positive for CD133 and negative for CD34 and CD45 could form floating cell clusters and could differentiate into neurons and glia. Following transplantation into rodents, these cells could migrate and differentiate into cells of multiple neural lineages *in vivo*. Selecting for $CD24^-$ cells further enriched for these abilities. Of interest is the fact that CD133 is not specific for neural cells, but is an antigen common to other stem cells, e.g., hematopoietic.

Finally Keyoung et al. (2001) transfected dissociated fetal human brain cells with retroviruses to express green fluorescent protein (GFP) coupled with nestin and musashi1 promoters, markers of immature neural progenitors. GFP was then expressed in uncommitted neuroepithelial cells that could also be isolated by FACS. It was demonstrated that this technique yielded cells that could self-renew, were multipotent, and could generate both neurons and glia. Similar techniques have been applied to isolating various neural progenitors from adult human brain biopsy specimens. While such isolation techniques may run the theoretical risk of toxicity from the GFP or retrovirus or altered gene expression (Martinez-Serrano et al., 2000; Roy et al., 2000), the use of such cells to date following transplantation in rodents has been safe and effective.

Recently, Nunes et al. (2003) developed a method for the isolation of multipotential neural progenitor cells from the subcortical white matter of the adult brain, which harbors a pool of glial progenitor cells. These cells can be isolated by FACS either transfection with GFP under the control of the CNP2 promoter, or A2B5-targeted immunotagging. Although these cells give rise largely to oligodendrocytes, neurons were also generated in low-density culture. Further, they demonstrated that glial progenitors include cells capable of neurogenesis, can be passaged as neurospheres *in vitro*, and generate functionally competent neurons and glia both *in vitro* and *in vivo*.

Human Neural Stem Cell Propagation

While genetic manipulation of human NSCs has probably been the most effective technique for insuring the propagation of engraftable clones without senescence, concerns always linger, heretofore not supported, that problems could emerge. In the case of v-myc, this gene is constitutively regulated by the cell in the manner in which it regulates cellular myc and following transplantation it is silenced (Flax et al., 1998; Rubio et al., 2000; Vescovi et al., 1999). In other words, normal cellular processes are used to regulate a gene that is likely to be pivotal in stem cell self-renewal. Thus the tumorigenic potential of these NSCs is likely minimal. However, to provide an extra level of assurance, it might be prudent to engineer cells with a CRE-loxP recombinase system for removing cell cycle regulatory or immortalizing genes just prior to or just following implantation (McKay, 1997; Westerman and Leboulch, 1996).

While the use of epigenetic techniques would seem to circumvent any such quandaries, these also have similar safety and efficacy concerns. The extended passage of NSCs by epigenetic means may induce alterations in growth properties including tumor formation following transplantation (Morshead et al., 2002). However, such problems have not been observed or reported as yet (Vescovi et al., 1999). The tumorigenic potential of human NSCs propagated in this manner is also likely to be minimal. However, senescence has been a problem in which a useful population of NSCs will cease self-renewing presenting another quandary.

Properties of Human Neural Stem Cells

We have developed and characterized a number of human NSC (hNSCs) clones from freshly dissociated fetal telencephalon using the protocols described above (Flax et al., 1998); some are v-myc-augmented and others are simply epigenetically propagated. These self-renewing clones exhibit classical NSC properties and grow as both neurospheres and adherent cells that express markers consistent with an undifferentiated state when proliferating. When these cells are transferred to serum-containing media on adherent substrata and exit from the cell cycle, they spontaneously differentiate into the 3 fundamental neural lineages: neurons, oligodendrocytes, and astrocytes. The latter cell type emerges most robustly when the hNSCs are cocultured with dissociated embryonic murine CNS tissue.

Following transplantation into germinal zones of the newborn mouse brain, the hNSCs participate in normal development and respond appropriately to developmental cues *in vivo* by exhibiting migration, integration into host parenchyma, and differentiation into multiple developmental- and regional-appropriate neural cell types (Flax et al., 1998; Rubio et al., 2000). Where and when neurogenesis and gliogenesis normally occur, the hNSCs also yield neurons and glia, respectively, intermixed non-disruptively with the progeny of host neural progenitors. Consistent with our definition of a true stem cell, the same clone of hNSCs can participate in forebrain development when implanted into the subventricular zone (SVZ), yet integrate into the cerebellum at the opposite end of the neuraxis when implanted into the external germinal layer (EGL). hNSCs that integrated into the SVZ followed 2 developmentally appropriate migratory routes: they either migrated tangentially to the olfactory bulbs via the rostral migratory stream where they became neurons, or more radially into the subcortical white matter and cortical parenchyma where they differentiated into oligodendroglia and astroglia. Following implantation into the cerebellar EGL, the same hNSC clone generating granule neurons. In fact, the hNSCs were capable of yielding granule neurons in mouse mutants that were deficient in this cell type (Flax et al., 1998), suggesting their potential for cell replacement therapies, much as murine NSCs did in an identical mouse model (Rosario et al., 1997).

Like rodent NSCs, engrafted hNSCs are also capable of expressing foreign transgenes within the parenchyma of host animals (Flax et al., 1998). This observation suggested the feasibility of using such cells for gene therapy in a range of pathological conditions. Indeed, NSCs constitutively express a number of useful gene products, including neurotrophic factors such as GDNF and lysosomal enzymes. These gene products are sufficiently abundant to cross-correct neurons and glia from mouse models of some lysosomal storage diseases, genetically-based neurodegenerative diseases typically caused by mutations in genes encoding lysosomal enzymes (Billinghurst et al., 1998; Flax et al., 1998).

To assess whether hNSCs might engraft and respect developmental cues in hosts that more closely resemble human anatomy and to help assess their ultimate potential utility in human disease hNSCs were transplanted under ultrasonic guidance into the cerebral ventricles of normal fetal Old World monkeys during corticogenesis (Figure 3). The hNSCs gained access to the primary germinal zone of the telencephelon, the ventricular zone (VZ; Park et al., 2002a). The hNSCs entered the VZ, they migrated throughout the primate brain, much as rodent NSCs did in the embyronic rodent brain (Brustle et al., 1998; Lacorazza et al., 1996). Interestingly, the progeny of a single hNSC clone distributed themselves into two subpopulations. One subpopulation contributed to corticogenesis via migration along radial glia to appropriate layers of the cortical plate and differentiated into lamina-appropriate neurons and glia. The second subpopulation remained undifferentiated and contributed to the SVZ, a secondary germinal zones, with occasional members interspersed throughout the brain parenchyma. These data suggest that an early genetic program allocates the progeny of NSCs either for immediate use in organogenesis or to pools for later use in the postdevelopmental brain (Ourednik et al., 2001). This method by which the hNSCs were introduced into the developing subhuman primate brain could be employed for in utero cellular therapy of human neurogenetic disorders. One could consider not only treating congenital disorders but also, altering the brain composition of patients harboring genes for neurodegenerative diseases that are not expressed until adulthood or middle-age but whose antenatal genetic diagnosis is possible, e.g., Huntington's Disease.

Other labs have also generated and characterized various human NSC lines (Rubio et al., 2000; Villa et al., 2000). Rubio et al. (2000) have developed the human NSC line HNSC.100, epigenetically expanded from cells derived from the diencephalic and telencephalic regions of a 10 week gestation age human embryo (Vescovi et al., 1999) and then immortalized with v-myc (Flax et al., 1998). At the time of initial publication, the HNSC.100 cell line had undergone over 250 population doublings without evidence of senescence. Following mitogen withdrawal, the cells exit the cell cycle and transition from an immature

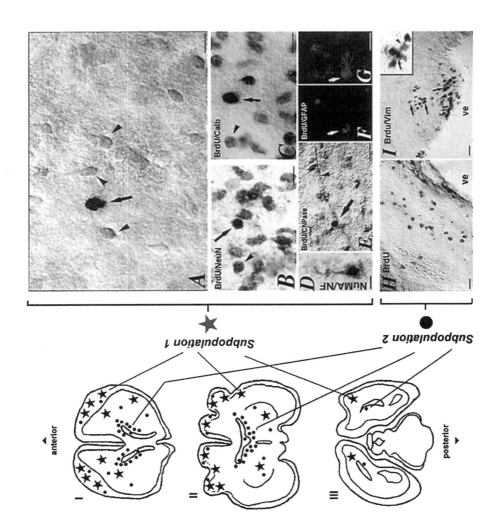

Figure 3. Segregation of fates of cloned hNSCs and their progeny into two subpopulations in the brains of developing Old World monkeys. Schematics (**left**) and photomicrographs (**right**) illustrating the distribution and properties of transplanted hNSCs (BrdU+) dispersed throughout and integrated into the ventricular zone (VZ). From there, clonally related hNSC-derived cells pursued one of two fates, as shown by immunocytochemical analysis (**A** to **I**). Donor cells that migrated outward from the VZ along radial glial fibers into the developing neocortex constituted subpopulation 2. The differentiated phenotypes of cells in subpopulation 1 (red stars) are pictured in panels **A** to **G**. (**A**) A hNSC-derived BrdU-positive cell (black nucleus, arrow), likely a neuron according to its size, morphology, large nucleus, and location, is intermingled with the monkey's neurons (arrowheads) in neocortical layers II and III. The neuronal identity of such donor-derived cells is confirmed in **B** to **D**. (**B**, **C**, and **E** to **G**) High-power photomicrographs of human donor-derived cells integrated into the monkey cortex double stained with antibodies against BrdU and cell type-specific markers: (**B**) NeuN and (**C**) calbindin for neurons (arrows, donor-derived cells; arrowheads, host-derived cells). (**E**) CNPase for oligodendroglia (arrow, BrdU-positive black nucleus in CNPase-positive brown cell; arrowhead indicates long process emanating from the soma). (**F** and **G**) Astroglia to BrdU (**F**); GFAP for astroglia in (**G**). The human origin of the cortical neurons is further confirmed in (**D**) where the human-specific nuclear marker NuMA (black nucleus) is colocalized in the same cell with neurofilament (NF) immunoreactivity (brown). Progeny from this same hNSC clone that constituted subpopulation 2 (blue dots) in the schematic and in **H** and **I** (arrows) remained mainly confined to the SVZ and stained only for an immature neural marker [vimentin (brown) colocalized with BrdU (black nucleus) better visualized in the inset (arrows); arrowhead indicates host vimentin-positive cell. Some members of subpopulation 2 were identified in the developing neocortex (blue dots) intermixed with differentiated cells. Abbreviations: ve, lateral cerebral ventricle; arrow, BrdU-positive donor-derived cell; arrowhead, BrdU-negative host-derived cell except in (**E**). Scale bars, 30 μm (**A**) - (**C**); 20 μm (**D**) - (**I**). Reprinted from *Science* (2001) 293:1820-1824.

phenotype (expression of vimentin and nestin) to a differentiated phenotype (morphologies and markers consistent with maturing neurons and glia). Similarly, following transplantation, these cells exit the cell cycle and differentiate (Rubio et al., 2000). They have also been engineered to express genes with therapeutic potential, such as tyrosine hydroxylase, the rate-limiting enzyme for dopamine synthesis (Villa et al., 2000).

Do the cell lines need to be well-defined clones or will uncloned populations work equally as well? Carpenter et al. (1999) and Fricker et al. (1999) have argued that it may not be essential to obtain pure clonal populations of hNSCs for therapy. They demonstrated that a mixture of multipotent and lineage-restricted neural progenitors isolated from an embryonic human fetus and expanded *in vitro* in the presence of EGF, FGF2 and LIF were engraftable in the adult rodent brain. These cells exhibited extensive migration, many along routes normally taken by endogenous neural precursor cells. These cells, too, yielded both glial and neuronal phenotypes. While, a multiclonal population of hNSCs is at risk for having a shifting compostion based on the vaying representation of any given clone at any given passage (usually the most proliferative clone), these authors reported that their progenitor populations could be expanded up to 10-million fold with no change in properties *in vivo*. Nevertheless, this caution should be kept in mind.

Utility in Transplantation

Table 1 lists the properties that are desirable in a human NSC line for therapeutic use. These properties include plasticity to accommodate to their region of engraftment: ability to migrate to regions of pathology, effectively and stably express genes of therapeutic importance, differentiate into multiple cell types in the damaged region, and make proper reconnections while without making inappropriate connections. Just as important, however, are safety requirements that must be satisfied prior to the clinical use of human neural stem or progenitor cells (Table 2 adapted from Martinez-Serrano et al., 2001), Although many of the characteristics listed have been satisfied by the various published hNSC lines, some issues still remain. Most prominent among these is insuring that a stable line of hNSCs can be propagated and expanded to yield inexhaustibly large numbers over unlimited amounts of time without reaching senescence yet without becoming neoplastically transformed. How that can be most safely and efficiently accomplished is an area of intense investigation. Can it be accomplished via carefully selected genes that operate on the cell cycle such as myc or by maintaining telomere length (telomerase), or by expansion with mitogens?

INVENTORY OF AVAILABLE EMBRYONIC STEM CELL (ESC) LINES AND THEIR POTENTIAL FOR NEURAL CELL FORMATION

The use of federal funds for research on human ESC lines has been restricted to those lines in existence prior to August 09, 2001. Table 3 (see page 261) is adapted from the National Institutes of Health registry of approved human ESC lines (http://escr.nih.gov). It provides the name and location of each distributor and whether they have met the eligibility criteria. The characterization of these cells is far from complete and each is in various stages of development. Only a few are available to investigators and the unique characteristics of each cell line are largely unknown. Probably only 3-4 of the listed lines has any utility for research. None are likely to be appropriate for clinical applications because of their need to be grown on mouse embryonic fibroblast feeder layers during key times in their maintenance. Other biological challenges must be tackled before ESC lines of any type can be used reliably, effectively, and safely in research or in clinical medicine. One challenge is the ability to direct ESCs down a particular desired lineage pathway to the exclusion of other non-desired pathways. A second challenge is excluding without exception the formation of teratomas, one of the defining characteristics of an ESC.

Generation of neural precursors from ESCs

Neural precursors have been generated *in vitro* from human ESCs (Reubinoff et al., 2000) and from human embryonic germ (EG) cells (Shamblott et al., 2001) by adapting protocols used for deriving neural cells from rodent ESCs. Similar attempts are underway to generate even more specialized neuronal cell types from human ESCs (Lee et al., 2000; Studer, 2001), e.g., dopaminergic and serotonergic neurons from mouse ESCs, by driving them through 5 stages (Figure 4). The difficulty of adapting these procedures to human ESCs reflects both the differences between mouse and human ESCs as well as the technical challenges of manipulating human cells.

Tropepe et al. (2001) have recently speculated that neural ectoderm is the default differentiation pathway for murine ESCs. That the same may pertain to human ESCs is suggested by reports of Reubinoff et al. (2000) and Zhang et al. (2001). The latter group demonstrated the spontaneous generation *in vitro* of human neural precursor cells from human ESC lines following prolonged cultivation (at least 3 weeks) at high density. Cellular aggregates within the ESC culture, more reminiscent of neural tubes than of embroid bodies, were formed that consisted of cells with complex morphologies. These cells were able to

Table 1. Desirable Neural Stem Cell Properties for CNS Gene Therapy and Repair

Properties	Description	Reference
Genetic manipulability	NSCs are easily transduced by most gene transfer methods, including most viral-mediated techniques.	Flax et al., 1998 Villa et al., 2000
Facile engraftability following simple implantation protocol	Easily engraftable by intra-ventricular injection or i.v. administration when intracranial pathology is present.	Villa et al., 2000 Rosario et al., 1997
Expression of therapeutic and tracking genes	Cells can be engineered to express both reporter genes and therapeutically important genes.	Flax et al., 1998 Villa et al., 2000 Aboody et al., 2000
Normal reintegration into host cytoarchitecture and circuitry	Differentiate to regionally-appropriate glial & neuronal phenotypes; important for re-establishing circuitry and/or repairing damage. Multiple cell types likely required to reconstitute damaged milieu.	Flax et al., 1998 Vescovi et al., 1999 Rosario et al., 1997
Migration capacity	Migrate along paths of endogenous stem cells as well as along non-stereotypical routes in order to "home in" on pathology.	Ourednik et al., 2001 Aboody et al., 2000 Fricker et al., 1999
Plasticity in vitro and in vivo	Ability to accommodate to multiple CNS regions and yield a variety of region-appropriate cell types to reconstitute the entire damaged milieu.	Bjornson et al., 1999 Flax et al., 1998 Rosario et al., 1997
Maintenance of transgene expression	Not only might neural cells express engineered neural gene products better and in a more regulated fashion than do nonneural cells, but certain gene products are intrinsically made by NSCs without the need for genetic engineering. NSCs can be engineered to carry multiple copies of a transgene thus minimizing the risk of transgene down regulation.	Flax et al., 1998
Ability to deliver multiple different transgenes	Multiple gene products can be inserted into a given NSC typically by multiple infections with a viral vector.	Villa et al., 2000

Table 1. (cont.) Desirable Neural Stem Cell Properties for CNS Gene Therapy and Repair

Properties	Description	Reference
Minimal negative effects	Through targeted administration, NSCs can selectively express gene products locally without necessarily affecting non-diseased cells. Alternatively, if desired, via administration to certain germinal zones, widespread dissemination of NSCs throughout the CNS can be achieved for addressing global pathologies. In such cases, the cells and gene product remain restricted to the CNS and cause no systemic side-effects.	Aboody et al., 2000 Fricker et al., 1999
Serve as producer cells for in vivo dissemination of viral vectors	Aid amplification of virus-mediated genes to large CNS regions.	Lynch et al., 1999
Tropism for and within regions of CNS degeneration	In presence of pathology, NSCs alter their migration and differentiation patterns towards replacement of diseased cells and repair of surrounding anatomy.	Aboody et al., 2000 Rosario et al., 1997
Immunotolerance	Within the same species, e.g., mouse NSCs transplanted into mouse hosts, even allografts of NSCs appear to be well-tolerated without evidence of immunorejection or the need for immunosuppression.	Snyder et al., 1992 Park et al., 2002b Ourednik et al., 2002 Aboody et al., 2000 Park et al., 2002a Lynch et al., 1999
In genetically augmented NSCs, downregulation of v-myc following transplantation to minimize likelihood of tumor formation	Following engraftment, genetically-augmented NSCs spontaneously silence v-myc expression; these NSCs never form in vivo and may even aid in tumor reduction.	Flax et al., 1998 Vescovi et al., 1999 Aboody et al., 2000

Table 2. Necessary Requirements for Insuring Safety of Neural Stem Cell Lines

- Absence of transformed cell characteristics

- Absence of contaminating microorganisms, viruses, and prions, excluding genetically introduced transgenes for propagation

- Normal karyotype

- Exit from the cell cycle following transplantation

- Ability to differentiate

- Avoidance of innervation into non-targeted region

- Introduction of a "safety" mechanism that can be easily targeted to eradicate cells if necessary in vivo

form floating vimentin- and nestin-positive cellular clusters following isolation and subculturing in serum-free media, and after replating on an adherent substratum, e.g., ornithine and laminin, they expressed neuronal markers (neurofilament, b_{III}-tubulin, NCAM, synaptophysin, MAP2a, MAP2b). The majority of neurons were glutamatergic with a smaller percentage positive for GABA and TH. Prolonged differentiation *in vitro* was necessary for GFAP immunoreactivity, and cells expressed O4 only when cultured in the presence of platelet-derived growth factor. After grafting into mice, clusters and incorporated cells were found in the majority of animals with phenotype properties similar to those seen in culture and they did not form teratomas.

Zhang et al. (2001) found that although these neurally-directed cells, like precursors from the brain, required FGF2 for proliferation, no additive effect was elicited by EGF or LIF, alone or in combination. This may suggest that proliferating ESC-derived neural precursors represent a more immature stage in development than precursor cells derived from the fetal brain. This is consistent with the suggestion that rodent EGF, and perhaps LIF, responsiveness is acquired at later stages of precursor cell differentiation (Kalyani et al., 1997; Tropepe et al., 1999).

Schuldiner et al. (2001) demonstrated a dose-dependent increase in the percent of cells expressing neurofilament protein following the application of retinoic acid and nerve growth factor (NGF) to dissociated embryoid bodies from the human ESC line H9. They found that the expression of dopamine D1 receptors and serotonin 2A and 5A receptors in differentiated but not untreated ESCs. They also detected the expression of dopa-decarboxylase, a key enzyme in the synthesis of both monoamines. *In vivo* evidence for these has not yet been reported.

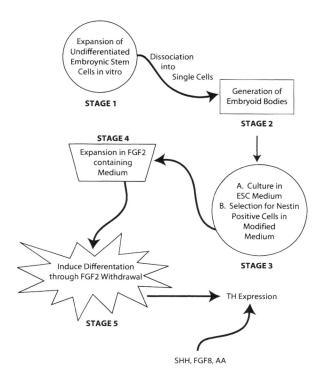

Figure 4 . ESC Differentiation into Neuronal Phenotypes. Following expansion of undifferentiated ESCs *in vitro*, embryoid bodies are generated. A defined medium is used to select for CNS stem cells that are later proliferated in the presence of the mitogen FGF2. Differentiation is induced following removal of the mitogen. The genes Pax2, Pax5, Wnt1, En1, and Nurr1 are detectable by stages 3 and 4. By stage 4 cells expressing the intermediate filament nestin are present. Expression of specific neuronal phenotypes is not observed until stage 5 including the expression of β_{III}-tubulin in 71% of the cells. Of these, 7% are positive for TH. This value is increased by supplementation with ascorbic acid , sonic hedgehog (SHH), and FGF8.

Future directions in the use of human ESCs will obviously entail trying to direct neural precursors down ever more specialized differentiation paths by attempting to mimic protocols used with mouse ES cells, e.g., dopaminergic neurons (Lee et al., 2000) or motor neurons (Wichterle et al., 2002). These protocols attempt to mimic developmental processes by applying proteins that have been identified to play a role in normal development. Improvements in techniques for culturing human ESCs are needed, e.g., growing the cells in monolayers (Ying et al., 2003), and without mouse embryonic fibroblast feeder layers.

CONCLUSION

In summary, emerging data suggest that a population of CNS progenitors or stem cells exists whose isolation, expansion, and transplantation may have clinical utility for cell replacement or molecular support therapy for some degenerative, developmental, and acquired insults to the CNS (Park et al., 2002a). The use of NSCs as graft material has helped reorient and broaden the scope of neural transplantation as a therapeutic intervention. Most neurologic diseases are not unifocal like Parkinsonism but are characterized by extensive, widespread, multifocal, or even global pathology, often requiring multiple repair strategies.

By virtue of their inherent biologic properties, NSCs seem to possess the multiple therapeutic capabilities demanded. At least three types of interventions are possible from the implantation of NSCs into the dysfunctional CNS. First, transplanted NSCs may differentiate into and replace damaged or degenerated neurons and/or glia. In some pathologic conditions, host factors may be released or expressed within the microenvironment that are detrimental to the survival or differentiation of donor progenitors. However, donor cells may be engineered *ex vivo* to be less sensitive to these factors, to secrete substances that might neutralize these forces or to express trophic agents to circumvent these problems. A second category of potential interventions is possible even for implanted cells that do not differentiate into the desired cell types. Donor NSCs may intrinsically provide factors, both diffusible and non-diffusible, that may enable the injured host to regenerate its own lost cells and/or neural circuitry. Finally, in the event that specific factors are not naturally produced in sufficient quantities a third strategy is possible. NSCs can be genetically engineered before transplantation to become resident "factories" for the sustained production of substances known to reactivate, mobilize, and recruit quiescent host progenitor cells; promote regeneration and further differentiation of immature nerve cells; attract ingrowth of host fibers; forestall degeneration resulting from insufficiency of a trophic factor, enzyme, or other factor; neutralize toxins; or allow cells to utilize alternative metabolic pathways. Because NSCs can incorporate into the host cytoarchitecture, they may prove to be more than vehicles for passive delivery of substances. The regulated release of certain substances through feedback loops may be reconstituted, as may the reformation of essential circuits. One multipotent NSC line may, in certain conditions, perform many of the above-mentioned functions even concurrently. Transplanting these cells, possibly in combination with other systemic or somatic cell therapies, may be a broadly applicable treatment for a number of neurological diseases (Park et al., 2002a).

The field of NSC biology is at a very early stage of development. Some of our suggestions are speculative, and much needs to be learned about the properties of such cells. While work is ongoing on the isolation, propagation, and transplantation of human NSCs, many important questions need to be addressed before it will be possible to use such cells in clinical applications (Park et al., 2002a). For instance, what factors insure the expansion, stability, engraftment, migration, and differentiation of transplanted NSCs? What variables dictate the efficiency of foreign gene expression by engrafted NSCs? What is required in a given disease for reversing progression and restoring function? When is the proper time to administer cells? What are the limits of reconstitution in the brain? Do donor-derived cells function normally? Through probing of these important questions, it is hoped that the biological properties of human NSCs can be combined with molecular engineering to restore normal functions in the repair of lesions in both inherited and acquired diseases affecting the CNS.

Implications of NSCs for the Field of Regenerative Medicine

The discovery of the existence of the neural stem cell as model for "programmed plasticity" and the potential that harnessing developmental processes may hold for addressing damage and disease, has heralded, a paradigm shift in the way medicine will approach therapy not only for the nervous system but for other organ systems as well.

The recognition of putative stem cells in other solid organ systems (Bonner-Weir and Sharma, 2002; Brittan and Wright, 2002; Forbes et al., 2002; Foster et al., 2002; Goldring et al., 2002; Hughes, 2002; Janes et al., 2002; Otto, 2002), and the ability of ESCs to give rise to such precursors (Reubinoff et al., 2000), suggests the broad scope of this approach: the harnessing of stem cells for the repair of many tissues, e.g., heart, liver, muscle, pancreas), in addition to the nervous sytem. While approaches to disease have heretofore focused on halting a specific pathology, stem cells could provide an additional approach by replacing defective components with more normal ones. There is also the possibility that tissue-resident stem cells from other organ systems can give rise to neural elements, may provide alternative sources for CNS repair. These observations remain controversial at present.

Questions have emerged as to whether one should effect repair through (a) the mobilization and recruitment of endogenous stem cells; (b) the extraction, expansion, and reimplantation of the patient's own stem cells; or (c) the implantation of exogenous well-characterized donor stem cells. The answers are not yet clear. Autografts of patient stem cells in situ expansion of native stem cells remains problematic. For example, stem cells derived from an individual with a genetic impairment, including Alzheimer's, Parkinson's, and

Huntington's diseases, may prove to be ineffective (Park et al., 2002a). An additional problem potentially is the fusion of a donor cell with a host cell (Wang et al., 2003) leading to misidentification of host vs. donor cells.

Future research will need to determine whether such cells can meet the gold standards of safety, efficiency, simplicity, and efficacy described in Tables 1 and 2. If these standards are met, it is quite conceivable that many different sources of stem cells, used singley or in combination, will ultimately be used for different purposes in treating a variety of diseases.

Table 3. NIH Embryonic Stem Cell Registry

Provider	NIH Code	Available	Characteristics	Reference(s)
BresaGen, Inc.				
Athens, Georgia	BG01 (hESBGN.01)	Yes	A	-
(http://www.bresagen.com)	BG02 (hESBGN.02)	Yes	A	-
	BG03 (hESBGN.03)	No	A	-
	BG04 (hESBGN.04)	No	A	-
CyThera, Inc.	CY12 (hES-1-2)	No	NC	-
San Diego, California	CY30 (hES-3-0)	No	NC	-
(http://www.cytheraco.com)	CY40 (hES-4.0)	No	NC	-
	CY51 (hES-5-1)	No	NC	-
	CY81 (hES-8-1)	No	NC	-
	CY82 (hES-8-2)	No	NC	-
	CY91 (hES-9-1)	No	NC	-
	CY92 (hES-9-2)	No	NC	-
	CY10 (hES-10)	No	NC	-
ES Cell International	ES01 (HES-1)	Yes	B	Reubinoff et al., 2000
Melbourne, Australia	ES02 (HES-2)	Yes	B	Reubinoff et al., 2000
(http://www.escellinternational.com)	ES03 (HES-3)	Yes	B	-
	ES04 (HES-4)	Yes	B	-
	ES05 (HES-5)	Yes	B	-
	ES06 (HES-6)	Yes	B	-

Table 3. (cont.) NIH Embryonic Stem Cell Registry

Provider	NIH Code	Available	Characteristics	Reference(s)
Geron Corporation Menlo Park, California (http://www.geron.com)	GE01 (H1)	Yes*	C	Carpenter et al., 1999 Thompson et al., 1984 Xu et al., 2001
	GE07 (H7)	Yes*	C	Carpenter et al., 1999 (Thompson et al., 1984 Xu et al., 2001
	GE09 (H9)	Yes*	C	Carpenter et al., 1999 Thompson et al., 1984 Xu et al., 2001 Amit et al., 2000
	GE13 (H13)	No	D	Thomson et al., 1998
	GE14 (H14)	Yes*	C	Thompson et al., 1984 Xu et al., 2001
	GE91 (H9.1)	Yes*	E	Amit et al., 2000
	GE92 (H9.2)	Yes*	E	Amit et al., 2000
Göteborg University Göteborg, Sweden	SA01 (Salgrenska 1)	No	NA	-
	SA02 (Salgrenska 2)	No	NA	-
	SA03 (Salgrenska 3)	No	NA	-
Karolinska Institute Stockholm, Sweden (http://www.ki.se)	KA08 (hICM8)	No	NA	-
	KA09 (hICM9)	No	NA	-
	KA40 (hICM40)	No	NA	-
	KA41 (hICM41)	No	NA	-
	KA42 (hICM42)	No	NA	-
	KA43 (hICM43)	No	NA	-

*Pursuant to Geron's agreement with the Wisconsin Alumni Research Foundation, Geron may transfer this cell line only to Geron collaborators. NA - Characterization *Not Available* at Time of Publication. NC - *Not Complete* at time of Publication. If no reference is cited, data was obtained only from http://escr.nih.gov.

Table 3. (cont.) NIH Embryonic Stem Cell Registry

Provider	NIH Code	Available	Characteristics	Reference(s)
Maria Biotech Co. Ltd.	MB01 (MB01)	No	F	-
Maria Infertility Hosp. Medical Inst.	MB02 (MB02)	No	G	-
Seoul, Korea	MB03 (MB03)	No	G	-
MizMedi Hosp.–Seoul Natl. Univ.	MI01 (Miz-hES1)	No	NA	-
Seoul, Korea				
Natl. Centre for Biol. Sciences	NC01 (FCNCBS1)	No	NA	-
Tata Inst. of Fund. Research	NC02 (FCNCBS2)	No	NA	-
Bangalore, India	NC03 (FCNCBS3)	No	NA	-
Pochon CHA University	CH01 (CHA-hES-1)	No	NA	-
Seoul, Korea	CH02 (CHA-hES-2)	No	NA	-
Reliance Life Sciences	RL05 (RLS ES 05)	No	NA	-
Mumbai, India	RL07 (RLS ES 07)	No	NA	-
(http://www.ril.com)	RL10 (RLS ES 10)	No	NA	-
	RL13 (RLS ES 13)	No	NA	-
	RL15 (RLS ES 15)	No	NA	-
	RL20 (RLS ES 20)	No	NA	-
	RL21 (RLS ES 21)	No	NA	-
Technion University	TE03 (I 3)	No	NA	-
Haifa, Israel	TE32 (I 3.2)	No	NA	-
(http://www.technion.ac.il)	TE33 (I 3.3)	No	NA	-
	TE04 (I 4)	No	NA	-
	TE06 (I 6)	No	NA	-
	TE62 (I 6.2)	No	NA	-
	TE07 (J 3)	No	NA	-
	TE72 (J 3.2)	No	NA	-
Univ. California, San Francisco	UC01 (HSF-1)	No	NA	-
CA, USA (http://escells.ucsf.edu)	UC06 (HSF-6)	Yes	H	-

A - Cells are positive for cell markers SSEA-3, SSEA-4, TRA 1-60, TRA 1-81, Oct-4, and alkaline phosphatase. Cells are negative for the cell marker SSEA1.

B - Cells are positive for cell markers Oct-4, SSEA-3 and -4, TRA-1-60 and 1-81, GCTM-2, TG343, GDF3, and alkaline phosphatase activity. Cells are negative for cell marker SSEA-1. Give rise to teratomas containing derivatives of three germ layers in SCID mice. Differentiate *in vitro* into extaembryonic and somatic cell lineages. Neural progenitor cells may be isolated from differentiating ES cell cultures and induced to form mature neurons. hES6 in progress for some of this characterization.

C - Cells were derived by J. Thomson at University of Wisconsin and are cell line WA01 (H1), WA07 (H7), WA09 (H9) and WA014 (H14). Cells can be maintained feeder-free. Cells express OCT-4 and hTERT, give rise to teratomas containing derivatives of three germ layers in SCID mice, and differentiate *in vitro* into derivatives of all three germ layers.

D - Cells were derived by J. Thomson at University of Wisconsin and are cell line WA013.

E - Clonal line derived from H9. Cells cloned at passage 29 and maintain stable karyotype and phenotype. Cells have telomerase activity. Cells give rise to teratomas containing derivatives of three germ layers in SCID mice.

F - Cells are positive for cell markers Oct-4, alkaline phosphatase activity and telomerase activity. Differentiate *in vitro* into mature neurons, glial cells, muscle cells and beating cardiomyocytes. AFP was also detected in this cells.

G - Cells are positive for cell markers SSEA-3, SSEA-4, Oct-4, alkaline phosphatase activity and telomerase activity. Cells are negative for the cell marker SSEA-1. Cell has normal karyotype (46 XX). Differentiation *in vitro* into neurons. Cells were cultured in feeder-free culture condition using Matrigel coated plate.

H - Cells have normal morphology and a normal female karyotype. Cells differentiate *in vitro* into all three germ layers. Cells are positive for cell marker Oct-4, SSEA-3, SSEA-4, and alkaline phosphatase. Cells are negative for SSEA-1.

I - Cells are positive for cell markers SSEA-3, SSEA-4, TRA-1-60, TRA-1-81, and alkaline phosphatase. Cells are negative for SSEA-1. Give rise to teratomas containing derivatives of three germ layers in SCID mice. Differentiate *in vitro*. Documented quality control includes karyotype analysis, DNA fingerprinting, mycoplasma testing, and human virus testing.

J - Cells are positive for cell markers SSEA-3, SSEA-4, TRA-1-60, TRA-1-81, and alkaline phosphatase. Cells are negative for SSEA-1. Give rise to teratomas containing derivatives of three germ layers in SCID mice. Differentiate *in vitro*.

REFERENCES

Aboody KS, Brown A, Rainov NG, Bower KA, Liu S, Yang W, Small JE, Herrlinger U, Ourednik V, Black PM, Breakefield XO, Snyder EY (2000) Neural stem cells display extensive tropism for pathology in adult brain: evidence from intracranial gliomas. Proc Natl Acad Sci U S A 97: 12846-12851.

Akerud P, Canals JM, Snyder EY, Arenas E (2001) Neuroprotection through delivery of glial cell line-derived neurotrophic factor by neural stem cells in a mouse model of Parkinson's disease. J Neurosci 21: 8108-8118.

Billinghurst LL, Taylor RM, Snyder EY (1998) Remyelination: cellular and gene therapy. Semin Pediatr Neurol 5: 211-228.

Bjornson CR, Rietze RL, Reynolds BA, Magli MC, Vescovi AL (1999) Turning brain into blood: a hematopoietic fate adopted by adult neural stem cells *in vivo*. Science 283: 534-537.

Bonner-Weir S, Sharma A (2002) Pancreatic stem cells. J Pathol 197: 519-526.

Brittan M, Wright NA (2002) Gastrointestinal stem cells. J Pathol 197: 492-509.

Brustle O, Choudhary K, Karram K, Huttner A, Murray K, Dubois-Dalcq M, McKay RD (1998) Chimeric brains generated by intraventricular transplantation of fetal human brain cells into embryonic rats. Nat Biotechnol 16: 1040-1044.

Buc-Caron MH (1995) Neuroepithelial progenitor cells explanted from human fetal brain proliferate and differentiate *in vitro*. Neurobiol Dis 2: 37-47.

Carpenter MK, Cui X, Hu ZY, Jackson J, Sherman S, Seiger A, Wahlberg LU (1999) *In vitro* expansion of a multipotent population of human neural progenitor cells. Exp Neurol 158: 265-278.

Castro RF, Jackson KA, Goodell MA, Robertson CS, Liu H, Shine HD (2002) Failure of bone marrow cells to transdifferentiate into neural cells *in vivo*. Science 297: 1299.

Clarke DL, Johansson CB, Wilbertz J, Veress B, Nilsson E, Karlstrom H, Lendahl U, Frisen J (2000) Generalized potential of adult neural stem cells. Science 288: 1660-1663.

D'Amour KA, Gage FH (2002) Are somatic stem cells pluripotent or lineage-restricted? Nat Med 8: 213-214.

Drago J, Murphy M, Carroll SM, Harvey RP, Bartlett PF (1991) Fibroblast growth factor-mediated proliferation of central nervous system precursors depends on endogenous production of insulin-like growth factor I. Proc Natl Acad Sci U S A 88: 2199-2203.

Flax JD, Aurora S, Yang C, Simonin C, Wills AM, Billinghurst LL, Jendoubi M, Sidman RL, Wolfe JH, Kim SU, Snyder EY (1998) Engraftable human neural stem cells respond to developmental cues, replace neurons, and express foreign genes. Nat Biotechnol 16: 1033-1039.

Forbes S, Vig P, Poulsom R, Thomas H, Alison M (2002) Hepatic stem cells. J Pathol 197: 510-518.

Foster CS, Dodson A, Karavana V, Smith PH, Ke Y (2002) Prostatic stem cells. J Pathol 197: 551-565.

Fricker RA, Carpenter MK, Winkler C, Greco C, Gates MA, Bjorklund A (1999) Site-specific migration and neuronal differentiation of human neural progenitor cells after transplantation in the adult rat brain. J Neurosci 19: 5990-6005.

Gage FH (2000) Mammalian neural stem cells. Science 287: 1433-1438.

Geschwind DH, Ou J, Easterday MC, Dougherty JD, Jackson RL, Chen Z, Antoine H, Terskikh A, Weissman IL, Nelson SF, Kornblum HI (2001) A genetic analysis of neural progenitor differentiation. Neuron 29: 325-339.

Goldring K, Partridge T, Watt D (2002) Muscle stem cells. J Pathol 197: 457-467.

Hughes S (2002) Cardiac stem cells. J Pathol 197: 468-478.

Janes SM, Lowell S, Hutter C (2002) Epidermal stem cells. J Pathol 197: 479-491.

Kalyani A, Hobson K, Rao MS (1997) Neuroepithelial stem cells from the embryonic spinal cord: isolation, characterization, and clonal analysis. Dev Biol 186: 202-223.

Keyoung HM, Roy NS, Benraiss A, Louissaint A, Jr., Suzuki A, Hashimoto M, Rashbaum WK, Okano H, Goldman SA (2001) High-yield selection and extraction of two promoter-defined phenotypes of neural stem cells from the fetal human brain. Nat Biotechnol 19: 843-850.

Kilpatrick TJ, Bartlett PF (1993) Cloning and growth of multipotential neural precursors: requirements for proliferation and differentiation. Neuron 10: 255-265.

Kitchens DL, Snyder EY, Gottlieb DI (1994) FGF and EGF are mitogens for immortalized neural progenitors. J Neurobiol 25: 797-807.

Kornblum HI, Raymon HK, Morrison RS, Cavanaugh KP, Bradshaw RA, Leslie FM (1990) Epidermal growth factor and basic fibroblast growth factor: effects on an overlapping population of neocortical neurons *in vitro*. Brain Res 535: 255-263.

Kruger GM, Morrison SJ (2002) Brain repair by endogenous progenitors. Cell 110: 399-402.

Lacorazza HD, Flax JD, Snyder EY, Jendoubi M (1996) Expression of human beta-hexosaminidase alpha-subunit gene (the gene defect of Tay-Sachs disease) in mouse brains upon engraftment of transduced progenitor cells. Nat Med 2: 424-429.

Lee SH, Lumelsky N, Studer L, Auerbach JM, McKay RD (2000) Efficient generation of midbrain and hindbrain neurons from mouse embryonic stem cells. Nat Biotechnol 18: 675-679.

Ling ZD, Potter ED, Lipton JW, Carvey PM (1998) Differentiation of mesencephalic progenitor cells into dopaminergic neurons by cytokines. Exp Neurol 149: 411-423.

Liu Y, Himes BT, Solowska J, Moul J, Chow SY, Park KI, Tessler A, Murray M, Snyder EY, Fischer I (1999) Intraspinal delivery of neurotrophin-3 using neural stem cells genetically modified by recombinant retrovirus. Exp Neurol 158: 9-26.

Lundberg C, Englund U, Trono D, Bjorklund A, Wictorin K (2002) Differentiation of the RN33B cell line into forebrain projection neurons after transplantation into the neonatal rat brain. Exp Neurol 175: 370-387.

Martinez-Serrano A, Rubio FJ, Navarro B, Bueno C, Villa A (2001) Human neural stem and progenitor cells: *in vitro* and *in vivo* properties, and potential for gene therapy and cell replacement in the CNS. Curr Gene Ther 1: 279-299.

Martinez-Serrano A, Villa A, Navarro B, Rubio FJ, Bueno C (2000) Human neural progenitor cells: better blue than green? Nat Med 6: 483-484.

Mayer-Proschel M, Kalyani AJ, Mujtaba T, Rao MS (1997) Isolation of lineage-restricted neuronal precursors from multipotent neuroepithelial stem cells. Neuron 19: 773-785.

McKay R (1997) Stem cells in the central nervous system. Science 276: 66-71.

Morrison SJ, White PM, Zock C, Anderson DJ (1999) Prospective identification, isolation by flow cytometry, and *in vivo* self-renewal of multipotent mammalian neural crest stem cells. Cell 96: 737-749.

Morshead CM, Benveniste P, Iscove NN, van der KD (2002) Hematopoietic competence is a rare property of neural stem cells that may depend on genetic and epigenetic alterations. Nat Med 8: 268-273.

Mujtaba T, Piper DR, Kalyani A, Groves AK, Lucero MT, Rao MS (1999) Lineage-restricted neural precursors can be isolated from both the mouse neural tube and cultured ES cells. Dev Biol 214: 113-127.

Noble M (2000) Can neural stem cells be used to track down and destroy migratory brain tumor cells while also providing a means of repairing tumor-associated damage? Proc Natl Acad Sci U S A 97: 12393-12395.

Nunes MC, Roy NS, Keyoung HM, Goodman RR, McKhann G, Jiang L, Kang J, Nedergaard M, Goldman SA (2003) Identification and isolation of multipotential neural progenitor cells from the subcortical white matter of the adult human brain. Nat Med 9: 439-447.

Onifer SM, Whittemore SR, Holets VR (1993) Variable morphological differentiation of a raphe-derived neuronal cell line following transplantation into the adult rat CNS. Exp Neurol 122: 130-142.

Otto WR (2002) Lung epithelial stem cells. J Pathol 197: 527-535.

Ourednik J, Ourednik V, Lynch WP, Schachner M, Snyder EY (2002) Neural stem cells display an inherent mechanism for rescuing dysfunctional neurons. Nat Biotechnol 20: 1103-1110.

Ourednik V, Ourednik J, Flax JD, Zawada WM, Hutt C, Yang C, Park KI, Kim SU, Sidman RL, Freed CR, Snyder EY (2001) Segregation of human neural stem cells in the developing primate forebrain. Science 293: 1820-1824.

Palmer TD, Schwartz PH, Taupin P, Kaspar B, Stein SA, Gage FH (2001) Cell culture. Progenitor cells from human brain after death. Nature 411: 42-43.

Park KI, Ourednik J, Ourednik V, Taylor RM, Aboody KS, Auguste KI, Lachyankar MB, Redmond DE, Snyder EY (2002a) Global gene and cell replacement strategies via stem cells. Gene Ther 9: 613-624.

Park KI, Teng YD, Snyder EY (2002b) The injured brain interacts reciprocally with neural stem cells supported by scaffolds to reconstitute lost tissue. Nat Biotechnol 20: 1111-1117.

Qian X, Shen Q, Goderie SK, He W, Capela A, Davis AA, Temple S (2000) Timing of CNS cell generation: a programmed sequence of neuron and glial cell production from isolated murine cortical stem cells. Neuron 28: 69-80.

Reubinoff BE, Pera MF, Fong CY, Trounson A, Bongso A (2000) Embryonic stem cell lines from human blastocysts: somatic differentiation *in vitro*. Nat Biotechnol 18: 399-404.

Rosario CM, Yandava BD, Kosaras B, Zurakowski D, Sidman RL, Snyder EY (1997) Differentiation of engrafted multipotent neural progenitors towards replacement of missing granule neurons in meander tail cerebellum may help determine the locus of mutant gene action. Development 124: 4213-4224.

Roy NS, Wang S, Jiang L, Kang J, Benraiss A, Harrison-Restelli C, Fraser RA, Couldwell WT, Kawaguchi A, Okano H, Nedergaard M, Goldman SA (2000) *In vitro* neurogenesis by progenitor cells isolated from the adult human hippocampus. Nat Med 6: 271-277.

Rubio FJ, Bueno C, Villa A, Navarro B, Martinez-Serrano A (2000) Genetically perpetuated human neural stem cells engraft and differentiate into the adult mammalian brain. Mol Cell Neurosci 16: 1-13.

Ryder EF, Snyder EY, Cepko CL (1990) Establishment and characterization of multipotent neural cell lines using retrovirus vector-mediated oncogene transfer. J Neurobiol 21: 356-375.

Sah D, Ray J, Gage FH (1997) Bipotent progenitor cell lines from the human CNS. Nat Biotechnol 15:574-580.

Schuldiner M, Eiges R, Eden A, Yanuka O, Itskovitz-Eldor J, Goldstein RS, Benvenisty N (2001) Induced neuronal differentiation of human embryonic stem cells. Brain Res 913: 201-205.

Shamblott MJ, Axelman J, Littlefield JW, Blumenthal PD, Huggins GR, Cui Y, Cheng L, Gearhart JD (2001) Human embryonic germ cell derivatives express a broad range of developmentally distinct markers and proliferate extensively *in vitro*. Proc Natl Acad Sci U S A 98: 113-118.

Snyder EY (1998) Neural stem cells: developmental lessons with therapeutic potential. The Neuroscientist 4(6):408-425.

Snyder EY, Deitcher DL, Walsh C, Arnold-Aldea S, Hartwieg EA, Cepko CL (1992) Multipotent neural cell lines can engraft and participate in development of mouse cerebellum. Cell 68: 33-51.

Snyder EY, Taylor RM, Wolfe JH (1995) Neural progenitor cell engraftment corrects lysosomal storage throughout the MPS VII mouse brain. Nature 374: 367-370.

Snyder EY, Yoon C, Flax JD, Macklis JD (1997) Multipotent neural precursors can differentiate toward replacement of neurons undergoing targeted apoptotic degeneration in adult mouse neocortex. Proc Natl Acad Sci U S A 94: 11663-11668.

Steindler DA (2002) Neural stem cells, scaffolds, and chaperones. Nat Biotechnol 20: 1091-1093.

Storch A, Paul G, Csete M, Boehm BO, Carvey PM, Kupsch A, Schwarz J (2001) Long-term proliferation and dopaminergic differentiation of human mesencephalic neural precursor cells. Exp Neurol 170: 317-325.

Studer L (2001) Stem cells with brainpower. Nat Biotechnol 19: 1117-1118.

Svendsen CN, Caldwell MA, Shen J, ter Borg MG, Rosser AE, Tyers P, Karmiol S, Dunnett SB (1997) Long-term survival of human central nervous system progenitor cells transplanted into a rat model of Parkinson's disease. Exp Neurol 148: 135-146.

Svendsen CN, Clarke DJ, Rosser AE, Dunnett SB (1996) Survival and differentiation of rat and human epidermal growth factor-responsive precursor cells following grafting into the lesioned adult central nervous system. Exp Neurol 137: 376-388.

Svendsen CN, ter Borg MG, Armstrong RJ, Rosser AE, Chandran S, Ostenfeld T, Caldwell MA (1998) A new method for the rapid and long term growth of human neural precursor cells. J Neurosci Methods 85: 141-152.

Temple S (2001) The development of neural stem cells. Nature 414: 112-117.

Teng YD, Lavik EB, Qu X, Park KI, Ourednik J, Zurakowski D, Langer R, Snyder EY (2002) Functional recovery following traumatic spinal cord injury mediated by a unique polymer scaffold seeded with neural stem cells. Proc Natl Acad Sci U S A 99: 3024-3029.

Thompson S, Stern PL, Webb M, Walsh FS, Engstrom W, Evans EP, Shi WK, Hopkins B, Graham CF (1984) Cloned human teratoma cells differentiate into neuron-like cells and other cell types in retinoic acid. J Cell Sci 72: 37-64.

Tropepe V, Hitoshi S, Sirard C, Mak TW, Rossant J, van der KD (2001) Direct neural fate specification from embryonic stem cells: a primitive mammalian neural stem cell stage acquired through a default mechanism. Neuron 30: 65-78.

Tropepe V, Sibilia M, Ciruna BG, Rossant J, Wagner EF, van der KD (1999) Distinct neural stem cells proliferate in response to EGF and FGF in the developing mouse telencephalon. Dev Biol 208: 166-188.

Uchida N, Buck DW, He D, Reitsma MJ, Masek M, Phan TV, Tsukamoto AS, Gage FH, Weissman IL (2000) Direct isolation of human central nervous system stem cells. Proc Natl Acad Sci U S A 97: 14720-14725.

Vescovi AL, Parati EA, Gritti A, Poulin P, Ferrario M, Wanke E, Frolichsthal-Schoeller P, Cova L, Arcellana-Panlilio M, Colombo A, Galli R (1999) Isolation and cloning of multipotential stem cells from the embryonic human CNS and establishment of transplantable human neural stem cell lines by epigenetic stimulation. Exp Neurol 156: 71-83.

Villa A, Snyder EY, Vescovi A, Martinez-Serrano A (2000) Establishment and properties of a growth factor-dependent, perpetual neural stem cell line from the human CNS. Exp Neurol 161: 67-84.

Wang X, Willenbring H, Akkari Y, Torimaru Y, Foster M, Al Dhalimy M, Lagasse E, Finegold M, Olson S, Grompe M (2003) Cell fusion is the principal source of bone-marrow-derived hepatocytes. Nature 422: 897-901.

Westerman KA, Leboulch P (1996) Reversible immortalization of mammalian cells mediated by retroviral transfer and site-specific recombination. Proc Natl Acad Sci U S A 93: 8971-8976.

Whittemore SR, White LA (1993) Target regulation of neuronal differentiation in a temperature-sensitive cell line derived from medullary raphe. Brain Res 615: 27-40.

Wichterle H, Lieberam I, Porter JA, Jessell TM (2002) Directed differentiation of embryonic stem cells into motor neurons. Cell 110: 385-397.

Yandava BD, Billinghurst LL, Snyder EY (1999) "Global" cell replacement is feasible via neural stem cell transplantation: evidence from the dysmyelinated shiverer mouse brain. Proc Natl Acad Sci U S A 96: 7029-7034.

Ying QL, Stavridis M, Griffiths D, Li M, Smith A (2003) Conversion of embryonic stem cells into neuroectodermal precursors in adherent monoculture. Nat Biotechnol 21: 183-186.

Zhang SC, Wernig M, Duncan ID, Brustle O, Thomson JA (2001) *In vitro* differentiation of transplantable neural precursors from human embryonic stem cells. Nat Biotechnol 19: 1129-1133.

Chapter 9

Isolation, Survival, Proliferation, and Differentiation of Human Neural Stem Cells

Beatriz Navarro, Ana Villa, Isabel Liste, Carlos Bueno and
Alberto Martínez-Serrano

INTRODUCTION

Neural stem cells are essential cellular elements for nervous system generation and maintenance (Anderson, 2001; Temple, 2001). During the last decade, an impressive amount of information has been generated regarding the basic *in vitro* and *in vivo* biology of neural stem cells (NSCs). A quick search in public databases on terms like survival, proliferation, and differentiation of NSCs immediately retrieves thousands of research articles and reviews published in the last two years on these topics. Although the majority of this research, particularly that one dealing with genetic analyses and behavior of NSCs in situations of neurodegeneration *in vivo*, has been obviously conducted in rodents, a growing knowledge about human NSCs (hNSCs) biology is also rapidly becoming available. The main objective of this chapter will be to summarize recent advances in our understanding of the biology of hNSCs, particularly in those aspects related to translation of basic research to potential therapeutic applications. We, in advance would like to apologize to many colleagues whose work can not be summarized here due to space constraints. Whenever needed, and due to the absence of information in the human setting in many respects, research in rodent systems will also be summarized. This chapter will be organized in a few sections aimed to cover isolation procedures of hNSCs, culturing or proliferation methods, and properties of the different cellular systems, factors influencing survival and differentiation (two aspects intimately linked), and finish with an account of recent transplantation experiments illustrating the survival, migration and differentiation capabilities of hNSCs when grafted into the fetal, neonatal and adult rodent and primate brain.

From: *Neural Stem Cells: Development and Transplantation*
Edited by: Jane E. Bottenstein © 2003 Kluwer Academic Publishers, Norwell, MA

ISOLATION AND PROPAGATION OF HUMAN NEURAL STEM CELLS (HNSCS)

Sources of hNSCs

The available tissue sources of cells showing genuine properties of hNSCs are mainly two: human embryonic stem cells (hES) and fresh human tissue derived from fetal, neonatal or adult nervous system samples (autopsies or biopsies). Alternative sources of neural precursors, progenitors, neurons or glia of human origin, like human umbilical cord, bone marrow or skin, will not be considered in this chapter (see Chapter 6 in this volume).

Human neural stem/precursor cells derived from hES cells

With the discovery of isolation, derivation and proliferation methods for hES cells back in 1998 by Thomson et al. (1998), a whole new field in cell and tissue bioengineering exploded. hES cells, due to their pluripotency (ability to generate differentiated cell types from all three embryonic germ layers) can be easily manipulated in culture to obtain other mature, differentiated cell types, neural cells included. The initial methods for the derivation of neural cells from hES cultures, described soon after the discovery of hES cells (Reubinoff et al. 2000, 2001; Zhang et al., 2001; Schuldiner et al., 2001; Carpenter et al., 2001; see also Chapter 5 in this volume), were all based on the generation of embryoid bodies (transient multicellular aggregates or structures, masses of cells growing as distinct entities in cultures of hES cells), which appear soon after culture conditions are modified (usually LIF removal, treatment with retinoic acid; Studer, 2001; Smith, 2001; Gottlieb, 2002). Isolation of these structures followed by serum removal and addition of neural stem cell mitogens (FGF2 and/ or EGF) leads to the generation of cells growing as floating aggregates, similar to the so-called neurospheres, and to the appearance/expression of multiple immature neural markers in the generated cells, particularly nestin. From this stage, and after plating those cell clumps as adherent cultures, the neural precursors continue their differentiation program yielding a mixed neural culture composed of different types of neurons (GABAergic, glutamatergic, serotoninergic, dopaminergic). Alternative methods, mainly applied to the generation of dopaminergic neurons from mouse and primate ES cells, have been developed on the basis of cocultures of ES cells onto a stromal feeder and inductive cell layer (Kawasaki et al., 2000, 2002). Information about the simultaneous generation of other (non-dopaminergic) neuronal phenotypes is limited in this case. Also, very recently, a new method for the conversion of mouse ES cells into neural precursors has been described, using a feeder-free but conditioned medium supplemented, adherent monolayer culture (Ying et al., 2003).

It is worth mentioning at this point that the isolation of human neural stem/precursor cells from hES cells is a poorly understood process, since experimental procedures to accomplish it have been mostly empirically derived (see comments by D'Amour and Gage, 2000; Studer, 2001; see also Chapter 5 in this volume). In addition, all systems reported so far require either the propagation or the neural induction of hES in cell culture systems based either in the coculture with feeder/inductive cell layers (fibroblasts or stromal cell feeder layers), or the use of conditioned medium. This lack of definition in medium composition and control upon operating mechanisms make the system poorly defined. Also, and whilst it is true that neural cells arise in these preparations (mostly neurons and astrocytes, but rarely oligodendrocytes), the induction of specific neuronal cell types remains a challenge. It should be highlighted too that, in spite of the positive evidence indicating that neural cells can be generated from hES cells, a substantial population of cells having phenotypes corresponding to other embryonic layers are usually also present. These findings pose important safety issues and limitations for future applications of these technologies in the clinical setting, which should be clarified in the near future.

Derivation of hNSCs from Fresh Human Tissue: Cell Strains and Cell Lines

Most methods employed for the derivation of hNSCs cultures from fresh tissue, already developed almost a decade ago, are based on the generation of floating aggregate cultures (the so called neurosphere cultures) using rodent cells (Reynolds and Weiss, 1992; Vescovi et al., 1993; Martínez-Serrano et al., 2001; Villa et al., 2001). In the case of human materials, post mortem tissue is most often used (from aborted human fetuses, or after autopsy of neonatal or adult specimens), but there are also examples of tissue obtained from biopsies after surgical interventions (see below).

Once tissue is available, it can be used as such or after purification/enrichment of the sample in cells with properties of neural stem/precursor cells (see below). The cultures afterwards can be grown following two principal schemes: A) the generation of floating neurosphere cultures which are subsequently passed by mechanical trituration or "chopping" (Svendsen et al., 1998) and where neural stem cell growth is achieved by the provision of mitogens (most commonly in the form of FGF2, EGF and/or LIF). Due to the continued action of these mitogens, the culture becomes progressively enriched in cells with the properties of neural precursors, and after a few passages, needed to eliminate primary postmitotic cells, it becomes a "cell strain" (a serially passaged culture of mortal cells). As discussed below, these neurosphere cultures have the main theoretical advantage of being propagated under the influence of epigenetic signals (mitogens added to the culture medium) and without the need for any genetic modification. This, of course, does not eliminate the possibility that genetic changes occurr in these cells: mutations, phenotypic changes, and so on, that may take place in any cell culture system. The main disadvantages of these cultures are their limited capacity for proliferation, and their continuously changing or evolving properties with time, together with the heterogeneity of cell types present in the neurospheres. B) Another method, based on the genetic perpetuation of the cells, has been also used for the generation of immortal cell lines of hNSCs (Martinez-Serrano et al., 2001; see Chapter 8 in this volume). In this case, a perpetuating gene (actually, v-myc for all reports published on human cells) is used to immortalize the cells in order to generate a true cell line. In spite of the presence of v-myc in these lines, the cells do not show any sign of transformation either *in vitro* or *in vivo*. It should be highlighted at this point that the use of perpetuating genes (v-myc in particular) results in conditional immortalization, since the cells are still dependent on mitogens for their growth (Villa et al., 2000), and readily differentiate after mitogen removal from the culture medium. The main advantages of cell lines are their vigorous growth, utility for further genetic modification and subcloning, their availability and banking, and the stability of their properties over time (as shown below).

hNSCs Isolation/Propagation Methods and Properties of the Resulting Cultures

Before progressing into the details of isolation procedures briefly, the culturing methods for human neural stem cells will be described. [For a detailed account of numerous methods and variants, see Martinez-Serrano et al. (2001) and LeBelle and Svendsen (2002)]. In general, hNSCs are cultured in a DMEM:F12 based medium, with an increased glucose concentration, and with some extra added buffering power (usually in the form of HEPES) to counteract the rapid acidification of the medium that normally occurs as the cells proliferate. N2 supplements (Bottenstein and Sato, 1979) are essential for cell growth, and in some cases 1% albumin and nonessential aminoacids are also added (although they may not be essential for cell proliferation). When albumin is used, it is recommended to obtain high quality, cell-culture tested material (e.g., Albumax from Life Technologies), since some batches of albumin may contain traces of reagents which may seriously compromise cell viability. For the particular case of neurosphere cultures, heparin is also recommended since it acts as a buffer for mitogens, and also reduces cell adherence to the plastic surface. Heparin, though, is clearly not indicated (and in fact unnecessary) for the maintenance of adherent cell lines. To this DMEM:F12-N2 basic medium, additional mitogens are usually added to enhance cell growth. The use of serum is normally not advised, since it may promote cell differentiation. The most common mitogens used are a combination of FGF2 and EGF (10-20 ng/ml), although each of them individually may act to stimulate proliferation of different subsets of cells. LIF or CNTF (10 ng/ml) have been used in some studies, since they seem to enhance cell proliferation and extend lifespan to a moderate extent (see below for details). Under these conditions, both neurospheres and cell lines proliferate efficiently.

For the case of cell lines, two thirds of the medium is replaced every 2-3 days, and once confluency is approached, cultures are trypsinized (avoiding the use of serum) and re-seeded at 1/5-1/10 splits. Considering the cell cycle length of cell lines (one to two days), the cells are normally passaged every 5-7 days.

The case of human neurospheres is markedly different from that of cell lines. Cultures are initiated from dissociated cell suspensions obtained from fresh tissue following a mild protease digestion, with papain or trypsin, and in the presence of DNaseI, and the cells are usually plated at a density of $1\text{-}2 \times 10^5$ cells /cm^2. After a few hours, the cells adhere to the plastic surface, and cultures initially look like any other primary neuronal culture. Following continuous mitogen stimulation over subsequent days, EGF-, FGF2-, or EGF/FGF2-responsive cells start to proliferate and initially form small clumps of cells that tend to detach from the plastic surface. Culture flasks are usually knocked everyday, to maximize the number of cells detaching from the plastic surface and forming suspension aggregates. Mitogens may be readded once or twice per week (or not, depending on laboratory routines). Floating cell aggregates are collected approximately 10 days following plating and mechanically triturated in a small volume of medium using siliconized and fire polished Pasteur pipettes of decreasing opening size, or, alternatively, regular automatic pipette tips until cell clumps are not visible to the naked eye. DNaseI may be included at this step, to minimize cell clumping during trituration, due to the genomic DNA released from dying cells. The yield, in most cases, is around 50%. Following centrifugation, the cells are again plated at $1\text{-}2 \times 10^5$ cells/ml in complete fresh medium. After 3-5 passages, cultures are mainly composed of floating aggregates of cells (neurospheres) originating from cells showing all the properties of an *in vitro* neural stem cell. Cellular composition in the spheres may vary with time, since the mitogens seem not to be able to counteract the propensity of the cells to differentiate, yielding a heterogeneous neural culture. A variant of trituration is the "chopping" method, in which spheres are sectioned into quarters, to preserve cell-to-cell contact (Svendsen et al., 1998).

Low oxygen conditions seem to slightly promote hNSCs growth, but are not really needed for the derivation of either neurosphere cultures or cell lines.

Even cultures initiated from human ventral mesencephalic tissue seem to behave exactly the same under low (5%) or conventional (20%, hyperoxic) oxygen conditions, in terms of proliferation and dopamine neuron generation (our unpublished results, but see Studer et al., 2000 and Storch et al., 2001).

Several interesting aspects regarding the properties of these proliferating hNSCs cultures have not been explored yet in detail. Among them, there is the issue of self-renewal. Self-renewal, in our opinion, is a concept that may better be reserved for *in vivo* neural stem cells, which are able to truly perpetuate their cellular population, possibly due to the combination of cell-intrinsic properties and environmental factors. Proliferation in culture may be seen as a different phenomenon, and equivalence between proliferation and self-renewal *in vitro* is far from clear. Basically, cells in culture are responding to strong mitogenic cocktails, and, as a result, proliferate. So, the cells are somehow forced to "self-renew". However, the changes that occur in culture with time, like the progressive decay in their potential to generate oligodendrocytes or dopaminergic neurons, speaks against a true self-renewal capacity of the cells *in vitro*. Proliferation does not mean self-renewal. Or, in other words, the simple observation that certain cells proliferate in culture while stimulated by powerful mitogens does not demonstrate a cell-intrinsic capacity for self-renewal, if we understand self-renewal as preservation of properties, and not merely as capacity for cellular proliferation.

Another two aspects which have not been analyzed yet, are 1) the occurrence of symmetric or asymmetric divisions in hNSC cultures, and 2) the issue of whether particular growth requirements may depend or not on the fetus age or the nervous system region from which the starting tissue is obtained. With regard to the type of cell division, it seems obvious that proliferating neural stem cells (either neurosphere-forming cells or cells of hNSC cell lines), as long as they proliferate and generate progeny with near-to-stable properties, must necessarily undergo some type of division allowing for the maintenance of a putative stem cell pool. Obviously, a neurosphere forming cell must have some capacity for self-renewal, but it is far from clear if such a cell undergoes symmetric or asymmetric divisions, how often, and how is that decision regulated. Initially, the neurosphere forming cells should undergo symmetric divisions, even if it is only to generate a new sphere, followed by asymmetric divisions, leading to the generation of more differentiated progeny. In the case of cell lines, it seems more plausible that these cells undergo symmetric divisions, continuously perpetuating cells endowed with homogeneous properties of a neural stem cell, until culture conditions are changed in order to promote its differentiation.

It is also important to discuss the consistent presence of undifferentiated neural stem-like cells in cultures treated with differentiating conditions. This is a common observation made in both neurosphere and cell line cultures, which consistently appear to contain a substantial percentage of cells that are undifferentiated. Alternatively, there could be an intrinsic program in the cells. These nondifferentiating cells in presumably "differentiated" cultures have not attracted a lot of attention from investigators, but they may be an excellent preparation to study basic aspects of homeostatic control of neural stem cell pools. In fact, the existence of a pool of cells which are refractory to differentiating conditions may very well constitute the proof for asymmetric divisions.

Coming back to the isolation of hNSCs, progress in recent years has lead to the development of a few methods that result in the generation of relatively pure cultures of hNSCs: 1) FACS sorting using cell surface markers, 2) FACS sorting using the expression of fluorescent reporter genes with specific promoters, 3) FACS sorting based on physical (size) properties, 4) Enrichment on the basis of specific mitogen treatment, and 5) immortalization of mitogen activated hNSCs. Not all these methods or combination of them have been applied to the various sources of hNSCs. (See Chapter 7 in this volume for a discussion of isolation and clonal analysis of neural stem cells.)

FACS based on surface markers

Fluorescence-activated cell sorting (FACS) has been successfully applied to rodent peripheral and central nervous system cells, and also to human CNS cells. The main advantage of FACS-based procedures is that the desired cells can be isolated at high purity from a bulk suspension, thereby providing a way to obtain hNSCs without the need for prolonged culturing and enrichment. Isolation of cells by FACS would, in theory, provide enough cells for experimentation or transplantation in a single step, as long as a source containing sufficient numbers of hNSCs is available. This aspect, particularly for the case of human materials, remains a hurdle, since the availability of such a tissue source for large-scale hNSC collection is unclear. The best documented example of FACS purification of hNSCs is based on use of the CD133 antigen in combination with other antigens (Uchida et al., 2000). Starting from fetal forebrain human tissue, the cell subpopulation showing the properties of self renewal and multipotency were also showed with high CD133 expression and absence of CD34, CD45 with negative-to-low expression of CD24 antigens. A fraction of the cells isolated following these criteria showed the capacity to self renew in neurosphere cultures and to generate neurons and glia upon differentiation. Even though data in their report and subsequent follow up studies (Tamaki et al., 2002) are compelling, other neurosphere and immortalized cell cultures of hNSCs fulfilling all the required criteria of neural stem cells, do not express CD133 at levels high enough for sorting, e.g., human forebrain-derived neurospheres described by Carpenter et al. (1999) and the human HNSC.100 cell line (unpublished observations). As previously discussed (Martinez-Serrano et al., 2001), the CD133+ population also contains cells with properties different from genuine hNSCs, since they show a very limited *in vivo* differentiation capacity, generating mostly glia after transplantation (Uchida et al., 2000; Tamaki et al., 2002).

A recent addition to the armamentarium of cell surface markers is the antigen called FA-1 (from fetal antigen-1, also known as delta-like, dlk) (Christophersen et al., 2002). This protein is expressed by cultured hNSCs, and it is also a marker of dopaminergic and other neurons in the adult mammalian brain. Therefore, it is being studied to develop useful methods for the isolation of not only hNSCs, but also of dopaminergic human progenitor cells for transplantation. Finally, tetraspanins and non-protein epitopes [CD9, CD15, CD81, CD95 (Fas) and GD2 ganglioside)], also present on hNSCs surfaces (Klassen et al., 2001), may additional further antigens helpful to refine FACS procedures for the purification of hNSCs,

Alternative methods used for the enrichment of hNSCs and other precursors or progenitors are based on the less discriminating combinations of cell-surface markers such as N-CAM+/A2B5− for enrichment of hNSCs (Haque et al., 2000) or the purification of glial restricted precursors from commercial sources of hNSC as an A2B5+/PSA-NCAM− subpopulation (Dietrich et al., 2002). In addition, oligodendrocyte progenitors have been isolated from samples obtained after surgical resections of white matter regions of the adult human brain, based solely on the expression of A2B5 antigen (Windrem et al., 2002). Physical parameters such as the size and cellular complexity of the presumptive neurosphere forming or neuronal progenitor cells have also been applied for both human and rodent neural precursor cells (Ostenfeld et al., 2002; Rietze et al., 2001).

Other methods, only used so far for rodent tissue, are 1) the purification of p75+/P0− neural crest stem cells (NCSCs) from the PNS (E14.5 rat sciatic nerve; Morrison et al., 1999), and 2) a negative selection procedure for FACS sorting of rat telencephalic neural stem cells (Maric et al., 2003). The applicability of these procedures for the sorting of hNSCs from heterogeneous populations of neural cells still remains to be determined.

FACS of specific populations of hNSCs, precursors and progenitors based on the expression of reporter fluorescent genes under the control of cellular- or phenotype-specific promoters.

This strategy represents a further step for the purification of specific cellular subtypes from a heterogeneous bulk preparation derived from human nervous tissue specimens. The concept behind it is very simple, since it is based on the transgenic expression of reporter genes (green fluorescent proteins, GFPs) under the regulatory control of phenotype specific promoters. In this manner, one would theoretically be able to target stem, precursor, progenitor, or mature cells, as long as the cell-type specificity of the promoter in the construct is preserved in a transgenic setting, which is not guaranteed). GFPs are the most commonly used reporter genes, in spite of the disadvantages this implies, due to their undesired effects on gene expression by the modified cells (Martinez-Serrano et al., 2000). Other alternatives, like the use of LacZ and fluorescent substrates for α-galactosidase, are much less explored.

Using a repertoire of promoters, fetal and adult human neural stem/precursor and progenitors have been isolated from different brain regions of samples obtained from biopsies or autopsies. Promoters used include those of the genes tubulin-α-1 for neurons, nestin for neural stem cells, musashi for immature precursors (also present in neuronal and astroglial progenitors), and CNPase for immature oligodendrocyte progenitors (Roy et al., 1999, 2000; Keyoung et al., 2001).

Enrichment of hNSCs/precursor cells through continued expansion of cell strains (neurosphere cultures).

Propagation procedures for the so-called neurosphere cultures were initially described over a decade ago, derived from studies done with rodent cellular preparations (Reynolds and Weiss, 1992; Vescovi et al., 1993). The adaptation of these methods for their use to culture hNSCs has been extensively reviewed elsewhere (Martinez-Serrano et al., 2001). The most recent and remarkable pieces of information related to human neurosphere cultures include the elucidation of LIF actions on the proliferation capacity of these cells, the proposal that low oxygen (normoxia) conditions in cell culture are more appropriate for culturing hNSCs than air oxygen tension (hyperoxia) and, more interestingly, a detailed description of the molecular and cellular properties of the neurosphere cultures in relation to their heterogeneity and limited survival in culture.

Human neurospheres are normally expanded under the mitogenic influence of FGF2 and/or EGF. Nowadays, and stemming from the ES field, LIF is quite often included in the cell culture medium to enhance cell division. LIF (or CNTF, which is exchangeable for LIF, since they share the same signaling subunit of the receptor) stimulates proliferation of human cells remarkably (Carpenter et al.,1999; Ostenfeld et al., 2000; Wright et al., 2002). In addition, LIF was expected to solve the problem of senescence of human neurosphere cultures through a putative dynamic regulation of telomerase activity levels (as proposed by Svendsen, 2000). Afterwards, however, LIF has been found to have no relevant influence on the levels of telomerase activity in long-term cultured neurospheres (Ostenfeld et al., 2000, Wright et al., 2002, Villa et al., submitted). In summary, LIF does not seem to be able to override the proliferation limit of human neurospheres beyond a certain number of cell divisions. It doubles the population doubling (PD) limit from around 40-50 PD (Ostenfeld et al., 2000) to a maximum of around 90 PD (Wright et al., 2002), but the cells finally enter senescence and the cultures become nonproductive, as expected of somatic, mortal cells. In all these studies, neurospheres were derived from the fetal human CNS.

Cell culture in low oxygen conditions, or the addition of erythropoietin (EPO) to neurosphere cultures, has been shown to enhance proliferation of rodent neural stem/progenitor cells from both the PNS and CNS (Morrison et al., 2000, Studer et al., 2000, Shingo et al., 2001). In contrast to regular cell culture conditions, growth in reduced levels of oxygen (3-5%), more similar to those present in the developing CNS, have been reported to enhance the mitogenic potential and also influence cell fate decisions of PNS neural crest stem cells by providing permissive conditions for catecholaminergic differentiation (Morrison et al., 2000). Reduced oxygen levels also enhance the survival, proliferation and catecholaminergic differentiation of rat embryonic ventral mesencephalic (VM) progenitor cells. Overall enhancement resulted in a 3-fold increase in TH$^+$ neuron production when compared to regular (hyperoxia) conditions, accounting for over one-half of the neuronally-differentiated cells (identified as α-III-tubulin$^+$ cells, Studer et al., 2000). A similar beneficial effect of culture in low oxygen conditions has been described for the proliferation and subsequent differentiation/dopaminergic induction of human VM progenitors (Storch et al., 2001).

Erythropoietin (EPO) endogenously produced by the mammalian brain as an hypoxia-inducible cytokine, also exerts profound actions on forebrain neural stem cells of rodent origin, both *in vitro* and *in vivo*, which may represent a parallel situation to that described under reduced oxygen levels (Shingo et al., 2001). In this case, EPO appeared to mimic the effects of moderate hypoxia on enhancement of neuron production by cultured mouse forebrain neurospheres, at the expense of multipotent progenitors.

The heterogeneity and senescence of neurosphere cultures has often been neglected. Heterogeneity of neurosphere cultures was postulated long ago (see discussion in Martínez-Serrano et al., 2001) on the basis of the expression of relatively mature neuronal and astroglial markers in neurosphere cultures, which contain a minority of cells having the potential for the generation of new neurospheres, i.e., cells able to "self-renew". The latter cells express markers of immature NSCs such as nestin and are the only ones that could be regarded as stem/precursor cells, operationally speaking, as long as the definition of the hNSC *in vivo* is not clarified. Neurosphere cells are surviving in a less than optimal growth medium (see below) that balances their intrinsic propensity for differentiation with the mitogenic stimuli provided by powerful growth factors such as FGF2, EGF, and LIF/CNTF.

Regarding the heterogeneity of neurosphere cultures, there are clear, detailed reports describing it at the molecular, ultrastructural, cellular, and functional (viability, proliferation, and lineage marker gene expression before and after differentiation) levels (Svendsen and Smith, 1999; Kukekov et al., 1999, 2002; Poltavtseva et al., 2001, 2002; Revishchin et al., 2001; Suslov et al., 2002). In fact, human neurospheres have been found to differ from each other even within a single clone of cells, in samples derived from embryos of the same gestation, from embryo to embryo, and even more, from lab to lab. Heterogeneity in the neurospheres could be caused, at least in part, by the absence of selective pressure in culture, i.e., cells can accumulate mutations freely or drift easily when cultured in bulk. It is also important to highlight that human neurosphere cells proliferate as long they manage to maintain a critical balance between their propensity to differentiate and the strong mitogenic stimuli present in their cell culture medium. In this scenario, it should not be so surprising to find that part of the cells in the spheres escape from the proliferative stimuli and progress to generate a repertoire of transit amplifying cells, which, in turn, easily give rise to fully differentiated neuronal and glial cells.

The issue of senescence of human neurosphere strains has become recently accepted in the field, since human neurosphere cultures derived from different sources have been consistently and systematically reported to show a proliferation limit in culture. Thus, neurospheres derived from hES cells (Reubinoff et al., 2000), from the fetal human CNS (Ostenfeld et al., 2000, 2001), or the neonatal or adult human CNS (Palmer et al., 2001) have all been reported to enter senescence, yielding nonproductive cultures after variable periods of time and population doublings (see Table 1).

In one case, Ostenfeld et al. (2000) reported that human neurosphere cultures gradually lose their initially low levels of telomerase activity with passaging, and suggested that this was causative of the reduction of their telomeres (assayed by telomere restriction fragment, TRF, length, which also include the length of the subtelomeric regions at the end of the chromosomes). On these grounds, the idea that the absence of telomerase was responsible for telomere shortening was put forward as a means to explain the proliferation limit commonly observed for human neurosphere cultures.

In our laboratory, we wanted to reexamine this hypothesis, and, to our surprise, we found that human neurosphere strains from different embryo ages, labs, and culturing conditions had, overall, rather long and heterogeneous, te-

Table 1. Proliferation limits of common sources of hNSCs grown as neurosphere strains

Cell Type	Limit	Population Doublings	References
hES cells	3-4 months	11-12	Reubinoff et al., 2001
Fetal CNS (E/F)	>1 year	40-50	Ostenfeld et al., 2000
Fetal CNS (E/F/L)	≈ 1.5-2 years	90	Wright et al., 2002
Neonatal CNS	4 months	10-20	Palmer et al., 2001
Adult CNS	4 months	10-20	Palmer et al., 2001

lomeres (Figure 1). Telomere shortening thus seemed not to be the cause of their poor proliferation properties and senescence. In all samples studied, a complete absence of telomerase activity was confirmed, in agreement with the results by Ostenfeld et al. (2001).

The presence of heterogeneous and long average TRF signals could be due to the existence of a subpopulation of cells in the neurosphere cultures with true stem cells properties which could be able to preserve the length of their telomeres (cells with long telomeres and telomerase positive), co-existing with other, more differentiated cells, with shortened telomeres. In order to clarify this point, we performed a Q-FISH (quantitative fluorescent in situ hybridization) assay to more precisely determine the actual length of the telomeric region at each chromosome end in the neurosphere samples (Figure 2). Q-FISH data confirmed that neurosphere cells did not show any relevant reduction of telomere size (neurosphere cells, 5.6-7.6 Kb range; immortalized cell lines, 1.8-3.4 Kb range). Q-FISH analysis, in addition, provided evidence demonstrating that no chromosomal free ends were present in human neurosphere cells in immortalized cell lines, nor were there any structural or numerical chromosomal aberrations. Furthermore, analysis of individual cellular metaphases ruled out the presence of a subpopulation of cells having telomeres of differing sizes, meaning that heterogeneity in telomere size occurs on a single cell basis, and is not a population-based phenomenon.

From these studies one can therefore conclude that neurosphere cells are able to maintain their telomeres at a reasonable length, and, therefore, suggesting a genetic crisis to explain their senescence has no strong basis. Interestingly, neurosphere cells preserve the end of their chromosomes in the absence of telomerase activity. This fact, in combination with the high heterogeneity found when analyzing individual telomere ends point to the activation of alternative lengthening of telomere pathways (ALT).

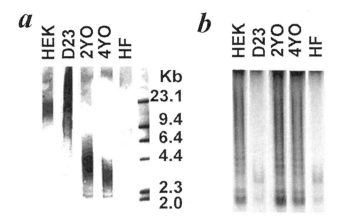

Figure 1. Telomere restriction fragment) (TRF) (a) and telomerase assays (b). Telomere length and telomerase activity in hNSCs and control cell lines. Representative results from 4-5 independent determinations as shown. **(a)** southern blot of telomere restriction fragment (TRF) assays of genomic DNA and **(b)** telomerase activity in cell lysates. Cells used were HEK293; neurospheres; D2 (line D) passage 23 (Vescovi et al. 1999); two- or four year-old (2YO, 4YO) HNSC.100 immortalized cells (Villa et al., 2000), and human skin biopsy derived primary fibroblasts (HF, passage 23) used as a negative control.

The argument presented above directly leads one to formulate the question of whether or not cells in the neurosphere cultures are actually showing or undergoing any senescence-related process. To start obtaining some evidence in this respect, we analyzed the expression of the well-accepted marker of cellular senescence, SA-β-Gal (standing for senescence-associated β-gal activity; Dimri et al., 1995). When different neurosphere cell strains were analyzed for this activity, we found a high percentage of cells (over 60%), both during proliferation and after differentiation conditions, staining positive for this marker (Figure 3; Villa et al., 2000; unpublished data). In contrast, immortalized cell lines did not show any stained cells.

Considering the present data, it is unreasonable to argue that human neurosphere strains have a limitation of their lifespan on the basis of a biological clock, that measures telomere length. Rather, the present data on telomere ends in human neurosphere cells (long and seemingly functional telomeres) are more consistent with the view that these cells are subjected to what has been loosely defined as SIPS (stress induced premature senescence), i.e., that culture conditions are suboptimal, imposing a continued stress on the cells.

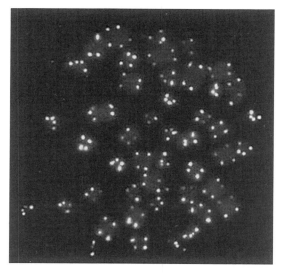

Figure 2. Q-FISH determination of telomere length. Pseudocolor image of chromosomes (blue) and telomere signal (yellow) at their ends in a metaphase spread of a human neurosphere cell. Note the high variability of the telomeric signals at different chromosomal ends, a manifestation of the heterogeneity of telomere lengths. Quantification of telomere lengths is given in the text.

Genetically perpetuated hNSCs lines

The field of genetically perpetuated lines of hNSCs that can generate functional neural cells has been recently and thoroughly reviewed (Vescovi and Snyder, 1999; Martinez-Serrano et al., 2001; Villa et al., 2001; 2002, Gottlieb et al., 2002). Therefore, the reader is referred to those original publications for detailed information on the available cell lines. In addition to the H1 and H6 cell lines described by Flax et al. (1998), the HNSC.100 (or hNS1) cell line described by Villa et al. (2000) and the newly generated hNS2 cells (unpublished data), another cell line called HB1-F3 has been recently described (Cho et al., 2000).

Interestingly, all the human cell lines reported so far have been perpetuated using v-myc. In the Villa et al. study (2000) other putative immortalizing genes were also studied for their efficiency, like the ^{ts}Tag from SV40, Bcl-X$_L$, and c-myc. c-Myc was the only one showing some effects in extending the lifespan of hNSCs, but failed to immortalize the cells. Another gene with putative immortalization capacity, telomerase reverse transcriptase (hTERT) has also been tried on hNSCs and reported to be inefficient (Pilcher et al., 2002). However, more recent reports indicate that hTERT is effective for human spinal cord neural precursors (Roy et al., 2002).

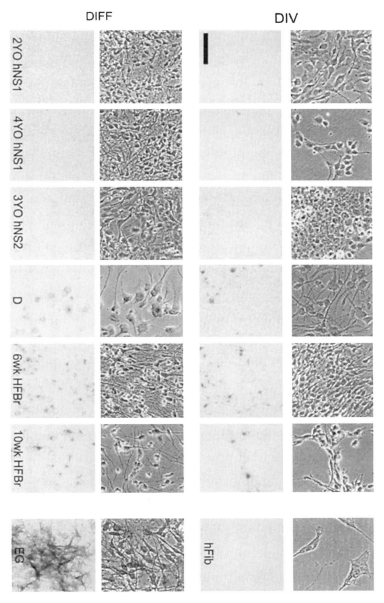

Figure 3. Senescence associated β-gal staining of hNSCs. Cell lines (2-year and 4-year old hNS1 and hNS2) and human neurosphere strains (line D, from diecephalon; and 6-week and 10-week HFBr, from human embryonic forebrain) were analyzed for SA-β-gal activity, during division (DIV) and differentiation (DIFF) conditions. Positive controls are rat ensheathing glia (EG) and presenescent human fibroblasts (hFib).

So far, all the immortalized cell lines described have been generated in a non-targeted way, meaning that the cells were not selected in culture for being endowed with any particular properties before their genetic modification. In addition, the transgene (v-myc) expression was not targeted to specific cell types since unspecific retroviral (MLV-based) vectors were used in all cases. The hNS1 cell line (Villa et al., 2000) was derived from a human neurosphere strain that was cultured under selective conditions for hNSC growth for one year. These cells were then cloned, generating a homogeneous cell line for further study.

Several aspects related to the basic biology of immortalized cell lines deserve further discussion: 1) the nature and properties of the immortalizing gene, 2) properties other than continued proliferation substantiating their immortal nature, 3) the immortal but not "transformed" nature of the cells, 4) their unaltered potential for terminal differentiation into functional neural cells, and 5) their stability over years in culture.

v-myc has been consistently reported to be the best gene for immortalization of hNSCs. However, its actions, underlying molecular mechanism, and efficiency at immortalizing hNSCs are very poorly understood. v-myc is a fusion protein of 110 Kd generated after rescue of chicken c-myc cDNA (exons 2 and 3) by an avian retrovirus. As a result, the coding region for active c-myc became fused to the gag sequence, resulting in a gag-myc entity, which, after processing and cleavage of its ends, results in a protein showing transforming activity on avian cells. However, there are no reports in the literature describing any experimental evidence on transformation of human cells by v-myc. Furthermore, the activation of two-to-three human oncogenes is needed to push cells into a transformed phenotype. Interestingly, the proto-oncogene, c-myc, failed to immortalize hNSCs (Villa et al., 2000, and other unpublished results obtained with forebrain and mesencephalic human tissue samples) and is also reported to be not sufficient to cause intraparenchymal tumors when transduced into rodent neural precursors *in vivo* (Fults et al., 2002).

Finally, a compelling and extensive body of evidence generated by several research groups working with immortalized cells demonstrates that v-myc cell lines do not form tumors, i.e., are not transformed. Box 1 describes some of the evidence generated by our group and that of Evan Snyder, working with different hNSCs lines (see Chapter 9 in this volume).

With regard to the immortal nature of v-myc hNSC lines, we know that continued proliferation over years and the ability to subclone is present. Recent investigation in our group has determined that v-myc, possibly through a direct interaction with myc-binding sites in the promoter for hTERT, results in the quick activation (<48hr) of telomerase activity in hNSCs (unpublished results). As a result, the telomeres of 2-and 4-year propagated hNS1 cells and of 3-year

Box 1. Evidence demonstrating the absence of transformation of v-myc perpetuated hNSCs

- No reported transforming activity of v-myc on human neural tissue in the scientific literature.

- Absence of growth in soft agar assays (Villa eat al., 2000).

- Their immortalization is conditional, since they are EGF/bFGF dependent. In the absence of these factors, and even in the presence of fetal bovine serum, the cells do not grow. After differentiation, the cells do not proliferate, as assessed by ^{3}H-thymidine incorporation in vitro or PCNA immunocytochemistry. The cell cycle length increases from 40 hours to over 30 days (Villa eat al., 2000).

- After in vitro differentiation, the cells:
  - achieved by growth factor withdrawal (Villa et al., 2000; Flax et al., 1998; Rubio et al., 2000).
  - stopped expressing nestin and vimentin, both in vitro and in vivo (Villa et al., 2000; Flax et al., 1998; Rubio et al., 2000).
  - reduced expression of v-myc, as demonstrated by RT-PCR (Villa et al., 2000; Flax et al., 1998; Rubio et al., 2000; unpublished data).
  - stopped expressing telomerase protein, and show no detectable telomerase activity (unpublished data) in contrast to transformed cells.
  - changed morphology and expressed neuronal and glial differentiation markers proper of terminally differentiated (and not just quiescent) cells (Villa et al., 2000; Flax et al., 1998; Rubio et al., 2000).
  - pool of cells in G2/S/M phases of the cell cycle profoundly decreased. This change was more clear-cut in genetically expanded (GE) cells than in human neurospheres (unpublished data).

- In vivo:
  - GE-hNSCs do not form tumors neither in fetal (Ourednik et al., 2001), neonatal (Flax et al., 1998) nor in adult immune-suppressed animals (Rubio et al., 2000; Aboody et al., 2000).
  - Transplanted GE-hNSCs downregulate nestin as soon as one week following implantation (Rubio et al., 2000).
  - Transplanted GE-hNSCs do not express PCNA, a marker of cycling cells (Rubio et al., 2000).
  - Transplanted GE-hNSCs differentiate to generate cells expressing mature neuronal and glial markers (Flax et al., 1998; Rubio et al., 2000; Ourednik et al., 2001).
  - Cell cycle duration, colony formation potential and neuron generation capability is unchanged over years in culture (unpublished data).

Reviews: Villa et al., 2001; Martinez-Serrano et al., 2001; Villa et al., 2002.

propagated hNS2 lines are short (around 2-3 Kb as determined by Q-FISH) but homogeneous, and are preserved in a stable manner for over hundreds of population doublings. No chromosomal free ends or chromosomal abnormalities (structural or numerical) were found in these cells lines. Furthermore, their clonogenic potential, cell cycle (^{3}H-thymidine incorporation), and rate of neuron generation (β-III-tubulin$^+$ neurons) were unaltered after 4 years in culture (unpublished data; see Figure 4).

The capacity for terminal differentiation of hNSCs was studied by flow cytometry, carrying neurosphere strains and cell lines in parallel. This cell cycle-exit study revealed that hNSCs cell lines efficiently exit the cell cycle. Following differentiation, 5% of the cells were arrested with a 2-4N DNA content. When expression of the cell cycle marker Ki-67 (late G1/S/G2/M phases) was studied, 10-15% of the total number of cells were positive for the marker, whereas neurosphere cells stain positive in up to 30% of the cells. Combining these data, it can be concluded that 7-9% of the differentiated cells from hNSC lines did not exit the cell cycle, remaining arrested in G1, whereas up to 25% of the cells in neurosphere cultures did not exit the cell cycle in order to terminally differ-

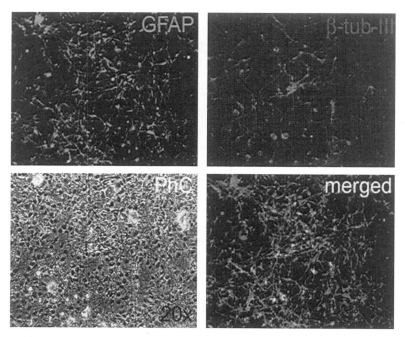

Figure 4. Neuron and glia generation by 4-year old hNS1 cells, differentiated in culture by mitogen withdrawal. The panels show astrogial cells (green, GFAP), neurons (red, β-III-tubulin), together with a phase contrast image and the merged image to illustrate the absence of doubly stained cells. This indicates that 4-year old cells retained their capability to make cell fate decisions along distinct neural lineages.

entiate. Therefore, immortalized hNSCs are not compromised in their capacity to exit the cell cycle and terminally differentiate when compared to neurosphere cultures, which have a larger population of cells unable to complete differentiation.

IN VIVO PERFORMANCE OF HUMAN NEURAL STEM/ PRECUSOR CELLS AFTER TRANSPLANTATION

Transplantation of hNSCs, either cell lines or strains, is not common in the scientific literature (see reviews in Martinez-Serrano et al., 2001; Le Belle and Svendsen, 2002). More recent studies describe integration, migration and differentiation of hNSC transplants. Several of these studies use fetal or neonatal mammalian recipients to fully reveal the plasticity and differentiation capacity of the cells. However, when hNSCs were grafted into the adult mammalian brain, the cells mostly generated astroglia, and only a few neurons differentiated in a region specific manner. When neurosphere cells were transplanted, neurons appearing in non-neurogenic regions may not be derived from neural stem cells, but rather from cells already committed for neuron generation (Martinez-Serrano et al., 2001; Fricker et al., 1999; Blackshaw and Cepko, 2002). While it is true that transplants of hNSCs perform much better in the developing CNS than in the adult brain, precise data from the adult brain will be most important for the future treatment of neurodegenerative diseases.

Perhaps the most exciting observation in recent transplantation experiments was that reported by Ourednik et al. (2001). They report that grafted hNSCs seem to be able to play a dual role *in vivo* when implanted into the cerebral ventricles of the fetal primate brain. A portion of the cells integrated into active germinal zones, joining the massive neurogenesis wave taking place at the time when the cells were transplanted. The transplanted cells contributed to corticogenesis, generating neurons and glia to the host brain. Of particular interest, they was described that some of the cells found final residence in regions or niches defined as secondary germinal zones, which persist after most brain development has taken place, i.e., the subventricular zone (SVZ). If these cells have the ability to choose between differentiation or quiescence, once could propose that NSCs appear to be able to autoregulate their population, in order to maintain self-renewal instead of becoming depleted following the generation of neuronal and glial progenitors. Three other studies have also contributed new data on the performance of human forebrain neural precursors from neurosphere strains grown with EGF/FGF2/LIF and grafted into the neonatal or adult rat brain (Englund et al., 2002a, b; Aleksandrova et al,. 2002).

In neonatal animals (Englund et al., 2002a), human neurosphere cells were implanted into the striatal and hippocampal formations, as well as into the SVZ.

Extensive migration, similar to that described by Armstrong et al. (2000) and Rosser et al. (2000) was observed. Also, region-specific differentiation was observed. This study, however, raises concern when it comes to translating these basic observations in experimental animals to the human brain, since an essentially uncontrolled cell migration occurred from the implantation sites. Whether or not this is a positive aspect for cell therapy remains to be clarified and discussed in depth, since colonization of remote brain areas by implants aimed at a circumscribed structure or region may not always be desirable.

In the adult rats (Englund et al., 2002b), migration patterns were observed 3.5 months following grafting. When the cells were grafted into the SVZ, they were found migrating as immature neuroblasts (doublecortin[+] cells, dcx), along the rostral migratory stream (RMS). In the hippocampus, possibly facilitated by the physical architecture and very small size of the subgranular and granular zones (SGZ, GZ, respectively), cells were found well integrated in the GZ after migrating a few micrometers from the implantation site. Two other migration paths appeared as diffuse, non-oriented cell migration, resulting in scattered cells radially dispersed in the neostriatum and hippocampus. These were also immature cells, staining positive for dcx. Finally, the authors reported extensive migration of the cells implanted in the striatum along white matter tracts, reaching regions far distant from the implantation site, (e.g., midbrain and frontal cortex). The authors stated that a large fraction of the grafted cells remained undifferentiated in a stem or progenitor cell stage. Whether or not this reflects a type of homeostatic control is far from clear. It can also be interpreted as a hurdle, since many of the grafted cells are unable to differentiate when implanted into non-germinal zones.

Aleksandrova et al. (2002) also describe the performance of implants of neurosphere cells into the young/adult rat brain. The cells were derived from 8-12 week old fetuses and were expanded in culture for 14 days prior to use, and the host brains were examined 10-20 days post-transplantation. Once more, profuse and nonoriented migration was observed with cells even reaching the contralateral hemisphere shortly after grafting. Many of them continued expressing markers of undifferentiated cells, such as vimentin.

Finally, FACs sorted and expanded hNSCs (CD133[+]) have been injected into neonatal NOD/SCID mice ventricles (Tamaki et al., 2002). GFP[+] cells were subsquently found integrated into the SVZ, RMS, olfactory bulb, hippocampus, but also at non neurogenic locations such as striatum, cerebellum and cortex. Differentiation was morphologically assessed and only a portion of the cells had elaborated processes and dendrites.

These recent studies confirm previous observations (Flax et al., 1998; Rubio et al., 2001; Fricker et al., 1999; Svendsen et al., 1996; Armstrong et al., 2000; Rosser et al., 2000) and indicate that hNSCs or progenitor cells can survive, extensively migrate, and integrate into the fetal, neonatal and adult mammalian brain. In neurogenic regions, some of the transplanted cells differentiate in a region-specific manner. However, a substantial portion of the implanted cells migrate in a diffuse manner, following permissive structures allowing for migration, like white matter tracts, and are found almost throughout the brain. Importantly, most of the cells that migrate extensively do not differentiate, remaining as nestin or vimentin-positive cells.

In spite of not being completely new findings, these observations reinforce the need for further experimentation and careful discussion. Demonstration of such excellent integration and migratory properties is of high importance on an experimental basis. However, when it comes to the design of future therapies for humans, it is far from clear that this extensive and uncontrolled migration, and limited differentiation of the cells will be acceptable. It is important to point out that some of the most devastating brain tumors, like high grade invasive gliomas, may arise from undifferentiated neural precursors (Bachoo et al., 2002).

Some recent transplantation reports have used pre-differentiated human neurosphere cells for grafting in the CNS. These can not be regarded as transplants of hNSCs or even precursors, but rather of neuronal or glial progenitors, and should be interpreted similarly to the use of primary neuronal tissue.

In a study dealing with the counteraction of apoptosis taking place after *in vitro* differentiation of human neurosphere cells, Caldwell et al. (2001) provided data on transplants of *in vitro* pre-differentiated cells. When these grafted cells were compared to their undifferentiated counterparts, they were found to generate large masses of cells with little evidence for migration away from the implantation site, as opposed to the smaller grafts of undifferentiated cells, which showed much higher migration capability. Therefore, it looks as if the immature properties of human neural precursors behave like a double-edged sword: When the inherent "stemness" and plasticity of the cells is exploited, more extensive migration is obtained, but differentiation becomes limited. Conversely, if the cells are pre-differentiated, migration and the capacity to colonize relative large areas of the brain, e.g., rat striatum, become limited.

A second report showed that *in vitro* pre-differentiated cells derived from human neurospheres can survive in the brain, and, at certain locations, probably due to the presence of specific neurotrophic factors and unknown microenvironmental factors, some populations of neurons survive better than others (Wu et al., 2002). The authors used a protocol including FGF2, herapin, and laminin, that resulted in a good rate of cholinergic neuron generation. The mechanism(s) regulating cholinergic differentiation were not investigated in detail. When these pre-differentiated (primed) cells were grafted to cholinergic and noncholinergic areas of the rat CNS, cholinergic neurons preferentially survived at locations that normally contain large cholinergic cell groups. These results were interpreted as a region-specific, instructive or inductive action on the grafted cells permitted by the priming method. However, the selective action of survival factors *in vivo* can not be excluded. Regardless what the operating mechanisms in this paradigm are, the fact is that at certain locations, cholinergic neurons were present at the time of sacrificing the animals. Clearly, further research will be needed to explain this phenomenon, and also to elucidate whether this priming/grafting design will work using other cell sources.

To end this section, we will briefly discuss the under-explored aspect of the capacity for functional integration and maturation of both neuronal and glial cells derived from NSC transplants. In terms of functional responses and the ability of implanted cells to sense and react to endogenous stimuli (physiological or pathological) in the recipient brain, a few experiments have started to provide evidence indicating that this is possible. For instance, rodent neural stem cells are reported to react like endogenous neurons after remote photic stimulation following grafting into the suprachiasmatic nucleus (Zlomanczuk et al., 2002). In two other reports, grafted rodent precursors have been found to fully differentiate, not only showing the desired morphology and expression of marker genes, but also the electrophysiological properties of endogenous local neurons. In one study, mouse ES cells were converted to dopaminergic neurons and grafted into the lesioned striatum of hemi-parkinsonian rats (Kim et al., 2002). In the second study, rat neural precursors of the immortalized (^{ts}Tag) line called RN33B, were grafted and analyzed electrophysiologically following integration and differentiation. Cortical grafted neurons were identified as having the correct set of electrical properties of functional pyramidal neurons (Englund et al., 2002c). Although these studies were conducted using rodent cells, they provide the long awaited evidence for functional integration of transplanted neural precursor cells into a recipient brain. It remains to be determined whether the human counterparts will behave similarly or not, although expectations remain high.

CONCLUDING REMARKS AND FUTURE PERSPECTIVES

Recent work has provided new and exciting knowledge of the biology of hNSCs *in vitro* and also following transplantation *in vivo*. Still, important issues remain to be investigated and resolved. In this chapter we have highlighted that improvements need to be made in cell culture conditions to avoid the expression of a senescent phenotype and to possibly extend the lifespan of cultured human neurosphere cells. Also, a better understanding of the biology of immortalizing genes, particularly v-myc, and that of the resultant cell lines, is needed.

An important set of unresolved issues *in vivo* is how to control the extensive and non-oriented cell migration shown by some neural stem/precursor cell preparations; how to stimulate terminal differentiation *in vivo*, to avoid the presence of large pools of undifferentiated cells; and how to improve the neurogenic capacity of hNSC transplants in non-neurogenic regions of the adult brain, such as the striatum. Finally, there is a need to investigate if the grafted cells have the potential to acquire electrical properties and integrate into the brain circuitry in a meaningful or relevant way in terms of both the newly generated glia and neurons.

These coming years will surely be very exciting, since most of these issues are currently being investigated by many groups. Therefore, expectations are high for hNSC cell therapies, although a great deal of research remains to be conducted to coax the tremendous potential and plasticity of neural stem and precursor cells into therapeutic realities.

Acknowledgements

C.B. is supported by an FPI grant from the Spanish Ministry of Science and Technology. Work at the author's laboratory was funded by grants from EU (BIO04-CT98-0530 and QLK3-CT-2001-02120), NsGene A/S (Denmark), Foundation La Caixa (Spain), and Spanish Ministry of Science and Technology (MCYT SAF2001-1038-C02-02). The institutional grant from Foundation Ramón Areces to the CBMSO is also gratefully acknowledged.

REFERENCES

Aboody KS, Brown A, Rainov NG, Bower KA, Liu S, Yang W, Small JE, Herrlinger U, Ourednik V, Black PMcL, Breakefield XO and Snyder EY (2000) Neural stem cells display extensive tropism for pathology in adult brain: Evidence from intracranial gliomas. Proc. Natl. Acad. Sci. USA, 97, 12846-12851.

Aleksandrova MA, Saburina IN, Poltavsteva RA, Revishchin AV, Korochkin LI and Sukhikh GT (2002) Behavior of human neural progenitor cells transplanted to rat brain. Dev. Brain Res., 134, 143-148.

Anderson DJ (2001) Stem cells and pattern formation in the nervous system: the possible versus the actual. Neuron 30, 19-35.

Armstrong RJ, watts C, Svendsen CN, Dunnett SB and Rosser AE (2000) Survival, neuronal differentiation, and fiber outgrowth of propagated human neural precursor grafts in an animal model of Huntington´s disease. Cell Transplant., 9, 55-64.

Bachoo RM, Maher EA, Ligon KL, Sharpless NE, Chan SS, You MJ, Tang Y, DeFrances J, Stover E, Weissleder R, Rowitch DH, Louis DN, DePinho RA. (2002) Epidermal growth factor receptor and Ink4a/Arf: convergent mechanisms governing terminal differentiation and transformation along the neural stem cell to astrocyte axis. Cancer Cell. 1, 269-77

Blackshaw S and Cepko CL (20002) Stem cells that know their place. Nat. Neurosci., 5, 1251-1252.

Bottenstein JE, Sato GH (1979) Growth of a rat neuroblastoma cell line in serum-free supplemented medium. Proc Natl Acad Sci USA 76: 514-517.

Caldwell MA, He X, Wilkie N et al. (2001) Growth factors regulate the survival and fate of cells derived from human neurospheres, Nat. Biotechnol. 19, 475–479.

Carpenter MK, Cui X, Hu ZY, Jackson J, Sherman S, Seiger A, Wahlberg LU (1999) *In vitro* expansion of a multipotent population of human neural progenitor cells. Exp Neurol. 158, 265-78.

Carpenter MK, Inokuma MS, Denham J, Mujtaba T, Chiu CP and Rao MS (2001) Enrichment of neurons and neural precursors from human embryonic stem cells. Exp. Neurol., 172, 383-397.

Cho T, Bae JH, Choi HB, Kim SS, McLarmon JG, Suh-Kim H, Kim SU and Min CK (2002) Human neural stem cells: electrophysiological properties of voltage-gated ion channels. Neuroreport 13, 1447-1452.

Christophersen N.S., P. Barraud, M. Gronborg, C.H. Jensen, B. Juliusson, L.U. Wahlberg. Expression of fetal antigen 1 in proliferating and dopaminergic differentiated stem and progenitor cultures derived from human CNS. Program No. 127.8. 2002 Abstract Viewer/Itinerary Planner. Washington, DC: Society for Neuroscience, 2002. Online.

D'Amour K and Gage FH (2000) New tools for human developmental biology. Nat. Biotech., 18, 381-382.

Dietrich J, Noble M and Mayer-Proschel M (2002) Characterization of A2B5+ glialprecursor cells from cryopreserved human fetal brain progenitor cells. Glia 40, 65-77.

Dimri GP, Lee X, Basile G, Acosta M, Scott G, Roskelley C, Medrano EE, Linskens M, Rubelj I, Pereira-Smith O, et al. (1995) A biomarker that identifies senescent human cells in culture and in aging skin *in vivo*. Proc Natl Acad Sci U S A. 92, 9363-9367.

Englund U, Bjorklund A, Wictorin K, Lindvall O and Kokaia M (2002 c) Grafted neural stem cells develop into functional pyramidal neurons and integrated into host cortical circuitry. Proc. Natl. Aad. Sci. USA, 99, 17089-17094.

Englund U, Bjorklund A, Wictorin K. (2002 b) Migration patterns and phenotypic differentiation of long-term expanded human neural progenitor cells after transplantation into the adult rat brain. Brain Res Dev Brain Res. 134, 123-141.

Englund U, Fricker-Gates RA, Lundberg C, Bjorklund A, Wictorin K. (2002 a) Transplantation of human neural progenitor cells into the neonatal rat brain: extensive migration and differentiation with long-distance axonal projections.Exp Neurol. 173, 1-21

Flax JD, Aurora S, Yang C, Simonin C, Wills AM, Billinghurst LL, Jendoubi MJ, Sidman RL, Wolfe JH, Kim SU and Snyder EY (1998) Engraftable human neural stem cells respond to developmental cues, replace neurons, and express foreign genes. Nat. Biotech., 16, 1033-1039.

Fricker, R. M., Carpenter, M. K., Winkler, C., Greco, C., Gates, M. A. and Björklund, A. (1999). Site-specific migration and differentiation of human neural progenitor cells after transplantation to the adult rat brain. J. Neurosci. 19: 5990–6005.

Fults D, Pedone C, Dai C and Holland EC (2002) MYC expression promotes the proliferation of neural progenitor cells in culture and *in vivo*. Neoplasia, 4, 32-39.

Gottlieb DI (2002) Large-scale sources of neural stem cells. Annu. Rev. Neurosci., 25, 381-407.

Haque NSK, Inokuma NS, Gold JD, Rao MS and Carpenter MK (2000) Transplantation of pluripotent human embryonic stem cells and their derivatives into the mammalian brain. Soc. Neurosci. Abstr. 327.8

Kawasaki H, Mizuseki K, Nishikawa S, Kaneko S, Kuwana Y, Nakanishi S, Nishikawa SI and Sacia Y (2000) Induction of midbrain dopaminergic neurons from ES cells by stromal cell-derived inducing activity. Neuron, 28, 31-40.

Kawasaki H, Suemori H, Mizuseki K, Wanatabe K, Urano F, Ichinose H, Haruta M, Takakashi M, Yoshikawa K, Nishikawa SI, Nakatsuji N abd Sasai Y (2002) Generation of dopaminergic neurons and pigmented epithelia from primate ES cells by stromal cell-derived inducing activity. Proc. Natl. Acad. Sci. USA 99, 1580-1585

Keyoung HM, Roy NS, Benraiss A, Louissaint A Jr, Suzuki A, Hashimoto M, Rashbaum WK, Okano H, Goldman SA (2001) High-yield selection and extraction of two promoter-defined phenotypes of neural stem cells from the fetal human brain. Nat Biotechnol , 19, 843-850

Kim JH, Auerbach JM, Rodriguez-Gomez JA, Velasco I, Gavin D, Lumelsky N, Lee SH, Nguyen J, Sanhez-Pernaute R, Bankiewicz K and McKay RDG (2002). Dopamine neurons derived from embryonic stem cells function in an animal model of Parkinson´s disease. Nature 418, 50-56.

Klassen H, Schwartz MR, Bailey AH and Young MJ (2001) Surface markers expressed by multipotent human andmouse neural progenitor cells include tetraspanins and non-protein epitopes. Neurosci. Lett., 312, 180-182.

Kukekov VG, Laywell ED, Suslov O, Davies K, Scheffler B, Thomas LB, O'Brien TF, Kusakabe M and Steindler DA (1999) Multipotent stem/progenitor cells with similar properties arise from two neurogenic regions of the adult human brain. Exp. Neurol., 156, 333-344.

Le Belle JE and Svendsen CN (2002). Stem cells for neurodegenerative disorders: where can we go from here?. BioDrugs 16 , 389-401.

Maric D, Maric I, Chang YH and Barker JL (2003) Prospective cell sorting of embryonic rat neural stem cells and neuronal and glial progenitors reveals selective effects of basic fibroblast growth factor and epidermal growth factor on self-renewal and differentiation. Proc. Natl. Acad. Sci. USA, 23, 240-251.

Martínez-Serrano A, Rubio FJ, Navarro B, Bueno C and Villa A (2001) Human neural stem and progenitor cells: *in vitro* and *in vivo* properties, and potential for gene therapy and cell replacement in the CNS. Curr. Gene Ther., 1, 279-299.

Martínez-Serrano A, Villa a, Navarro B, Rubio FJ and Bueno C (2000) Human neural progenitor cells: better blue than green? Nat Medicine, 6, 483-484.

Morrison SJ, Csete M, Groves AK, Melega W, Wold B and Anderson DJ (2000) Culture in reduced levels of oxygen promotes clonogenic sympathoadrenal differentiation by isolated neural crest stem cells. J. Neurosci., 20, 7370-7376.

Morrison SJ, white PM, Zock C and Anderson DJ (1999) Prospective identification, isolation by flow-cytometry, and *in vivo* self-renewal of multipotent mammalian neural crest stem cells. Cell 96, 737-749.

Ostenfeld T , Joly E , Tai YT , Peters A , Caldwell M, Jauniaux E and Svendsen CN (2002) Regional specification of rodent and human neurospheres. Dev. Brain Res. 134, 43–55.

Ostenfeld T, Cladwell MA, Prowse KR, Linskens MH, Jauniaux E and Svendsen CN (2000) Human neural precursor cells express low levels of telomerase *in vitro* and show diminishing cell proliferation with extensive axonal outgrowth following trasnplantation. Exp. Neurol., 164, 215-226.

Ourednik V, Ourednik J, Flax JD, et al. (2001) .Segregation of human neural stem cells in the developing primate forebrain. Science 293, 1820-1824.

Palmer TD, Schwartz PH, Taupin P et al. (2001) Progenitor cells from human brain after death. Nature 411, 42-43.

Pilcher HR, Rodriguez T, Majudmar S, Mann V, Dong J Sinden J et al. (2002)) Immortalization of human neural stem cells using large T-antigen and hTERT. Dev. Brain Res., 134, A19-A48. P6-17.

Poltavtseva RA, Marey MV, Aleksandrova MA, Revishcin AV, Korochkin LI and Sukhikh GT (2002) Evaluation of progenitor cell cultures from human embryos for neurotransplantation. Dev. Brain Res., 134, 149-154.

Poltavtseva RA, Rzhaninova AA, Revishchin AV, Aleksandrova MA, Korochkin LI, Repin VS and Sukhikh GT (2001) *In vitro* development of neural progenitor cells from human embryos. Bull. Exo. Biol. Med., 132, 1000-1003.

Reubinoff B.E., Itsykon P., Turetsky T., Pera M.F., Reinhartz E., Itzik A. and Ben-Hur T. (2001) Neural progenitors from human embryonic stem cells. Nat Biotech., 19, 1134-1140.

Reubinoff BE, Pera MF, Fong CY, Trounson A and Bongso A (2000) Embrionic stem cell lines from human blastocysts: somatic differentiation *in vitro*. Nat. Biotech., 18, 399-404

Revishchin AV, Poltavtseva RA, Marei MV, Aleksandrova MA, Viktorov IV, Korochkin LI and Sukhikh GT (2001) Structure of cll clusters formed incultures of dissociated human embryonic brain. Bull. Exp. Biol. Med., 132, 856-860.

Reynolds BA and Weiss S (1992) Generation of neurons and astrocytes from isolated cells from the adult mammalian central nervous system. Science 255, 1707-1710.

Rietze RL, Valcanis H, Brooker GF, Thomas T, Voss AK and Bartlett PF (2001) Purification of a pluripotent neural stem cell from the adult mouse brain. Nature, 412, 736-739.

Rosser AE, Tyers P and Dunnett SB (2000) The morphological development of neurons derived from EGF- and FGF-2-driven human CNS precursors depends on their site of integration in the neonatal rat brain. Eur. J. Neurosci., 12, 2405-2413.

Roy NS, Benraiss A, Wang S, Fraser RAR, Goodman R, Couldwell WT, Nedergaard M, Kawaguchi A, Okano H and Goldman SA (2000) Promoter-targeted selection and isolation of neural progenitor cells from the adult human ventricular zone. J. Neurosci. Res., 59, 321-331.

Roy NS, Benraiss A, Wang S, Fraser RAR, Goodman R, Couldwell WT, Nedergaard M, Kawaguchi A, Okano H and Goldman SA (2000) Promoter-targeted selection and isolation of neural progenitor cells from the adult human ventricular zone. J. Neurosci. Res., 59, 321-331.

Roy NS, Nakano T, Keyoung H, Carpenter M, Jiang L, Kang J, Nedergaard M and Goldman SA (2002) Telomerase-immortalization of neuronally-restricted progenitors derived from the fetal human spinal cord. Program No. 825.6. 2002 Abstract Viewer/Itinerary Planner. Washington, DC: Society for Neuroscience, 2002. Online.

Roy NS, Wang S, Jiang L, Kang J, Benraiss A, Harrison-Restelli C, Fraser RAR, Couldwell WT, Kawaguchi A, Okano H, Nedergaard M and Goldman SA (2000) *In vitro* neurogenesis by progenitor cells isolated from the adult human hippocampus. Nat. Medicine, 6, 271-277.

Rubio FJ, Bueno C, Villa A, Navarrro B and Martínez-Serrano A (2000) Genetically perpetuated Human Neural Stem Cells engraft and differentiate into the adult mammalian brain Mol. Cell. Neuroscience, 16, 1-13.

Schuldiner M, Eiges R, Eden A, Yanuka O, Itskovitz-Eldor J, Goldstein RS and Benvenisty N (2001) Induced neuronal differentiation of human embryonic stem cells.. Brain Res. 913, 201-205.

Shingo T, Sorokan ST, Shimazaki T and Weiss S (2001) Erythropoietin regulates the *in vitro* and *in vivo* production of neuronal progenitors by mammalian forebrain neural stem cells. J. Neurosci., 21, 9733-9743.

Smith AG (2001) Embryo-derived stem cells: of mice and men. Annu. Rev. Cell. Dev. Biol., 17, 435-462.

Storch A, Paul G, Csete M, Boehm BO, Carvey PM, Kupsch A and Schwarz J (2001) Long-term proliferation and dopaminergic differentiation of human mesencephalic neural precursor cells. Exp. Neurol., 170, 317-325.

Studer L (2001) Stem cells with brain power. Nat. Biotech., 19, 1117-1118.

Studer L, Csete M, Lee SH, Kabbani N, Walikonis J, Wold B, McKay R. (2000) Enhanced proliferation, survival, and dopaminergic differentiation of CNS precursors in lowered oxygen. J Neurosci. 2000 Oct 1;20(19):7377-83.

Suslov ON, Kukekov VG, Ignatova TN and Steindler DA(2002) Neural stem cell heterogeneity demonstrated by molecular phenotyping of clonal neurospheres. Proc. Natl. Acad. Sci. USA, 99, 14506-14511.

Suslov ON, Kukekov VG, Ignatova TN, Steindler DA (2002) Neural stem cell heterogeneity demonstrated by molecular phenotyping of clonal neurospheres. Proc Natl Acad Sci U S A 99, 14506-11

Svendsen CN (2000) Controlling the growth and differentiation of human neural stem cells. Symposium: Recent advances in neural stem cell technology. 30th Soc. for Neurosci. Meeting, New Orleans 3-10 November 2000.

Svendsen CN and Smith AG (1999) New prospects for human stem-cell therapy in the nervous system. Trends Neurosci., 22, 357-364.

Svendsen CN, Clarke DJ, Rosser AE and Dunnett SB (1996) Survival and differentiation of rat and human epidermal growth factor-responsive precursor cells following grafting into the lesioned adutl central nervous system. Exp. Neurol., 137, 376-388.

Svendsen CN, ter Borg MG, Armstrong RJE et al. (1998) A new method for the rapid and long term growth of human neural precursor cells. J. Nsci. Meth., 85, 141-152.

Tamaki S, Exkert K, He D, Sutton R, Doshe M, Jain G, Tushinski R, Reitsma M, Harris B, Tsukamoto A, Gage F, Weissman I and Uchida N (2002) Engraftment of sorted/expanded human central nervous sytem stem cells from fetal brain. J. Nsci. Res., 69, 976-986.

Temple S (2001) The development of neural stem cells. Nature 414, 112-117.

Thomson JA, Itskovitz-Eldor J, Shapiro SS, Waknitz MA, Swiergiel JJ, Marshall VS, Jones JM. (1998) Embryonic stem cell lines derived from human blastocysts. Science 282, 1145-1147

Uchida N, Buck DW, He D, Reitsma MJ, Masek M, Phan TV, Tsukamoto AS, Gage FH and Weissman IL (2000) Direct isolation of human central nervous system stem cells. Proc. Natl. Acad. Sci. USA, 97, 14720-14725.

Vescovi Al and Snyder EY (1999) Establishment and properties of neural stem cell clones: plasticity *in vitro* and *in vivo*. Brain Pathol. 9, 569-598.

Vescovi AL, Parati EA, Gritti A, Poulin P, Ferrario M, Wanke E, Frölichsthal-Schoeller P, Cova L, Arcellana-Panlilio M, Colombo A and Galli R (1999) Isolation and cloning of multipotential stem cells from the embryonic human CNS and establishement of transplantable human neural stem cell lines by epigenetic stimulation. Exp. Neurology, 156, 71-83.

Villa A, Navarro B and Martínez-Serrano A (2002) Genetic perpetuation of *in vitro* expanded human neural stem cells (HNSCs): cellular properties and therapeutic potential. Brain Res Bull, 57, 789-794.

Villa A, Rubio FJ, Navarro B, Bueno C and Martínez-Serrano A (2001) Human neural stem cells *in vitro*. A focus on their isolation and perpetuation. Biomed. Pharmacother., 55, 91-5.

Villa A, Snyder EY, Vescovi, A et al. (2000) Establishment and properties of a growth factor dependent, perpetual neural stem cell line from the human CNS. Exp. Neurol., 161, 67-84.

Windrem MS, Roy NS, Wang J, Nunes M, Benraiss A, Goodman R, McKhann GM 2nd, Goldman SA. (2002) Progenitor cells derived from the adult human subcortical white matter disperse and differentiate as oligodendrocytes within demyelinated lesions of the rat brain. J Neurosci Res., 69, 966-75

Wright L.S., J. Li, S. Klein, K. Wallace, J.A. Johnson, C.N. Svendsen. Increased neurogenesis and GFAP expression in human neurospheres following LIF treatment: A micro-array study. Program No. 329.3. 2002 Abstract Viewer/Itinerary Planner. Washington, DC: Society for Neuroscience, 2002. Online.

Wright LS, Li J, Klein S, Wallace K, Johson JA and Svendsen CN (2002) Increased neurogenesis and GFAP expression in human neurospheres following LIF treatment: A micro-array study. Program No. 329.3. 2002 Abstract Viewer/Itinerary Planner. Washington, DC: Society for Neuroscience, 2002. Online.

Wu P, Tarasenko YI, Gu Y, Huamg LYM, Coggeshall RE and Yu Y (2002) Region-specific generation of cholinergic neurons from fetal human neural stem cells grafted in adult rat. Nat. Neurosci., 5, 1271-1278.

Ying QL, Stavridis M, Griffiths D, Li M and Smith A (2003) Conversión of embryonic stem cells into neuroectodermal precursors in adherent monoculture. Nat. Biotech., 21, 183-186.

Zhang S.-C., Wernig M., Duncan I.D., Brüstle O. and Thomson J.A. (2001) *In vitro* differentiation of transplantable neural precursors from human embryonic stem cells. Nat Biotech., 19, 1129-1133.

Zlomanczuk P, Mrugala M, de la Iglesia HO, Ourednik V, Quesenberry PJ, Snyder EY, Schwartz WJ (2002) Transplanted clonal neural stem-like cells respond to remote photic stimulation following incorporation within the suprachiasmatic nucleus. Exp Neurol 174, 162-168.

Chapter 10

Neuronal Replacement by Transplantation

Daniel J. Guillaume and Su-Chun Zhang

The vertebrate central nervous system (CNS) is structured during development by sequential placement and wiring of various types of cells at particular locations. In the mature brain and spinal cord, the anatomical neuronal network is relatively stable with little neuronal replacement although functional plasticity is ongoing throughout life (Nottebohm, 2002). Hence, the mammalian brain and spinal cord, unlike many other tissues, has very limited capacity to replace neurons, particularly projection neurons, and to rewire neuronal circuitry through the use of endogenous stem cells in neurological disorders. One strategy to circumvent this is by utilizing exogenous cells as a source of replacement and repair of neuronal circuitry. Transplantation of neural progenitor cells in animal models of neurological disorders has shown survival of grafted cells and contribution to functional recovery in some instances. However, neuronal replacement therapy remains a distant goal. Major hurdles include the lack of effective donor cells and difficulty in remodeling the non-neurogenic adult CNS environment. Recent stem cell technology offers hope for generating potentially effective donor cells. Current efforts are focused on identifying which lineage position of the progenitors is ideal for transplantation, and whether the grafted progenitors can differentiate into the desired phenotypes and produce functional connections with recipient cells, subsequently contributing to functional recovery. Equally important is the recreation of a neurogenic environment in the adult CNS to promote the survival, migration, differentiation, and functional integration of the grafted neural progenitor cells. Further animal studies in both areas will be necessary to translate neural transplantation to a future therapy for various neurological conditions.

NEURONAL SPECIFICATION DURING DEVELOPMENT

Development of the mammalian nervous system begins with a sheet of columnar neuroepithelial cells called the neural plate. The neuroepithelial cells are initially specified from naïve embryonic stem cells during early embryogenesis. In mouse, it occurs around day 7 and in human, it takes place at the end

From: *Neural Stem Cells: Development and Transplantation*
Edited by: Jane E. Bottenstein © 2003 Kluwer Academic Publishers, Norwell, MA

of the third gestation week. These neuroepithelial cells are often called neural stem cells, as they generate neurons, astrocytes, and oligodendrocytes in a coordinated temporal and spatial order. This definition of neural stem cells is, however, overly simplified. Neuroepithelial cells at a particular location of the neural plate will ultimately give rise to a specific set of glia and neurons with unique connections (Jessell, 2000). Even the neuroepithelial cells at a given location of the neural plate may have different fates over time (Temple, 2001). Ventral neuroepithelial cells in the spinal cord generate motoneurons first but primarily oligodendrocytes at a later stage (Novitch et al., 2001; Zhou et al., 2002). Hence, neuroepithelial cells, or neural stem cells, likely differ from each other depending upon when and where the cells are located or isolated.

During embryogenesis, neuroepithelial cells migrate away from the ventricular zone and begin to differentiate into neurons first, followed by glial cells. This stereotyped order of neural differentiation is also observed *in vitro* where mouse cortical neural stem cells in clonal cultures produce neurons and then glia (Qian et al., 2000). This stereotypic temporal pattern of neural differentiation is also true for the neuroepithelial cells that are generated *in vitro* from embroynic stem (ES) cells, even though the neuroepithelia have not been in the brain (Zhang et al., 2001). Within the neuronal population, projection neurons are born first and interneurons later. Hence, neural stem cells isolated from early neural plate may generate projection neurons but differentiate predominantly into interneurons after expansion for a period of time. This is clearly illustrated by the failure to generate dopamine neurons from expanded midbrain neural stem/progenitor cells (Studer et al., 1998; Ostenfeld et al., 2002). Similarly, precursor cells from the subventricular zone will no longer generate projection neurons by birth (Lim et al., 1997). This temporal shift of the differentiation potential of neural stem/progenitor cells is often overlooked. Although neural stem cells differentiate into neurons, astrocytes, and oligodendrocytes, one needs to look further into the subtypes of neurons and glia produced. This hierarchy of cell lineage development has been regarded as the consequence of a cell autonomous mechanism of fate determination.

The fate of neuroepithelial cells is also influenced by cell-extrinsic factors. One major characteristic of neuroepithelial cells is their positional identity. Regionalization of neuroepithelial cells is achieved via responsiveness of neural precursor cells to morphogen gradients such as sonic hedgehog (SHH), bone morphogenetic proteins (BMPs), fibroblast growth factors (FGFs), retinoic acid (RA), and wnt proteins that control dorsal-ventral and rostral-caudal fate. Neuroepithelial cells in the ventral spinal cord are patterned to motoneurons and interneurons in response to discrete concentrations of SHH (Jessell, 2000; Briscoe et al., 1999). Alteration of local SHH concentration may promote the differentiation of one type of neuron, often at the expense of others (Briscoe et al., 1999). In culture, naïve neuroepithelial cells, particularly those generated from mouse ES cells, can be directed to a specific neuronal fate depending on the presence of morphogens. In the presence of FGF8 and SHH which confer midbrain dopamine neuron identity, ES-derived neuroepithelial cells differentiate predominantly into dopamine neurons (Lee et al., 2000). In contrast, in response to caudalizing signals such as RA, the same type of cells generates spinal cord motoneurons (Wichterle et al., 2002). Thus, environmental cues play an important role in directing the fate of neuroepithelial cells. Nevertheless, neuroepithelial cells that have passed the birthday of a particular cell type and/or are derived from ectopic regions may become unresponsive to the same set of signals for generation of this particular type of cells, illustrating the importance of coupling the cell-intrinsic program and environmental signals.

During development, neuroepithelial cells migrate, exit the cell cycle, and differentiate into neurons and glia. Hence, the number of neural stem cells decreases progressively. In the embroynic day 10 (E10) rat spinal cord, about 50% of the neural plate cells are regarded as neural stem cells. This portion drops to 10% at E12 and 1% by postnatal day 0 (P0) (Kalyani et al., 1998). In adults, neural stem cells further decrease in number and localize mainly to limited areas such as the subventricular zone and the hippocampal dentate gyrus, although fewer cycling putative stem cells can be found elsewhere in the CNS (for review see Magavi and Macklis, 2001). Moreover, the differentiation potential of these neural stem cells is likely different from those during development. While they can generate neurons and glia *in vitro*, adult neural stem cells, such as those of the hippocampal dentate gyrus and the subventricular zone, generate only interneurons and glia under most circumstances (for exception, see Magavi et al., 2000). Consequently, the adult mammalian brain and spinal cord, has very limited capacity to replace neurons and to rewire neuronal circuitry through the use of endogenous stem cells in neurological disorders. In many cases it may require replacement of lost neurons with exogenous cells, through transplantation, together with the recreation of the environment to support the survival, differentiation, and integration of grafted cells into a neuronal circuitry.

NEURONAL REPLACEMENT: THE DONOR

What Is The Ideal Donor?

Different diseases affect different types or groups of cells. Hence, the cells used for replacement depend upon the affected population. In general, the ideal donor cells should be easy to obtain or produce in a large numbers (efficiency); effective in cellular or molecular replacement (efficacy); tolerant to immune rejection; and resistant to generation of aberrant tissues (safety). For neuronal replacement therapy in the nervous system, another critical requirement is functional integration of grafted cells into the existing circuitry. Presently, no cells or tissues meet all of these criteria. In order to obtain a large number of donor cells, they will likely require expansion *in vitro* before transplantation. Expandable cells are generally immature, or progenitor cells. This raises the questions of whether and how the progenitor cells differentiate into the cells of a therapeutic choice. Any *in vitro* expanded cells may run into a risk of aberrant growth following grafting. Hence, the selection of donor cells will likely be the balanced consideration of these traits of individual cell types (Table 1).

In addition to replacing damaged cells, donor cells may also act as a vehicle for delivering or supplementing missing molecules. While many types of cells may fulfill the task, cells of neural origin are likely better than non-neural cells for neurological diseases. Neural stem cells and their derivatives may sus-

Table 1. Potential Donor Cells

Cell type	Donor source	Advantages	Disadvantages
Embryonic stem cells	Embryo	Unlimited expansion in undifferentiated state, enormous differentiation potential	Difficulty in directing in vitro differentiation, potential of teratoma formation
Neural stem/ progenitor cells	Fetal brain	Long-term in vitro expansion, multiple differentiation potential, safe	Lack of ability to differentiate into projection neurons
Neuronal-restricted progenitor cells	Embryo or fetal brain	Predictable differentiation to neuronal phenotype, safe	Restricted neuronal differentiation potential, very limited expandability
Primary neural cells	Fetal brain	Predictable phenotypic differentiation	Lack of expandability, restricted differentiation potential, require stage-specific embryonic source, mixed/inconsistent cell types
Adult stem cells	Adult "self" neural or nonneural	Avoids ethical and immunological constraints, expandable in culture, possible autologous transplant	Greatly restricted potential in neural differentiation, questionable functional differentiation and integration
Xenotransplant cells	Non-human	Readily available, possible genetic modification	Tendency of immune rejection, possible zoonoses

tain neural transgene expression whereas neural genes in non-neural cells may be easily down-regulated. Products expressed intrinsically by the grafted vehicle neural cells may promote the therapeutic effects of the engineered cells.

Source Of Donor

Primary tissues or cells

Primary fetal cells are generally committed to a particular phenotype, corresponding to their site of origin. After implantation, they complete their maturation process through a cell autonomous mechanism. This has been the case in clinical trials in Parkinson's Disease patients using fetal mesencephalic tissues (Bjorklund & Lindvall, 2000). One major advantage of this approach is that the phenotypic differentiation of grafted cells is predictable. This principle is perhaps critical in developing donor cells from stem cells for transplant therapy in adults, where the environment is unlikely to be as permissive as in embryonic stages to direct uncommitted cells to a specific phenotype (see below). The disadvantage is that these cells have a limited capacity to expand to a large quantity. Experimentally, these cells must be prepared from a stage specific embryo. Clinically, there are numerous practical constraints that limit the use of primary fetal brain tissues in cell transplantation beyond the ethical concerns (Freed, 2002). This explains why the use of primary cells is unlikely a general transplant therapy for neurodegenerative disorders including Parkinson's Disease.

Neural stem/progenitor cells

In contrast to primary cells, neural stem/progenitor cells that are isolated from embryonic brain and spinal cord can be expanded in culture for a prolonged period of time using genetic or epigenetic approaches. The expanded progenitors are able to further differentiate into neurons, astrocytes and oligodendrocytes, although the capacity of neuronal differentiation appears to decline over time (Svendsen & Caldwell, 2000). Current approaches for maintaining and expanding neural stem/progenitors, either the adherent (Palmer et al., 1997) or the non-adherent "neurosphere" cultures (Reynolds et al., 1992), are yet ideal to maintain or expand truly undifferentiated neural stem cells or cells that are still plastic enough to be directed to various neuronal lineages. In reality, these cultures are mixtures of largely progenitors and a small proportion of generic neural stem cells. For example, less than 1% of the cells in the neurospheres isolated from E8.5 mouse neural plate can generate secondary neurospheres (Tropepe et al., 1997).

From a developmental perspective, neural stem/progenitor cells are often confined to a specific regional identity at the time of isolation given that neuroepithelia are specified to rostro-caudal and dorsal-ventral domains during early development. They are further developmentally restricted over time in culture. This means that neural stem/progenitor cells isolated after neural tube formation are likely confined within certain lineages of regional and temporal identity. This is illustrated by the fact that neural progenitor cells isolated from different regions of the neuraxis retain marker gene expression of a particular region (Hitoshi et al., 2002; Ostenfeld et al., 2002). Using an embryonic intraparenchymal transplantation approach, Olsson et al. (1997) demonstrated that midbrain/hindbrain precursor cells isolated from E13.5 have a decreased level of heterotopic integration compared to those derived from E10.5, suggesting that neuroepithelial cells increasingly restrict their potential during embryonic development. For therapeutic purpose, it may offer a window of opportunity for deriving cells that match the cell replacement requirement. However it limits the possibility of differentiating these progenitors into other lineages. Nevertheless, progenitor cells within a certain window of development, appear to retain the plasticity to give rise to cells of other lineages. For example, coculture of midbrain/hindbrain neural stem cells with ventral forebrain explant induces the expression ventral forebrain markers (Hitoshi et al., 2002). Similarly, neural precursor cells isolated from the hippocampal dentate gyrus and grafted into the rostral migratory stream, migrated into the olfactory bulb and differentiated into tyrosine hydroxylase-positive neurons, a non-hippocampal phenotype (Suhonen et al., 1996). Two strategies could be developed to complement the deficit. One is to develop methodology that can selectively maintain and expand generic neural stem cells, similar to that for maintaining embryonic stem cells. The other is to identify signals that, *in vitro*, induce the neural stem/progenitors to become cells of a particular regional identity (see section on ES cells).

In addition to epigenetic stimulation of neural stem/progenitor cell proliferation, various genetic means have been applied to propagate neural stem cells. The most extensively used approach is immortalization with oncogenes such as v-myc. The oncogene-immortalized cells are kept in cell cycle yet they remain responsive to growth factors for proliferation. The clonal line C17.2, engineered by overexpression of v-myc in cerebellar granular cells (Snyder et al., 1992), appears to be trapped in a synchronized proliferative state. Yet, they are capable of differentiating into a wide variety of neurons and glial cells *in vitro*. Remarkably, these cells can differentiate into neuronal and glial cell types depending on where they are placed. When these cells are implanted into the dysmyelinated brain of shiverer mice, the cells differentiate into oligodendrocytes and produce myelin sheaths (Yandava et al., 1999). When the cells are grafted into the injured brain or spinal cord, these cells differentiate into glia and neurons that are appropriate to the region (Vescovi & Snyder, 1996). These results suggest that the environment largely controls the fate of neural stem cells. They also suggest that the cells apparently "forgot" their cerebellar origin. It is possible that the genetic manipulation reverses the developmental program to a more primitive state. The remarkably broad spectrum and high degree of differentiation is so far unmatched by any types of nongenetically modified neural stem cells. The uniformity of cellular characteristics and the remarkable differentiation potential of C17.2 cells might suggest that non lineage committed neuroepithelial cells, if similarly generated and maintained by nongenetic modifications, may have a similar potential. This also raises the question of whether a common neural stem cell or a more committed neuronal progenitor is preferred for cell replacement therapy (see below).

Embryonic stem cells

Neural stem/progenitor cells can be generated not only from fetal brain tissues, but also from their precursors, embryonic stem (ES) cells. ES cells are derived from the inner cell mass of preimplantation embryos at the blastocyst stage (Evans & Kaufman, 1981; Thomson et al., 1998). They are capable of almost unlimited proliferation in an undifferentiated state yet retain the potential to differentiate into many, if not all, cell and tissue types of the body. Methodologically, current approaches using feeder cells with or without the addition of cytokines such as leukemia inhibitory factor (LIF) allow us to selectively maintain mouse ES cells in a synchronized, undifferentiated state, in contrast to the methods available for propagating neural stem/progenitor cells. This means that ES cells can be trapped in the cell cycle for a long period, whereas most neural stem cells will progress into a progenitor stage in current culture systems. Hence, ES cells can provide a large quantity of consistent starting materials for deriving the desired cells for cell therapy. A major challenge is how to teach the naïve ES cells to choose a neural fate. Based on what we have learned from neural induction and neural patterning in animals, we are now able to efficiently differentiate mouse and human ES cells to neuroepithelial cells (Bain et al., 1995; Okabe et al., 1996; Zhang et al., 2001). The ES-derived neuroepithelial cells can be further differentiated into specialized neurons such as dopamine neurons and motoneurons (Lee et al., 2000; Wichterle et al., 2002). These *in vitro* generated neurons exhibit electrical properties (Finley et al., 1996) and make connections with endogenous cells, with subsequent contribution to functional recovery (Kim et al., 2002).

Another problem with the use of ES cells in cell therapy is the tendency for teratoma formation. This is due mainly to the presence of undifferentiated

ES cells in the graft (Benninger et al., 2000; Bjorklund et al., 2002). Present differentiation protocols often employ aggregated ES cell culture (embryoid body formation) and some cells in the embryoid body remain undifferentiated even in prolonged culture (Billon et al., 2002). Use of this mixture for grafting would run the risk of growth of unwanted tissues or teratoma formation by contaminating, undifferentiated ES cells. Another possible reason is that cells generated from ES cells *in vitro* may be specified but not fully committed to a neural fate. Hence, these ES-derived cells may still be plastic enough to choose alternative fates under certain conditions. Strategies can be designed and techniques may be developed to overcome this problem. One method is to positively sort out the target cells using cell surface markers or using cell type specific transcription factors through homologous recombination or the use of promoters (Li et al., 1998; Wichterle et al., 2002; Zwaka & Thomson, 2003). Another method is to remove the pluripotent stem cells by using stem cell surface molecules or similar genetic manipulations.

Although ES cells may provide a source of cells for neuronal replacement, another barrier that needs to be addressed is immune rejection. One way to overcome this problem is to reprogram patients' somatic cells (therapeutic cloning). In this way, DNA of a patient's cells, e.g., skin cells or fibroblasts, can be reprogrammed to a developmental process by introducing them into an oocyte. The egg containing the transferred material will develop into a blastocyst in which the inner cell mass can be cultured as ES cells. The technology for nuclear reprogramming is in place for many species (Wilmut et al., 1997; Munsie et al., 2000). Theoretically, human cells can be reprogrammed in a similar manner. ES cells generated in this way should have similar advantages and disadvantages as those of regular ES cells. Because the ES cells are genetically matched for all nuclear genes of the patient, the cells would be immunologically compatible with the patient.

Adult stem cells

In light of the ethical and immunological constraints surrounding the use of embryo and fetal tissue-derived neural cells, stem cells generated from adult neural and non-neural tissues may be preferable. As discussed above, stem cells exist in the adult, including the CNS, albeit in a smaller number and perhaps also with different potentials. Neural stem/progenitor cells have been isolated from brains of adult animals and humans (Palmer et al., 1997; Gage et al., 1998; Kukekov et al., 1999) and expanded in culture. These cells appear to have astrocytic traits (Laywell et al., 2000). These neural progenitors, like their embryonic counterparts, can differentiate into neurons and glia both *in vitro* and following transplantation into the brain (Suhonen et al., 1996; Shihabuddin et al.,

2000; Herrera et al., 1999). Progenitor cells isolated from adult spinal cord tissue may differentiate into hippocampal neurons after transplantation (Shihabuddin et al., 2000) and hippocampal progenitors can contribute to neurons in the olfactory bulb (Suhonen et al., 1996). In both cases, however, the adult progenitors mainly contribute to interneruons. This occurs when the progenitor cells are placed in neurogenic regions such as dentate gyrus or rostral migratory stream. If placed in nonneurogenic regions, these same progenitors differentiate primarily into glial cells (Suhonen et al., 1996). Studies thus far suggest that the adult neural stem cells have a lesser degree of plasticity to generate a variety of neuronal types.

Adult stem cells from tissues other than the nervous system appear also able to generate neuron-like cells in experimental settings. For example, stem cells or precursor cells isolated from skin or bone marrow can differentiate into neuronal-like cells *in vitro* and in some cases following neural transplantation (Toma et al., 2001; Jiang et al., 2002). This phenomenon has generally been regarded as transdifferentiation, i.e., fate shift across the lineage boundary (see Chapter 6 in this volume). Cross-lineage differentiation would make it possible to have an autologous graft from the patient's own body and thus offers a new prospect for a more flexible source of cells. However, this phenomenon appears to be very rare and to happen in somewhat unusual conditions. For example, the multiple-lineage differentiation potential appears after bone marrow stromal cells are expanded in culture for a long time (Jiang et al., 2002). Some of the observations may be attributed to fusion with recipient cells and/or transformation (Morshead et al., 2002; Ying et al., 2002). There is presently no demonstration that the transdifferentiated neurons are functional.

Xenotransplant

The use of xenografts, such as fetal porcine cells, avoids the ethical issues surrounding human fetal tissue. However, xenotransplantation runs the risk of immune rejection. One possible way to overcome this is to engineer the donor cells with immune competent molecules to shield immune rejection. Most studies to date have used fetal pig neural cells, and these cells were utilized in small clinical trials. A post-mortem histological study of fetal pig neural cells placed into the striatum of a Parkinson's patient revealed graft survival and axon extension of several millimeters 7 months later (Deacon et al., 1997). Neural precursors, obtained from embryonic pig, have been transplanted into the 6-OHDA-lesioned rat model of Parkinson's disease (Armstrong et al., 2002). Many cells differentiated into neurons and extended axons from the substantia nigra throughout the striatum. Although there was evidence of synaptic connections, no improvement in function was noted. Overall, survival and clinical benefits are uncertain with xenotransplantation. There is also a risk of transmitting pathogens from animals to humans.

Choice Of Donor Cells

Selection of donor cells depends on not only the cells themselves but also the nature of the disease. The primary consideration regarding choice of donor cells is efficacy, i.e., whether the donor cells can effectively replace the diseased cells in an anatomical (circuitry) and functional manner. Production efficiency and safety are also important factors that would render cell replacement a realistic therapy or not. The nature of a disease dictates the selection of donor cells, particularly the lineage position of the cells. In neurological traumas in which many types of cells are damaged, e.g., spinal cord injury and ischemia, uncommitted neural stem cells would be the preferred donors for transplant. These multipotential neural stem cells, under the influence of environmental signals present following injury, may hopefully differentiate into the variety of cell types that are lost. For neurodegenerative diseases in which a specific type of neuron is damaged or lost, lineage committed neuronal progenitor cells are likely to be the ideal donors (see Table 1).

Lineage progression from ES cells to neurons is a process of progressive fate restriction, in which the proliferation, migration, and differentiation potential decreases. In contrast, the safety of donor cells increases (Figure 1A). In practice, terminally differentiated neurons are non-dividing process-bearing cells. They are thus not well suited for transplantation purposes due to poor cell survival in the cell preparation and transplantation processes. Thus, cells used for transplantation are generally immature. Neural stem cells would ideally suit that need given their proliferation and multi-lineage differentiation potential. In reality, however, neural stem cells isolated from the brain or spinal cord and expanded in culture have an increasing tendency to differentiate into glial cells over neurons (Svendsen and Caldwell, 2000, Jori et al., 2003). The adult CNS environment also tends to favor glial differentiation. Thus, neural stem cells grafted into the adult brain often end up with predominantly glial progeny with few neurons (Rubio et al., 2000). This suggests that the neural stem cells need to be induced to a neuronal fate before implantation, or an environmental cue needs to be in place to guide their differentiation into neurons in the brain (see Table 1).

An ideal neuronal progenitor cell would differentiate into a variety of neuron types but not glial cells, and they would still retain a certain degree of cell proliferation and migration potential. Following transplantation, they would differentiate and mature into neurons that are appropriate to the brain region where the cells are placed. A neuronal progenitor, also known as a neuronal-

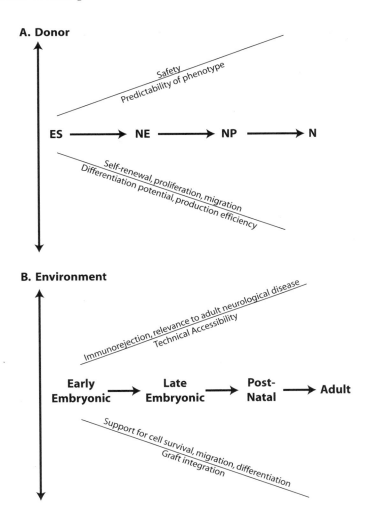

Figure 1. A. Relationship between cell behaviors and lineage position of a cell. As a cell progresses from an embryonic stem (ES) cell to a neuron, production efficiency (*in vitro* expansion capability) and ability to self-renewal, proliferate, migrate, and differentiate progressively decreases, while neuronal phenotype predictability and safety of the transplant increases. **B. Relation between donor cells and the environment where the cells are placed.** From embryonic to adult life, the neurogenic environment shifts to a non-neurogenic niche. Hence, environmental signals that support cell survival, migration, differentiation, and graft integration decrease, while technical accessibility improves, immunological rejection is more likely, and the pathological conditions mimicked in animal models become more relevant to adult neurological diseases.

restricted progenitor (NRP), has been isolated from rat neural tubes based on its expression of the embryonic neural cell adhesion molecule (E-NCAM, Mayer-Proschel et al., 1997). The sorted cells express neuronal markers such as β_{III}-tubulin and MAP2 and differentiate into only neurons but not glia *in vitro* and after transplantation into brains (Mayer-Proschel et al., 1997; Kalyani et al., 1998; Yang et al., 2000; Han et al., 2002). *In vitro*, the NRPs can differentiate into neurons with various neurotransmitter phenotypes (Kalyani et al., 1998). It is unclear however, whether the NRPs from a particular region of the neural tube will differentiate into different types of neurons depending upon the brain areas in which they are grafted. Cells in the nervous system, unlike those of the hematopoietic system, are endowed with positional identities during early development. NRPs derived from embryonic spinal cord express choline acetyl transferase when grafted into the occipital cortex where the host neurons do not normally express this marker (Yang et al., 2000), suggesting that the maturation of NRPs is cell-intrinsic. The positional identity of neural progenitor cells adds complexity to the selection of donor cells for neural transplantation. This could require different kinds of NRPs from various CNS regions for individual disease conditions. It is hence essential to understand the signaling pathway(s) that leads to a particular neuronal fate and the phenotypes and/or genotypes the progenitor presents. In this way, a specific neuronal progenitor may be induced, isolated, and/or expanded for cell replacement in an individual condition. Unfortunately, we know little as yet of the nature of these neuronal progenitors.

In theory, neuronal progenitors are more restricted than their precursors, neural stem cells, or neuroepithelial cells. Neural stem cells isolated from brain are already specified or restricted to the fate of their origin. In contrast, neuroepithelial cells differentiated from ES cells appear to be much more plastic. Mouse ES cell-derived neuroepithelial cells may be directed to become midbrain dopamine neurons or spinal cord motoneurons depending upon positional information. Cells treated with sonic hedgehog that induces a ventral neural fate, and FGF8 that determines midbrain/hindbrain identity will differentiate into midbrain dopamine neurons (Lee et al., 2000). These ventralized neuroepithelial cells, however, can take on a spinal cord motoneuronal fate when treated with a caudalizing factor, retinoic acid (Wichterle et al., 2002). Since we know little about the transient neuronal progenitor population, current practice is to partially induce the neuroepithelial cells toward certain fates such as dopamine neurons or motoneurons to bias the fate choice before transplantation. Alternatively, specialized neuronal progenitors may be enriched or purified before transplantation using genetic means (Li et al., 1998; Kim et al., 2002; Wichterle et al., 2002,). Not only the neurotransmitter phenotype, but also the positional identity of a neuron type are critical in order to achieve a functional connection with target cells. For example, mesencephalic dopamine neurons can make functional synapses with striatal neurons after transplantation whereas dopamine neurons derived from hypothalamus or arcuate nucleus cannot make functional connections with striatal neurons (Abrous et al., 1988; Hudson et al., 1994). There will be a need for a variety of motoneurons of specific segmental identities to match a specific muscle. This puts emphasis on the need for understanding the patterning of neuronal types during development and of the development of strategies for directing the differentiation of neuronal cells with particular neurotransmitter and positional identities. In any case, specialized neuronal progenitors are ideal for replacing neurons in individual neurodegenerative diseases. ES cells are likely the best source because of their ability to renew themselves continuously and to generate a neuroepithelial intermediate that is plastic enough to give rise to a variety of neuronal types.

NEURONAL REPLACEMENT: THE NICHE

Success of neural transplantation depends not only on the donor cells but also on the niche in which the cells are placed. There are two environments for grafted cells: 1) a developing immature CNS in which neurogenesis is ongoing and 2) a mature stable CNS where neurogenesis has largely ceased with the exception of restricted brain regions. Both recipient environments have been exploited to investigate the differentiation potential of neural progenitors, functional integration of grafted cells, and therapeutic potential of neural precursors in individual neurological conditions (Figure 1B).

Fate of grafted neural precursor cells in the developing CNS

Transplantation into the embryonic brain allows exploration of stem cell participation during CNS development and investigation of how the survival, migration, differentiation, and integration of grafted neural precursors are regulated by extrinsic factors (Brustle, 1998). In this model, cells are injected into the cerebral ventricle of embryonic rat brain between E15 and E18. At this time, active neurogenesis is still taking place in many brain areas and the ventricular system is technically accessible (Brustle et al., 1995; Campbell et al., 1995; Fishell, 1995). Grafted neural progenitor cells, derived from various developing brain regions as well as from mouse ES cells, migrate to widespread areas of the brain and incorporate into host brain parenchyma. Furthermore, the grafted neural precursors appear to acquire region-specific neuronal phenotypes. For example, striatal neural precursors incorporate into the host cortex and acquire the morphology and axonal projection resembling those of cortical neurons, whereas they retain striatal neuronal identity when placed back into the striatum (Fishell, 1995). These observations suggest that their fate differentiation is largely influenced by position-specific cues. A similar observation using human neural stem cell transplantation into embryonic monkey brains has shown that grafted cells migrate along radial glia to temporally appropriate layers of the cortical plate and differentiate into lamina-appropriate neurons or glia (Ourednik et al., 2001). Hence, embryonic transplantation provides a useful model to examine the neural lineage potential of cells such as ES-derived cells and those from other species including human.

Transplantation into embryonic mouse or rat brains is limited to late embryonic stages due to the technical difficulty of manipulating early mammalian embryos. This hinders the delineation of the differentiation potential of neural stem/progenitors into a wide range of neurons, particularly those generated at an early stage of development. This difficulty can be overcome by the use of chick embryos. The developmental stages of chicks are well defined and the large size of these embryos are amenable to various manipulations. In particular, the primordial nervous system develops within 24 hours of embryonic life, which leaves a large embryonic window to examine the developmental potential of neural precursor cells. Using this system, Wichterle et al. (2002) have demonstrated that immature motoneurons generated from mouse ES cells survive, mature, and make connections with their target muscles following transplantation into the ventral neural tube. Such an exquisite demonstration would be difficult to achieve in mouse embryos at the stage of motoneuron development. Thus, the chick embryos offer a superb recipient to examine early neural lineage potential of stem cells and cells from other species (Goldstein et al., 2002).

While injection of cells into embryonic mouse or rat brain is still technically difficult, transplantation into neonatal mouse or rat brain is more achievable. With few exceptions (Englund et al., 2002), cells are injected into the cerebral ventricles of 1-2 day old mice or rats. Again, cells derived from various sources such as immortalized cell lines (Snyder et al., 1992; Englund et al., 2002), neural stem cells (Flax et al., 1998; Tamaki et al., 2002), or human ES-derived neuroepithelial cells (Zhang et al., 2001), can migrate and incorporate into both neurogenic and nonneurogenic regions. However, by comparison with transplantation into embryonic brains, neural precursors injected into the neonatal cerebral ventricles distribute largely to the subventricular and/or neurogenic areas. For example, human ES cell-derived neuroepithelial cells, following transplantation into cerebral ventricles, migrate and incorporate primarily into the hippocampus, rostral migratory stream, and olfactory bulb. To a lesser degree, some cells do incorporate into nonneurogenic cortical regions (Zhang et al., 2001). Human ES cell-derived neural epithelial cells, identified by prelabeling with green fluorescent protein, differentiate into both neurons and glia, although the glial differentiation occurs much later (unpublished data; Figure

2). This suggests that neurogenic signals are critical for migration and differentiation of grafted neural precursors.

Transplantation into embryonic or neonatal recipients provides a proof-of-concept model for examining the potential of neural precursor cells to participate in normal development. The extensive migration, and particularly the region-appropriate cell type differentiation, suggest that neural precursors do have multilineage potential and that their fate is largely dictated by the signals in a specific brain region at a particular development stage. In other words, neural precursor cells can recapitulate normal developmental processes if given an appropriate environment. This has clinical relevance in hereditary CNS disorders in which some neurogenic signals may be present to promote the incorporation of grafted neural progenitor cells. It also suggests that recreation of the

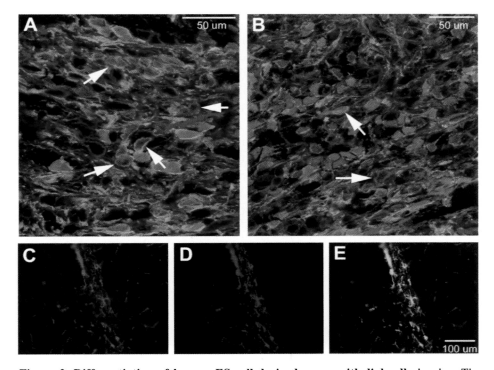

Figure 2. Differentiation of human ES cell-derived neuroepithelial cells *in vivo*. The neuroepithelial cells, transfected to express green fluorescent protein (GFP), were transplanted into newborn immunodeficient SCID mice through intraventricular injection. One month after implantation, many of the grafted GFP⁺ cells are positively labeled with β_{III}-tubulin (red in **A**) and fewer cells are labeled with MAP2 (red in **B**). Arrows indicate the GFP⁺ cells that colabel with β_{III}-tubulin in (**A**) or MAP2 (**B**). After 3 months, glial differentiation is observed as shown by the expression of GFAP (**D**) in some of the GFP-expressing cells (**C**). (**E**) is a merged image of (**C**) and (**D**).

neurogenic environment will be essential in order for neural stem cells to adopt a desired neuronal fate in mature brain and spinal cord.

Fate of grafted neural precursor cells in the adult CNS

Differentiation in neurogenic regions of the adult CNS

In contrast to the embryonic brain, the adult CNS environment lacks the developmental cues for directing region-specific cell migration and neuronal differentiation with the exception of neurogenic areas such as the subventricular zone and hippocampal dentate gyrus. Hence, multipotent neural stem/progenitor cells implanted into the subventricular zone can migrate along the rostral migratory stream to the olfactory bulb and differentiate into interneurons (Fricker et al., 1999; Yang et al., 2000). Similarly, adult hippocampal progenitors acquire local neuronal identities and express the neurotransmitter phenotypes found in the olfactory bulb when implanted into the neurogenic adult rat subventricular zone or rostral migratory stream, but they fail to produce neurons when transplanted into the nonneurogenic cerebellum (Suhonen et al., 1996). Implantation of neurosphere cells (neural stem/progenitor cells) into the nonneurogenic adult rat striatum or substantial nigra results in differentiation of transplanted stem cells primarily into astrocytes (Rubio et al., 2000) with only a fraction becoming neurons (Winkler et al., 1998; Rosser et al., 1997, Cao et al., 2001, Chow et al., 2000). These observations corroborate findings from developmental studies that neural stem/progenitor cells need an appropriate niche to develop into region-specific neuronal types. While the normal adult CNS lacks such a developmental cue, an injured CNS may recapitulate, at least partially, the developmental process to facilitate repair. Presently, it is not known to what degree and for how long the regenerative signals are present following injury or with disease. Hence, for neuronal replacement therapy in an adult environment, it may be preferable to have different types of neural progenitor cells for specific neurological conditions.

Neural transplantation into injured and degenerated adult CNS

Adult CNS conditions that are potential candidates for transplant therapy are: (1) those that affect a specific CNS region such as stroke or spinal cord injury, destroying many cell types in anatomically restricted areas, or (2) those conditions that affect a specific population of neuronal cells in a restricted brain region, such as Parkinson's Disease or Huntington's Disease, destroying certain tracts or nuclei with specific neuroanatomical and neurochemical characteristics (Rossi & Cattaneo, 2002). Diseases that affect diffuse areas of the CNS such as Alzheimer's Disease involve a neurodegenerative process, that affects many cell types in many regions. In the latter case, it would be difficult to restore the anatomical and functional network, comprised of many neuronal types connecting to each other in unique patterns through neural transplantation. In these instances, it would seem more pragmatic to protect native cells through trophic support or to encourage endogenous stem cells to migrate and replace the diseased cells (See Chapter 12 in this volume).

Transplantation into confined CNS areas of damage

Damage to the CNS in a confined region is most often caused by trauma to the spinal cord or brain, or ischemia due to thromboembolism or hypoperfusion/cardiac arrest. Cells in the affected territory, including neurons and glial cells, undergo necrosis and apoptosis. Beyond the primary attack, there is often an inflammatory reaction, which exacerbates the damage by expanding the injury to surrounding areas. Neural transplantation has been applied to such a regional condition in animal models of ischemia and spinal cord injury. Two ischemic animal models have been useful in exploring various restorative strategies: (1) the global hypoxic-ischemic model causing selective neuronal death in the hippocampal CA1 region due to temporary heart arrest and (2) focal ischemia in the neocortex and lateral striatum created by middle cerebral artery occlusion. Transplantation of neural progenitor cells from fetal rat striatum (Borlongan et al., 1998), adult hippocampus (Toda et al., 2001), or human carcinoma cell lines (Borlongan et al., 1998; Barami et al., 2001) into these animal models has shown that the grafted cells survive and that some cells differentiate into cells expressing neuronal markers. Some of the behavioral deficits are also alleviated or reversed in the grafted animals. Similarly, injection of human bone marrow stromal cells intravenously into ischemic rats results in behavioral recovery with a small population of grafted cells in the brain expressing neuronal makers (Chen et al., 2002). However, there is little evidence to support restoration of connectivity or functional integration of implanted cells into the host's circuitry. Some of the effects appear to be achieved via protective/trophic roles of grafted cells.

Spinal cord injury is another devastating neurological condition without an effective therapy. It can be largely mimicked in rats by a computerized weight drop system to create a contusion injury to the spinal cord (Basso et al., 1995). Transplantation of neural progenitor cells into the lesion leads to a certain degree of functional recovery (McDonald et al., 1999; Coumans et al., 2001; Ogawa et al., 2002). Histological analyses reveal little evidence of reconnection of the severed axons in the lesion by grafted cells. Instead, the functional recovery seems attributed to trophic support for the survival of injured cells and a substratum effect on the sprouting and outgrowth of endogenous axons. Another functional contribution is likely attributed to remyelination of naked but still connected axons by grafted cells surrounding the lesion (McDonald et al., 1999; Liu et al., 2000).

Given the damage to multiple cell types in these regional neurological conditions, cells with multiple lineage differentiation potential, such as neural stem cells, would be ideal donor cells for transplantation. The hope is that the injured CNS environment will reactivate some of the developmental programs and that these multipotent cells will differentiate into the missing neuronal and glial cells and reconstitute the injured tissue. Available evidence indicates that the environment remains largely gliogenic and only a small fraction of cells differentiate into neuronal elements (McDonald et al., 1999; Cao et al., 2001). This would suggest that even in the regional condition that affects multiple cell types, a combination of committed neurons (e.g., motoneurons or serotonergic neurons) and oligodendrocytes instead of undifferentiated neural stem cells, might be more desirable. In addition, pattern formation involves sequential and coordinated placement and wiring of specific cell types during development. Thus, a scaffold placed in the regional condition might be helpful in the reconstruction of the neural tissue (Park et al., 2002).

Reinnervation of the target and reconstruction of the neural circuit

Diseases that affect a specific cell lineage and/or pathway in a confined brain region are top candidates for cell replacement therapy. In Parkinson's Disease (PD), a special group of neurons in the substantia nigra of the midbrain degenerate. These neurons project to and make connections with neurons in the striatum and release a specific neurotransmitter, dopamine. Hence, replacement of cells that produce dopamine could conceivably correct the locomotive dysfunction caused by lack of dopamine. Indeed, transplantation of immature dopamine-producing neurons that are derived from embryonic tissues (Perlow et al., 1979; Bjorklund & Stenevi, 1979) or generated from mouse ES cells (Kim et al., 2002) into the striatum of PD rat models results in release of dopamine from the grafted cells with concomitant locomotive functional recovery (Bjorklund & Lindvall, 2000). Clinical trials with the transplantation of immature dopamine neurons from fetal human midbrain tissues lead to the relief of motor symptoms in some patients (Lindvall et al., 1994; Piccini et al., 1999; Freed et al., 2001).

Huntington's Disease (HD) is an autosomal dominant neurodegenerative condition that results in the preferential loss of the GABAergic projecting medium-sized spiny neurons in the striatum (Li, 1999). Transplantation of GABAergic neurons into the striatum appears to restore some neural function in both animal models and patients (Isacson et al., 1984; Freeman et al., 2000; Bachoud-Levi et al., 2000). In both PD and HD, cells that contribute to functional recovery are immature dopamine- or GABA-producing neurons. They are committed to the dopamine or GABA neuronal fate yet they are immature and ideal for transplantation. After grafting, they mature according their own program, independent of host environment. If, however, multipotential neural stem cells are transplanted into the striatum of PD or HD animals, the grafted cells differentiate into mainly astrocytes with few neurons and scarce dopamine neurons (Svendsen et al., 1997; Ostenfeld et al., 2001). Thus, for neuronal replacement therapy in neurodegenerative diseases that affect a particular population of cells, lineage committed neuronal progenitors are the ideal donor.

Another neurodegenerative disease, amyotrophic lateral sclerosis (ALS, or Lou Gehrig's disease), also affects a specific group of cells, i.e., motoneurons. However, in ALS, motoneurons throughout the brain and spinal cord are damaged, although to varying degrees. This would require a multiple transplant strategy or at least long-distance migration of grafted cells into the right places and differentiation into the right types of motoneurons. Furthermore motoneurons in the spinal cord have to penetrate the CNS/PNS boundary, travel a long distance, match the muscle cells, and innervate the atrophied muscle cells. Thus, a series of hurdles need to be overcome in order to make cell replacement therapy a realistic goal for ALS. Progress has been made in the induction and purification of immature motoneurons from mouse embryonic stem cells (Wichterle et al., 2002). These ES cell-generated motoneurons can indeed innervate muscle cells following transplantation into early embryonic chick neural tube. Key issues are the modification of the CNS/PNS boundary to be permissive to the penetration of axons, recreation of a permissive axonal pathway, and reactivation of the atrophied muscle cells. For clinical application, the generation of human motoneurons is a critical step.

Homotopic vs. heterotopic graft

Cells in the nervous system communicate with each other through direct and often distant synaptic contact. Hence, degeneration of a group of neurons in one brain region results in denervation of target cells in another area. In Parkinson's Disease (PD), for example, degeneration of dopamine neurons in the substantia nigra of the midbrain leads to denervation of neurons in the striatum. Neuronal replacement therapy aims to reinnervate the target and/or to reconstruct the neuronal circuitry both anatomically and functionally. This raises the issue of where to place the cells and how to facilitate reconnections with targets. Ideally, cells should be placed in the region where they normally belong, the so-called homotopic graft. However, the adult CNS environment lacks the cues to guide long-distance axonal growth to the target. Hence, dopamine neurons grafted into the substantia nigra of a PD rat model fail to extend axons to and innervate striatal neurons. Another problem associated with homotopic transplantation is that the transplanted cells could become the target of neurodegenerative assault. To bypass this problem, cells can be placed within, or very close to the target, the so-called heterotopic transplant. In the case of PD, cells are often transplanted into the striatum. In this way, dopamine positive fibers form a network with neurons in the striatum and functional recovery ensues through the release of diffusible neurotransmitters in a paracrine manner (Bjorklund & Lindvall, 2000). The major problem of ectopic transplant is that the grafted cells are not participants in the feedback neural circuitry. For example, dopamine neurons placed ectopically in the striatum lack some of the major afferent inputs that normally regulate the activity of dopamine neurons in the substantia nigra. Hence, although they innervate the target striatal neurons, the release of dopamine from the ectopically implanted cells is not regulated, as it is when the cells are in the midbrain. Nevertheless, there is some evidence to indicate that the intrastriatal dopamine neurons receive afferent connections from the host cerebral cortex and striatum (Fisher et al., 1991), suggesting partial integration of the grafted cells into the neuronal circuitry. This explains why the functional recovery is often incomplete.

In the case of Huntington's Disease (HD) cells can be directly implanted into the striatum (homotopic) without the concern that the disease will adversely affect the donor cells, as they would not contain the HD mutation. In a postmortem clinical study, neuronal protein aggregates of mutated huntingtin, typical of HD pathology, did not appear in the tissue resulting from human fetal cells implanted into a patient with HD (Freeman et al., 2000). Rodent striatal cells transplanted into a rodent HD model have been shown to survive, produce GABA, and improve neurologic function (Isacson et al., 1984, Campbell et al., 1993, Deckel et al., 1983). Human donor cells obtained from the ventricular eminence (Sanberg et al., 1997; Freeman et al., 2000) have also displayed connectivity with host brain and behavioral improvement in a rodent recipient. In primates, striatal allografts and xenografts survive and improve motor and cognitive deficits (Hantraye et al., 1992, Kendall et al., 1998, Palfi et al., 1998). Similar to PD, the grafted cells in the striatum need to reconnect the afferent fibers from the cortex and send projections to the globus pallidus to reestablish the cortico-striato-pallidal circuit.

Recreation of the niche

Discussion thus far suggests that a neurogenic niche needs to be recreated in order to achieve functional neuronal replacement by grafted progenitor cells.

It is generally believed that neurogenic cues may be at least partially reactivated after degeneration or injury to compensate for damage. This is suggested by the fact that the rate of neuronal production in the SVZ is elevated after cytotoxic treatment (Doetsch et al., 1999), ischemia (Jin et al., 2001), psychosocial stress, mechanical injury (Gould et al 1997; Gould and Tanapat, 1997), or seizures (Bengzon et al., 1997; Parent et al., 1997). These insults elicit the upregulation of FGF2 (Yoshimura et al., 2001), which may boost neurogenic processes. FGF2 knockout mice do not show increased neuronal production after seizures or ischemia (Yoshimura et al., 2001), suggesting that FGF2 is an important neurogenic factor in this process as it is during development. Nevertheless, transplantation of neural precursors into these injured environments leads to primarily glial differentiation, suggesting that the reactivated neurogenic program may be compromised by the concomitant and/or subsequent activation of the gliogenic program. There is presently a lack of understanding of how a neurogenic niche can be activated and maintained in the adult CNS.

Insights into how neurogenic activity is maintained in normal adult brain may help us modify the adult CNS niche to promote neuronal differentiation by grafted precursor cells. In the subventricular zone (SVZ), differentiation of neural stem cells may be partially controlled by noggin and bone morphogenetic protein (BMP). Noggin in the SVZ is thought to drive progenitor cell differentiation towards neuronal fates, whereas BMP, expressed by SVZ progenitors has gliogenic activity. BMP, but not noggin, is expressed in the adult striatum, leading to differentiation into glia of most transplanted cells. With noggin overexpression in the striatum, many donor cells acquire neuronal identities, but still retain features of olfactory bulb interneurons (Lim et al., 2000). In the rat hippocampus, SHH appears to be critical in maintaining the neural progenitor pool during adulthood (Lai et al., 2003).

Under a very special condition in which selective neuronal apoptosis is induced by chromophore-targeted laser photolysis without scar tissue formation, Magavi et al. (2000) observed regeneration of projection neurons in nonneurogenic cerebral cortex. Insights from this study may have relevance to endogenous repair after a synchronized apoptosis of multiple cells such as in ischemia, as demonstrated (Nakatomi et al., 2002).

FUNCTIONAL INTEGRATION INTO EXISTING CIRCUITRY

The goal of neuronal replacement is the functional integration of grafted cells into the existing circuitry and subsequent improvement in neurological function. In order to achieve this, neural stem/progenitor cells must migrate to (or be placed in) the appropriate position, differentiate into region-specific neuronal types, project axons to and make synaptic connections with the right targets, as well as receive inputs from presynaptic elements of the circuitry. As discussed above, grafted neural precursors can indeed differentiate into neurons appropriate to the brain region, especially in the developing brain or neurogenic regions of the adult brain. To confirm that the graft-derived neurons are functional, several parameters are often measured, including retrograde tracing, electrophysiological recording, and ultrastructural evidence of synapses between grafted cells and host neurons. Transplanted neuronal progenitor cells from embryonic rat tissues into the cerebral ventricles of E19 rats migrate and differentiate into neurons. Electrophysiological analysis indicates that some of the graft-derived neurons can generate an action potential (Auerbach et al., 2000), suggesting that the implanted neurons are functional. In another study, immortalized neural progenitor cells (cell line RN33B), following transplantation into neonatal rat cortex and hippocampus, differentiate into neurons that are morphologically similar to host cells in the regions. Retrograde tracing analysis indicates that the grafted neurons in the cortex make connection with thalamic neurons and those in the hippocampus form connections with neurons in the contralateral hippocampus. Electrophysiological recording of brain slices confirms that the grafted GFP-labeled neurons fire action potentials in response to evoked stimuli in the targets (Englund et al., 2002). This study demonstrates that grafted neurons can form functional connection with host neurons. Another way of detecting functional integration into neuronal circuitry is the expression of immediate early genes following neural stimulation. Implantation of neural stem cells derived from neonatal mouse cerebellum (C17.2 cell line) into embryonic mouse suprachiasmatic nucleus results in differentiation of grafted cells to neurons that produce vasopressin (characteristic of SCN neurons) and react to photic stimulation with expression of c-Fos protein. This regional and physiological appropriate response to stimulation suggests that exogenous stem cells can integrate in an appropriate fashion (Zlomanczuk et al., 2002).

Taken together, evidence for functional integration of grafted cells into adult brain is at present sketchy. This is probably due to the lack of neurogenic signals that promote the survival, differentiation, and synaptic formation. *In vitro*, retrovirally labeled neural progenitors from the hippocampus can migrate extensively, differentiate into neurons, and form functional synapses with cultured neonatal neurons or neurons in hippocampal slices (Song et al., 2002; van Praag et al., 2002). Recently, it has been shown that dopamine neurons generated from mouse ES cells form functional neurons following transplantation into the denervated striatum (Kim et al., 2002). Apparently, more studies with appropriate animal models are needed to address whether neurons grafted into the adult CNS can be integrated into functional neuronal circuitry.

NEURONAL REPLACEMENT: FROM ANIMALS TO HUMAN

Experimental studies with neural transplantation have demonstrated that implanted neurons or their precursors can survive and make connections with host neurons, and sometimes contribute to functional recovery in animal models of neurological injuries and degeneration. These studies have led to clinical trials of neural transplantation for Parkinson's Disease and Huntington's Disease (Bjoklund & Lindvall, 2000). However, there are still major issues to be solved before neural transplantation becomes a therapy for neurological injuries or degeneration. A reliable and renewable source of donor cells is necessary. The establishment of human ES cells (Thomson et al., 1998) and the application of stem cell technology brings optimism to this aspect. Major efforts are being applied to direct the stem cells to specialized neural fates. It has recently been shown that human ES cells can be directed to neuroectodermal cells that are able to give rise to a variety of neuronal and glial cell types (Zhang et al., 2001). Given that embryonic induction and lineage differentiation is a conserved process across species, it is likely that specialized neurons such as dopamine neurons and motoneurons can be induced from human stem cells, as has been shown in mouse ES cells. Another major issue with regard to the use of human ES cells is proof of safety. This may require removing undifferentiated cells or positively selecting for the desired cells before transplantation. In addition, recent success of homologous recombination in human ES cells (Zwaka & Thomson, 2003) moves stem cells one step closer to clinical application.

The survival and differentiation of grafted cells will require halting the disease process and reactivating developmental signals. Thus, recreation of a permissive environment in the host CNS is another challenge to the success of neural transplantation therapy. This has not been studied in detail. Similarly, further studies are needed to explore the functional integration of grafted neurons.

Acknowledgements

Studies in our laboratory have been supported by the NIH (RR016588-01, NSO 45926-01), Michael J. Fox Foundation, National ALS Association, and Myelin Project.

REFERENCES

Abrous N, Guy J, Vigny A, Calas A, Le Moal M, Herman JP (1988) Development of intracerebral dopaminergic grafts: a combined immunohistochemical and autoradiographic study of its time course and environmental influences. J. Comp Neurol 273:26-41.

Armstrong RJ, Hurelbrink CB, Tyers P, Ratcliffe EL, Richards A, Dunnett SB, Rosser AE, Barker RA (2002) The potential for circuit reconstruction by expanded neural precursor cells explored through porcine xenografts in a rat model of Parkinson's disease. Exp Neurol 175: 98-111.

Auerbach JM, Eiden MV, McKay RD (2000) Transplanted CNS stem cells form functional synapses *in vivo*. Eur J Neurosci 12:1696-704.

Bachoud-Levi A-C, Remy P, Nguyen J-P, Brugieres P, Lefaucheur J-P, Bourdet C, Baudic S, Gaura V, Maison P, Haddad B, Boisse M-F, Grandmougin T, Jeny R, Bartolomeo P, Barba GD, Degos J-D, Lisovoski F, Ergis A-M, pailhous E, Cesaro P, Hantraye P, Peschanski M (2000) Motor and cognitive improvements in patients with Huntington's disease after neural transplantation. Lancet 9:1975-1979.

Bain G, Kitchens D, Yao M, Huettner JE, Gottlieb DI (1995) Embryonic stem cells express neuronal properties *in vitro*. Dev Biol 168:342-357.

Barami K, Hao HN, Lotoczky GA, Diaz FG, Lyman WD (2001) Transplantation of human fetal brain cells into ischemic lesions of adult gerbil hippocampus. J Neurosurg 95:308-15.

Basso DM, Beattie MS, Bresnahan JC (1995) A sensitive and reliable locomotor rating scale for open field testing in rats. J Neurotrauma 12:1-21.

Bengzon J, Kokaia Z, Elmer E, Nanobashvili A, Kokaia M, Lindvall O (1997) Apoptosis and proliferation of dentate gyrus neurons after single and intermittent limbic seizures. Proc Natl Acad Sci USA 94:10432-10437.

Benninger Y, Marino S, Hardegger R, Weissmann C, Aguzzi A, Brandner S (2000) Differentiation and histological analysis of embryonic stem cell-derived neural transplants in mice. Brain Pathol 10:330-341.

Billon N, Jolicoeur C, Ying QL, Smith A, Raff M (2002) Normal timing of oligodendrocyte development from genetically engineered, lineage-selectable mouse ES cells. J Cell Sci 115:3657-3665.

Bjorklund A, Stenevi U (1979) Reconstruction of the nigrostriatal dopamine pathway by intracerebral nigral transplants. Brain Res 177:555-560.

Bjorklund A, Lindvall O (2000) Cell replacement therapies for central nervous system disorders. Nat Neurosci 3:537-544.

Bjorklund LM, Sanchez-Pernaute R, Chung S, Andersson T, Chen IY, McNaught KS, Brownell AL, Jenkins BG, Wahlestedt C, Kim KS, Isacson O (2002) Embryonic stem cells develop into functional dopaminergic neurons after transplantation in a Parkinson rat model. Proc Natl Acad Sci USA 99:2344-2349.

Borlongan CV, Tajima Y, Trojanowski JQ, Lee VM, Sanberg PR (1998) Cerebral ischemia and CNS transplantation: differential effects of grafted fetal rat striatal cells and human neurons derived from a clonal cell line. NeuroReport 9:3703-3709.

Borlongan CV, Tajima Y, Trojanowski JQ, Lee VM, Sanberg PR (1998) Transplantation of cryopreserved human embryonal carcinoma-derived neurons (NT2N cells) promotes functional recovery in ischemic rats. Exp Neurol 149:310-321.

Briscoe J, Sussel L, Serup P, Hartigan-O'Connor D, Jessell TM, Rubenstein JL, Ericson J (1999) Homeobox gene Nkx2.2 and specification of neuronal identity by graded Sonic hedgehog signaling. Nature 398:622-627.

Brustle O, Maskos U, McKay RD (1995) Host-guided migration allows targeted introduction of neurons into the embryonic brain. Neuron 15:1275-1285.

Brustle O, Choudhary K, Karram K, Huttner A, Murray K, Dubois-Dalcq M, McKay RD (1998) Chimeric brains generated by intraventricular transplantation of fetal human brain cells into embryonic rats. Nat Biotech 16:1040-1044.

Brustle O (1999) Building brains: neural chimeras in the study of nervous system development and repair. Brain Pathol 9:527-545.

Campbell K, Kalen P, Wictorin K, Lundberg C, Mandel RJ, Bjorklund A (1993) Characterization of GABA release from intrastriatal striatal transplants: dependence on host-derived afferents. Neurosci 53:403-415.

Campbell K, Olsson M, Bjorklund A (1995) Regional incorporation and site-specific differentiation of striatal precursors transplanted to the embryonic forebrain ventricle. Neuron 15:1259-1273.

Cao QL, Zhang YP, Howard RM, Walters WM, Tsoulfas P, Whittemore SR (2001) Pluripotent stem cells engrafted into the normal or lesioned adult rat spinal cord are restricted to a glial lineage. Exp Neurol 167:48-58.

Chen YL, Chen XG, Wang L, Gautam SC, Xu YX, Katakowski M, Zhang LJ, Lu M, Janakiraman N, Chopp M (2002) Human marrow stromal cell therapy for stroke in rat: Neurotrophins and functional recovery. Neurol 59:514-523.

Chow SY, Moul J, Tobias CA, Himes BT, Liu Y, Obrocka M, Hodge L, Tessler A, Fischer I (2000) Characterization and intraspinal grafting of EGF/bFGF-dependent neurospheres derived from embryonic rat spinal cord. Brain Res 874:87-106.

Coumans JV, Lin TT, Dai HN, MacArthur L, McAtee M,, Nash C, Bregman BS (2001) Axonal regeneration and functional recovery after complete spinal cord transection in rats by delayed treatment with transplants and neurotrophins. J Neurosci 21:9334-9344.

Davidson BL, Allen ED, Kozarsky KF, Wilson JM, Roessler BJ (1993) A model system for *in vivo* gene transfer into the central nervous system using an adenoviral vector. Nat Genet 3:219-223.

Deacon T, Schumacker J, Dinsmore J, Thomas C, Palmer P, Kott S, Edge A, Penney D, Kassissieh S, Dempsey P, Isacson O (1997) Histological evidence of fetal pig neural cell survival after transplantation into a patient with Parkinson's disease. Nat Med 3:350-353.

Deckel AW, Robinson RG, Coyle JT, Sanberg PR (1983) Reversal of long-term locomotor abnormalities in the kainic acid model of Huntington's disease by day 18 fetal striatal implants. Eur J Pharmacol 93:287-288.

Doetsch F, Caille I, Lim DA, Garcia-Verdugo JM, Alvarez-Buylla A (1999) Subventricular zone astrocytes are neural stem cells in the adult mammalian brain. Cell 97:703-716.

Englund U, Bjorklund A, Wictorin K, Lindvall O, Kokaia M (2002) Grafted neural stem cells develop into functional pyramidal neurons and integrate into host cortical circuitry. Proc Natl Acad Sci USA 99:17089-17094.

Evans MJ, Kaufman MH (1981) Establishment in culture of pluripotential cells from mouse embryos. Nature 292:154-156.

Finley, M.E, Kulkarni, N, Huettner, JE (1996) Synapse formation and establishment of neuronal polarity by P19 embryonic carcinoma cells and embryonic stem cells. J. Neurosci 16:1056-1065.

Fishell G (1995) Striatal precursors adopt cortical identities in response to local cue. Development 121:803-812.

Fisher LJ, Young SJ, Tepper JM, Groves PM, Gage FH (1991) Electrophysiological characteristics of cells within mesencephalon suspension grafts. Neurosci 40:109-122.

Flax JD, Aurora S, Yang C, Simonin C, Wills AM, Billinghurst LL, Jendoubi M, Sidman RL, Wolfe JH, Kim SU, Snyder EY (1998) Engraftable human neural stem cells respond to developmental cues, replace neurons, and express foreign genes. Nat Biotech 16:1033-1039.

Freed CR (2002) Will embryonic stem cells be a useful source of dopamine neurons for transplant into patients with Parkinson's disease? Proc Natl Adad Sci USA 99:1755-1757.

Freed CR, Greene PE, Breeze RE, Tsai WY, DuMouchel W, Kao R, Dillon S, Winfield H, Culver S, Trojanowski JQ, Eidelberg D, Fahn S (2001) Transplantation of embryonic dopamine neurons for severe Parkinson's disease. N Engl J Med 344:710-719.

Freeman TB, Hauser RA, Sanberg PR, Saporta S (2000) Neural transplantation for the treatment of Huntington's disease. Prog Brain Res 127:405-411.

Freeman TB, Cicchetti F, Hauser RA, Deacon TW, Li XJ, Hersch SM, Nauert GM, Sanberg PR, Kordower JH, Saporta S, Isacson O (2000) Transplanted fetal striatum in Huntington's disease: phenotypic development and lack of pathology. Proc Natl Acad Sci USA 97:13877-13882.

Fricker RA, Carpenter MK, Winkler C, Greco C, Gates MA, Bjorklund A (1999) Site-specific migration and neuronal differentiation of human neural progenitor cells after transplantation in the adult rat brain. J Neurosci 19:5990-6005.

Gage FH, Kempermann G, Palmer TD, Peterson DA, Ray J (1998) Multipotent progenitor cells in the adult dentate gyrus. J Neurobiol 36:249-66.

Goldstein RS, Drukker M, Reubinoff BE, Benvenisty N (2002) Integration and differentiation of human embryonic stem cells transplanted to the chick embryo. Dev Dyn 225:80-86.

Gould E, McEwen BS, Tanapat P, Galea LA, Fuchs E (1997) Neurogenesis in the dentate gyrus of the adult tree shrew is regulated by psychosocial stress and NMDA receptor activation. J Neurosci 17:2492-2498.

Gould E, Tanapat P (1997) Lesion-induced proliferation of neuronal progenitors in the dentate gyrus of the adult rat. Neuroscience 80:427-436.

Han SS, Kang DY, Mujtaba T, Rao MS, Fischer I (2002) Grafted lineage-restricted precursors differentiate exclusively into neurons in the adult spinal cord. Exp Neurol 177:360-375.

Hantraye P, Riche D, Maziere M, Isacson O (1992) Intrastriatal transplantation of cross-species fetal striatal cells reduces abnormal movements in a primate model of Huntington disease. Proc Natl Acad Sci USA 89:4187-4191.

Herrera DG, Garcia-Verdugo JM, Alvarez-Buylla A (1999) Adult-derived neural precursors transplanted into multiple regions in the adult brain. Ann Neurol 46:867-877.

Hida H, Hashimoto M, Fujimoto I, Nakajima K, Shimano Y, Nagatsu T, Mikoshiba K, Nishino H (1999) Dopa-producing astrocytes generated by adenoviral transduction of human tyrosine hydroxylase gene: *in vitro* study and transplantation to hemiparkinsonian model rats. Neurosci Res 35:101-112.

Hitoshi S, Tropepe V, Ekker M, van der Kooy D (2002) Neural stem cell lineages are regionally specified, but not committed, within distinct compartments of the developing brain. Development 129:233-244.

Hudson JL, Bickford P, Johansson M, Hoffer BJ, Stromberg I (1994) Target and neurotransmitter specificity of fetal central nervous system transplants: importance for functional reinnervation. J Neurosci 14:283-290.

Issacson O, Brundin P, Kelly PA, Gage FH, Bjorklund A (1984) Functional neuronal replacement by grafted striatal neurons in the ibotenic acid-lesioned rat striatum. Nature 311:458-460.

Jessell TM (2000) Neuronal specification in the spinal cord: inductive signals and transcriptional codes. Nat Rev Genet 1:20-29.

Jiang Y, Jahagirdar BN, Reinhardt RL, Schwartz RE, Keene CD, Ortiz-Gonzalez XR, Reyes M, Lenvik T, Lund T, Blackstad M, Du J, Aldrich S, Lisberg A, Low WC, Largaespada DA, Verfaillie CM (2002) Pluripotency of mesenchymal stem cells derived from adult marrow. Nature 418:41-49.

Jin K, Minami M, Lan JQ, Mao XO, Batteur S, Simon RP, Greenberg DA (2001) Neurogenesis in dentate subgranular zone and rostral subventricular zone after focal cerebral ischemia in the rat. Proc Natl Acad Sci USA 98:4710-4715.

Jori FP, Galderisi U, Piegari E, Cippollaro M, Cascino A, Peluso G, Cotrufo R, Giordana A, Melone MAB (2003) EGF-responsive rat neural stem cells: molecular follow-up of neuron and astrocyte differentiation *in vitro*. J Cell Physiol 195:220-233.

Kalyani AJ, Piper D, Mujtaba T, Lucero MT, Rao MS (1998) Spinal cord neuronal precursors generate multiple neuronal phenotypes in culture. J Neurosci 18:7856-7868.

Kendall AL, Rayment FD, Torres EM, Baker HF, Ridley RM, Dunnett SB (1998) Functional integration of striatal allografts in a primate model of Huntington's disease. Nat Med 4:727-729.

Kim JH, Auerbach JM, Rodriguez-Gomez JA, Velasco I, Gavin D, Lumelsky N, Lee SH, Nguyen J, Sanchez-Pernaute R, Bankiewicz K, McKay R (2002) Dopamine neurons derived from embryonic stem cells function in an animal model of Parkinson's disease. Nature 418:50-56.

Kukekov VG, Laywell ED, Suslov O, Davies K, Scheffler B, Thomas LB, O'Brien TF, Kusakabe M, Steindler DA (1999) Multipotent stem/progenitor cells with similar properties arise from two neurogenic regions of adult human brain. Exp Neurol 156:333-344.

Lai K, Kaspar BK, Gage FH, Schaffer DV (2003) Sonic hedgehog regulates adult neural progenitor proliferation *in vitro* and *in vivo*. Nat Neurosci 6:21-27.

Laywell ED, Rakic P, Kukekov VG, Holland EC, Steindler DA (2000) Identification of a multipotent astrocytic stem cell in the immature and adult mouse brain. Proc Natl Acad Sci USA 97:13883-13888.

Li, M, Pevny L, Lovell-Badge R, Smith A (1998) Generation of purified neural precursors from embryonic stem cells by lineage selection. Curr Biol 8:971-977.

Li XJ (1999) The early cellular pathology of Huntington's disease. Mol Neurobiol 20:111-124.

Lee SH, Lumelsky N, Studer L, Auerbach JM, McKay RD (2000) Efficient generation of midbrain and hindbrain neurons from mouse embryonic stem cells. Nat Biotech 18:675-679.

Lim DA, Fishell GJ, Alvarez-Buylla A (1997) Postnatal mouse subventricular zone neuronal precursors can migrate and differentiate within multiple levels of the developing neuraxis. Proc Natl Acad Sci USA 94:14832-14836.

Lim DA, Tramontin AD, Trevejo JM, Herrera DG, Garcia-Verdugo JM, Alvarez-Buylla A (2000) Noggin antagonizes BMP signaling to create a niche for adult neurogenesis. Neuron 28:713-726.

Lindvall O, Sawle G, Widner H, Rothwell JC, Bjorklund A, Brooks D, Brundin P, Frackowiak R, Marsden CD, Odin P, et al (1994) Evidence for long-term survival and function of dopaminergic grafts in progressive Parkinson's disease. Ann Neurol 35:172-180.

Liu S, Qu Y, Stewart TJ, Howard MJ, Chakrabortty S, Holekamp TF, McDonald JW (2000) Embryonic stem cells differentiate into oligodendrocytes and myelinate in culture and after spinal cord transplantation. Proc Natl Acad Sci USA 97:6126-6131.

Magavi SS, Leavitt BR, Macklis JD (2000) Induction of neurogenesis in the neocortex of adult mice. Nature 405:951-955.

Magavi SS, Macklis JD (2001) Manipulation of neural precursors in situ: induction of neurogenesis in the neocortex of adult mice. Neuropsychopharm 25: 816-835.

Mayer-Proschel M, Kalyani AJ, Mujtaba T, Rao MS (1997) Isolation of lineage-restricted neuronal precursors from multipotent neuroepithelial stem cells. Neuron 19:773-785.

McDonald JW, Liu Z, Qu Y, Liu S, Mickey SK, Turetsky D, Gottlieb DI, Choi DW (1999) Transplanted embryonic stem cells survive, differentiate and promote recovery in injured rat spinal cord. Nat Med 5:1410-1412.

Morshead CM, Benveniste P, Iscove NN, van der Kooy D (2002) Hematopoietic competence is a rare property of neural stem cells that may depend on genetic and epigenetic alterations. Nat Med 8:268-273.

Munsie MJ, Michalska AE, O'Brien CM, Trounson AO, Pera MF, Mountford PS (2000) Isolation of pluripotent embryonic stem cells from reprogrammed adult mouse somatic cell nuclei. Curr Biol 10:989-992.

Nakatomi H, Kuriu T, Okabe S, Yamamoto S, Hatano O, Kawahara N, Tamura A, Kirino T, Nakafuku M (2002) Regeneration of hippocampal pyramidal neurons after ischemic brain injury by recruitment of endogenous neural progenitors. Cell 110:429-441.

Nottebohm F (2002) Neuronal replacement in adult brain. Brain Res Bull 57:737-749.

Novitch BG, Chen AI, Jessell TM (2001) Coordinate regulation of motor neuron subtype identity and pan-neuronal properties by the bHLH repressor Olig2. Neuron 31:773-789.

Ogawa Y, Sawamoto K, Miyata T, Miyao S, Watanabe M, Nakamura M, Bregman BS, Koike M, Uchiyama Y, Toyama Y, Okano H (2002) Transplantation of *in vitro*-expanded fetal neural progenitor cells results in neurogenesis and functional recovery after spinal cord contusion injury in adult rats. J Neurosci Res 69:925-933.

Okabe S, Forsberg-Nilsson K, Spiro AC, Segal M, McKay RDG (1996) Development of neuronal precursor cells and functional postmitotic neurons from embryonic stem cells *in vitro*. Mech Dev 59:89-102.

Olsson M, Campbell K, Turnbull DH (1997) Specification of mouse telencephalic and mid-hindbrain progenitors following heterotopic ultrasound-guided embryonic transplantation. Neuron 19:761-772.

Ostenfeld T, Joly E, Tai YT, Peters A, Caldwell M, Jauniaux E, Svendsen CN (2002) Regional specification of rodent and human neurospheres. Dev Brain Res 134:43-55.

Ourednik V, Ourednik J, Flax JD, Zawada WM, Hutt C, Yang C, Park KI, Kim SU, Sidman RL, Freed CR, Snyder EY (2001) Segregation of human neural stem cells in the developing primate forebrain. Science 293:1820-1824.

Palfi S, Conde F, Riche D, Brouillet E, Dautry C, Mittoux V, Chibois A, Peschanski M, Hantraye P (1998) Fetal striatal allografts reverse cognitive deficits in a primate model of Huntington's disease. Nat Med 4:963-966.

Palmer TD, Takahashi J, Gage FH (1997) The adult rat hippocampus contains primordial neural stem cells. Mol Cell Neurosci 8:389-404.

Parent JM, Yu TW, Leibowitz RT, Geschwind DH, Sloviter RS, Lowenstein DH (1997) Dentate granule cell neurogenesis is increased by seizures and contributes to aberrant network reorganization in the adult rat hippocampus. J Neurosci 17:3727-3738.

Park KI, Teng YD, Snyder EY (2002) The injured brain interacts reciprocally with neural stem cells supported by scaffolds to reconstitute lost tissue. Nat Biotech 20:1111-1117.

Perlow MJ, Freed WJ, Hoffer BJ, Seiger A, Olson L, Wyatt RJ (1979) Brain grafts reduce motor abnormalities produced by destruction of nigrostriatal dopamine system. Science 204:643-647.

Piccini P, Brooks DJ, Bjorklund A, Gunn RN, Grasby PM, Rimoldi O, Brundin P, Hagell P, Rehncrona S, Widner H, Lindvall O (1999) Dopamine release from nigral transplants visualized *in vivo* in a Parkinson's patient. Nat Neurosci 2:1137-1140.

Qian X, Shen Q, Goderie SK, He W, Capela A, Davis AA, Temple S (2000) Timing of CNS cell generation: a programmed sequence of neuron and glial cell production from isolated murine cortical stem cells. Neuron 28:69-80.

Reynolds BA, Tetzlaff W, Weiss S (1992) A multipotent EGF-responsive striatal embryonic progenitor cell produces neurons and astrocytes. J Neurosci 12:4565-4574.

Richardson WD (2001) Hedgehog-dependent oligodendrocyte lineage specification in the telencephalon. Development 128:2545-2554.

Rosser AE, Tyers P, ter Borg M, Dunnett SB, Svendsen CN (1997) Co-expression of MAP-2 and GFAP in cells developing from rat EGF responsive precursor cells. Brain Res Dev Brain Res 98:291-295.

Rossi F, Cattaneo E (2002) Opinion: neural stem cell therapy for neurological diseases: dreams and reality. Nat Rev Neurosci 3:401-409.

Rubio FJ, Bueno C, Villa A, Navarro B, Martinez-Serrano A (2000) Genetically perpetuated human neural stem cells engraft and differentiate into the adult mammalian brain. Mol Cell Neurosci 16:1-13.

Sanberg PR, Borlongan CV, Koutouzis TK, Norgren RB Jr, Cahill DW, Freeman TB (1997) Human fetal striatal transplantation in an excitotoxic lesioned model of Huntington's disease. Ann NY Acad Sci 831:452-460.

Shihabuddin LS, Horner PJ, Ray J, Gage FH (2000) Adult spinal cord stem cells generate neurons after transplantation in the adult dentate gyrus. J Neurosci 20:8727-8735.

Snyder EY, Deitcher DL, Walsh C, Arnold-Aldea S, Hartwieg EA, Cepko CL (1992) Multipotent neural cell lines can engraft and participate in development of mouse cerebellum. Cell 68:33-55.

Song HJ, Stevens CF, Gage FH (2002) Neural stem cells from adult hippocampus develop essential properties of functional CNS neurons. Nat Neurosci 5:438-445.

Studer L, Tabar V, McKay RD (1998) Transplantation of expanded mesencephalic precursors leads to recovery in parkinsonian rats. Nat Neurosci 1:290-295.

Suhonen JO, Peterson DA, Ray J, Gage FH (1996) Differentiation of adult hippocampus-derived progenitors into olfactory neurons *in vivo*. Nature 383:624-627.

Svendsen CN, Caldwell MA, Shen J, ter Borg MG, Rosser AE, Tyers P, Karmiol S, Dunnett SB (1997) Long-term survival of human central nervous system progenitor cells transplanted into a rat model of Parkinson's disease. Exp Neurol 148:135-146.

Svendsen CN, Caldwell MA (2000) Neural stem cells in the developing central nervous system: implications for cell therapy through transplantation, in SB dunnett and A. Bjorklund (Eds.), Progress in Brain Research, Vol 127,13-34.

Takayama H, Ray J, Raymon HK, Baird A, Hogg J, Fisher LJ, Gage FH (1995) Basic fibroblast growth factor increases dopaminergic graft survival and function in a rat model of Parkinson's disease. Nat Med 1:53-58.

Tamaki S, Eckert K, He D, Sutton R, Doshe M, Jain G, Tushinski R, Reitsma M, Harris B, Tsukamoto A, Gage F, Weissman I, Uchida N (2002) Engraftment of sorted/expanded human central nervous system stem cells from fetal brain. J Neurosci Res 69:976-986.

Tekki-Kessaris N, Woodruff R, Hall AC, Gaffield W, Kimura S, Stiles CD, Rowitch DH,

Temple S (2001) The development of neural stem cells. Nature 414:112-117.

Toda H, Takahashi J, Iwakami N, Kimura T, Hoki S, Mozumi-Kitamura K, Ono S, Hashimoto N (2001) Grafting neural stem cells improved the impaired spatial recognition in ischemic rats. Neurosci Lett 316:9-12.

Toma JG, Akhavan M, Fernandes KJ, Barnabe-Heider F, Sadikot A, Kaplan DR, Miller FD (2001) Isolation of multipotent adult stem cells from the dermis of mammalian skin. Nat Cell Biol 3:778-784.

Thomson JA, Itskovitz-Eldor J, Shapiro SS, Waknitz MA, Swiergiel JJ, Marshall VS, Jones JM (1998) Embryonic stem cell lines derived from human blastocysts. Science 282:1145-1147.

Tropepe V, Craig CG, Morshead CM, van der Kooy D (1997) Transforming growth factor-alpha null and senescent mice show decreased neural progenitor cell proliferation in the forebrain subependyma. J Neurosci 17:7850-7859.

van Praag H, Schinder AF, Christie BR, Toni N, Palmer TD, Gage FH (2002) Functional neurogenesis in the adult hippocampus. Nature 415:1030-1034.

Vescovi AL, Snyder EY (1999) Establishment and properties of neural stem cell clones: plasticity *in vitro* and *in vivo*. Brain Pathol 9:569-598.

Wichterle H, Lieberam I, Porter JA, Jessell TM (2002) Directed differentiation of embryonic stem cells into motor neurons. Cell 110:385-397.

Wilmut I, Schnieke AE, McWhir J, Kind AJ, Campbell KH (1997) Viable offspring derived from fetal and adult mammalian cells. Nature 385:810-813.

Winkler C, Fricker RA, Gates MA, Olsson M, Hammang JP, Carpenter MK, Bjorklund A (1998) Incorporation and glial differentiation of mouse EGF-responsive neural progenitor cells after transplantation into the embryonic rat brain. Mol Cell Neurosci 11:99-116.

Yang H, Mujtaba T, Venkatraman G, Wu YY, Rao MS, Luskin MB (2000) Region-specific differentiation of neural tube-derived neuronal restricted progenitor cells after heterotopic transplantation. Proc Natl Acad Sci USA 97:13366-13371.

Yandava BD, Billinghurst LL, Snyder EY (1999) "Global" cell replacement is feasible via neural stem cell transplantation: evidence from the dysmyelinated shiverer mouse brain. Proc Natl Acad Sci USA 96:7029-7034.

Ying QL, Nichols J, Evans EP, Smith AG (2002) Changing potency by spontaneous fusion. Nature 416:545-548.

Yoshimura S, Takagi Y, Harada J, Teramoto T, Thomas SS, Waeber C, Bakowska JC, Breakefield XO, Moskowitz MA (2001) FGF-2 regulation of neurogenesis in adult hippocampus after brain injury. Proc Natl Acad Sci USA 98:5874-5479.

Zhang SC, Wernig M, Duncan ID, Brustle O, Thomson JA (2001) *In vitro* differentiation of transplantable neural precursors from human embryonic stem cells. Nat Biotech 19:1129-1133.

Zhou Q, Anderson DJ (2002) The bHLH transcription factors OLIG2 and OLIG1 couple neuronal and glial subtype specification. Cell 109:61-73.

Zlomanczuk P, Mrugala M, de la Iglesia HO, Ourednik V, Quesenberry PJ, Snyder EY, Schwartz WJ (2002) Transplanted clonal neural stem-like cells respond to remote photic stimulation following incorporation within the suprachiasmatic nucleus. Exp Neurol 174:162-168.

Zwaka TP, Thomson JA (2003) Homologous recombination in human embryonic stem cells. Nat Biotech 21:319-321.

Chapter 11

Derivation of Myelin-forming Cells for Transplantation Repair of the CNS

Ian D. Duncan and Yoichi Kondo

INTRODUCTION

There is great current interest in the feasibility of remyelinating the central nervous system (CNS) by transplantation of cells of the oligodendrocyte lineage, or other myelin-forming cells. For the last 20 years, glial cell transplantation has been extensively used to explore interactions between grafted and endogenous cells and how cells, transplanted as allografts or xenografts, are able to ensheath and myelinate foreign axons (Duncan et al., 1988). The extensive myelination achieved in many of these studies by the transplanted cells led to the consideration that this approach might be used therapeutically in human myelin disorders (Blakemore et al., 1996; Duncan et al., 1997; Blakemore and Franklin, 1999; Duncan, 2001).

A key consideration is the choice of cell to be used in transplantation that will generate sufficient oligodendrocytes or other myelin-forming cells for repair. We are now able to take cells from their primitive beginnings in the form of embryonic stem (ES) cells or neural stem cells from rodent sources and generate sufficient oligodendrocytes *in vitro* and *in vivo* that will myelinate large areas of the CNS of experimental animals. In this chapter we will briefly discuss the models which these cells are tested in, the choice of cells that might be used therapeutically, and the clinical disorders they might be used in. Finally we will discuss how to evaluate their success following transplantation in terms of structural repair and restoration of function.

ANIMAL MODELS

There are now a wide variety of models in which transplantation of myelin forming cells has been performed. In general, the model chosen depends upon the questions to be asked and the disease to be targeted. A major division of the models used is into 1) the myelin mutants where myelin is absent or

From: *Neural Stem Cells: Development and Transplantation*
Edited by: Jane E. Bottenstein © 2003 Kluwer Academic Publishers, Norwell, MA

markedly reduced, 2) or focal models of demyelination, where injection of myelinotoxic chemicals are used. In the case of the mutants, transplant studies have been performed on the jimpy (jp) (Lachapelle et al., 1990), quaking (qk) (Duncan et al., 1981), and shiverer (shi) (Gumpel et al., 1985; Gansmüller et al., 1991; Warrington et al., 1993; Mitome et al., 2001) mice and on the myelin deficient (md) (Duncan et al., 1988; Tontsch et al., 1994; Rosenbluth et al., 1990), Long Evans shaker (les) (Zhang et al., 2003) and taiep (Duncan and Zhang, unpublished data) rats. The shaking (sh) pup, a canine X-linked mutant, has provided an excellent model in which to devise strategies of allograft transplantation that can scale-up the size of repair, similar to what may be required in humans. As the sh pup lives well into adulthood, it provides a model in which long-term, adult transplants can be performed (Archer et al., 1997). In general terms, the greater the range of models that mimic a specific human disease, the better the chances for significant animal data relative to that disease. For example, the models of Krabbe's Disease extend from the twitcher (twi) mouse to a non-human primate and in the case of Pelizaeus Merzbacher Disease (PMD), from mice, to canine models.

To study focal acute lesions that may be associated with clinical deficits, injection of a myelinotoxic chemical such as lysophosphatidyl choline or ethidium bromide is the method of choice. Prior irradiation of the area is used to inhibit endogenous remyelination (Blakemore and Franklin, 1991; Blakemore et al., 1995b; Honmou et al., 1996). Areas of focal myelin loss can be created either in the spinal cord or brain. It is essential however that such lesions are well characterized prior to grafting cells and that there has not been a significant loss of axons thus allowing remyelination to occur. Such focal lesions are also useful in testing the ability of oligodendrocyte progenitors (OPCs) or other myelinating cells to migrate through normal neuropil to reach areas of myelin loss.

CHOICE OF DONOR CELLS

The choice of cells used to repair focal areas of myelin loss or more generalized regions of dysmyelination is complex and as yet, undecided. In general, the primary requirements of such cells are that they can migrate, divide in a controlled fashion and ensheath and myelinate axons. In addition, cells should interact favorably with other cells of the CNS and if possible be nonimmunogenic. Finally, cells should be readily accessible and producible in sufficient number without ethical or technical difficulties. The latter point may mean that cells need also be expanded in culture prior to transplantation.

Once cells with these criteria are selected, it should also be considered whether they can interact with axons in varying pathological milieus. Examples of this include the ability of the transplanted cells to migrate and myelinate axons in areas of astrocytic hypertrophy (gliosis) and their ability to survive and function in an inflammatory background. It is not known whether all candidate cell types will be able to ensheath chronically demyelinated or dysmyelinated axons; further experimental data will determine this.

At present, therefore, there is a wide variety of cell types that could fit the above description. The first cell to be considered is the endogenous myelinating cell of the CNS, the oligodendrocyte, or a cell at any stage along its developmental pathway that can be coaxed into differentiating into this cell. Consideration should also be given to whether stem cells or progenitors that give rise to oligodendrocytes can be generated from the adult CNS as well as from embryonic or neonatal sources. Secondly, Schwann cells which are known to be able to myelinate CNS axons and that are frequently seen in the spinal cord in a variety of disorders must be considered. A cell that may have properties of both Schwann cells and CNS cells, the olfactory ensheathing cell (OEC) is the final possible myelin producing cell of the nervous system. Finally, and most contentiously, are stem cells from other tissues that can be coaxed to become myelinating cells prior to transplantation. The issue of transdifferentiation of such stem cells remains a hotly debated subject (see Chapter 6 in this volume).

Oligodendrocyte lineage cells

It is axiomatic that the cell used to replace lost or dysfunctional oligodendrocytes will be the oligodendrocyte itself. As the cell that normally populates and interacts with all other cells of the CNS, it seems the obvious choice. In the developing CNS, oligodendrocytes arise from neuroepithelial precursors and differentiate through many stages *in vivo* (in a similar fashion to that detailed *in vitro*; Figure 1). The sites of origin of these cells have been vigorously debated, especially in the brain where more than one site has been identified (Olivier et al., 2001; Woodruff et al., 2001; Qi et al., 2002). In the spinal cord, the major site of origin of cells that give rise to oligodendrocytes is in the ventral cord, adjacent to the central canal (Pringle and Richardson, 1993; Richardson et al., 1997). While OPCs were originally identified *in vivo* by their labeling with probes to the platelet-derived growth factor-alpha receptor (PDGFαR), more recently an array of markers and transcription factors expressed early, and in some cases transiently, have been identified (Kessaris et al., 2001; Rowitch et al., 2002; Zhou et al., 2001; Takebayashi et al., 2002). There is not overall agreement on the time of expression of these markers or their uniqueness for cells of the oligodendrocyte lineage. Nonetheless, they will eventually provide a better understanding of the stages of development that occur from neural stem cells to the bona fide OPC.

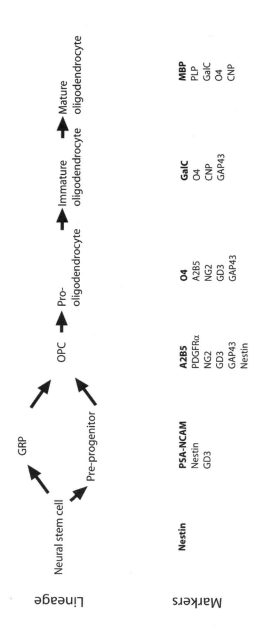

Figure 1: Progression through the oligodendrocyte lineage from the neural stem cell to mature oligodendrocyte. The developmental lineage (top) and markers of these stages (bottom) is based upon combined data from many labs and is still evolving. Neural stem cells mature and give rise to a cell known as a glial restricted precursor (GRP). These cells and oliodendrocytes progenitor cells (OPCs) then express a wide variety of markers including some transcription factors.

The key features of cells selected for brain repair are that the cells are migratory and mitotic. A number of general principles apply here. In regard to motility, as cells develop more processes they become less migratory and stop dividing as they associate, ensheath, and myelinate axons. Thus, it would seem that cells of the earlier lineage that are motile and mitotic would be most successful for transplantation. Indeed, while preparations containing predominantly mature oligodendrocytes can myelinate md rat axons (Duncan et al., 1992), other work suggests that OPCs are more effective (Warrington et al., 1993; Archer et al., 1997).

Initial studies on transplanting cell preparations used a mixed preparation of glia derived from the brain or spinal cord that contained OPCs and more mature cells (Blakemore and Crang, 1988; Duncan et al., 1988; Gumpel et al., 1983). Isolation of more purified collections of oligodendroglia have utilized the standard technique described by McCarthy and de Vellis (1980) or more recently used growth factor expansion to generate pure populations of progenitors or cell lines. One of the most utilized cell lines has been CG4 (central glial -4) produced by Louis et al. (1992). Using the isolation approach of McCarthy and de Vellis, they established pure cultures of oligodendrocytes that in the presence of medium conditioned by the B104 neuroblastoma cell line gave rise to a proliferative progenitor. It had been shown previously that B104 conditioned media maintained neonatal and adult OPCs as nondifferentiated, dividing cells, although the factors responsible for this have not been identified (Hunter and Bottenstein, 1990, 1991). CG4 cells remained proliferative and undifferentiated until removal of B104 medium whereupon 98% of the cells differentiate into oligodendrocytes (Louis et al., 1992). We and others have used the CG-4 cell line at early passage as a source of myelinating oligodendrocytes (Tontsch et al., 1994; Franklin et al., 1996a). However, at later passages (P20-25) it may lose its ability to myelinate *in vivo*, yet remain highly mitotic. Thus, a more reproducible means of creating a regular supply of such cells has been explored.

We examined the ability of neural stem cells grown as free floating collections of cells or neurospheres (Reynolds and Weiss, 1992) to produce oligodendrocytes. When the neurospheres were plated and growth factors withdrawn, they produced a minor percentage of oligodendrocytes compared to neurons and astrocytes. However, when these cells were transplanted into the md rat, extensive myelination ensued suggesting that the myelin deficient environment had promoted an oligodendrocyte differentiation (Hammang et al., 1997). Transplantation of neurospheres has also been described by others, and these neural stem cells gave rise to oligodendrocytes *in vivo* (Ader et al., 2000; Mitome et al., 2001).

However this source of cells relies on the differentiation *in vivo* of neural stem cells in oligodendrocytes after transplantation and may not be the most efficient method of preparing purified cells for repair. Baron-Van Evercooren and colleagues used a combination of the prior methods to derive purified, floating collections of OPCs derived from newborn rat cerebral hemispheres, grown in the presence of B104 medium, that they called oligospheres (Avellana-Adalid et al., 1996; Vitry et al., 1999). These cells proficiently myelinated axons on transplantation. We modified this technique using neurospheres as the source of cells. By gradually substituting EGF and FGF2 for B104 medium, we were able to produce highly purified collections of OPCs that myelinated mutant axons after transplantation (Figure 2; Zhang et al., 1998b). It is also possible to generate such cells from the adult rodent brain (Zhang et al., 1999b) and from other species, including canine (Zhang et al., 1998a) and porcine (Smith and Blakemore, 2000) brain.

The source of human oligodendrocyte progenitors, however, remains more problematic. It has been shown that human oligodendrocytes from fetal brain are capable of myelinating rodent axons on transplantation (Gumpel et al., 1987; Seilhean et al., 1996). When derived from adult human brain as a mixed glial preparation, these cells did not myelinate demyelinated axons in the adult rat spinal cord (Targett et al., 1996). However, when more purified populations of OPCs were isolated, these cells could extensively myelinate demyelinated axons in the adult rat corpus callosum (Windrem et al., 2002). The differences in results of these two studies are likely due to the purity of OPCs in the latter study.

We have also explored the myelinating potential of neural stem cells derived from human fetal brain. It is possible to culture these cells as neurospheres in the presence of EGF and FGF2, and like rodent neurospheres, they produce a low percentage of oligodendrocytes (Zhang et al., 1999a). Transplantation of these human neurospheres has not led to myelination of md rat axons, two weeks after transplantation, although the cells survived but did not apparently differentiate (Zhang and Duncan, unpublished data). In attempts to produce larger numbers of progenitors, we have been unable to generate human oligospheres using techniques the described above.

Embryonic stem (ES) cells

The demonstration that human ES cells can be grown and propagated in culture (Thomson et al., 1998; Shamblott et al., 1998) has led to the hope that these pluripotent cells will become the cornerstone of transplant repair for many organs. ES cells were first isolated from the mouse blastocyst and have now been grown from nonhuman primate (Thomson et al., 1995) as well as human sources (Thomson et al., 1998). It has been clearly shown that mouse ES cells can be coaxed in culture to give rise to glial progenitors and subsequent oligodendrocytes and astrocytes using sequential combinations of growth factors

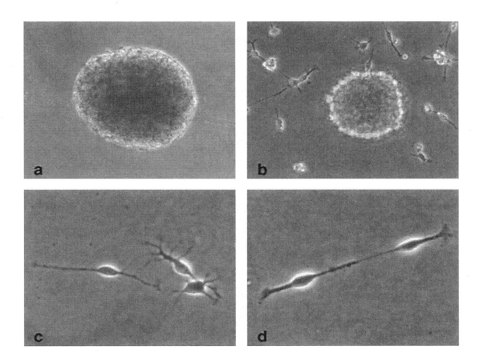

Figure 2. Different growth factors maintain neural stem cells as undifferentiated cells, or direct differentiation towards oligodendrocyte progenitor cells (OPCs). Canine neurospheres (**a**) and oligospheres (**b**) grown in serum-free neurosphere medium (Zhang et al., 1998a) supplemented with FGF2 plus EGF, and FGF2 plus PDGF-AA, respectively. The spheres were derived from the striatum of a one week-old pup, and were cultured for 5 days. Note the floating neurosphere in (**a**) with all cells remaining in the sphere. In contrast, in the oligosphere (**b**) some cells have migrated out and adhere to the dish, mostly with the morphology of OPCs, suggesting that PDGF has driven the cells toward an oligodendroglial differentiation. This can be seen more clearly in (**c**) and (**d**). In (**c**) a classic bipolar OPC is adjacent to two other OPCs that have begun to develop additional processes. In (**d**) a dividing progenitor produces two bipolar OPCs.

(Brustle et al., 1999). Approximately 30% of the cells are oligodendrocytes and astrocytes. Transplantation of ES cell-derived glial progenitors into the md rat resulted in extensive myelination by mouse oligodendrocytes that developed from these progenitors. Mouse ES cells maintained and differentiated in the so-called 4⁻/4⁺ retinoic acid protocol (Bain et al., 1995) were transplanted as neurally differentiated cells into adult rat spinal cord injuries (McDonald et al., 2000). Improvement in clinical function was ascribed to the differentiation of many of these cells into oligodendrocytes, although no evidence was provided that the cells were producing myelin. In a second study from the same group, a method of purification for oligodendrocytes was described (Liu et al., 2000). Free-floating populations of these cells growing in conditioned medium derived from the cultures at an earlier stage were generated. These oligospheres (a term originally coined by Avellane-Adalid et al., 1996) contained a high percentage of cells labeled with the early oligodendrocyte marker (O4). These cells matured in culture to express later oligodendrocyte markers and on transplantation into shi mice, made myelin basic protein (MBP)-positive (Liu et al., 2000).

Studies with human ES cells have so far not produced the significant oligodendrocyte differentiation described for mouse cells. Zhang et al. (2001) devised a method of purification of neuroepithelial cells from embryoid bodies grown in the presence of FGF2. With time, clustering of cells into rosettes occurred and these structures eventually occupied much of the sphere. Dispase treatment of these cells after plating led to rosette retraction, leaving adherent cells behind. These clumps of cells were pelleted and contained purified neuroepithelial cells as identified by immunolabelling for nestin, musashi1, and PSA-NCAM. FGF2 enhanced division of these cells, while LIF and PDGF have a lesser effect. When grown in culture, these neural stem cells gave rise first to neurons of variable types and eventually to astrocytes, with few oligodendrocytes being produced in medium containing PDGF. Transplantation of these cells into newborn mice also showed that they would differentiate into neurons and astrocytes, but not oligodendrocytes. No tumor formation was seen up to 8 weeks after transplantation. In a similar study by Reubinoff et al., (2000), a simpler method of neural differentiation from human ES cells was used, maintaining ES cells without passage or replenishing the feeder cells, which led to spontaneous neural differentiation. Clumps of cells presumed to be neural precursors were replated and grown in medium containing EGF and FGF-2. These cells develop into varied neuronal phenotypes *in vitro* and when transplanted into newborn mice, cells migrated and differentiated into neurons, astrocytes, and oligodendrocytes. The latter is a notable difference from the report by Zhang et al. (2001). As noted by Studer (2001), definitive evidence of significant and functional oligodendrocyte differentiation has not been shown as yet.

Therefore the ability to produce significant numbers of oligodendrocytes from human ES cells is lacking. The reasons for this in contrast to mouse ES cells is not known but may relate to the lack of differentiation cues present in the culture system, or that the temporal course of development of human oligodendrocytes is longer than for other neural cells. Futher studies of growth factor effects on human ES cells (Schuldiner et al., 2000) may help to identify the culture conditions required to promote oligodendrocyte development.

Schwann cells

These cells remain as a second key cell type that could be used to myelinate or remyelinate the CNS. While Schwann cells myelinate only a single internode and have been evolutionarily displaced from the CNS in favor of the oligoden-drocyte which can myelinate multiple axons, nonetheless they migrate into the CNS in many neuropathological conditions and can myelinate CNS axons (Duncan and Hoffman, 1997). The ability to biopsy a patient's own peripheral nerve to isolate Schwann cells and expand them *in vitro* (Morrissey et al., 1995; Tennekoon et al., 1995), makes this cell a serious candidate for transplant re-pair. In addition, it has been shown that remyelination resulting from either spontaneous Schwann cell invasion or transplantation will restore nerve con-duction (Felts and Smith, 1992; Honmou et al., 1996).

Successful remyelination by Schwann cells may however be prevented by their failure to interact positively with host astrocytes. As gliosis is a key fea-ture of many myelin disorders, this could be a major problem if astrocytes in-hibit the migratory ability of transplanted Schwann cells (Franklin and Blakemore, 1993). However, there is not total agreement in the literature re-garding Schwann cell-astrocyte interactions, although it is clear that Schwann cell myelination is always seen in areas that are GFAP-deficient (Duncan and Hoffman, 1997; Shields et al., 2000). The question remains as to whether Schwann cells can only myelinate axons in areas deficient in astrocytes or that Schwann cell invasion/development in the CNS results in astrocyte retraction. Finally, it is also not known whether Schwann cells could be called on to myelinate the entire CNS in disorders characterized by global myelin absence or loss, given their 1:1 relationship with axons. Perhaps such repair may be limited to focal, anatomically strategic lesions that occur at sites that give rise to clinical deficits.

Olfactory ensheathing cells

Olfactory ensheathing cells (OECs) are unique glia that ensheathe or sur-round small diameter axons of the olfactory nerve and also the nerve fiber layer of the olfactory bulb, and support neurogenesis in the olfactory system (re-viewed by Ramon-Cueto and Avila, 1998). They share the properties of both Schwann cells and astrocytes (Doucette, 1990). For instance, while OECs sup-port axonal regrowth like Schwann cells, they form a glial limitans at the PNS-CNS transition zone as do astrocytes (Ramon-Cueto and Valverde, 1995). Al-though OECs do not normally form myelin, upon transplantation into the CNS, OECs isolated from rat (Franklin et al., 1996b), canine (Smith et al., 2002), or human (Barnett et al., 2000; Kato et al., 2000) sources remyelinated axons of the demyelinated rat spinal cord and improved axonal conduction (Imaizumi et al., 1998).

On transplantation into the rat CNS, OECs of rat or human origin myelinate axons in a manner similar to Schwann cells, associating with single axons and making P0 (a peripheral myelin protein) positive myelin (Barnett et al., 2000; Franklin et al., 1996b). Porcine OECs transplanted into traumatic lesions of the spinal cord also make peripheral myelin (Imaizumi et al., 2000a). Not all OECs however differentiate into Schwann cell-like cells. Both rat and human OECs also give rise to a cell that appears astrocyte-like (Barnett et al., 2000; Franklin et al., 1996b). A potential advantage of transplanting OECs compared to Schwann cells is their ability to interact with host astrocytes (Franklin and Barnett, 1997; Franklin et al., 1996b). Such interactions have been explored in cell culture, comparing Schwann cell and OEC behavior when cocultured with astrocytes (Lakatos et al., 2000). A number of differences were seen, including free intermingling of OECs and astrocytes but not Schwann cells that promoted astrocytic hypertrophy. While human Schwann cells can be obtained from peripheral nerve biopsy, OECs can also be isolated from the periphery, e.g., the lamina propria (Au and Roskams, 2003), suggesting that autologous transplantation might be possible.

However, results questioning the ability of OECs to myelinate axons have recently been published. Rat OECs highly purified by immunopanning with the p75 antibody did not myelinate the axons of dorsal root ganglion neurons *in vitro*. Instead, OECs produced long, flattened sheets that separated axons into bundles (Plant et al., 2002). However, the axons in the cultures appeared smaller than those myelinated by Schwann cells, and this may be a contributing factor in their failure to myelinate. When such purified OECs or Schwann cells were transplanted into the contused rat spinal cord, only Schwann cells significantly promoted axon sparing/regeneration and improvement in hind limb locomotor performance (Takami et al., 2002). These data have raised important questions about OECs, and unequivocal proof of the origin of transplanted cells and those cells that differentiate from them should answer these questions.

Non-neural cells

Lack of availability of sufficient numbers of human OPCs for transplantation has led to consideration of alternative sources of myelinating cells such as Schwann cells or OECs. Myelinating cells derived from non-neural tissue might even be more appealing because of ease of collection, *in vitro* manipulation and autologous transplantation. Mesenchymal and hematopoietic stem cells derived from bone marrow have been of particular interest. The ability of such stem cells to transdifferentiate into a neural lineage is discussed in detail elsewhere in this book (Chapter 6), but brief mention will be made of data suggesting development of such cells into oligodendrocytes or Schwann cells. Mesenchymal stem cells (MSCs) from adult marrow have been reported to give rise to galactocerebroside (GalC) positive oligodendrocytes *in vitro* (Jiang et al., 2002).

When these cells were injected into a blastocyst it was suggested that cells seen in the chimera in the corpus callosum were oligodendrocytes. When similar cells were transplanted into the penumbra of infarcts in the rat brain, some MSCs expressed GalC (Zhao et al., 2002), although this is technically a difficult antibody to use as a marker of oligodendrocytes *in vivo*. Enriched adult mouse bone marrow-derived hematopoietic progenitor cells were transplanted into newborn mouse brain and cell differentiation followed, indicated by markers of neurons and glia. Using the O4 antibody, they reported that up to 50% of cells were β-galactosidase/O4-positive. Bone marrow stromal cells injected either directly into demyelinating lesions (Akiyama et al., 2002b) or intravenously (Akiyama et al., 2002a) were shown to survive *in vivo*, express β-galactosidase, and differentiate into cells similar to Schwann cells. The myelin made by these cells restored conduction to remyelinated axons (Akiyama et al., 2002a, 2002b). Finally, human cord blood cells were reported to give rise to oligodendrocytes *in vitro* and these cells were purported to express DM-20, the minor isoform of the proteolipid protein gene (Buzanska et al., 2002).

These data are intriguing yet incomplete at present. It will be important to show that such cell differentiation is reproducible in different laboratories, that large numbers of unequivocally identifiable cells are produced, and that these cells are functional, i.e., can make significant quantities of myelin after transplantation.

Evaluation of Transplantation

Myelination

In general, success of transplantation of OPCs or any other myelinating cell is judged on a) cell survival, b) migration, and c) myelination. In addition, an increase in cell number through controlled division following transplantation may also be required to repair large areas of the CNS.

Cell survival. Transplantation of practically any cell into the CNS is followed by death of a proportion of these cells. We have shown that up to 50% of OPCs die within 24 hours of transplantation into the md rat spinal cord (Zhang et al., 1999c). However, this figure may vary depending on the pathologic milieu into which the cells are transplanted. It is critical that the transplanted cells are labeled so that demonstration of their survival (and later differentiation and myelination) and distinction from host cells is unequivocal. Methods of labeling OPCs, Schwann cells, or OECs have been presented and reviewed elsewhere (Duncan, 1996; Blakemore et al., 1995a). In brief, cells can be generated to express β-galactosidase or green fluorescent protein (GFP), either by viral transduction or by transgenic approaches (Iwashita et al., 2000; Tontsch et al.,

1994; Windrem et al., 2002; Mitome et al., 2001). Vital dyes such as fast-blue, bisbenzimide (Hoechst 33258; Gansmuller et al., 1991) or DiI (Baron-Van Evercooren et al., 1996) have been used to label OPCs. Other techniques include labeling dividing cells with BrdU prior to transplantation or in situ hybridization using a probe to the Y-chromosome to detect male cells transplanted into female hosts (O'Leary and Blakemore, 1997b). While each may have its advantages, the best technique is to double label to allow the identification of both the grafted cells and the myelin they produce.

Migration. To accurately determine the extent of cell migration after transplantation, one needs to be assured that cells are labeled and can be clearly identified *in vivo*. A second caveat is that there is a difference between active migration in the parenchyma of the CNS versus passive movement. Passive dispersion of cells can arise as a result of the actual act of injection of a cell suspension (Lipsitz et al., 1995), and the extent of this may vary in the brain and spinal cord and in different pathological environments. Secondly, unintentional injection of cells into the ventricular system at any level can result in their widespread dispersion throughout the brain and spinal cord. Judgment of migration in a single microscopic section may be inaccurate as this represents only a single moment in time. It may be necessary to evaluate a large group of animals injected similarly, at different time points after surgery, to see whether there is evidence of progressive migration away from the site of focal implantation. We have attempted to evaluate migration *in vivo* by labeling transplanted OPCs with iron nanoparticles in the md rat spinal cord. The spinal cord was excised two weeks later and the cord was examined by magnetic resonance imaging (MRI). There was complete overlap of the MRI signal with PLP-positive myelin (Bulte et al., 1999). Using a similar protocol in the Long Evans shaker rat, individual animals were sequentially followed by MRI. In this case there was an overlap between β-galactosidase-positive cells, PLP immunolabelling and the MRI signal (Bulte et al., 2001). *In vitro* MRI imaging of transplanted OPCs has also been demonstrated by others (Franklin et al., 1999). Labeling of cells with iron particles, allowing their MRI detection *in vivo,* perhaps along with MRI evidence of an increase in myelin formation are powerful tools in studying migration and survival of cells in the intact animal.

Myelination. It is our view that this is best judged in 1μm sections or on EM. Toluidine blue 1μm sections provide unequivocal evidence of myelination by labeled cells in recipients where there is little or no host myelin such as in the md rat (Figure 3). Immunolabeling for a missing protein in the myelin mutants, i.e., PLP in the md rat (Duncan et al., 1988; Zhang et al., 1998b) or MBP in the shi mouse (Mitome et al., 2001) is unequivocal proof of the origin of the myelin. In the case of Schwann cell or even OEC transplants, the myelin will be P0-positive, definitive proof of the cell origin, provided the donor cells are ac-

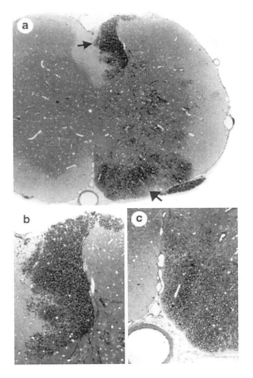

Figure 3: Focal transplantation results in extensive myelination at the site of implantation. An *md* rat was transplanted at 7 days of age with a dissociated cell preparation from the spinal cord of a normal rat and perfused 11 weeks later. In (**a**) myelin produced by the transplanted cells can be seen in the dorsal and ventral (right) columns (arrows). The gray matter on the right side of the cord is also myelinated. On higher power in (**b**) and (**c**) large areas of myelin can be seen adjacent to areas in the dorsal column (**b**) and the left ventral column (**c**), that have remained nonmyelinated. Myelination by the transplanted cells is seen many millimeters rostral and caudal to the segment shown in (**a**). (Modified with permission from Zhang and Duncan, 1999).

curately labeled and can be distinguished from host cells. In the case of focal injection of myelinotoxic chemicals in areas that are irradiated, there is no endogenous repair by oligodendrocytes or Schwann cells, therefore the transplanted cells must be responsible for remyelination of such areas. In the case of Schwann cells, this is only definitive if these cells are clearly identified *in vivo* by a known label, since Schwann cells inevitably invade the spinal cord in most lesions (Duncan and Hoffman, 1997). More recently such studies have utilized genetically labeled cells as definitive proof of the origin of the remyelinating cells (Iwashita et al., 2000).

Functional Recovery

While myelination of areas of non-myelination or demyelination by transplanted cells suggests that nerve fiber function would be restored, it requires formal proof. This was first established in detailed electrophysiological studies of conduction through patches of myelin made by transplanted cells by (Utzschneider et al., 1994). They showed that nerve conduction velocity was restored nearly to normal in a transplant-derived myelinated area of the dorsal column in the neonatal md rat. In further experiments, they tested the physiologic responses of remyelinated axons in the dorsal column of rats that had been demyelinated by injection of ethidium bromide and remyelinated by rat or human Schwann cells, (Imaizumi et al., 2000b; Kohama et al., 2001) and rat or human OECs (Imaizumi et al., 2000b). In all preparations they demonstrated restoration of conduction in remyelinated fibers. They have also explored the feasibility of differentiation of bone marrow stromal cells differentiating into a myelinating cell when transplanted directly into focally demyelinated lesions (Akiyama et al., 2002b) or by intravenous injection (Akiyama et al., 2002a). In both experiments, axons were remyelinated by cells with a Schwann cell phenotype with restoration of conduction velocity. These experiments suggest collectively that focal repair following transplantation leads to restoration of function.

The final test of such cell replacement therapy must be in behavioral recovery after transplantation. Such studies have been rare but there is convincing evidence that focal remyelination of ethidium bromide-induced demyelination of the cervical spinal cord by transplanted OPCs can restore function (Jeffery et al., 1999). Locomotor activity as judged by the ability to traverse a wooden beam, was regained in rats in which cervical cord lesions were repaired by grafted cells. While these results are encouraging, the 'holy grail' is to restore function in the CNS with multiple lesions (e.g. multiple sclerosis) or even in those animals/humans with a global absence or loss of myelin (e.g. the leukodystrophies). One such experiment has suggested that this may be possible. Transplantation of the C17.2 immortalized mouse cell line, C17.2 into the newborn shi mouse led to widespread incorporation of the cells in the brain and a loss of tremor in some of the affected mice (Yandava et al., 1999). However, in a later study using the same recipient, even more extensive myelin formation by transplanted cells did not lead to any behavioral response (Mitome et al., 2001). While the first results were encouraging, the difference in results between the two studies is unexplained. In addition, it is extremely unlikely that human clinical trials would use immortalized cells.

Site and timing of engraftment

In the spinal cord, placement of a micropipette into the quadrants can lead to deposition of cells in the dorsal, lateral, or ventral columns. At the same time, deposition of cells in the adjacent gray matter, or their migration there from the white matter, can lead to almost total myelination/remyelination of that spinal cord segment (Archer et al., 1997). In the brain, targeting of specific white matter tracts such as the corpus callosum or internal capsule, requires stereotactic approaches. Stereotaxis is essential for precise localization of cells such as in transplantation into the superior cerebellar peduncle (Shields et al., 2000).

Such precise targeting could be employed in diseases such as MS where focal demyelination of structures such as the cerebellar peduncles cause severe clinical deficits.

In addition to grafting cells into brain parenchyma, the ventricular system has been used as a conduit for the dispersion of cells throughout the brain. The lateral ventricle has been targeted in many studies of transplantation of neural stem cells (Brustle et al., 1997; Yandava et al., 1999). Transplantation into the ventricles of the neonatal mouse or rat has led to the extensive dispersion of cells throughout the brain. It is not entirely clear how the cells cross the ependymal lining and penetrate the brain. It is also not clear whether certain cell types will reproducibly follow set pathways of migration. In a study of transplantation of OPCs into the lateral ventricle of the fetal md rat, we found cells scattered variably throughout gray and white matter four weeks later, including the corpus callosum, inferior colliculus, hippocampus, and thalamus (Learish et al., 1999). Cells were also observed in the olfactory bulb, cerebellum, brain stem, cerebral cortex, hypothalamus, and the optic nerve and chiasm. Some of these cells differentiated into myelinating oligodendrocytes but others remained as undifferentiated cells in areas such as the molecular layer of the cerebellum. Finally, some cells differentiated into astrocytes, although fewer than those becoming myelinating cells. It is unclear what directs such cell migration or if it is random. Certainly, some cells appear to follow the rostral migratory stream to the olfactory bulb as do neural stem cells similarly transplanted. A better understanding of the cues that direct migration of transplanted OPCs in the developing brain may help promote more widespread migration. A recent study by Mitome et al. (2001) has been most encouraging in regard to achievement of significant myelination by grafted cells. When neurospheres were injected into the lateral ventricle and cisterna magna of shi mice at postnatal day 0 and 2, extensive myelination of the corpus collosum and ventral hippocampal fissures as well as other areas of the brain was found (Mitome et al., 2001). These data suggest that significant parts of whole white matter tracts may be repaired by cell transplantation, and importantly shows that injections at multiple times may help to achieve this.

The timing of transplantation of OPCs may be critical to the chances of myelinating or remyelinating focal, multifocal, or large-scale areas of the CNS. Much of the work on the myelin mutants has been carried out in neonates where it is likely that myelination resulting from transplanted cells occurs in competition with any host myelination. At fetal and neonatal times, cues for myelin forming cells to ensheath axons must be at their greatest. From a clinical perspective, transplanting cells during this timeframe would be important in the childhood myelin disorders or leukodystrophies. Pelizaeus Merzbacher disease is a representative disorder and advances have been made using its animal models, the jimpy mouse, md rat, and shaking pup (Archer et al., 1997). The greatest challenge however is in successful transplantation in the adult CNS where plasticity may be less and the cues for myelination diminished. In addition, reactive changes in the neuropil such as gliosis may be inhibitory factors in myelin formation by transplanted cells. Despite these caveats, much of the work on repair of focally demyelinated lesions has been performed in adult rats. We have also successfully transplanted a mixed glial preparation into the adult sh pup at 9 months of age, a time at which there is prominent gliosis (Archer et al., 1997). While transplantation into the parenchyma of adults leads to myelination, cells grafted into the ventricles of mutants or normal animals may not become incorporated into the brain. However, if this is carried out in adult rats with experimental allergic encephalomyelitis, cells are indeed able to migrate from the ventricles into the brain (Ben-Hur et al., 2003). It may be that important migratory stimuli are found in such lesions, perhaps a significant observation for MS where many major lesions are located at periventricular sites.

MIGRATION AND PROLIFERATION: METHODS FOR ENHANCING THEM

A major goal of transplantation of myelinating cells is to have them repair large and dispersed areas of demyelination, or in the case of the inherited disorders, be capable of global replacement. Thus transplanted cells must be migratory and be able to divide *in vivo* in a controlled manner, as only a finite number of cells can be implanted. In regard to migration, studies on the oligodendrocyte lineage show that only progenitors or pre-progenitors have the ability to migrate, at least *in vitro*. As cells differentiate and become multipolar, migration ceases. It is known that migration of OPCs must occur in the brain and spinal cord during development for normal myelination to occur.

Promotion of migration of transplanted OPCs may be enhanced by a better understanding of the molecules expressed both by the cells and by surrounding tissues. Molecules known to promote migration of OPCs *in vitro* include PDGF and FGF2 (Armstrong et al., 1990; Milner et al., 1997) while tenascin-c is known to inhibit migration (Kiernan et al., 1996). OPC migration may also be influenced by the expression of receptors for semaphorin or netrin (Spassky et al., 2002). Both of these molecules are known to act as chemoattractants or repellants. It is also important to understand why OPCs stop migrating and ensheath axons. Certainly, if they differentiate and become multipolar, migration would cease. Expression of the chemokine receptor CXCR2 by OPCs may be a key to positioning cells to myelinate the CNS (Tsai et al., 2002). The integrin family almost certainly plays a key role in OPC migration through their interaction with the extracellular matrix. In particular, expression of $\alpha v \beta 1$ integrin plays a role in cell migration (Milner et al., 1996; Buttery et al., 1999).

As yet, there have been no studies exploring the migratory behavior of transplanted cells induced to overexpress molecules thought to promote migration. Any expression of such molecules however must be compatible with oligodencrocyte function, that is, cells must stop migrating, mature, ensheath axons and make myelin. Indirect evidence of transplanted cells being influenced by chemotactic stimuli comes from experiments where OPCs were transplanted in the md rat spinal cord some distance from transplanted B104 cells (Milward et al., 2000). The latter produce factors known to maintain OPCs dividing and undifferentiated (Hunter and Bottenstein, 1990, 1991). In this study we showed that the transplanted OPCs migrated selectively toward the B-104 cells suggesting the latter produce chemotactic factors.

A key question regarding migration of transplanted cells is whether they will be able to migrate through areas of normal myelination to reach dispersed lesions as in MS, and whether abnormalities of the milieu such as inflammation or gliosis will influence this. Work of Blakemore and Franklin suggests that transplanted OPCs have a limited ability to survive in the normal CNS and migrate toward focal lesions, although prior X-irradiation of the neuropil may enhance this (Franklin et al., 1996a; O'Leary and Blakemore, 1997a; Chari and Blakemore, 2002). In contrast, others have suggested that OPCs have considerable ability to migrate through areas of normal myelin (Baron-Van Evercooren et al., 1996). The differences noted may relate to the different models used. Interestingly, inflammation may promote the spread of transplanted cells, both in the spinal cord (Tourbah et al., 1997) and brain (Ben-Hur et al., 2003). The latter study used transplantation into the ventricles as a means of disseminating cells. We and others have used the ventricular cavities as a means of promoting widespread integration of OPCs into the brain (Learish et al., 1999; Wu et al., 2002; Mitome et al., 2001). As noted before, injections on two occasions into the ventricular system may promote more extensive migration and repair (Mitome et al., 2001). Most recently it has been shown that neural stem cells may be disseminated widely throughout the CNS in mice with experimental allergic encephalomyelitis when cells are injected intrathecally or intravenously (Pluchino et al., 2003).

A second method of extending myelination by transplanted cells is to increase the number of cells that either survive grafting or divide. Many OPCs die on transplantation (Zhang et al., 1999c). We have shown that more OPCs divide when cografted with B104 cells (Milward et al., 2000), and it is known that PDGF produced by ectopically transplanted cells increases the number of OPCs in the optic nerve (Barres et al., 1992) and spinal cord (Björklund and Lindvall, 2000). Indirect evidence that transplanted cells will divide with time, giving rise to more oligodendrocytes and hence increased myelination, was noted when neural stem cells were injected into a mutant mouse (Ader et al., 2001). Thus, these combined data indicate the potential for OPC migration and division to occur and be enhanced *in vivo* following transplantation.

FUTURE PERSPECTIVES

There now has been extensive experimentation on glial cell transplantation that suggests that is a safe technique that results in extensive repair and, in some cases, improvement or return of function. Two critical issues remain to be resolved, however, in using this approach in human myelin disorders, although they may not hinder initial clinical trials from being performed. Indeed a small Phase I trial of autologous Schwann cell transplantation in MS patients has already been performed confirming the safety of the procedure. There is still uncertainty however about the best human cell to be used. While ES cells may be the eventual solution to the issue of cell numbers, sufficient oligodendrocytes from these cells have not been generated *in vitro* or *in vivo* compared to mouse ES cells. Likewise, human neural stem cells grown as neurospheres have generated more oligodendrocytes than ES cells, but insufficient for large scale repair, although FACS sorting of dissociated cell preparations may provide the purity and number of cells required. Schwann cells and OECs remain as promising cell sources. In summary, it would appear beneficial to continue to explore all cell source options while more is learned about oligodendrocyte differentiation from human ES cells.

The disease or diseases to be targeted also remains a question of considerable debate. A start could be made in the repair of focal areas of myelin, such as in the spinal cord of some MS patients (combined with medical therapy to block or lessen ongoing disease) if there was consensus about the right cell to implant. The greater challenge is in replacing oligodendrocytes at multiple sites or along the entire neuroaxis. Success with focal lesions however will be a huge step in moving towards global repair.

Acknowledgements

Studies noted here from our laboratory have been supported by NIH (R01 NS33710-03), the Myelin Project, the Oscar Rennebohm Foundation, the Elizabeth Elser Doolittle Charitable Trust, and the Roddis Foundation.

REFERENCES

Ader M, Meng J, Schachner M, Bartsch U (2000) Formation of myelin after transplantation of neural precursor cells into the retina of young postnatal mice. GLIA 301-310.

Ader M, Schachner M, Bartsch U (2001) Transplantation of neural precursor cells into the dysmyelinated CNS of mutant mice deficient in the myelin-associated glycoprotein and Fyn tyrosine kinase. Eur J Neurosci 14: 561-566.

Akiyama Y, Radtke C, Honmou O, Kocsis JD (2002a) Remyelination of the spinal cord following intravenous delivery of bone marrow cells. GLIA 39: 229-236.

Akiyama Y, Radtke C, Kocsis JD (2002b) Remyelination of the rat spinal cord by transplantation of identified bone marrow stromal cells. J Neurosci 22: 6623-6630.

Archer DR, Cuddon PA, Lipsitz D, Duncan ID (1997) Myelination of the canine central nervous system by glial cell transplantation: a model for repair of human myelin disease. Nature Med 3: 54-59.

Armstrong RC, Harvath L, Dubois-Dalcq ME (1990) Type 1 astrocytes and oligodendrocyte-type 2 astrocyte glial progenitors migrate toward distinct molecules. J Neurosci Res 27: 400-407.

Au E, Roskams AJ (2003) Olfactory ensheathing cells of the lamina propria *in vivo* and *in vitro*. GLIA 41: 224-236.

Avellana-Adalid V, Nait-Oumesmar B, Lachapelle F, Baron-Van Evercooren A (1996) Expansion of rat oligodendrocyte progenitors into proliferative "oligospheres" that retain differentiation potential. J Neurosci Res 45: 558-570.

Bain G, Kitchens K, Yao M, Huettner JE, Gottlieb DI (1995) Embryonic stem cells express neuronal properties *in vitro*. Developmental Biology 168: 342-357.

Barnett SC, Alexander CL, Iwashita Y, Gilson JM, Crowther J, Clark L, Dunn LT, Papanastassiou V, Kennedy PGE, Franklin RJM (2000) Identification of a human olfactory ensheathing cell that can effect transplant-mediated remyelination of demyelinated CNS axons. Brain 123: 1581-1588.

Baron-Van Evercooren A, Avellana-Adalid V, Ben Younes-Chennoufi A, Gansmuller A, Nait-Oumesmar B, Vignais L (1996) Cell-cell interactions during the migration of myelin-forming cells transplanted in the demyelinated spinal cord. GLIA 16: 147-164.

Barres BA, Hart IK, Coles HSR, Burne JF, Voyvodic JT, Richardson WD, Raff MC (1992) Cell death and control of cell survival in the oligodendrocyte lineage. Cell 70: 31-46.

Ben-Hur T, Einstein O, Mizrachi-Kol R, Ben-Menachem O, Reinhartz E, Karussis D, Abramsky O (2003) Transplanted multipotential neural precursor cells migrate into the inflamed white matter in response to experimental autoimmune encephalomyelitis. GLIA 41: 73-80.

Björklund A, Lindvall O (2000) Cell replacement therapies for central nervous system disorders. Nature Neuroscience 3: 537-544.

Blakemore WF, Crang A (1988) Extensive oligodendrocyte remyelination following injection of cultured central nervous system cells into demyelinating lesions in adult central nervous system. Dev Neurosci 10: 1-11.

Blakemore WF, Crang AJ, Franklin RJM (1995a) Transplantation of glial cells. In: Neuroglial Cells (Ransom BR, Kettenmann H, eds), pp 869-882. Cambridge: Oxford University Press.

Blakemore WF, Franklin RJM (1991) Transplantation of glial cells into the CNS. TINS 14:323-327.

Blakemore WF, Franklin RJM (1999) Transplantation options for therapeutic CNS remyelination. Cell Transplantation.

Blakemore WF, Franklin RJM, Noble M (1996) Glial cell transplantation and the repair of demyelinating lesions. In: Glial Cell Development. Basic principles and clinical relevance. (Jessen KR, Richardson WD, eds), pp 209-220. Oxford: BIOS Scientific.

Blakemore WF, Olby NJ, Franklin RJM (1995b) The use of transplanted glial cells to reconstruct glial environments in the CNS. Brain Pathol 5: 443-450.

Brustle O, Cunningham M, Tabar V, Studer L (1997) Experimental transplantation in the embryonic, neonatal, and adult mammalian brain. In: Current Protocols in Neuroscience (Crawley J, Gerfen C, McKay RDG, Rogawski M, Sibley D, Skolnick P, eds), pp 3.10.11-13.10.28. New York: John Wiley.

Brustle O, Jones E, Learish R, Karran K, Chaudhary K, Weistler O, Duncan ID, McKay RDG (1999) Myelin-repair by transplantation of embryonic stem cell-derived glial precursors. Science.

Bulte JWM, Douglas T, Witwer B, Zhang S-C, Strable E, Lewis BK, Zywicke H, Miller B, van Gelderen P, Moskowitz BM, Duncan ID, Frank JA (2001) Magnetodendrimers allow endosomal magnetic labeling and *in vivo* tracking of stem cells. Nat Biotechnol 19: 1141-1147.

Bulte JWM, Zhang S-C, van Gelderen P, Herynek V, Jordan EK, Duncan ID, Frank JA (1999) Neurotransplantation of magnetically labeled oligodendrocyte progenitors: MR tracking of cell migration and myelination. Proc Natl Acad Sci.

Buttery PC, Mallawaarachchi CM, Milner R, Doherty P, ffrench-Constant C (1999) Mapping regions of the b1 integrin cytoplasmic domain involved in migration and survival in primary oligodendrocyte precursors using cell-permeable homeopeptides. Biochem Biophys Res Commun 259: 121-127.

Buzanska L, Machaj EJK, Zablocka B, Pojda Z, Domanska-Janik K (2002) Human cord blood-derived cells attain neuronal and glial features *in vitro*. J Cell Sci 115: 2131-2138.

Chari DM, Blakemore W (2002) New insights into remyelination failure in multiple sclerosis: implications for glial cell transplantation. Mult Scler 8: 271-277.

Doucette R (1990) Glial influences on axonal growth in the primary olfactory system. GLIA 3: 433-449.

Duncan ID (1996) Glial cell transplantation and remyelination of the CNS. Neuropathol Appl Neurobiol 22: 87-100.

Duncan ID (2001) Strategies for repair in MS: the potential role of glial-cell transplantation. In: Multiple Sclerosis: Tissue Destruction and Repair (Kappos L, ed), pp 25-32. Martin Dunitz Publishers.

Duncan ID, Aguayo AJ, Bunge RP, Wood PM (1981) Transplantation of *in vitro* cultures of rat Schwann cells into the mouse spinal cord. J Neurol Sci 41: 241-252.

Duncan ID, Grever WE, Zhang S-C (1997) Repair of myelin disease: strategies and progress in animal models. Molecular Medicine Today 3: 554-561.

Duncan ID, Hammang JP, Jackson KF, Wood PM, Bunge RP, Langford LA (1988) Transplantation of oligodendrocytes and Schwann cells into the spinal cord of the myelin-deficient rat. J Neurocytol 17: 351-360.

Duncan ID, Hoffman RL (1997) Schwann cell invasion of the central nervous system of the myelin mutants. J Anat 190: 35-49.

Duncan ID, Paino C, Archer DR, Wood PM (1992) Functional capacities of transplanted cell-sorted adult oligodendrocytes. Dev Neurosci 14: 114-122.

Felts PA, Smith KJ (1992) Conduction properties of central nerve fibers remyelinated by Schwann cells. Brain Res 574: 178-192.

Franklin RJM, Barnett SC (1997) Do olfactory glia have advantages over Schwann cells for CNS repair. J Neurosci Res 50: 665-672.

Franklin RJM, Bayley SA, Blakemore WF (1996a) Transplanted CG4 cells (an oligodendrocyte progenitor cell line) survive, migrate, and contribute to repair of areas of demyelination in X-irradiated and damaged spinal cord but not in normal spinal cord. Exp Neurol 137: 263-276.

Franklin RJM, Blakemore WF (1993) Requirements for Schwann cell migration within CNS environments: a viewpoint. Int J Devl Neurosci 11: 641-649.

Franklin RJM, Blaschuk KL, Bearchell MC, Prestoz LLC, Setzu A, Brindle KM, ffrench-Constant C (1999) Magnetic resonance imaging of transplanted oligodendrocyte precursors in the rat brain. NeuroReport 10: 3961-3965.

Franklin RJM, Gilson JM, Franceschini IA, Barnett SC (1996b) Schwann cell-like myelination following transplantation of an olfactory bulb-ensheathing cell line into areas of demyelination in the adult CNS. GLIA 17: 217-224.

Gansmüller A, Clerin E, Krüger F, Gumpel M, Lachapelle F (1991) Tracing transplanted oligodendrocytes during migration and maturation in the shiverer mouse brain. GLIA 4: 580-590.

Gansmuller A, Clerin E, Kruger F, Gumpel M, Lachapelle F (1991) Tracing transplanted oligodendrocytes during migration and maturation in the shiverer mouse brain. GLIA 4: 580-590.

Gumpel M, Baumann N, Raoul M, Jacque C (1983) Survival and differentiation of oligodendrocytes from neural tissue transplanted into new-born mouse brain. Neurosci Letters 37: 307-311.

Gumpel M, Lachapelle F, Baumann N (1985) Central nervous tissue transplantation into mouse brain: differentiation of myelin from transplanted oligodendrocytes. In: Neural Grafting in the Mammalian CNS (Björklund A, Stenevi U, eds), pp 151-158. Elsevier Science Publishers, B.V.

Gumpel M, Lachapelle F, Gansmüller A, Baulac M, Baron-Van Evercooren A, Baumann N (1987) Transplantation of human embryonic oligodendrocytes into shiverer brain. Ann N Y Acad Sci 495: 71-85.

Hammang JP, Archer DR, Duncan ID (1997) Myelination following transplantation of EGF-responsive neural stem cells into a myelin-deficient environment. Exp Neurol 147: 84-95.

Honmou O, Kocsis JD, Waxman SG, Felts PA (1996) Restoration of normal conduction properties in demyelinated spinal cord axons in the adult rat by transplantation of exogenous Schwann cells. J Neurosci 16: 3199-3208.

Hunter S, Bottenstein JE (1990) Growth factor responses of enriched bipotential glial progenitors. Dev Brain Res 54: 235-248.

Hunter SF, Bottenstein JE (1991) O-2A glial progenitors from mature brain respond to CNS neuronal cell line-derived growth factors. J Neurosci Res 28: 574-582.

Imaizumi T, Lankford KL, Burton WV, Fodor WL, Kocsis JD (2000a) Xenotransplantation of transgenic pig olfactory ensheathing cells promotes axonal regeneration in rat spinal cord. Nature Biotechnology.

Imaizumi T, Lankford KL, Kocsis JD (2000b) Transplantation of olfactory ensheathing cells or Schwann cells restores rapid and secure conduction across the transected spinal cord. Brain Res 854: 70-78.

Imaizumi T, Lankford KL, Waxman SG, Greer CA, Kocsis JD (1998) Transplanted olfactory ensheathing cells remyelinate and enhance axonal conduction in the demyelinated dorsal columns of the rat spinal cord. J Neurosci 18: 6176-6185.

Iwashita Y, Fawcett JW, Crang AJ, Franklin RJM, Blakemore WF (2000) Schwann cells transplanted into normal and x-irradiated adult white matter do not migrate extensively and show poor long-term survival. Exp Neurol 164: 292-302.

Jeffery ND, Crang AJ, O'Leary MT, Hodge SJ, Blakemore WF (1999) Behavioural consequences of oligodendrocyte progenitor cell transplantation into experimental demyelinating lesions in the rat spinal cord. Eur J Neurosci 11: 1508-1514.

Jiang Y, Vaessen B, Lenvik T, Blackstad M, Reyes M, Verfaillie CM (2002) Multipotent progenitor cells can be isolated from postnatal murine bone marrow, muscle, and brain. Exp Hematol 30: 896-904.

Kato T, Honmou O, Uede T, Hashi K, Kocsis JD (2000) Transplantation of human olfactory ensheathing cells elicits remyelination of demyelinated rat spinal cord. GLIA 209-218.

Kessaris N, Pringle N, Richardson WD (2001) Ventral neurogenesis and the neuron-glial switch. Neuron 31: 677-680.

Kiernan BW, Götz B, Faissner A, ffrench-Constant C (1996) Tenascin-C inhibits oligodendrocyte precursor cell migration by both adhesion-dependent and adhesion-independent mechanisms. Mol. Cell. Neurosci. 7:322-335.

Kohama I, Lankford KL, Preiningerova J, White FA, Vollmer TL, Kocsis JD (2001) Transplantation of cryopreserved adult human Schwann cells enhances axonal conduction in demyelinated spinal cord. J Neurosci 21: 944-950.

Lachapelle F, Lapie P, Nussbaum JL, Gumpel M (1990) Immunohistochemical studies on cross-transplantations between jimpy, shiverer, and normal newborn mice. J Neurosci Res 27: 324-331.

Lakatos A, Franklin RJM, Barnett SC (2000) Olfactory ensheathing cells and Schwann cells differ in their *in vitro* interactions with astrocytes. GLIA 32: 214-225.

Learish RD, Brustle O, Zhang S-C, Duncan ID (1999) Widespread myelination following intraventricular transplantation of oligodendrocyte progenitors into the cerebral ventricle of embryonic myelin-deficient rats. Ann Neurol 46: 716-722.

Lipsitz D, Archer DR, Duncan ID (1995) Acute dispersion of glial cells following transplantation into the myelin deficient rat spinal cord. GLIA 14: 237-242.

Liu S, Stewart TJ, Howard MJ, Chakrabortty S, Holekamp TF, McDonald JW (2000) Embryonic stem cells differentiate into oligodendrocytes and myelinate in culture and after spinal cord transplantation. PNAS 97: 6126-6131.

Louis JC, Magal E, Muir D, Manthorpe M, Varon S (1992) CG-4, a new bipotential glial cell line from rat brain, is capable of differentiating *in vitro* into either mature oligodendrocytes or type-2 astrocytes. J Neurosci Res 31: 193-204.

McCarthy KD, de Vellis J (1980) Preparation of Separate Astroglial and Oligodendroglial Cell Cultures From Rat Cerebral Tissue. J Cell Biol 85: 890-902.

McDonald JW, Liu XZ, Qu Y, Liu S, Mickey SK, Turetsky D, Gottlieb DI, Choi DW (1999) Transplanted embryonic stem cells survive, differentiate and promote recovery in injured rat spinal column. Nat Med 5: 1410-1412.

Milner R, Anderson HJ, Rippon RF, McKay J, Franklin RJM, Marchionni MA, Reynolds R, ffrench-Constant C (1997) Contrasting effects of mitogenic growth factors on oligodendrocyte precursor cell migration. GLIA 19: 85-90.

Milner R, Edwards G, Streuli C, ffrench-Constant C (1996) A role in migration for the avb1 integrin expressed on oligodendrocyte precursors. J Neurosci 16: 7240-7252.

Milward EA, Zhang S-C, Zhao M, Lundberg C, Ge B, Goetz BD, Duncan ID (2000) Enhanced proliferation and directed migration of oligodendroglial progenitors co-grafted with growth factor-secreting cells. GLIA 32: 264-270.

Mitome M, Low HP, van den Pol A, Nunnari JJ, Wolf MK, Billings-Gagliardi S, Schwartz WJ (2001) Towards the reconstruction of central nervous system white matter using neural precursor cells. Brain 124: 2147-2161.

Morrissey TK, Levi ADO, Nuijens A, Sliwkowski MX, Bunge RP (1995) Axon-induced mitogenesis of human Schwann cells involves heregulin and p185[erbB2]. Proc Natl Acad Sci USA 92: 1431-1435.

O'Leary MT, Blakemore WF (1997a) Oligodendrocyte precursors survive poorly and do not migrate following transplantation into the normal adult central nervous system. J Neurosci Res 48: 159-167.

O'Leary MT, Blakemore WF (1997b) Use of a rat Y chromosome probe to determine the long-term survival of glial cells transplanted into areas of CNS demyelination. J Neurocytol 26: 191-206.

Olivier C, Cobos I, Villegas EMP, Spassky N, Zalc B, Martinez S, Thomas JL (2001) Monofocal origin of telencephalic oligodendrocytes in the anterior entopeduncular area of the chick embryo. Development 128: 1757-1769.

Plant GW, Currier PF, Cuervo EP, Bates ML, Pressman Y, Bunge MB, Wood PM (2002) Purified adult ensheathing glia fail to myelinate axons under culture conditions that enable Schwann cells to form myelin. J Neurosci 22: 6083-6091.

Pluchino S, Quattrini A, Brambilla E, Gritti A, Salani G, Dina G, Galli R, Del Carro U, Amadio S, Bergami A, Furlan R, Comi G, Vescovi AL, Martino G (2003) Injection of adult neurospheres induces recovery in a chronic model of multiple sclerosis. Nature 422: 688-694.

Pringle NP, Richardson WD (1993) A singularity of PDGF alpha-receptor expression in the dorsoventral axis of the neural tube may define the origin of the oligodendrocyte lineage. Development 117: 525-533.

Qi Y, Stapp D, Qiu M (2002) Origin and molecular specification of oligodendrocytes in the telencephalon. Trends in Neurosciences 25: 223-225.

Ramon-Cueto A, Avila J (1998) Olfactory ensheathing glia: properties and function. Brain Res Bull 46: 175-187.

Ramon-Cueto A, Valverde F (1995) Olfactory bulb ensheathing glia: a unique cell type with axonal growth-promoting properties. GLIA 14:163-173.

Reubinoff BE, Pera MF, Fong C-Y, Trounson A, Bongso A (2000) Embryonic stem cell lines from human blastocysts: somatic differentiation *in vitro*. Nature Biotechnol 18: 399-404.

Reynolds BA, Weiss S (1992) Generation of neurons and astrocytes from isolated cells of the adult mammalian central nervous system. Science 255: 1707-1710.

Richardson WD, Pringle NP, Yu W-P, Hall AC (1997) Origins of spinal cord oligodendrocytes: Possible developmental and evoluntionary relationships with motor neurons. Dev Neurosci 19: 58-68.

Rosenbluth J, Hasegawa M, Shirasaki N, Rosen CL, Liu Z (1990) Myelin formation following transplantation of normal fetal glia into myelin-deficient rat spinal cord. J Neurocytol 19: 718-730.

Rowitch DH, Lu QR, Kessaris N, Richardson WD (2002) An 'oligarchy' rules neural development. Trends in Neurosciences 25: 417-422.

Schuldiner M, Yanuka O, Itskovitz-Eldor J, Melton DA, Benvenisty N (2000) Effects of eight growth factors on the differentiation of cells derived from human embryonic stem cells. PNAS 97: 11307-11312.

Seilhean D, Gansmüller A, Baron-Van Evercooren A, Gumpel M, Lachapelle F (1996) Myelination by transplanted human and mouse central nervous system tissue after long-term cryopreservation. Acta Neuropathol (Berl) 91: 82-88.

Shamblott MJ, Axelman J, Wang S, Bugg EM, Littlefield JW, Donovan PJ, Blumenthal PD, Huggins GR, Gearhart JD (1998) Derivation of pluripotent stem cells from cultured human primordial germ cells. PNAS 95: 13726-13731.

Shields SA, Blakemore WF, Franklin RJM (2000) Schwann cell remyelination is restricted to astrocyte-deficient areas after transplantation into demyelinated adult rat brain. J Neurosci Res 60: 571-578.

Smith PM, Blakemore WF (2000) Porcine neural progenitors require commitment to the oligodendrocyte lineage prior to transplantation in order to achieve significant remyelination of demyelinated lesions in the adult CNS. Eur J Neurosci 12: 2414-2424.

Smith PM, Lakatos A, Barnett SC, Jeffery, ND, Franklin RJM (2002) Cryopreserved cells isolated from the adult canine olfactory bulb are capable of extensive remyelination following transplantation into the adult rat CNS. Exp Neurol 176:402-406.

Spassky N, de Castro F, Le Bras B, Heydon K, Quéraud-LeSaux F, Bloch-Gallego E, Chédotal A, Zalc B, Thomas JL (2002) Directional guidance of oligodendroglial migration by class 3 semaphorins and netrin-1. J Neurosci 22: 5992-6004.

Studer L (2001) Stem cells with brainpower. Nature Biotechnol 19: 1117-1118.

Takami T, Oudega M, Bates ML, Wood PM, Kleitman N, Bunge MB (2002) Schwann cell but not olfactory ensheathing glia transplants improve hindlimb locomotor performance in the moderately contused adult rat thoracic spinal cord. Journal of Neuroscience 22: 6670-6681.

Takebayashi H, Nabeshima Y, Yoshida S, Chisaka O, Ikenaka K (2002) The basic helix-loop-helix factor Olig2 is essential for the development of motoneuron and oligodendrocyte lineages. Curr Biol 12: 1157-1163.

Targett MP, Sussman J, Scolding N, O'Leary MT, Compston DAS, Blakemore WF (1996) Failure to achieve remyelination of demyelinated rat axons following transplantation of glial cells obtained from the adult human brain. Neuropathol Appl Neurobiol 22: 199-206.

Tennekoon GI, Lerner MA, Kirk C, Rutkowski JL (1995) Purification and expansion of human Schwann cells *in vitro*. Nature Med 1: 80-83.

Thomson JA, Itskovitz-Eldor J, Shapiro SS, Waknitz MA, Swiergiel JJ, Marshall VS, Jones JM (1998) Embryonic stem cell lines derived from human blastocysts. Science 282: 1145-1147.

Thomson JA, Kalishman J, Golos TG, Durning M, Harris CP, Becker RA, Hearn JP (1995) Isolation of a primate embryonic stem cell line. PNAS 92: 7844-7848.

Tontsch U, Archer DR, Dubois-Dalcq M, Duncan ID (1994) Transplantation of an oligodendrocyte cell line leading to extensive myelination. PNAS 91: 11616-11620.

Tourbah A, Linnington C, Bachelin C, Avellana-Adalid V, Wekerle H, Baron-Van Evercooren A (1997) Inflammation promotes survival and migration of the CG4 oligodendrocyte progenitors transplanted in the spinal gord of both inflammatory and demyelinated EAE rats. J Neurosci Res 50: 853-861.

Tsai HH, Frost E, To V, Robinson S, ffrench-Constant C, Geertman R, Ransohoff RM, Miller RH (2002) The chemokine receptor CXCR2 controls positioning of oligodendrocyte precursors in developing spinal cord by arresting their migration. Cell 110: 373-383.

Utzschneider DA, Archer DR, Kocsis JD, Waxman SG, Duncan ID (1994) Transplantation of glial cells enhances action potential conduction of amyelinated spinal cord axons in the myelin-deficient rat. PNAS 91: 53-57.

Vitry S, Avellana-Adalid V, Hardy R, Lachapelle F, Baron-Van Evercooren A (1999) Mouse oligospheres: From pre-progenitors to functional oligodendrocytes. J Neurosci Res 735-751.

Warrington AE, Barbarese E, Pfeiffer SE (1993) Differential myelinogenic capacity of specific developmental stages of the oligodendrocyte lineage upon transplantation into hypomyelinating hosts. J Neurosci Res 34: 1-13.

Windrem MS, Roy NS, Wang J, Nunes M, Benraiss A, Goodman R, McKhann GM, II, Goldman SA (2002) Progenitor cells derived from the adult human subcortical white matter disperse and differentiate as oligodendrocytes within demyelinated lesions of the rat brain. J Neurosci Res 69: 966-975.

Woodruff RH, Tekki-Kessaris N, Stiles CD, Rowitch DH, Richardson WD (2001) Oligodendrocyte development in the spinal cord and telencephalon: common themes and new perspectives. Int J Dev Neurosci 19: 379-385.

Wu SF, Suzuki Y, Kitada M, Kataoka K, Kitaura M, Chou H, Nishimura Y, Ide C (2002) New method for transplantation of neurosphere cells into injured spinal cord through cerebrospinal fluid in rat. Neurosci Lett 318: 81-84.

Yandava BD, Billinghurst LL, Snyder EY (1999) "Global" cell replacement is feasible via neural stem cell transplantation: Evidence from the dysmyelinated shiverer mouse brain. Proc Natl Acad Sci USA 96: 7029-7034.

Zhang S-C, Duncan ID (1999) Remyelination and restoration of axonal function by glial cell transplantation. In: Functional Neural Transplantation (Dunnett SB, Bjorklund A, eds), Amsterdam: Elsevier.

Zhang S-C, Ge B, Duncan ID (1999a) Tracing human oligodendroglial development *in vitro*. J Neurosci Res. 59:421-429.

Zhang S-C, Ge B, Duncan ID (1999b) Adult brain retains the potential to generate oligodendroglial progenitors with extensive myelination capacity. PNAS 96: 4089-4094.

Zhang S-C, Goetz BD, Duncan ID (2003) Suppression of activated microglia promotes survival and function of transplanted oligodendroglial progenitors. GLIA 41: 191-198.

Zhang S-C, Lipsitz D, Duncan ID (1998a) Self-renewing canine oligodendroglial progenitor expanded as oligospheres. J Neurosci Res 54: 181-190.

Zhang S-C, Lundberg C, Lipsitz D, O'Connor LT, Duncan ID (1998b) Generation of oligodendroglial progenitors from neural stem cells. J Neurocytol 27: 475-489.

Zhang, S.-C., Wagner, D., and Duncan, I. D. Acute death of grafted oligodendroglial progenitors. Soc for Neurosci 25. 1999c.

Zhang S-C, Wernig M, Duncan ID, Brüstle O, Thomson JA (2001) *In vitro* differentiation of transplantable neural precursors from human embryonic stem cells. Nature Biotech 19:1129-1133.

Zhao LR, Duan WM, Reyes M, Keene CD, Verfaillie CM, Low WC (2002) Human bone marrow stem cells exhibit neural phenotypes and ameliorate neurological deficits after grafting into the ischemic brain of rats. Exp Neurol 174: 11-20.

Zhou Q, Choi G, Anderson DJ (2001) The bHLH transcription factor Olig2 promotes oligodendrocyte differentiation in collaboration with Nkx2.2. Neuron 31: 791-807.

Chapter 12

Induction of Adult Cortical Neurogenesis From Neural Precursors In Situ

Paola Arlotta, Jinhui Chen, Sanjay S. P. Magavi and Jeffrey D. Macklis

ADULT NEUROGENESIS

Until very recently, the relative lack of recovery from CNS injury and neurodegenerative disease resulted in the entire field reaching the conclusion that neurogenesis does not occur in the adult mammalian brain. A series of groundbreaking results from different groups over the last four decades have contradicted this classical view and provided strong evidence that neurogenesis, the birth of new neurons, does extend past embryonic and fetal stages of development and occurs, although with limitations, also in the adult brain.

Joseph Altman and colleagues were the first to use techniques sensitive enough to detect the ongoing cell division that occurs in adult brain and published evidence that neurogenesis constitutively occurs in the hippocampus (Altman and Das, 1965) and olfactory bulb (Altman, 1969) of the adult mammalian brain. However, the absence of neuron-specific immunocytochemical markers at the time resulted in the identification of putatively newborn neurons being made on purely morphological criteria, which led to a widespread lack of acceptance of these results.

Only a decade ago, technical advances including the use of cell type-specific markers to clearly identify newborn neurons, allowed two independent groups to more definitively show that precursor cells isolated from the forebrain can differentiate into neurons *in vitro* (Reynolds and Weiss, 1992; Richards et al., 1992). These results led to an explosion of research in the field. Normally occurring neurogenesis in the olfactory bulb, olfactory epithelium (see Barber, 1982; Crews and Hunter, 1994), and hippocampus have now been well-characterized in the adult mammalian brain.

Olfactory Bulb Neurogenesis

Adult olfactory bulb neurogenesis has been most extensively studied in the rodent, though there is *in vitro* (Kirschenbaum et al., 1994; Pincus et al.,

From: *Neural Stem Cells: Development and Transplantation*
Edited by: Jane E. Bottenstein © 2003 Kluwer Academic Publishers, Norwell, MA

1998) and *in vivo* (Bernier et al., 2000) evidence suggesting that such neuronal precursors exist in humans. Several experiments show that the precursors that contribute to olfactory bulb neurogenesis reside in the anterior portion of the subventricular zone (SVZ; Luskin, 1993), sometimes called the subependymal zone (SEZ). When retroviruses or tritiated thymidine (Lois and Alvarez-Buylla, 1994), vital dyes (Lois and Alvarez-Buylla, 1994; Doetsch and Alvarez-Buylla, 1996), or virally labeled SVZ/SEZ cells (Luskin and Boone, 1994; Doetsch and Alvarez-Buylla, 1996) are microinjected into the anterior portion of the SVZ/ SEZ of postnatal animals, labeled cells are eventually found in the olfactory bulb. Upon reaching the olfactory bulb, these labeled neurons differentiate into interneurons specific to the olfactory bulb, olfactory granule cells and peri-glom- erular cells. To reach the olfactory bulb, the neuroblasts undergo tangential chain migration within sheaths of slowly dividing astrocytes (Garcia-Verdugo et al., 1998) along the rostral migratory stream (RMS) (Rousselot et al., 1995). Once in the olfactory bulb, the neurons migrate along radial glia away from the RMS and differentiate into interneurons that participate in functional synaptic circuitry (Carlen et al., 2002).

Of considerable interest have been the factors that contribute to the direc- tion of migration of the neuroblasts, as well as factors involved in initiating and controlling migration itself. *In vitro* experiments show that caudal septum ex- plants secrete a diffusible factor, possibly the molecule Slit (Mason et al., 2001) that repels olfactory bulb neural precursors (Hu and Rutishauser, 1996). Con- sistent with the idea that SVZ/SEZ precursor migration is directed by repulsion is the finding that SVZ/SEZ precursors migrate anteriorly along the RMS even in the absence of the olfactory bulb (Kirschenbaum et al., 1999). The tangential migration of the cells seems to be at least partially mediated by PSA-NCAM, expressed by the neuroblasts themselves (Hu et al., 1996). This may be modi- fied by tenascin and chondroitin sulfate proteoglycans that are located near the SVZ/SEZ (Gates et al., 1995). Understanding the factors that contribute to nor- mal SVZ/SEZ precursor migration could be important in developing approaches to induce such precursors to migrate to injured or degenerating regions of the brain.

Although the identity of the adult multipotent neural precursors in the SVZ/SEZ is still somewhat controversial (discussed later in this chapter), a num- ber of experiments have been performed to manipulate their fate and examine their potential, both *in vitro* and *in vivo*. These results will guide attempts to manipulate endogenous precursors for brain repair. *In vitro*, SVZ/SEZ precur- sors have been exposed to a number of factors to determine their responses. Generally, precursor cells have been removed from the brain, dissociated, and cultured in epidermal growth factor (EGF) and/or basic fibroblast growth factor 2 (FGF2). The EGF and/or FGF2 are then removed, and the cells are exposed to growth factors of interest. The details of this process, including the particular regions the cells are derived from, the media they are plated in, and the substrata they are plated on, can have significant effects on the fate of the precursors (Ahmed et al., 1995; Gritti et al., 1995; Kirschenbaum and Goldman, 1995; Gritti et al., 1996; Gross et al., 1996; Goldman et al., 1997; Williams et al., 1997; Arsenijevic and Weiss, 1998; Gritti et al., 1999; Whittemore et al., 1999; Gritti et al., 2002). The effects of several growth factors have also been tested *in vivo*, to investigate their effects under physiological conditions. Intracerebroventricularly (ICV) infused EGF or transforming growth factor (TGF)-α induce a dramatic increase in SVZ/SEZ precursor proliferation, and FGF2 induces a smaller increase in proliferation (Craig et al., 1996; Kuhn et al., 1997; Nakatomi et al., 2002). Even subcutaneously delivered FGF2 can induce the proliferation of SVZ/SEZ precursors in adult animals (Wagner et al., 1999). But despite the fact that newborn, mitogen-induced cells disperse into regions of the brain surrounding the ventricles, it is generally accepted that none of the newborn cells differentiate into neurons in these non-neurogenic regions (Kuhn et al., 1997). Interestingly, it has been recently reported that olfactory bulb neurogenesis increases during pregnancy and even following mating behavior in rodents, and it is mediated by the hormone prolactin (Shingo et al., 2003). These results further suggest that it may be possible to use growth factors/hormones to manipulate adult endogenous precursors *in vivo* for brain repair.

Several reports have attempted to establish the differentiation potential of SVZ/SEZ multipotent precursors, but these have yielded conflicting results. Postnatal mouse SVZ/SEZ precursors can differentiate into neurons in a number of regions in the developing neuraxis (Lim et al., 1997), while their fate is more limited to astroglia when they are transplanted into adult brain (Herrera et al., 1999). Adult mouse SVZ/SEZ precursors injected intravenously into sublethally irradiated mice have been reported to differentiate into hematopoetic cells, interpreted as demonstrating the broad potential of neural precursors for differentiation and interlineage "transdifferentiation"(Bjornson et al., 1999; Vescovi et al., 2002). However, it is possible that a contaminant blood stem cell or a chance transformation of cultured SVZ/SEZ cells led to a single transformant precursor accounting for this finding. It has also been reported that labeled multipotent neural precursors derived from adult mouse and transplanted into stage 4 chick embryos or developing mouse morulae or blastocysts, can integrate into the heart, liver, and intestine, and express proteins specific for each of these sites (Clarke et al., 2000). Moreover, acutely isolated and clonally derived neural precursors/stem cells have been shown to produce skeletal myotubes *in vitro* and *in vivo* following transplantation into adult animals (Galli et al., 2000). Though these results are not entirely unambiguous in view of recent reports of fusion between embryonic stem cells and mature cells (Terada et al., 2002;

Wurmser and Gage, 2002; Ying et al., 2002), adult multipotent neural precursors may not be totipotent, but they appear to be capable of differentiating into a wide variety of cell types under appropriate conditions, even if with low frequency.

These results indicate that the local cellular and molecular environment in which SVZ/SEZ neural precursors are located can play a significant role in their differentiation. Providing the cellular and molecular signals for appropriate differentiation and integration of new neurons will be critical for neuronal replacement therapies in which endogenous neural precursors are either transplanted or manipulated in situ.

Hippocampal Neurogenesis

Neurogenesis in the adult hippocampus has been extensively studied and described *in vivo* in adult rodents (Altman and Das, 1965), monkeys (Gould et al., 1998; Gould et al., 1999b; Kornack and Rakic, 1999) and humans (Eriksson et al., 1998). Newborn cells destined to become neurons are generated along the innermost aspect of the granule cell layer, the subgranular zone, of the dentate gyrus of the adult hippocampus. The cells migrate a short distance into the granule cell layer, send dendrites into the molecular layer of the hippocampus, and send their axons into the CA3 region of the hippocampus (Stanfield and Trice, 1988; Hastings and Gould, 1999; Markakis and Gage, 1999). Adult-born hippocampal granule neurons are morphologically indistinguishable from surrounding granule neurons (Gage et al., 1998; Kempermann et al., 2003) and appear to be functional (Song et al., 2002; van Praag et al., 2002). Hippocampal precursors cells are studied *in vitro* much like SVZ/SEZ precursors. Hippocampal precursors proliferate in response to FGF2 and can differentiate into astroglia, oligodendroglia, and neurons *in vitro* (Gage et al., 1998). Brain-derived neurotrophic factor (BDNF) increases both neuronal survival and differentiation, while neurotrophin-3 (NT3), neurotrophin-4/5 (NT4/5), and ciliary neurotrophic factor (CNTF) have more limited effects (Lowenstein and Arsenault, 1996). Further demonstrating the existence of precursors in the adult human, multipotent precursors derived from the adult human brain can be cultured *in vitro* (Kukekov et al., 1999; Roy et al., 2000).

Hippocampal neurogenesis occurs throughout adulthood, but declines with age (Kuhn et al., 1996). This age-related decline could be due to a depletion of multipotent precursors with time, a change in precursor cell properties, or a change in the levels of molecular factors that influence neurogenesis. Understanding what causes this age-related decrease in neurogenesis may be important in assessing the potential utility of future neuronal replacement therapies based on manipulation of endogenous precursors. Although aged rats have dramatically lower levels of neurogenesis than young rats, adrenalectomized aged rats have levels of neurogenesis very similar to those of young adrenalecto-mized rats (Cameron and McKay, 1999; Montaron et al., 1999). These results suggest that it is at least partially increased corticosteroids, which are produced by the adrenal glands, and not a decrease in the number of multipotent precur-sors, that leads to age-related decreases in neurogenesis. Events occurring in the hippocampus dramatically demonstrate that behavior and environment can have a quite direct influence on the brain's microcircuitry. Animals living in an enriched environment containing toys, running wheels, and social stimulation contain more surviving newborn neurons in their hippocampus than control mice living in standard cages (Kempermann et al., 1997). An intriguing, but completely speculative, idea that has been advanced by some in this field is that the processes mediating these effects on neurogenesis may underlie some of the benefits that physical and social therapies provide for patients with stroke and brain injury.

Some of the molecular mechanisms that mediate behavioral influences on hippocampal neurogenesis have begun to be elucidated (Aberg et al., 2000; Trejo et al., 2001). Stress increases systemic adrenal steroid levels and reduces hippocampal neurogenesis (Tanapat et al., 1998). Intriguingly, some antide-pressant medications appear to increase neurogenesis (Chen et al., 2000; Malberg et al., 2000). Together, these results demonstrate that adult neurogenesis can be modified by systemic signals, suggesting that modifying such systemic signals, and not only local ones, may be useful in developing potential future neuronal replacement therapies involving manipulation of endogenous precursors (Kempermann et al., 2000).

Adult hippocampal multipotent precursors can adopt a variety of fates *in vivo*, suggesting that they may be able to appropriately integrate into neuronal microcircuitry outside of the dentate gyrus of the hippocampus. Hippocampal precursors transplanted into neurogenic regions of the brain can differentiate into neurons, while precursors transplanted into non-neurogenic regions do not differentiate into neurons at all. Adult rat hippocampal precursors transplanted into the rostral migratory stream migrate to the olfactory bulb and differentiate into a neuronal subtype not found in the hippocampus, tyrosine-hydroxylase-positive neurons (Suhonen et al., 1996). However, although adult hippocampal precursors transplanted into the retina can adopt neuronal fates and extend neurites, they do not differentiate into photoreceptors, demonstrating at least conditional limitation of their differentiation fate potential (Nishida et al., 2000; Young et al., 2000). These findings demonstrate the importance of the local cellular and molecular microenvironment in determining the fate of multipotent precursors. These results also highlight that, although adult hippocampal precursors can adopt a variety of neuronal fates, they may not be able to adopt every neuronal fate.

Identity of adult neural precursors

The effort to identify the neural precursors that contribute to olfactory bulb neurogenesis has generated a great deal of controversy. Research took an interesting turn with the provocative discovery that "glia-like" cells, e.g. cells showing phenotypic and antigenic features of glia, including cytoplasmic glycogen inclusions and expression of the intermediate filament protein GFAP, glial acidic fibrillary protein, from the adult SVZ/SEZ can give rise to neurons and may therefore be neural precursors/stem cells (Doetsch et al., 1999). In a series of elegant experiments, Alvarez-Buylla and colleagues showed that some SVZ astrocytes in the adult mammalian brain are able to form multipotent neurospheres *in vitro* (considered by many a property of neural stem cells) and are the source of new neurons of the olfactory bulb. These data strongly suggest that at least a subset of SVZ astrocytic cells are neural precursors. They selectively labeled SVZ astrocytic cells and traced the label via selective viral infection using a receptor for an avian leukosis virus under the control of the GFAP promoter in transgenic mice, to newly generated neurons of the olfactory bulb (Doetsch et al., 1999). Similarly, GFAP-positive astrocyte-like cells in the subgranular layer (SGL) of the adult hippocampus (the other major site of adult neurogenesis) divide and generate new neurons (Seri et al., 2001). These data support the idea that a small subset of astrocyte-like cells have the properties of neural precursors.

In contrast with these results, Frisen's group published experiments reporting that ependymal cells lining the lumenal surface of the adult ventricular zone are adult neural precursors (Johansson et al., 1999). These results contrast with those of Doetsch et al. (1999), which reported that ependymal cells did not divide *in vivo* or under their culture conditions *in vitro*. A parallel report from van der Kooy's lab reported that, while ependymal cells divide as spheres *in vitro*, they do not posses multipotential precursor properties (Chiasson et al., 1999). It remains to be established whether experimental differences between the three groups account for their different results (reviewed in Barres, 1999). The idea that some subset of astrocyte-like cells can behave as neural precursors is supported by later studies suggesting that 1-10% of astrocytes isolated from several regions of the early postnatal brain (SVZ, cerebral cortex, cerebellum, spinal cord) and grown as monolayers, are able to form "neurospheres" that can give rise to both neurons and glia *in vitro* (Laywell et al., 2000). Moreover, retrovirally mediated expression of exogenous genes can drive some postnatal astrocyte-like cells to become neurons *in vitro* (Heins et al., 2002). Interestingly, there are developmental and spatial constraints to astrocyte multipotency, which is normally restricted to the first two postnatal weeks in mice, with the apparent exception of a subset of astrocyte-like cells from the SVZ, which appear to retain their multipotency during adulthood (Doetsch et al., 1999). This could mean that such cells in young animals may still be immature and retain their precursor attributes initially, until approximately P10-11 (Laywell et al., 2000). However, it may be that the neurogenic environment of the SVZ/SEZ can support/ and maintain the multipotency of a subset of astrocyte-like precursors, even in the adult brain.

Although the idea that glial cells could be neural precursors may sound unusual, it is supported by previous studies (long ignored) that in lizards and newts a special class of radial ependymoglia that extend from the ventricular zone (VZ) lumen to the pial surface, can divide and give rise to both neurons and glia after injury, critical for spinal cord regeneration (Chernoff, 1996). In line with these observations, similar radial glial cells are present in the adult brain of the canary in regions of active neurogenesis, suggesting that they may be the progenitors of newly generated neurons (Alvarez-Buylla et al., 1990). Moreover, results from our laboratory show that adult astrocytes from the cerebral cortex can dedifferentiate into transitional radial glia in response to signals that induce neurogenesis in transplanted immature precursors (Leavitt et al., 1999). In general, the notion that cells of the radial glial lineage may have features of neural precursors during development and thus be able to generate both glia and neurons is not new (Malatesta et al., 2000; Noctor et al., 2001; Noctor et al., 2002). We may imagine adult astrocyte-like cells as precursors with broad potential in neurogenic regions of the CNS (SVZ/SEZ and dentate gyrus) and more restricted potential in non-neurogenic regions, e.g. cerebral cortex, spinal cord. If this is true, one future challenge will be to understand whether astrocyte-like cells from non-neurogenic regions can be induced to assume multipotential neural precursor properties, toward the goal of neuronal repopulation and CNS repair in situ.

The location of adult mammalian multipotent precursors

If adult multipotent precursors were limited to the two neurogenic regions of the brain, the olfactory bulb and hippocampal dentate gyrus, it would severely limit the potential of neuronal replacement therapies based on in situ manipulation of endogenous precursors. However, adult multipotent precursors are not limited to the olfactory epithelium, anterior SVZ/SEZ, and hippocampus of the adult brain; they have been isolated and cultured *in vitro* from caudal portions of the SVZ/SEZ, septum (Palmer et al., 1995), striatum (Palmer et al., 1995), substantia nigra (Lie et al., 2002), cortex (Palmer et al., 1999), optic nerve (Palmer et al., 1999), spinal cord (Weiss et al., 1996; Shihabuddin et al., 1997), and retina (Tropepe et al., 2000). The precursors derived from all these regions can self-renew and differentiate into neurons, astroglia, and oligodendroglia *in vitro*, but they can normally differentiate only into glia or die *in vivo*. This would suggest that differences in local cues in different regions of the brain *in vivo* control and limit the fates of endogenous precursors. Precursors isolated from each region have different requirements for their proliferation and differentiation. Understanding the similarities and differences between the properties of multipotent precursors derived from different regions of the brain will be instrumental in developing neuronal replacement therapies based on manipulation of endogenous precursors.

CORTICAL NEUROGENESIS

Constitutive Neurogenesis of the Neocortex

The vast majority of studies, including our own, investigating potential neurogenesis in the neocortex of the well-studied rodent brain do not find normally occurring, constitutive adult cortical neurogenesis (Magavi et al., 2000; Benraiss et al., 2000). However, three studies report extremely low level, nonpurposefully induced neurogenesis in specific regions of the neocortex of adult primates, sometimes after electrophysiologic recording in those regions (Gould et al., 1999a; Gould et al., 2001) and in the visual cortex of adult rats (Kaplan, 1981). Our group and other groups have not been able to reproduce these findings in rodents or primates (Kornack and Rakic, 2001), so it is still unclear whether perhaps an extremely low level of neurogenesis occurs normally in the neocortex of any mammals, or whether there are technical explanations for false-positive results using specific experimental techniques.

Induction Cortical Neurogenesis

Multipotent precursors from several regions of the adult brain have a broad potential and can differentiate into at least three different cell types, astroglia, oligodendroglia, and neurons, given an appropriate *in vitro* or *in vivo* environment. This led us to explore the fate of multipotent precursors in an adult cortical environment that has been manipulated to support neurogenesis.

Our lab has previously shown that cortex undergoing synchronous degeneration of apoptotic projection neurons (Sheen et al., 1992; Macklis, 1993; Madison and Macklis, 1993; Sheen and Macklis, 1994; Scharff et al., 2000) forms an instructive environment that can guide the differentiation of transplanted cells. Immature neurons or multipotent neural precursors transplanted into targeted cortex migrate selectively to layers of cortex undergoing degeneration of projection neurons (Macklis, 1993; Sheen and Macklis, 1995; Hernit-Grant and Macklis, 1996), differentiate into projection neurons (Macklis, 1993; Sheen and Macklis, 1995; Hernit-Grant and Macklis, 1996; Snyder et al., 1997; Leavitt et al., 1999; Shin et al., 2000; Fricker-Gates et al., 2002), receive afferent synapses (Macklis, 1993; Snyder et al., 1997; Shin et al., 2000; Fricker-Gates et al., 2002), express appropriate neurotransmitters and receptors (Shin et al., 2000; Fricker-Gates et al., 2002), and re-form appropriate long-distance connections to the original contralateral targets of the degenerating neurons (Hernit-Grant and Macklis, 1996; Shin et al., 2000; Fricker-Gates et al., 2002). Together, these results suggested to us that cortex undergoing targeted apoptotic degeneration could direct endogenous multipotent precursors to integrate into adult cortical microcircuitry.

Based on the results outlined above, we investigated the fate of endogenous multipotent precursors in cortex undergoing targeted apoptotic degeneration (Magavi et al., 2000). In these experiments, we addressed (1) the question of whether the normal absence of constitutive cortical neurogenesis reflects an intrinsic limitation of the endogenous neural precursors' potential, or rather a lack of appropriate microenvironmental signals necessary for neuronal differentiation and/or survival. We also asked the question of (2) whether endogenous neural precursors could potentially be manipulated in situ, toward future neuronal replacement therapies. Finally, (3) we tried to understand if the same molecular signals can support/induce endogenous precursors to differentiate into different categories of neurons appropriate for the region of cortex targeted. The ability to repopulate the brain with the correct type of neurons for a specific region will almost certainly be a pre-requisite to functional replacement of lost neurons from endogenous multipotent precursors.

We found that endogenous multipotent precursors, normally located in the adult brain, could be induced to differentiate into neurons in the adult mammalian neocortex. Moreover, the same sequence/combination of signals could support differentiation of two distinct populations of projection neurons, i.e. corticothalamic projection neurons of layer 6 of cortex (Magavi et al., 2000) and corticospinal projection neurons of layer 5 of cortex (Chen et al., unpublished observations). In the first study, we induced synchronous apoptotic degeneration (Macklis, 1993; Sheen and Macklis, 1995) of cortico-thalamic neurons in layer 6 of anterior cortex (Figure 1) and examined the fates of dividing cells within the cortex, using BrdU and markers of progressive neuronal differentiation. BrdU$^+$ newborn cells expressed NeuN, a mature neuronal marker, and survived at least 28 weeks. Subsets of BrdU$^+$ precursors expressed doublecortin (dcx), a protein expressed exclusively in immature, usually migrating neurons

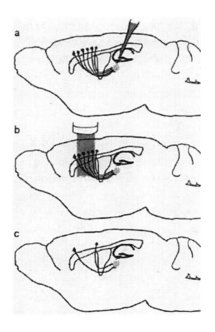

Figure 1. Targeted apoptotic neurodegeneration. Biophysically targeted apoptotic neurodegeneration produces highly specific cell death of selected projection neurons within defined regions of the neocortex. Degeneration results from the photoactivation of retrogradely transported nanospheres carrying the chromophore chlorin e6. (**a**) Nanospheres carrying chlorin e6 are injected into the axonal terminal fields of the targeted projection neurons. The nanospheres are retrogradely transported to the somata of the projection neurons and stored nontoxically within neuronal lysosomes. (**b**) The projection neurons are then exposed to long wavelength light, which specifically induces the chlorin e6 to produce singlet oxygen, and induces apoptosis of neurons containing chlorin e6. (**c**) Since neither the chlorin e6 nor the long-wavelength light cause apoptosis by themselves, only neurons both selectively labeled with the chromophore and located in the controlled light path undergo apoptotic degeneration

(Francis et al., 1999; Gleeson et al., 1999), and hu, an early neuronal marker (Marusich et al., 1994; Barami et al., 1995). We observed no new neurons in control, intact cortex. Some newborn neurons had pyramidal neuron morphology (large, 10-15 μm diameter somata with apical processes; see Figure 1A-D from Magavi et al., 2000) characteristic of neurons that give rise to long-distance projections. Retrograde labeling from thalamus demonstrated that newborn, BrdU+ neurons can form long-distance corticothalamic connections. Together, these results demonstrate that endogenous neural precursors can be induced in situ to differentiate in a layer- and region-specific manner into cortical neurons appropriate for the region of cortex targeted, survive for many months, and form appropriate long-distance connections in the adult mammalian brain.

To further understand the source of these newborn neurons, we examined where newly born cells were located in experimental and control animals. In experimental cortex, a small population of newborn cells located within the cortex of experimental animals may originate from precursors located within cortex itself, whereas a predominant population, which was not present in control animals, appeared to originate in or near the subventricular zone (SVZ). At two weeks, pairs of BrdU+ but non-neuronal cells, apparently daughters of the same precursor, were found throughout both control and experimental cortex (Figure 1I from Magavi et al., 2000). It is possible that some of these newborn cells are the daughters of cortically located adult multipotent neural precursors that have been described *in vitro* (Marmur et al., 1998; Palmer et al., 1999).

We investigated the early differentiation and migration of the newborn neurons using a marker of early postmitotic neurons, doublecortin (dcx; exclusively expressed in immature migrating and differentiating neurons; Francis et al., 1999; Gleeson et al., 1999). In experimental mice only, newborn BrdU$^+$/dcx$^+$ neurons with appeared to migrate from the SVZ through the corpus callosum and into targeted regions of cortex (Figure 2 from Magavi et al., 2000). No BrdU$^+$/dcx$^+$ cells were found in the corpus callosum or cortex of control animals. Newly born BrdU$^+$/dcx$^+$ neurons within the corpus callosum displayed morphologies characteristic of migrating neurons, while newborn dcx$^+$ neurons located in deep layers of cortex displayed more complex morphologies with apical processes that suggest their further differentiation. These results demonstrate the progressive differentiation of endogenous precursors into mature neurons. The location of dcx$^+$ neurons suggests that at least some of the newborn neurons that form in targeted cortex are derived directly from SVZ precursors. However, these data do not rule out the possibility that other precursors contribute to neurogenesis. Further understanding of the source of the cells that can contribute to induced neurogenesis could be critical for determining whether endogenous precursors can potentially form the basis of effective neuronal replacement therapies.

The continued differentiation of these newborn neurons was examined using antibodies to hu, an RNA-binding protein that begins to be expressed in neuronal nuclei and somata soon after differentiation (Marusich et al., 1994; Barami et al., 1995; Figure 3 from Magavi et al., 2000), and by expression of NeuN, a mature neuronal marker (Figure 2). The expression of these markers further confirms induced cortical neuron differentiation by endogenous neural precursors. These newborn cortical neurons, derived from endogenous precursors, can reform long distance projections. We injected the retrograde label FluoroGold into the same thalamic sites as the original nanosphere injections. At 9 weeks, we observed newborn BrdU$^+$ neurons retrogradely labeled with FluoroGold that had large nuclei and large cell bodies denoting pyramidal projection neuron morphology (Figure 3). These results show that endogenous precursors that differentiate into mature neurons can establish appropriate, long-distance cortico-thalamic connections in the adult brain.

Taken together, our results demonstrate that endogenous neural precursors can be induced to differentiate into neocortical neurons in a layer- and region-specific manner and reform appropriate corticothalamic connections in regions of adult mammalian neocortex that do not normally undergo neurogenesis. The same microenvironment that supports the migration, neuronal differentiation, and appropriate axonal extension of transplanted neuroblasts and precursors also supports and instructs the neuronal differentiation and axon extension of endogenous precursors. More recently, we have applied similar

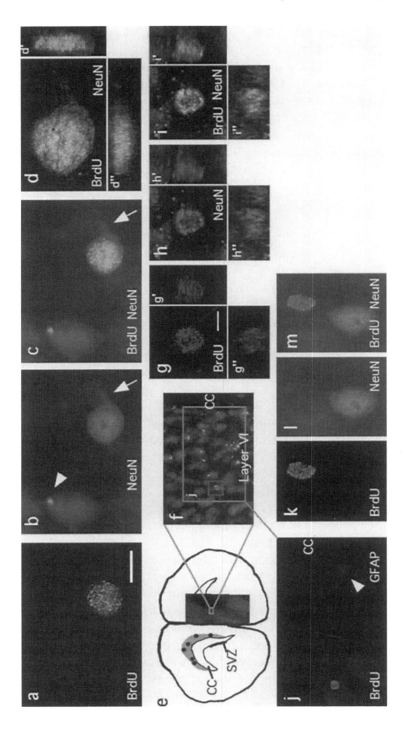

Figure 2. Induction of cortical neuronal differentiation in endogenous neural precursors. Newborn, BrdU⁺ cells can be induced to differentiate into neurons expressing NeuN, a mature neuronal marker, in regions of cortex undergoing targeted apoptotic degeneration of corticothalamic neurons. **a.** A large, densely stained BrdU⁺ nucleus (red) 28 weeks after induction of apoptosis. Scale bar: 10 μm. **b.** The BrdU⁺ neuron, lower right, shows typical NeuN staining, with strong nuclear and weaker cytoplasmic labeling. Apical dendrite (arrow). The neuron at left in a different focal plane remains from the original nanosphere-targeted neuronal population; a lysosome containing nanospheres is indicated (arrowhead). No BrdU⁺/NeuN⁺ newborn neurons contained nanospheres, demonstrating that they are not original, targeted neurons. **c.** Overlay of **a** and **b. d.** Confocal images combined to produce a 3-dimensional reconstruction of this newborn neuron. **d', d''.** cell viewed along the x-axis and y-axis, respectively, demonstrating co-localization of BrdU and NeuN. **e.** Left, camera lucida showing location of BrdU⁺/NeuN⁺ cells (dots) within the targeted region (gray). Right, a sample newborn neuron in cortical layer VI; corpus callosum (CC); subventricular zone (SVZ). **f.** Higher magnification view of layer VI shows a BrdU⁺ newborn neuron (red box). **g-i.** Confocal 3-d reconstruction of red boxed region. **g', h', i'.** X-axis. **g'', h'', i''.** Y-axis. **g.** BrdU (red), indicating the cell underwent mitosis during the 2 weeks following induction of apoptosis. Scale bar: 5 μm **h.** NeuN (green). **i.** Merged image of **g** and **h.** BrdU (red) and NeuN (green) labeling are coincident in all three dimensions. **j.** The BrdU⁺/NeuN⁺ neuron (red) is GFAP-negative (arrowhead). **k, l, m.** The presence of BrdU⁺/NeuN⁻ and BrdU⁻/NeuN⁺ cells demonstrate that the double labeling protocol is specific. Image from same section as **a-d. k.** BrdU (red) and **l.** NeuN (green) do not show cross-reactivity. **m.** Overlay of **k** and **l.** [from Magavi et al., Nature 405:951-955 (2002)]

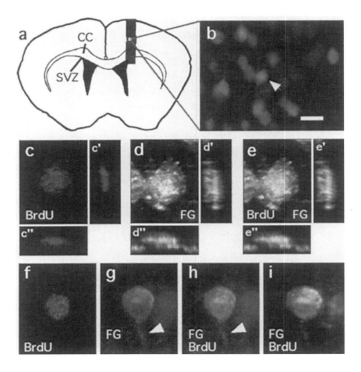

Figure 3. Newborn, BrdU⁺ neurons in cortex extend long-distance projections to the thalamus. a. Camera lucida of a BrdU⁺ (red) / FluoroGold⁺ (FG⁺) (white) retrogradely labeled neuron in layer VI of the cortex at 9 weeks. Corpus callosum (CC) **b.** Higher magnification shows the newborn corticothalamic neuron (arrow) and FG⁺/BrdU⁻ original neurons. Scale bar: 20 μm. **c-e.** Confocal 3-dimensional reconstruction of the neuron. **c', d', e'.** X-axis. **c'', d'', e''.** Y-axis. **c.** BrdU⁺ nucleus, indicating the cell underwent mitosis in the two weeks following induction of apoptosis. **d.** FG⁺ soma, indicating that this neuron projects to the thalamus. **e.** Merged image, confirming double-labeling in three dimensions. **f.** BrdU⁺ (red) nucleus of an adult-born FG⁺ cortico-thalamic neuron. **g.** FG⁺ (blue) cell body with labeled axon (arrow). **h.** Overlay of **f** and **g**. **i.** Confocal microscopy confirms double labeling. [from Magavi et al., Nature 405:951-955 (2002)]

approaches to layer 5 corticospinal projection neurons and find that endogenous precursors are capable of differentiating into such projection neurons, which then become anatomically integrated into neocortical circuitry.

Together, these results indicate that it is possible to induce neurogenesis of neocortical projection neurons de novo, even in the highly inhibitory environment of the adult neocortex, from endogenous neural precursors in situ without transplantation. Elucidation and manipulation of the relevant molecular controls over induced neurogenesis may allow the future development of novel neuronal replacement therapies for CNS injury and neurodegenerative disease.

CONCLUSION

Recent research suggests that it may be possible to manipulate endogenous neural precursors in situ to undergo neurogenesis in the adult brain, toward future neuronal replacement therapy for neurodegenerative diseases and other CNS injuries. Multipotent precursors, capable of differentiating into astroglia, oligodendroglia, and neurons, exist in many regions of the adult brain. These precursors have considerable plasticity, and, although they may have limitations in their integration into some host sites, they are capable of differentiation into neurons appropriate for a wide variety of recipient regions when heterotopically transplanted. Many adult precursors are capable of migrating long distances, using both tangential and radial forms of migration. Endogenous adult neural precursors are also capable of extending axons significant distances through the adult brain. In addition, *in vitro* and *in vivo* experiments have begun to elucidate the responses of endogenous precursors to both growth factors and behavioral manipulations, and are beginning to provide key information toward manipulation of their proliferation and differentiation. Recent experiments from our lab have shown that, under appropriate conditions, endogenous precursors can be induced and controlled to differentiate into neurons, extend long distance axonal projections, and survive for long periods in regions of the adult brain that do not normally undergo neurogenesis. These results indicate that there exists a sequence and combination of molecular signals by which neurogenesis can be induced in the adult mammalian cerebral cortex, where it does not normally occur.

Together, these data suggest that neuronal replacement therapies based on manipulation of endogenous precursors may be possible in the future. However, several questions must be answered before neuronal replacement therapies using endogenous precursors become a reality. The multiple signals that are responsible for endogenous precursor division, migration, differentiation, axon extension, and survival will need to be elucidated in order for such therapies to be developed efficiently (Catapano et al., 2001). Given the cellular complexity of the brain with several different types of neurons and glia, the ability to induce differentiation of the correct kind of neurons appropriate to a specific brain site will most likely be a prerequisite for successful neuronal replacement therapies. Finally, newborn neurons must be able to establish the appropriate connections, often to distant targets, and respond to functional input in a physiological manner in order to achieve functional integration.

These challenges also exist for neuronal replacement therapies based on transplantation of precursors, since donor cells, whatever their source, must interact with the mature CNS's environment in order to integrate into the brain. In addition, it remains an open question whether potential therapies manipulating endogenous precursors in situ would be necessarily limited to portions of the brain near adult neurogenic regions. However, even if multipotent precursors are not located in very high numbers outside of neurogenic regions of the brain, it may be possible to induce them to proliferate from the smaller numbers that are more widely distributed throughout the neuraxis. Potentially, it may be possible to induce the repopulation of diseased mammalian brain via the specific activation and instructive control over endogenous neural precursors along desired neuronal lineages.

A distinct and intriguing issue relates to the reason why neurogenesis occurs at all in the adult brain. In other words, we really do not know very much about the function that adult born neurons may have. Most hypotheses about the function of these neurons are at this stage speculative. Nonetheless, it has been proposed that in the case of olfactory bulb neurogenesis, newborn interneurons may play a part in sharpening the response of neighboring cells to odors (Gheusi et al., 2000). More recently, Weiss's group has reported that olfactory bulb adult neurogenesis may be important to the ability of a rodent mother to recognize and nurture her offspring, as it is shown that a release of prolactin in pregnancy is associated with increased olfactory bulb neurogenesis (Shingo et al., 2003). Ongoing work in our lab directly addresses this issue, and has uncovered a unique, experience-dependent function of newborn olfactory granule neurons Magavi et al., unpublished observations). The second main region of ongoing constitutive neurogenesis in the adult is the dentate gyrus of the hippocampus. Given the fact that the hippocampus is critical to learning and memory functions, there has been a great deal of interest in understanding dentate gyrus neurogenesis and in finding ways to boost it. Although much progress has been made, we still do not know what newborn neurons of the dentate gyrus really do—in fact, the role of the dentate gyrus is poorly understood overall, and its role in memory formation/storage, if any, is not at all clear. The idea that there may be a direct link between adult hippocampal neurogenesis and enhanced learning abilities, although quite intriguing, has been so far just correlative. Considering the speed of progress achieved in this field during the past decade, the future holds great promise to offer significant insight into these and other critical issues of this field.

Acknowledgements

This work was partially supported by grants from the NIH (NS41590, NS45523, HD28478, MRRC HD18655), Alzheimer's Association, Human Frontiers Science Program, and National Science Foundation to J.D.M. P.A. was supported by a Wills Foundation fellowship. S.S.M. was partially supported by an NIH predoctoral training grant and fellowships from the Leopold Schepp

Foundation and the Lefler Foundation. We would like to thank Farren B. Briggs for outstanding technical assistance in preparing this chapter. Parts of this chapter were updated and modified from a similar review article by the same authors for a different readership (Arlotta et al., 2003).

REFERENCES

Aberg MA, Aberg ND, Hedbacker H, Oscarsson J, Eriksson PS (2000) Peripheral infusion of IGF-I selectively induces neurogenesis in the adult rat hippocampus. J Neurosci 20:2896-2903.

Ahmed S, Reynolds BA, Weiss S (1995) BDNF enhances the differentiation but not the survival of CNS stem cell-derived neuronal precursors. J Neurosci 15:5765-5778.

Altman J (1969) Autoradiographic and histological studies of postnatal neurogenesis. IV. Cell proliferation and migration in the anterior forebrain, with special reference to persisting neurogenesis in the olfactory bulb. J Comp Neurol 137:433-457.

Altman J, Das GD (1965) Autoradiographic and histological evidence of postnatal hippocampal neurogenesis in rats. J Comp Neurol 124:319-335.

Alvarez-Buylla A, Theelen M, Nottebohm F (1990) Proliferation "hot spots" in adult avian ventricular zone reveal radial cell division. Neuron 5:101-109.

Arlotta P, Magavi SS, Macklis JD. (2003) Molecular manipulation of neural precursors in situ: induction of adult cortical neurogenesis. Exp Gerontol. 38(1-2):173-82.

Arsenijevic Y, Weiss S (1998) Insulin-like growth factor-I is a differentiation factor for postmitotic CNS stem cell-derived neuronal precursors: distinct actions from those of brain-derived neurotrophic factor. J Neurosci 18:2118-2128.

Barami K, Iversen K, Furneaux H, Goldman SA (1995) Hu protein as an early marker of neuronal phenotypic differentiation by subependymal zone cells of the adult songbird forebrain. J Neurobiol 28:82-101.

Barber PC (1982) Neurogenesis and regeneration in the primary olfactory pathway of mammals. Bibliotheca Anatomica:12-25.

Barres BA (1999) A new role for glia: generation of neurons! Cell 97:667-670.

Benraiss A, Chmielnicki E, Lerner K, Roh D, Goldman SA. (2001) Adenoviral brain-derived neurotrophic factor induces both neostriatal and olfactory neuronal recruitment from endogenous progenitor cells in the adult forebrain. J Neurosci. 21(17):6718-31.

Bernier PJ, Vinet J, Cossette M, Parent A (2001) Characterization of the subventricular zone of the adult human brain: evidence for the involvement of Bcl-2. Neurosci Res 37:67-78.

Bjornson CR, Rietze RL, Reynolds BA, Magli MC, Vescovi AL (1999) Turning brain into blood: a hematopoietic fate adopted by neural stem cells *in vivo*. Science 283:534-537.

Cameron HA, McKay RD (1999) Restoring production of hippocampal neurons in old age. Nat Neurosci 2:894-897.

Carlen M, Cassidy RM, Brismar H, Smith GA, Enquist LW, Frisen J (2002) Functional intergration of adult-born neurons. Cur Biol 12:606-608.

Catapano LA, Arnold MW, Perez FA, Macklis JD (2001) Specific neurotrophic factors support the survival of cortical projection neurons at distinct stages of development. J Neurosci 21:8863-8872.

Chen G, Rajkowska G, Du F, Seraji-Bozorgzad N, Manji HK (2000) Enhancement of hippocampal neurogenesis by lithium. J Neurochem 75:1729-1734.

Chernoff EA (1996) Spinal cord regeneration: a phenomenon unique to urodeles? Int J Dev Biol 40:823-831.

Chiasson BJ, Tropepe V, Morshead CM, van der Kooy D (1999) Adult mammalian forebrain ependymal and subependymal cells demonstrate proliferative potential, but only subependymal cells have neural stem cell characteristics. J Neurosci 19:4462-4471.

Clarke DL, Johansson CB, Wilbertz J, Veress B, Nilsson E, Karlstrom H, Lendahl U, Frisen J (2000) Generalized potential of adult neural stem cells. [see comments]. Science 288:1660-1663.

Craig CG, Tropepe V, Morshead CM, Reynolds BA, Weiss S, van der Kooy D (1996) *In vivo* growth factor expansion of endogenous subependymal neural precursor cell populations in the adult mouse brain. J Neurosci 16:2649-2658.

Crews L, Hunter D (1994) Neurogenesis in the olfactory epithelium. Persp Dev Neurobiol 2:151-161.

Doetsch F, Alvarez-Buylla A (1996) Network of tangential pathways for neuronal migration in adult mammalian brain. Proc Natl Acad Sci USA 93:14895-14900.

Doetsch F, Caille I, Lim DA, Garcia-Verdugo JM, Alvarez-Buylla A (1999) Subventricular zone astrocytes are neural stem cells in the adult mammalian brain. Cell 97:703-716.

Eriksson PS, Perfilieva E, Bjork-Eriksson T, Alborn AM, Nordborg C, Peterson DA, Gage FH (1998) Neurogenesis in the adult human hippocampus [see comments]. Nat Med 4:1313-1317.

Francis F, Koulakoff A, Boucher D, Chafey P, Schaar B, Vinet MC, Friocourt G, McDonnell N, Reiner O, Kahn A, McConnell SK, Berwald-Netter Y, Denoulet P, Chelly J (1999) Doublecortin is a developmentally regulated, microtubule-associated protein expressed in migrating and differentiating neurons. Neuron 23:247-256.

Fricker-Gates RA, Shin JJ, Tai CC, Catapano LA, Macklis JD (2002) Late-stage immature neocortical neurons reconstruct interhemispheric connections and form synaptic contacts with increased efficiency in adult mouse cortex undergoing targeted neurodegeneration. J Neurosci 22:4045-4056.

Gage FH, Kempermann G, Palmer TD, Peterson DA, Ray J (1998) Multipotent progenitor cells in the adult dentate gyrus. J Neurobiol 36:249-266.

Galli R, Borello U, Gritti A, Minasi MG, Bjornson C, Coletta M, Mora M, De Angelis MG, Fiocco R, Cossu G, Vescovi AL (2000) Skeletal myogenic potential of human and mouse neural stem cells. Nat Neurosci 3:986-991.

Garcia-Verdugo JM, Doetsch F, Wichterle H, Lim DA, Alvarez-Buylla A (1998) Architecture and cell types of the adult subventricular zone: in search of the stem cells. J Neurobiol 36:234-248.

Gates MA, Thomas LB, Howard EM, Laywell ED, Sajin B, Faissner A, Gotz B, Silver J, Steindler DA (1995) Cell and molecular analysis of the developing and adult mouse subventricular zone of the cerebral hemispheres. J Comp Neurol 361:249-266.

Gheusi G, Cremer H, McLean H, Chazal G, Vincent JD, Lledo PM (2000) Importance of newly generated neurons in the adult olfactory bulb for odor discrimination. Proc Natl Acad Sci USA 97:1823-1828.

Gleeson JG, Lin PT, Flanagan LA, Walsh CA (1999) Doublecortin is a microtubule-associated protein and is expressed widely by migrating neurons. Neuron 23:257-271.

Goldman SA, Kirschenbaum B, Harrison-Restelli C, Thaler HT (1997) Neuronal precursors of the adult rat subependymal zone persist into senescence, with no decline in spatial extent or response to BDNF. J Neurobiol 32:554-566.

Gould E, Reeves AJ, Graziano MSA, Gross CG (1999a) Neurogenesis in the Neocortex of Adult Primates. Science 286:548-552.

Gould E, Vail N, Wagers M, Gross CG (2001) Adult-generated hippocampal and neocortical neurons in macaques have a transient existence. Proc Natl Acad Sci USA 98:10910-10917.

Gould E, Tanapat P, McEwen BS, Flugge G, Fuchs E (1998) Proliferation of granule cell precursors in the dentate gyrus of adult monkeys is diminished by stress. Proc Natl Acad Sci S A 95:3168-3171.

Gould E, Reeves AJ, Fallah M, Tanapat P, Gross CG, Fuchs E (1999b) Hippocampal neurogenesis in adult Old World primates. ProcNatl Acad Sci USA96:5263-5267.

Gritti A, Vescovi AL, Galli R (2002) Adult neural stem cells: plasticity and developmental potential. J Physiol 96:81-90.

Gritti A, Cova L, Parati EA, Galli R, Vescovi AL (1995) Basic fibroblast growth factor supports the proliferation of epidermal growth factor-generated neuronal precursor cells of the adult mouse CNS. Neurosci Lett 185:151-154.

Gritti A, Frolichsthal-Schoeller P, Galli R, Parati EA, Cova L, Pagano SF, Bjornson CR, Vescovi AL (1999) Epidermal and fibroblast growth factors behave as mitogenic regulators for a single multipotent stem cell-like population from the subventricular region of the adult mouse forebrain. J Neurosci 19:3287-3297.

Gritti A, Parati EA, Cova L, Frolichsthal P, Galli R, Wanke E, Faravelli L, Morassutti DJ, Roisen F, Nickel DD, Vescovi AL (1996) Multipotential stem cells from the adult mouse brain proliferate and self-renew in response to basic fibroblast growth factor. J Neurosci 16:1091-1100.

Gross RE, Mehler MF, Mabie PC, Zang Z, Santschi L, Kessler JA (1996) Bone morphogenetic proteins promote astroglial lineage commitment by mammalian subventricular zone progenitor cells. Neuron 17:595-606.

Hastings NB, Gould E (1999) Rapid extension of axons into the CA3 region by adult-generated granule cells [published erratum appears in J Comp Neurol 1999 Dec 6;415(1):144]. J Comp Neurol 413:146-154.

Heins N, Malatesta P, Cecconi F, Nakafuku M, Tucker KL, Hack MA, Chapouton P, Barde YA, Gotz M (2002) Glial cells generate neurons: the role of the transcription factor Pax6. Nat Neurosci 5:308-315.

Hernit-Grant CS, Macklis JD (1996) Embryonic neurons transplanted to regions of targeted photolytic cell death in adult mouse somatosensory cortex re-form specific callosal projections. Exp Neurol 139:131-142.

Herrera DG, Garcia-Verdugo JM, Alvarez-Buylla A (1999) Adult-derived neural precursors transplanted into multiple regions in the adult brain. Ann Neurol 46:867-877.

Hu H, Rutishauser U (1996) A septum-derived chemorepulsive factor for migrating olfactory interneuron precursors. Neuron 16:933-940.

Hu H, Tomasiewicz H, Magnuson T, Rutishauser U (1996) The role of polysialic acid in migration of olfactory bulb interneuron precursors in the subventricular zone. Neuron 16:735-743.

Johansson CB, Momma S, Clarke DL, Risling M, Lendahl U, Frisen J (1999) Identification of a neural stem cell in the adult mammalian central nervous system. Cell 96:25-34.

Kaplan MS (1981) Neurogenesis in the 3-month-old rat visual cortex. J Comp Neurology 195:323-338.

Kempermann G, Kuhn HG, Gage FH (1997) More hippocampal neurons in adult mice living in an enriched environment. Nature 386:493-495.

Kempermann G, van Praag H, Gage FH (2000) Activity-dependent regulation of neuronal plasticity and self repair. Prog Brain Res 127:35-48.

Kempermann G, Gast D, Kroneberg G, Yamaguchi M, Gage FH (2003) Early determination and long-term persistence of adult-generated new neurons in the hippocampus of mice. Development 130:391-399.

Kirschenbaum B, Goldman SA (1995) Brain-derived neurotrophic factor promotes the survival of neurons arising from the adult rat forebrain subependymal zone. Proc Natl Acad Sci USA 92:210-214.

Kirschenbaum B, Doetsch F, Lois C, Alvarez-Buylla A (1999) Adult subventricular zone neuronal precursors continue to proliferate and migrate in the absence of the olfactory bulb. J Neurosci 19:2171-2180.

Kirschenbaum B, Nedergaard M, Preuss A, Barami K, Fraser RA, Goldman SA (1994) *In vitro* neuronal production and differentiation by precursor cells derived from the adult human forebrain. Cer Cortex 4:576-589.

Kornack DR, Rakic P (1999) Continuation of neurogenesis in the hippocampus of the adult macaque monkey. Proc Natl Acad Sci USA 96:5768-5773.

Kornack DR, Rakic P (2001) Cell proliferation without neurogenesis in adult primate neocortex. Science 294:2127-2130.

Kuhn HG, Dickinson-Anson H, Gage FH (1996) Neurogenesis in the dentate gyrus of the adult rat: age-related decrease of neuronal progenitor proliferation. J Neurosci 16:2027-2033.

Kuhn HG, Winkler J, Kempermann G, Thal LJ, Gage FH (1997) Epidermal growth factor and fibroblast growth factor-2 have different effects on neural progenitors in the adult rat brain. J Neurosci 17:5820-5829.

Kukekov VG, Laywell ED, Suslov O, Davies K, Scheffler B, Thomas LB, O'Brien TF, Kusakabe M, Steindler DA (1999) Multipotent stem/progenitor cells with similar properties arise from two neurogenic regions of adult human brain. Exp Neurol 156:333-344.

Laywell ED, Rakic P, Kukekov VG, Holland EC, Steindler DA (2000) Identification of a multipotent astrocytic stem cell in the immature and adult mouse brain. Proc Natl Acad Sci USA 97:13883-13888.

Leavitt BR, Hernit-Grant CS, Macklis JD (1999) Mature astrocytes transform into transitional radial glia within adult mouse neocortex that supports directed migration of transplanted immature neurons. Exp Neurol 157:43-57.

Lie DC, Dziewczapolski G, Willhoite AR, Kaspar BK, Shults CW, Gage FH (2002) The adult substania nigra contains progenitor cells with neurogenic potential. J Neurosci 22:6639-6649.

Lim DA, Fishell GJ, Alvarez-Buylla A (1997) Postnatal mouse subventricular zone neuronal precursors can migrate and differentiate within multiple levels of the developing neuraxis. Proc Natl Acad Sci USA 94:14832-14836.

Lois C, Alvarez-Buylla A (1994) Long-distance neuronal migration in the adult mammalian brain. Science 264:1145-1148.

Lowenstein DH, Arsenault L (1996) The effects of growth factors on the survival and differentiation of cultured dentate gyrus neurons. J Neurosci 16:1759-1769.

Luskin MB (1993) Restricted proliferation and migration of postnatally generated neurons derived from the forebrain subventricular zone. Neuron 11:173-189.

Luskin MB, Boone MS (1994) Rate and pattern of migration of lineally-related olfactory bulb interneurons generated postnatally in the subventricular zone of the rat. Chem Senses 19:695-714.

Macklis JD (1993) Transplanted neocortical neurons migrate selectively into regions of neuronal degeneration produced by chromophore-targeted laser photolysis. J Neurosci 13:3848-3863.

Madison R, Macklis JD (1993) Noninvasively induced degeneration of neocortical pyramidal neurons *in vivo*: selective targeting by laser activation of retrogradely transported photolytic chromophore. Exp Neurol 121:153-159.

Magavi SS, Leavitt BR, Macklis JD (2000) Induction of neurogenesis in the neocortex of adult mice [see comments]. Nature 405:951-955.

Malatesta P, Hartfuss E, Gotz M (2000) Isolation of radial glial cells by fluorescent-activated cell sorting reveals a neuronal lineage. Development 127:5253-5263.

Malberg JE, Eisch AJ, Nestler EJ, Duman RS (2000) Chronic antidepressant treatment increases neurogenesis in adult rat hippocampus. J Neurosci (Online) 20:9104-9110.

Markakis EA, Gage FH (1999) Adult-generated neurons in the dentate gyrus send axonal projections to field CA3 and are surrounded by synaptic vesicles. J Comp Neurol 406:449-460.

Marmur R, Mabie PC, Gokhan S, Song Q, Kessler JA, Mehler MF (1998) Isolation and developmental characterization of cerebral cortical multipotent progenitors. Dev Biol (Orlando) 204:577-591.

Marusich MF, Furneaux HM, Henion PD, Weston JA (1994) Hu neuronal proteins are expressed in proliferating neurogenic cells. J Neurobiol 25:143-155.

Mason HA, Ito S, Corfas G (2001) Extracellular signals that regulate the tangential migration of olfactory bulb neuronal precursors: inducers, inhibitors, and repellents. J Neurosci 21:7654-7663.

Montaron MF, Petry KG, Rodriguez JJ, Marinelli M, Aurousseau C, Rougon G, Le Moal M, Abrous DN (1999) Adrenalectomy increases neurogenesis but not PSA-NCAM expression in aged dentate gyrus. Eur J Neurosci 11:1479-1485.

Nakatomi H, Kuriu T, Okabe S, Yamamoto S, Hatano O, Kawahara N, Tamura A, Kirino T, Nakafuku M (2002) Regeneration of hippocampal pyramidal neurons after ischemic brain injury by recruitment of endogenous neural progenitors. Cell 110:429-441.

Nishida A, Takahashi M, Tanihara H, Nakano I, Takahashi JB, Mizoguchi A, Ide C, Honda Y (2000) Incorporation and differentiation of hippocampus-derived neural stem cells transplanted in injured adult rat retina. Invest Ophthalmol Vis Sci 41:4268-4274.

Noctor SC, Flint AC, Weissman TA, Dammerman RS, Kriegstein AR (2001) Neurons derived from radial glial cells establish radial units in neocortex. Nature 409:714-720.

Noctor SC, Flint AC, Weissman TA, Wong WS, Clinton BK, Kriegstein AR (2002) Dividing precursor cells of the embryonic cortical ventricular zone have morphological and molecular characteristics of radial glia. J Neurosci 22:3161-3173.

Palmer TD, Ray J, Gage FH (1995) FGF-2-responsive neuronal progenitors reside in proliferative and quiescent regions of the adult rodent brain. Mol Cell Neurosci 6:474-486.

Palmer TD, Markakis EA, Willhoite AR, Safar F, Gage FH (1999) Fibroblast growth factor-2 activates a latent neurogenic program in neural stem cells from diverse regions of the adult CNS. J Neurosci 19:8487-8497.

Pincus DW, Keyoung HM, Harrison-Restelli C, Goodman RR, Fraser RA, Edgar M, Sakakibara S, Okano H, Nedergaard M, Goldman SA (1998) Fibroblast growth factor-2/brain-derived neurotrophic factor-associated maturation of new neurons generated from adult human subependymal cells. Ann Neurol 43:576-585.

Reynolds BA, Weiss S (1992) Generation of neurons and astrocytes from isolated cells of the adult mammalian central nervous system. Science 255:1707-1710.

Richards LJ, Kilpatrick TJ, Bartlett PF (1992) De novo generation of neuronal cells from the adult mouse brain. Proc Natl Acad Sci USA 89:8591-8595.

Rousselot P, Lois C, Alvarez-Buylla A (1995) Embryonic (PSA) N-CAM reveals chains of migrating neuroblasts between the lateral ventricle and the olfactory bulb of adult mice. J Comp Neurol 351:51-61.

Roy NS, Wang S, Jiang L, Kang J, Benraiss A, Harrison-Restelli C, Fraser RA, Couldwell WT, Kawaguchi A, Okano H, Nedergaard M, Goldman SA (2000) *In vitro* neurogenesis by progenitor cells isolated from the adult human hippocampus [see comments]. Nat Med 6:271-277.

Scharff C, Kirn JR, Grossman M, Macklis JD, Nottebohm F (2000) Targeted neuronal death affects neuronal replacement and vocal behavior in adult songbirds. [see comments]. Neuron 25:481-492.

Seri B, Garcia-Verdugo JM, McEwen BS, Alvarez-Buylla A (2001) Astrocytes give rise to new neurons in the adult mammalian hippocampus. J Neurosci 21:7153-7160.

Sheen VL, Macklis JD (1994) Apoptotic mechanisms in targeted neuronal cell death by chromophore-activated photolysis. Exp Neurol 130:67-81.

Sheen VL, Macklis JD (1995) Targeted neocortical cell death in adult mice guides migration and differentiation of transplanted embryonic neurons. J Neurosci 15:8378-8392.

Sheen VL, Dreyer EB, Macklis JD (1992) Calcium-mediated neuronal degeneration following singlet oxygen production. NeuroReport 3:705-708.

Shihabuddin LS, Ray J, Gage FH (1997) FGF-2 is sufficient to isolate progenitors found in the adult mammalian spinal cord. Exp Neurol 148:577-586.

Shin JJ, Fricker-Gates RA, Perez FA, Leavitt BR, Zurakowski D, Macklis JD (2000) Transplanted neuroblasts differentiate appropriately into projection neurons with correct neurotransmitter and receptor phenotype in neocortex undergoing targeted projection neuron degeneration. J Neurosci (Online) 20:7404-7416.

Shingo T, Gregg C, Enwere E, Fujikawa H, Hassam R, Geary C, Cross JC, Weiss S (2003) Pregnancy-stimulated neurogenesis in the adult female forebrain mediated by prolactin. Science 299:117-120.

Snyder EY, Yoon C, Flax JD, Macklis JD (1997) Multipotent neural precursors can differentiate toward replacement of neurons undergoing targeted apoptotic degeneration in adult mouse neocortex. Proc Natl Acad Sci USA 94:11663-11668.

Song HJ, Stevens CF, Gage FH (2002) Neural stem cells from adult hippocampus develop essntial properties of functional CNS neurons. NatNeurosci 5:438-445.

Stanfield BB, Trice JE (1988) Evidence that granule cells generated in the dentate gyrus of adult rats extend axonal projections. Exp Brain Res 72:399-406.

Suhonen JO, Peterson DA, Ray J, Gage FH (1996) Differentiation of adult hippocampus-derived progenitors into olfactory neurons *in vivo*. Nature 383:624-627.

Tanapat P, Galea LA, Gould E (1998) Stress inhibits the proliferation of granule cell precursors in the developing dentate gyrus. Intl J Dev Neurosci 16:235-239.

Terada N, Hamazaki T, Oka M, Hoki M, Mastalerz DM, Nakano Y, Meyer EM, Morel L, Petersen BE, Scott EW (2002) Bone marrow cells adopt the phenotype of other cells by spontaneous cell fusion. Nature 416:542-545.

Trejo JL, Carro E, Torres-Aleman I (2001) Circulating Insulin-Like Growth Factor I Mediates Exercise-Induced Increases in the Number on New neurons in the Adult Hippocampus. J Neurosci 21:1628-1634.

Tropepe V, Coles BL, Chiasson BJ, Horsford DJ, Elia AJ, McInnes RR, van der Kooy D (2000) Retinal stem cells in the adult mammalian eye. Science 287:2032-2036.

van Praag H, Schinder AF, Christie BR, Toni N, Palmer TD, Gage FH (2002) Functional neurogenesis in the adult hippocampus. Nature 415:1030-1034.

Vescovi A, Gritti A, Cossu G, Galli R (2002) Neural stem cells: plasticity and their transdifferentiation potential. Cells Tiss Org 171:64-76.

Wagner JP, Black IB, DiCicco-Bloom E (1999) Stimulation of neonatal and adult brain neurogenesis by subcutaneous injection of basic fibroblast growth factor. J Neurosci 19:6006-6016.

Weiss S, Dunne C, Hewson J, Wohl C, Wheatley M, Peterson AC, Reynolds BA (1996) Multipotent CNS stem cells are present in the adult mammalian spinal cord and ventricular neuroaxis. J Neurosci 16:7599-7609.

Whittemore SR, Morassutti DJ, Walters WM, Liu RH, Magnuson DS (1999) Mitogen and substrate differentially affect the lineage restriction of adult rat subventricular zone neural precursor cell populations. Exp Cell Res 252:75-95.

Williams BP, Park JK, Alberta JA, Muhlebach SG, Hwang GY, Roberts TM, Stiles CD (1997) A PDGF-regulated immediate early gene response initiates neuronal differentiation in ventricular zone progenitor cells. Neuron 18:553-562.

Wurmser AE, Gage FH (2002) Stem cells: cell fusion causes confusion. Nature 416:485-487.

Ying QL, Nichols J, Evans EP, Smith AG (2002) Changing potency by spontaneous fusion. Nature 416:545-548.

Young MJ, Ray J, Whiteley SJ, Klassen H, Gage FH (2000) Neuronal differentiation and morphological integration of hippocampal progenitor cells transplanted to the retina of immature and mature dystrophic rats. Mol Cell Neurosci 16:197-205.

Chapter 13

Neural Stem Cells for Cellular Therapy in Humans

Mary B. Newman, Thomas B. Freeman, Cyndy D. Davis and Paul R. Sanberg

INTRODUCTION

The transplantation of neural stem cells (NSCs) has emerged in the last two decades as a viable alternative treatment for traumatic brain injury, neurodegenerative diseases, and certain neurological disorders. The driving force behind the use of NSCs for repair or treatment of injured/damaged brain has been the successful use of human fetal ventral mesencephalic tissue (dopaminergic neurons) in some Parkinson's Disease (PD) patients. A possible alternative to the use of fetal tissue for cell-based therapies is the procurement of NSCs from human embryonic or adult postmortem brain. NSCs may also be derived from other sources, including pluripotent embryonic carcinoma (EC), embryonic germ (EG), embryonic stem (ES), haematopoietic stem (from bone marrow or umbilical cord blood) and embryonic/adult brain cells from other species. All of these have their advantages and disadvantages as a cell source for use in cellular therapies.

The ideal candidates for neural transplantation would be neural stem/progenitor cells that could easily be obtained with little or no ethical concerns, maintained and expanded *in vitro* until needed, induced to differentiate into all three neural cell types (neurons, astrocytes, oligodendrocytes), selectively enhanced for certain genotypic or phenotypic properties, have a low immune-rejection possibility for the recipient, could survive and integrate within the host brain, and have no potential for tumorigenicity. The field of cellular therapy has made tremendous progress, particularly in the isolation and characterization of neural stem cells for their possible use in the clinical environment. Critical findings from clinical neural transplantation trials, challenges that remain, and new directions in relation to clinical applications will be discussed in this chapter.

From: *Neural Stem Cells: Development and Transplantation*
Edited by: Jane E. Bottenstein © 2003 Kluwer Academic Publishers, Norwell, MA

CELLULAR THERAPY FOR NEUROLOGICAL DISORDERS

The administration of NSCs may provide therapeutic benefits in several ways: (1) the replacement of specific cells, such as dopaminergic cells for PD, (2) the delivery or endogenous stimulation of neurotrophins, cytokines, or growth factors that could aid in the repair of the injured brain, (3) the stimulation of host brain neural stem cells to proliferate and aid in repair (see Chapter 12 in this volume), (4) the regeneration of tissue, (5) the delivery of genetically modified neural stem cells to the host (see Chapter 14 in this volume), and (6) a combination of these strategies. Depending on the disease or injury different strategies may be chosen (Table 1). Because a single cell type may not be able to perform all of these functions, more than one cell type might be necessary (Carpenter et al., 2003). Moreover, it is likely that greater therapeutic benefits will be realized when cells are co-transplanted with other cell types or with growth factors. Such a synergistic effect is evident from previous clinical and preclinical studies. In a PD clinical trial, transplanted fetal ventral mesencephalic dopaminergic neurons survived and integrated better within the host tissue when co-transplanted with embryonic striatal cells (Meyer et al., 1995). In addition, graft survival has also been enhanced by co-treatment with glial-derived neurotrophic factor (for review see Hoffer and Olson, 1997; Yurek and Fletcher-Turner, 1999; Mendez et al., 2000b). Dopamine neurons also demonstrated improved survival rates when co-transplated with testis-derived Sertoli cells (Willing et al., 1999). Therefore, the utilization of several transplantation strategies may act synergistically and prove to be more beneficial to recovery of the patient.

Prior to considering cellular therapy as an option, the type of disease or injury must be identified, the pathology understood, the target area identified, and the appropriate treatment strategy selected (Table 1). For example, in the case of PD, the progression of the disease is characterized by selective degeneration of substantia nigra dopaminergic neurons resulting in the loss of dopamine in the striatum. The strategy in this disorder is to replace dopaminergic neurons (cell-specific) in the nigrostriatal pathway (area identified). This disease is unlike cerebral ischemia or traumatic brain injury in which a regional area of tissue is damaged and all neural cell types are more than likely required for repair. Similarly, in the case of Huntington's Disease (HD), in which there is a massive loss of medium spiny GABAergic projection neurons of the caudate and putamen and to a lesser degree other neurotransmitter phenotypes, the replacement of several neuronal cell types is needed in an identified area.

Currently, no established criteria for cell-based therapies exist; however, factors influencing the therapeutic effects and survival of grafted cells have

Table 1. Strategies for Cell-Based Therapies

Type	Disorder	Cells Required	Target
Cell-Specific	Multiple sclerosis	Oligodendrocytes	Remyelination of axons
Regional	Traumatic brain injury	All neural cells	Regeneration of tissue in damaged area
Cell-specific/area identified	Stroke	All neural cells	Regeneration of tissue in ischemic area
Area identified/multiple cells	Parkinson's Disease	Dopaminergic (DA) neurons	Replace DA neurons in putamen or nigrostrlatal pathway
	Huntington's Disease	Medium spiny neurons	Regeneration of striatal tissues
		Medium aspiny neurons	Replacement of primarily GABA neurons, followed by enkephalin and substance P neurons

Treatment Strategies

- Cell replacement
- Tissue regeneration
- Delivery of neurotropic factors
- Stimulation of endogenous neurotrophins
- Genetically modified cells
- Mobilization of host brain neural stem cells
- Revascularization of injury area
- Supply or stimulation of myelinating cells
- Combination of strategies

n closely examined. A host of previous studies have provided knowledge regarding the general issues that should be addressed in cell-based therapies (Table 2). We will address the practical use and implications of neural transplantation-based therapies, the choice of donor cell phenotypes before and after engraftment, tumorigenicity and graft rejection, and clinical functional recovery assessments for certain neurodegenerative diseases and CNS injuries.

Neural Tissue Transplantation in Parkinson's and Huntington's Disease

Cell replacement is no longer considered an idealistic approach in the treatment central nervous sytem (CNS) disorders, but rather as a potential therapeutic alternative. PD has paved the way for the use of cellular therapies in other neurodegenerative diseases, neurological disorders, and traumatic brain injury, in the CNS. The earliest clinical transplantation trials were initiated in Sweden

Table 2. General Issues for Cell-Based Therapies

- Degree of damage to host environment; stage of the disease
- Source of the cells (consistent and renewable)
- Purity of the cells (homogenous population)
- Maturity or age of the cells
- Quantity of cells to be transplanted
- Ability of the cells to migrate to or away from injured area or not to migrate
- Potential of cells to differentiate into appropriate phenotypes before or after grafting
- Programmed cell death
- Culture conditions of cells before transplantation
- Need for immunosuppression in CNS grafts
- Cells ability to integrate and form functional connections within host environment
- Ability of the cells to respond to endogenous signals in appropriate manner
- Potential for tumorigenicity of grafted cells
- Method of delivery of cells
- Standardization or reproducibility of transplants
- Immunomatching between host and donor; genotyping
- Atypical circuit reconstruction

Note: Depending on the particular disease or disorder, more specific issues may arise. Individual patient issues that may affect the choice of cell-based therapy also need be recognized.

using adrenal medullary grafts in the 1980s (Backlund et al., 1985; Lindvall et al., 1987) and then in Mexico (Madrazo et al., 1987), in which there were reported benefits to PD patients that received the grafts. However, the engraftment of adrenal medullary tissues was short lived due to non-replicable findings. The use of fetal tissue for transplantation was pioneered by the work of Elizabeth Dunn in 1904 at the University of Chicago who demonstrated that these grafts survived in animals (Borlongan et al., 2000). However, not until eighty-five years later was human fetal ventral mesencephalic tissue transplanted in PD patients, in which successful amelioration of symptoms was shown (Lindvall et al., 1989). This group was the first to report the clinical benefits of fetal neural tissue transplanted into the putamen of PD patients and with graft survival and dopamine synthesis evaluated by positron emission tomography (PET; Lindvall et al., 1990a; Lindvall et al., 1990b). However, only minor changes in fluoro-dopa (f-dopa) uptake were observed (see below). Since that time, approximately 400 PD patients have undergone fetal ventral mesencephalic tissue transplantation.

In 1995, a PD patient who had received fetal dopaminergic neurons died of causes unrelated to surgery. This was the first case study to report direct evidence for graft survival, clinical improvement, and increased f-dopa (Freeman et al., 1995b; Kordower et al., 1995; Kordower et al., 1996). The postmortem brain showed that dopaminergic grafts survived with neurititic outgrowths and formed synaptic connections within host tissue. Clinical improvements were positively correlated with increased f-dopa and graft survival with reinnervated host tissue.

Other clinical trials have also reported benefits of grafted fetal tissue/cells (Wenning et al., 1997; Hagell et al., 1999; Hauser et al., 1999; Brundin et al., 2000; Mendez et al., 2000a; Mendez et al., 2002). Together, these studies were encouraging and provided hope not only for PD patients but also for all patients suffering from other CNS disorders/diseases. However, less promising results were observed in a more recent double-blind, sham surgery-controlled clinical trial performed by Freed and colleagues (2001). In this study, PD patients between the ages of 34 and 75 were transplanted with cultured mesencephalic tissue from four embryos or they received shamsurgery, with a hole drilled in the skull but with no penetration of the dura. They reported significant improvements, but only in the younger population of patients (p = 0.01 for scores on Unified Parkinson's Disease Rating Scale and p = 0.006 for the Schwab and England scores). Furthermore, dyskinesias recurred in 15% of the patients after the first year. This was an important study that addressed the issue of efficacy in regards to the first surgical control trial.. However, the lack of significant improvement reported in the older population (or late stage of disease; see discussion below) of this study has brought some doubt and uncertainty to the future direction of the neural transplantation field (Sanberg et al., 2002).

Recently, results from another study, a randomized placebo-controlled trial, have been reported (Freeman et al., 2003). Transplanted recipients were randomized to receive either a "low dose" (one donor per side) or "high dose" (four donors per side) of fetal dopaminergic tissue transplanted into the postcommissural putamen. Similar to the Freed et al. (2001) study, the control consisted of recipients that underwent a sham operation without needle penetration into the brain. Significant improvement ($p < 0.05$) was seen in the group that received four donors per side at six months after transplantation, but this benefit was lost following discontinuation of cyclosporin immunosuppression. For this study, the predetermined primary endpoint was the change in the Unified PD Rating Score (UPDRS) on motor scores. Two years after transplantation, there was no significant treatment effect in recipients receiving either one or four donors/side when compared to the control group ($p = 0.24$). Subjects in the placebo and one donor groups deteriorated by 9.4 ± 4.25 and 3.5 ± 4.23 respectively, whereas, those in the four donor groups improved by 0.72 ± 4.05 points. Posthoc stratification based on disease severity suggested a treatment effect at two years in patients that initially presented with moderate symptoms. Those receiving four donors per side improved by 1.5 ± 4.2 points while those in the placebo group deteriorated by 21.4 ± 4.3 points ($p = 0.005$). Importantly, 56% of transplanted patients developed dyskinesias that persisted following withdrawal of dopaminergic medication. Three of these patients had significant disability from these graft-induced dyskinesias. Striatal fluoro-dopa uptake was significantly increased following transplantation and robust graft survival was observed postmortem.

Several important lessons have been learned from these two recent placebo-controlled trials of human fetal tissue transplantation for the treatment of PD. Both trials had negative primary end points. These trials, therefore, clearly demonstrate the necessity of testing all reconstructive therapies utilizing controlled trials (Freeman et al., 1999). The cause of dyskinesias generated by grafts must be better understood before future trials go forward. It appears that this problem is not strictly a dose-related phenomena, as these dyskinesias occurred in patients that received both "low" and "high" doses of dopaminergic tissue. The loss of benefit following withdrawal of immunosuppression raises the possibility that a subclinical rejection process affects grafts, although this has never been previously described. Immune markers in graft sites were reported by Kordower et al. (1997a) but the significance of this remains unknown. Furthermore, the dyskinesias developed following discontinuation of cyclosporin as well. This suggests that long-term use of immunosuppression may be warranted. In addition, solid (as opposed to suspension) grafts contain mesenchymal vasculature from the donor, which may induce a stronger immunologic response from the graft recipient. A review of the pivotal trial by Olanow et al. (2001), combined with reevaluation of the data in the study by Freed et al., (2001) corroborates the finding that subjects with earlier stage disease are more likely to benefit from neural transplantation. Also, human as well as rodent fetal grafts produce up to three logs of variability in survival between grafts, even when identical methods were used (Kordower et al., 1997b; Karlsson, 2001). Together these findings suggest that a more reliable cell source with less inconsistency and better graft survival is important for minimizing the variability in outcomes seen in these fundamental trials.

Neural transplantation of human fetal tissue has also used to treat other diseases besides PD. For example, Huntington's Disease (HD) is another progressive and fatal neurodegenerative disease, which has been linked to a mutation in the huntingtin gene (a polyglutamine repeat in the N-terminal region) (Huntington's Disease Collaborative Research Group, 1993; Freeman et al., 2000a,c). Although less in number then PD clinical trials, transplantation of human fetal striatal grafts in HD patients has been reported (Bachoud-Levi et al., 1999; Hauser et al., 1999; Bachoud-Levi et al., 2000; Fink et al., 2000; Hauser et al., 2002a). The most recent report is by Hauser et al. (2002b) in which the lateral ventricular eminence from two to eight fetal striata were transplanted bilaterally into the striatum of seven HD patients. Each patient served as their own control, in which baseline measures were used in comparisons for efficacy of transplants using the United HD Rating Scale (UHDRS) and other neurolopsychological tests. Although no significant improvements or worsening of symptoms was observed (possibly due to the small sample size and open label design (a study in which the investigator knows the treatment condition that each patient received), this study demonstrated that neural grafts survived at least18 months after transplantation. Although grafts were placed in the striatum in HD patients, the neural degeneration in this disease is widespread. Therefore, long-term follow up of these graft recipients is particularly important.

Comparisons between clinical trials are often difficult due to differences in the transplantation protocols (Olanow et al., 1996). The variability in these trial protocols included: stage of the disease, age of recipient, use of immunosuppression, tissue/cell procurement and handling procedures, surgical delivery procedures, target area selected for the grafting, and clinical recovery assessment methods used. However, the studies discussed here and other similar clinical studies provide key insights to the critical factors that should be considered for the strategic framework for any type of cellular transplantation therapy. The critical factors addressed include, but are not limited to, the ethical issues surrounding aborted fetuses, availability of fetal tissue at the age required for transplantation, the technical difficulties in isolation, and recovery of fragmented brain fetal tissue, and low survival rate of transplanted neurons. In the case of PD, fetal midbrain dopamine cells from postconception age 6–9 weeks have a greater survival rate following transplantation than other developmental ages (Freeman et al., 1995a). However, a low survivability of only 5–10% of the total population transplanted has been reported (Kordower et al., 1996; Freed et al., 2001), in which 6–7 human fetuses are required to provide a sufficient number of surviving dopaminergic neurons after transplantation (Clarkson, 2001). In HD, neural tissue transplantations have used a range of 1–8 fetuses, between the gestation ages of 7.5–9 weeks, however, the percentage of surviving transplanted neurons has not been determined in these studies (Kopyov et al., 1998; Bachoud-Levi et al., 2000; Freeman et al., 2000c; Hauser et al., 2002b).

The ethical issues surrounding the use of fetal tissue, the shear number of fetuses required for transplantation, variability in survival between grafts, low survival rate of cells, and the procurement, handling and storage difficulties all contribute to the need for a more reliable source of cells, such as those derived from a NSC line.

Transplantation of Neural Stem Cells in Animals

The studies above describing the treatment of PD and HD with fetal tissue emphasizes the importance and need to establish stable NSC lines to be used in cellular transplantation therapies. The most realistic approach would be to culture human NSCs, which then may be immortalized, propagated *in vitro*, and prepared as needed. In addition to PD and HD, several other diseases or disorders, such as Multiple Sclerosis, stroke, and traumatic brain injury (TBI) would benefit from NSC line(s).

NSCs have been commonly defined as multipotent stem cells with the potential to self-renew and to differentiate into all three neural cell types: neurons, astrocytes and oligodendrocytes. Control over the isolation, selection, expansion, and differentiation of cultured NSCs for cell transplantation is the aim of most clinical researchers, regardless of the source of cells. The ability to direct the NSCs to either a heterogeneous or homogeneous population in accordance with a particular disease would be ideal for cellular transplantation.

The past decade has shown that NSCs from human and animals can be isolated, immortalized, and used for cellular repair in animal models of injury/disease (Gage, 2000; Lindvall and Hagell, 2001; Price and Williams, 2001). Studies using experimental animal models highlight the far-reaching possibilities that may exist with the use and further development of NSCs. In one study, NSCs isolated from striata and ventral mesencephala of BalbC mice fetuses (E14-15) and cultured with epidermal growth factor (EGF) were transplanted into myelin-deficient rat spinal cords and subsequently differentiated into oligodendrocytes, as determined by morphological examination and ultrastructural characterization. The host environment induced apparent lineage selection, because the isolated NSCs differentiated into all three neural cell types if maintained in culture. In addition, histological examination showed that some animals exhibited myelination in previously deficient areas, as determined by toluidine blue staining, thus demonstrating that NSCs from mice integrate and are functional within the host environment (Hammang et al., 1997).

Myelination in a more "global" manner has been reported using a NSC line (C17.2) derived from neonatal mouse cerebellum (Yandava et al., 1999). For discussions of the use of NSCs to generate myelin see Rogister et al. (1999a,b and Chapter 11 in this volume). In another intriguing study, a conditionally immortalized rat hippocampal embryonic cell line (HiB5) labeled with tritiated thymidine was used in a study to examine the effectiveness of NSCs to treat traumatic brain injury (TBI). In this study, TBI rats were transplanted with either HiB5 cells or HiB5 cells that were transduced to produce nerve growth factor (NGF) with a retroviral vector coded for mouse NGF. Interestingly, animals transplanted with either donor type demonstrated marked improvement in neuromotor functions and spatial learning tasks when compared to controls. A significant reduction in hippocampal CA3 cell death was only observed in the rats that received the NGF transfected cells (Philips et al., 2001). Although this was an important finding, the authors failed to phenotypically characterize the grafted cells. Nontransfected HiB5 grafted cells, according to previous reports, have been shown to differentiate into multiple cell types, and mainly immature glia (Martinez-Serrano et al., 1995; Lundberg et al., 1997; Martinez-Serrano and Bjorklund, 1997). NSCs have also been used with polymer scaffolding in an animal model of spinal cord injury. In this study, a murine NSC line (C17.2) was maintained in serum-free medium before transplantation.. Rats that received NCS-seeded scaffolds showed significant improvement in functional recovery when compared to the control group (cells alone or lesions alone). To identify C17.2 NSCs, a mouse specific antibody (M2) was used alsong with specific phenotypic makers. Tissue preservation and regeneration was most likely responsible for the observed behavioral improvement (Teng et al., 2002). In addition, NSCs in the scaffolds had less glial fibrillary acidic protein (GFAP) and were highly immunopositive for nestin, which suggests they were not contributing significantly to glial scarring and were mostly undifferentiated. This, however, is unlike the NSCs that were transplanted alone, as they were more immunopositive for GFAP. This study emphasizes an additional use for NSC lines, and how environmental factors may influence their functionality and phenotype. Other studies have demonstrated that multipotent neural cell lines may be engrafted and integrate into the developing mouse cerebellum (Snyder et al., 1992) as well as the neonatal rat brain (Englund et al., 2002). In addition, NSCs engineered to release GDNF prevented the further degeneration of dopaminergic neurons and improved behavioral motor function in the 6-hydroxydopamine rodent model of PD (Akerud et al., 2001). Moreover, NSC lines have yielded region-specific cell types when transplanted into lesioned striatum, cortex, or neocortex of the adult rodent brain (Snyder et al., 1997). The studies discussed here are a few of the many that exemplify the numerous potential uses of NSC lines in cell-based therapies (see Chapter 8 in this volume).

ASSESSMENT OF DONOR CELL PHENOTYPE

The selection and choice of the donor cells is dependent on several factors. First, the disease, trauma, or disorder that donor cells will be used to treat must be considered. For example, PD would require the use of cells with a dopamine phenotype or NSCs with the potential of differentiating into dopaminergic cells *in vivo*. In the case of traumatic brain injury or stroke, progenitor stem cells that are capable of producing all three neural cell types would be necessary. Reconstruction of the brain tissue would require the NSCs to have a larger repertoire of cell types and the ability to affect the host environment in more than one way, such as stimulation of host NSCs and neurotrophic factors and/or the delivery of trophic factors to aid repair.

Second, the mechanisms by which the donor cells will deliver their therapeutic benefits to the recipient once transplanted, must be considered. There are several theoretical ways in which NSCs may aid in the repair or lessen the symptoms of recipient's disease or injury. NSCs may replace lost cells, supply needed neurotrophic factors, stimulate endogenous stem cells, or deliver a gene product. Much needed pre-clinical studies utilizing different animal models of disease are needed in order to determine the best mechanism that will deliver the greatest therapeutic benefits. The origin, treatment of cells in culture, and gene therapy greatly influence the mechanisms of action of NSCs. Therefore, the type of treatment needed for a given disease or injury must be resolved before the donor cells and handling of the cells are determined.

Third, the age of the recipient or the progression of the disease can restrict the possibility of cellular therapy. In both humans and animals, fewer transplanted neurons are found within mature recipients when compared to younger or developing recipients (Rubio et al., 2000; Martinez-Serrano et al., 2001; Villa et al., 2002). As discussed earlier, functional recovery scores have shown that recipients with less advanced stages of PD who received embryonic dopamine neurons had improved clinical outcomes when compared to those in more advance stages of the disease. In the Freed et al. (2001) study, patients of age 60 or less displayed better overall symptom improvement scores when compared to those patients greater than age 60 or sham surgery patients. The younger patients at baseline levels had improved scores during "on" medication periods compared to the older patients. This was not true for the baseline of the "off" medication period, indicating that the stage of neurodegeneration is a key factor in the patient response to both types of therapeutic treatment. In addition, the importance of the progression (stage) of the disease was demonstrated in three HD patients who suffered from postoperative subdural hemorrhage (SDH) after receiving human fetal striatal grafts. In this study, researchers agreed that the SDHs were related to the degree of cerebral atrophy and that future transplantation should be performed in patients with early stages of disease and with less atrophy (Hauser et al., 2002b).

To date, most transplantation has been in patients with advanced stages of disease, although neural transplantation at a less progressive stage of the disease may be more beneficial. Additionally, in the case of PD, it may take up to 3 years for patients to receive the optimal benefits from grafts (Lindvall et al., 1994; Freeman et al., 2000a). Moreover, a decrease in neuritic outgrowth from grafts in older recipients was observed (Gage et al., 1983; Crutcher, 1990). Furthermore, trophic support of existing host dopaminergic neurons provided by the neural grafts may decrease the progression of the disease, which has been demonstrated in animal models. Lastly, it could be argued that patients with advanced disease provide a degenerative host environment, which may be resistant to neural repair or the incorporation of the neural graft, depending upon the degree of degeneration.

Fourth, the cell source, culturing conditions, and purification of the donor cells are areas of concern. Embryonic tissue, of both allogeneic and xenogeneic has been used in clinical trials for both PD and HD. However, in the PD patients that received porcine fetal mesencephalic cell suspensions, the benefits were less pronounced than or allografts. Furthermore, f-dopa uptake results were not consistent with graft survival, and no clinical improvements were observed in HD patients that had received xenografts (Fink et al., 2000). Although there was no overall improvement observed, the xenografts were well tolerated and no adverse effects were determined. It is likely that the lack of clinical improvement in this study was due to the use of xenogeneic porcine cells. NSCs have been isolated from both developing and mature rodent brain (Johe et al., 1996; Svendsen et al., 1996; Zigova et al., 1998) and human brain (Sah et al., 1997; Svendsen et al., 1997; Flax et al., 1998; Villa et al., 2000). Expansion of these cells has been achieved by both genetic and epigenetic means. Although such cells upon transplantation into the developing rat brain (Brustle et al., 1998; Flax et al., 1998), normal (Fricker et al., 1999; Vescovi et al., 1999a), or lesioned (Svendsen et al., 1999; Vescovi et al., 1999b) adult rat brain have shown the potential for survival and repopulation, the resulting cells are at various stages of differentiation and of a heterogeneous nature (Gage, 1998; Kukekov et al.,

1999; Scheffler et al., 1999). While suitable for some cell transplantation techniques, the lack of homogenity is not suitable for transplantations that require a specific neural cell type or phenotype, e.g., a dopaminergic neuron. Regardless of the origin of NSCs, an underlying difficulty is the small yield of neurons that is observed upon transplantation in both neurogenic and nonneurogenic brain regions (Rubio et al., 2000; Martinez-Serrano et al., 2001; Villa et al., 2002). NSCs that have been isolated from different species, the different ages of the donor, and the culturing conditions employed all contribute to the incongruent findings reported. These and other studies emphasize the need for further experimentation that optimizes culturing conditions for the desired phenotype of the cells required. In addition, the potential use of NSCs for cell-based therapies raises the question as to what differentiation stage (Gage, 2000; Freed, 2002) the cells should be at the time of transplantation. Whether more restricted precursor cells or more differentiated neural cells are required must be evaluated according to the disease and therapeutic method selected.

The process of inducing NSCs to differentiate into a desired neural cell type is much simpler than deriving a particular phenotype, e.g., dopaminergic. The difficulty lies in achieving a pure culture. For example, a study culturing mouse ES cells to induce dopaminergic neurons, observed that only 23 % of the cells exhibited a dopaminergic phenotype (Lee et al., 2000). If more restricted stem cells are used such as NSCs, the results vary between a low percentage (Carpenter et al., 1999; Ostenfeld et al., 2000) to very high percentages (Carvey et al., 2001) of dopaminergic neurons, and the true phenotype is often ambiguous (Arenas, 2002).

There are great difficulties in the establishment of specific phenotypes and properties of cultured cells. A combination of several factors must be used to induce a particular phenotype. Regardless of the cell types desired, the cell of interest must be indistinguishable from the native cell within the endogenous environment. Moreover, functionality and integration must be examined, beyond the neuronal appearance and expression of differentiated properties of the cell. In the case of neurons, this would include the ability of the cell to express the correct morphological features along with a panel of phenotypic markers; synthesis, release, and uptake of neurotransmitters; integration within the native environment; survival; ability to make functional synapses with other cells; and responses like an endogenous neuron to environmental signals (Freed, 2002).

Identification of Neural Stem Cells

A common way of sorting cells is by the use of antibodies to cell surface antigens, followed by manually sorting the cells, immunopanning, or fluorescence activated cell sorting. An alternative method of selecting stem cells, which has been used to isolate human hematopoietic stem cells (Jones et al., 1995; Storms et al., 1999), is use of the enzyme aldehyde dehydrogenase (ALDH). Because ALDH is an intracellular enzyme and well expressed in hematopoietic stem cells, by using a fluorescent substrate for this enzyme, researchers are able to sort cells, flow cytometry, to achieve a highly viable pure population of stem cells. This technique alleviates the problem of marking dead or dying cells that can occur with the use of monoclonal antibodies. However, to date this method has not been used to select or sort NSCs. Several different markers have been used to isolate NSCs, which is dependent upon the tissue source and age of donor. NSCs derived from murine (Redies et al., 1991; Mujtaba et al., 1999) and human ES cell cultures (Thomson et al., 1998; Carpenter et al., 2001) or human fetal brain (Dietrich et al., 2002) have been selected by expression of surface marker A2B5, which recognizes an epitope common to neurons, neuroendocrine, and glial cells, and the polysialylated form of neural cell adhesion molecule (PSA-NCAM). These immature cells, depending on the cell-based therapy required, may be propagated to supply more committed neural progenitors, to a specific phenotype, or could be used with no further differentiation if desired.

Cells isolated from human fetal brain and spinal cord (week 12) that are CD133$^+$ (primitive stem and progenitor cell marker), CD34$^-$ (hematopoietic stem and progenitor cell marker), CD45$^-$ (leukocyte marker), and CD24$^{-/lo}$ (B cell marker), can generate neurospheres, self-renew, and differentiate into neurons and glia (Uchida et al., 2000; Tamaki et al., 2002; see Chapter 7 in this volume). Uchida et al. (2000) also demonstrated that both CD34 and CD45 positive cells do not initiate neurospheres. Likewise, neurospheres do not express either of these markers, whereas the CD133 marker was reported to be expressed on 90–95% of the neurosphere cells. Currently, CD133 and A2B5 appear to be the stem cell markers of choice in isolating NSCs from fetal brain or embryonic stem cells. Whether more or less restricted progenitor cells or more differentiated or committed neural cells are required for cell-based therapies must be evaluated according to the disease and therapeutic model selected.

The use of positive and negative markers from the hematopoietic field may be useful in isolating neural stem cells from other cell types. At present, there are no adequate markers available to identify more primitive cells within neural tissue. In addition, there are marked differences among neural stem cells isolated and cultured from human fetuses even of the same gestation age. Poltavtseva et al. (2002) have reported such differences in the phenotype, development, and viability of these neural cells, all of which could influence their usefulness as a transplantable cell source. Therefore, careful characterization of each population of cells is required before *in vivo* studies are initiated.

Nestin, an intermediate filament protein, is a neural precursor cell marker expressed in rat (Lendahl et al., 1990), mice (Reynolds and Weiss, 1992; Palmer et al., 1999), and humans (Johansson et al., 1999; Shih et al., 2001; Poltavtseva et al., 2002). Although nestin has been used extensively to determine cells of neural lineage, this marker has also been found in endothelial cells, embryonic tissue, glioblastoma mulitforme, and melanoma (Shih et al., 2001; Mokry and Nemecek, 1998); therefore, it is not specific to the CNS. Most researchers agree that NSCs should also express EGF and FGF2 receptors, because neurospheres require EGF and FGF for cell division. NSCs isolated from both fetal and adult human brain express nestin, EGFR and FGFR1, which is dependent on the culturing conditions. These studies conclude that NSCs from cultured human fetal and adult neurospheres that are EGF and FGF2 responsive differentiate into all three neural cell types: neurons, astrocytes, and oligodendrocytes (Kukekov et al., 1999; Shih et al., 2001). Also, transcription factors may be used to identify the neural stem cell, e.g., sox1 and sox2. Sox1 is confined to the neuroepithelium of the neural plate and sox 2 is found in the neural crest, indicating that it is expressed earlier then sox1 (Svendsen et al., 1999). Li et al. (1998) used a strategy of lineage selection to isolate mouse ES cells based on positive labeling for β-galactosidase activity in which the reporter gene βgeo had been incorporated in the sox2 gene by homologous recombination. The cells were then further purified by G418 selection, which eliminated the negative sox2 cells, resulting in neuroepithelial progenitor cells.

Determining that a mature cell is from a specific lineage both *in vitro* and *in vivo* is often difficult. The standard practice of cell identification has been based on the use of single antibodies, such as, GFAP for astrocytes, GalC for oligodendrocytes, and TuJ1 for neurons. For cell-based therapies, it will become important to identify the progeny that will develop from NSCs and what factors will influence their cell fate (Gage, 2000).

Directing the Differentiation of Neural Stem Cells

Prior to transplantation, stem cells may be directed to a tissue-specific cell type or phenotype-specific cell in two ways. The first is epigenetic expansion, which is the culturing of the cells over a period of time accompanied by supplementation of the medium with growth factors, cytokines, or other chemical substances that will direct the cells towards a specific phenotype. This method attempts to mimic the endogenous environment, which directs the stem cells to a tissue-specific type during development. The second method is based on gene fusion (genetic perpetuation). For example, Klug et al. (1996) successfully transfected pluripotent ES cells with the alpha-cardiac myosin heavy chain promotor. The cultured cells differentiated into cardiomyocytes (>99%) after G418 selection. This antibiotic-resistance gene method may be used to allow only desired cell types to survive. The principle behind this is the attachment of a selected marker to the promoter sequence specific to the chosen tissue type. The expansion of NSCs by epigenetic methods has yielded heterogeneous population of cells at various stages of differentiation (Kukekov et al., 1999; Scheffler et al., 1999; Svendsen and Smith, 1999; Gage, 2000). Therefore, this method may not provide the optimal source of NSCs for transplantation. Genetically expanded cell lines may be more advantageous due to their ability to generate a homogeneous cell population.

Rodent clonal NSC lines have been successful in cell transplantation experiments for neurodegenerative diseases and the delivery of gene therapy (Whittemore and Snyder, 1996; Gage and Christen, 1997; Martinez-Serrano et al., 2001). The success of these transplants may be due to the stability of the cell lines as well as the survival and integration of these cells in both the young, developing, and aged rodent brain. The success observed with rodent clonal NSC lines has encouraged the search for human equivalents (Villa et al., 2000). Indeed, human NSC lines have been established by using either or both epigenetic and genetic propagation (Rubio et al., 2000; Villa et al., 2000). Although the techniques used to derive the NSC lines are complex and differ from laboratory to laboratory, there are basic principles that are congruent. Once cells from the tissue of interest have been isolated, they are typically cultured in medium containing one or more mitogens (i.e., EFG, FGF2). Differentiation of the cells is induced by withdrawing the mitogens or by the addition of other factors. Phenotypic identification is determined by the use of antibodies directed against antigens specific for different neural cell types. Both EGF and FGF2 are reported to be essential for propagating human NSC lines (Rubio et al., 2000; Villa et al., 2000) and are not necessary for rodent NSCs, indicating a clear difference between human and rodent stem cells. Embryonic stem cells can be maintained *in vitro* in the presence of leukemia inhibitory factor (LIF) and these cells will produce neural stem cells when supplemented with FGF2 in culture (Tropepe et al., 2001). Both EGF and FGF2 will keep neural stem cells in a proliferative state (Gritti et al., 1999; Ookura et al., 2002), while brain derived neurotrophic factor (BDNF) will induce neural progenitors to differentiate (Pincus et al., 1998; Benoit et al., 2001).

Some stem cells have been reported to transdifferentiate, or give rise to a different stem cell with distinct properties, which may be an alternative to NSCs isolated from human embryonic, adult, or the postmortem brain (see Chapter 6 in this volume). Kondo and Raff (2000) have demonstrated that oligodendrocyte progenitor cells transdifferentiate to pluripotent stem cells and give rise to neurons. However, there is a need to show that a single stem cell from one tissue source can differentiate into another tissue type or more than one tissue type. Even though stem cells may be isolated from a specific tissue, there could be a mixed population of stem cells present (Kuehnle and Goodell, 2002; Ogawa et al., 2002). This was observed in stem cells that were originally determined to be muscle progenitor cells and were thought to transdifferentiatie into haematopoietic cells; however, it was subsequently determined that, haematopoietic stem cells resided in the muscle tissue that was isolated. More surface markers are needed to isolate stem cells from specific organs or origins (see Chapter 3 in this volume) similar to those reported for haematopoietic stem cells. Additional markers are slowly emerging and should allow further analysis of NSCs and transdifferentiated NSCs.

While it is possible to direct the differentiation of pluripotent stem cells and multipotent NSCs *in vitro*, we are only at the initial stages of learning how to direct stem cells to particular phenotypes that might be useful *in vivo*. Some studies have reported the phenotype before and after transplantation of NSCs, however, the methods needed to direct cell differentiation both in culture and posttransplant is not currently available, but is necessary to comprehend the potential of these cells.

SOURCES AND MIGRATION OF STEM CELLS

Sources of Human Neural Stem Cells

In addition to the embryonic stem cell and the neural stem cell obtained from the CNS, other sources for human NSCs exist. Possible alternative sources are adult stem cells, bone marrow cells, NT2 cells, and umbilical cord blood cells. While the adult stem cells hold promise, there are limitations with these cells: only minute quantities are available, cells have not been isolated from all tissues, they are difficult to isolate and collect, and once isolated, they will require expansion to sufficient numbers in culture. These limitations restrict their practical use given our current technology. Alternatively, transplantation of bone marrow stem cells has been used clinically with excellent results. The main problems regarding the use of bone marrow cells include, the matching of compatible host to donor, invasiveness of the procedure in order to harvest marrow, recovery of the donor, and disparity between donor cells and recipient (graft versus host disease) which necessitates immunosuppression before and after transplantation (except when the transplantation is being used to treat immune deficiency diseases). NT2 cells (also called Ntera2/D) are a clonal cell line, which can be induced to differentiate into postmitotic neuronal cells through exposure to retinoic acid. The resulting differentiated cells are called NT2N or hNT neurons. Although NT2 cells are derived from a teratocarcinoma cell line and are classified as pluripotent human embryonal carcinoma (EC) cells

(Andrews, 1984; Pleasure et al., 1992), they have been extensively tested for safety, toxicity, and tumorigenicity and have been approved by the FDA for clinical testing (Sanberg et al., 2002). In addition, these cells have been used in preclinical and clinical trials for the treatment of stroke. However, NT2N cells have primarily been directed toward the dopamerigic phenotype (Zigova et al., 2000) and have not been actively directed to other neuronal phenotypes. Whether NT2 cells are capable of being multipotent is not currently known. In addition, these cells are transplanted after they have differentiated and have limited migratory ability if any. Therefore, these cells are limited in their ability to treat/ repair a wide variety of CNS injuries or degenerative diseases.

Another possible source of NSCs is from human umbilical cord blood (HUCB). These cells contain many properties that make them ideal for use in cell-based therapies. HUCB contains a heterogeneous population of cells rich in haematopoietic stem and progenitor cells with possible pluripotent characteristics (Broxmeyer, 1996; Mayani & Lansdorp, 1998). Although HUCB cells have been extensively studied and used in the treatment of various nonmalignant and malignant haematopoietic diseases (Lu, Shen & Broxmeyer, 1996), properties that are important for cell-based therapy of injured and degenerating tissue are just beginning to be evaluated.

The use of HUCB cells offers numerous advantages. First, cord blood offers few ethical concerns. Second, HUCB is easy to obtain without jeopardizing the mother or infant. Third, cryopreservation does not seem to affect the viability of the stem or progenitor cells. Cryopreserved cord blood has been stored up to 4 years and used in successful clinical transplantation (Broxmeyer, 1998). In addition, *in vitro* studies show that cord blood stored for 10 years has a high viability of immature progenitor cells that do not differ in proliferation or differentiation capacities from HUCB stored for shorter periods (Broxmeyer & Cooper, 1997). Fourth, HUCB yields higher numbers of haematopoietic progenitor cells with a higher proliferation rate and expansion potential than those of adult bone marrow (Hows, et al 1992; Cardoso et al 1993). Fifth, the clinical use of HUCB exhibits a low incidence of graft-vs-host disease (GVHD) rejection when compared to that of adult bone morrow (Gluckman et al., 1997; Wagner et al., 1992), even in children that received one antigen HLA-mismatch (Wagner et al., 1992; Wagner et al., 1995). The immaturity of cord blood cells could be postulated as the reason for this low rejection rate (Newman et al., in press), suggesting that immunosuppressive therapy may not be necessary, which would be important clinically.

In addition to the above advantages, both human bone marrow and umbilical cord blood cells have been shown *in vitro* and *in vivo* to differentiate into neural cell types (Sanchez-Ramos, 2002; Newman et al., in press). Human cord blood cells *in vitro* differentiate into neurons and astrocytes when exposed to all-trans-retinoic acid and nerve growth factor (Sanchez-Ramos et al., 2001), beta-mercaptoethanol (Ha et al., 2001), or when plated in a low glucose and acidic medium that is supplemented with FGF2 and EGF (Bickness et al., 2002). Interestingly, only one published study to date has shown HUCB-derived cells can express the oligodendrocyte marker GalC after negative selection for CD34 and CD45 cells and exposure to retinoic acid plus brain derived neurotrophic factor (Buzanska et al., 2002). One of the first studies demonstrating neural markers after HUCB cell transplantation was performed in collaboration with Dr. Chopp at the Henry Ford Health Center and our laboratory. In this study, rats were subjected to middle cerebral artery occlusion (MCAO) and received a tail vein injection of HUCB cells after the stroke. No reduction in lesion size was observed; however, HUCB cells survived and were found in the ipsilateral cortex and striatum of the damaged brain, and some cells were shown to have the neuronal markers NeuN and MAP2, the astrocytic marker GFAP, and the endothelial cell marker FVIII (Chen et al., 2001). A second study performed by the same two groups, using a traumatic brain injury (TBI) model, showed the migration of HUCB cells to the parenchyma of the damaged brain (Lu et al., 2002), and after 28 days *in vivo* immunocytochemistry demonstrated that the cord blood cells expressed the neuronal markers NeuN and MAP2. In addition, the TBI rats showed functional improvement with reduced deficits when compared to control animals by utilizing the rotarod for motor skills and with the Modified Neurological Severity Score (mNSS) for neurological assessment. A third study from our laboratory used HUCB cells transplanted into rat pup (1 day old) subventricular zone. Cells positive for the TuJ1 marker were observed although the majority showed an astrocytic-like phenotype and morphology (Zigova et al., 2002). Overall, HUCB cells embody several properties which encourage their utilization beyond the haematopoietic field and may prove to be valuable in the cell transplantation field.

Migration or Homing of Neural Stem Cells

One of the issues to consider when examining the feasibility of a cell type for cell-based therapy is the ability of that cell to migrate to the site of degeneration or injury (Table 2). We know that stem and progenitor cells migrate from "organ to organ" during embryogenesis through adulthood (Broxmeyer, 1998), and we are just beginning to understand the factors and circumstances that drive or direct these cells to their target. Embryonic stem cells proliferate and their progeny undergo a process of progressive lineage restriction to finally generate differentiated cells that form the mature tissue. Stem cell proliferation, differentiation, and migration are regulated by a strict timetable. In addition, the involvement of the extracellular matrix (ECM) is vital for cellular migration, along with the complex signaling of several cytokines, integrins, selectins, and adhesion molecules/receptors, which induce a cascade of intracellular and extracellular events that participate in the directing of cell migration.

Until recently, only embryonic stem cells were thought to be pluripotent. However, a certain degree of plasticity is now recognized within adult stem cells, in that some of these cells are capable of phenotypic changes and assist in regeneration of distal tissue (Blau, et al., 2001). Moreover, adult neural stem cells that reside within the subventricular zone (subependymal layer) of the forebrain, migrate a great distance within the rostral migratory stream to reach the olfactory bulb and regenerate this population of cells (Gritti et al., 2002; Peretto et al., 1999). The external granular proliferative layer in the dentate gyrus of the hippocampus is known to give rise to both astrocytes and neurons (Kuhn et al., 1996; Okano et al., 1993). In addition, neurogenesis has been demonstrated in animal models after global ischemia (Liu et al., 1998) and induced seizures (Parent et al., 1997). Moreover, human adult neural stem cells have been shown to migrate to the site of damage after being transplanted into animal models of neuronal injury (Akiyama et al., 2001; Herrera et al., 1999; Kurimoto, et al., 2001).

Human haematopoietic progenitor stem cells migrate to the bone marrow after transplantation and during fetal development. These cells also migrate from bone marrow to the peripheral blood in response to cytokines (Imai et al., 1998; Kim & Broxmeyer, 1998). Clinically, the cytokines granulocyte-colony stimulating factor (G-CSF) and granulocyte-macrophage colony stimulating factors (GM-CSF) have been used to elevate levels of haematopoietic stem and progenitor cells in peripheral blood and bone marrow in patients with hematological malignancy (Siena et al., 2000; To et al., 1997). The mobilization of stem cells within the stromal layer of bone marrow and peripheral blood is a complex interaction of numerous cytokines (Horuk, 2001; Pelus et al., in press).

HUCB cells, have been tested in compartmentalized culture dishes to determine the ability of these cells to migrate toward soluable extracts of homogenized tissue of normal and injured brain. HUCB cells migrate toward the developing rodent brain, especially, the extracts of the striatum of neonates when compared to old rats and controls (Newman et al, 2002). In addition, these cells also migrate toward ischemic extracts 24 hours after a stroke (Chen et al., 2001). Moreover, there appears to be a time-dependence in the migration of HUCB cells to ischemic extracts. More cells migrate at 72 hours to both hippocampal and ischemic extracts of adult rats when compared to other time points and controls (Newman et al., 2003). Our laboratory is currently investigating this phenomenon along with the induction factors responsible for the migration of these cells. In addition, very little is known regarding the migratory properties of NSCs.

The stage of differentiation is one of the main concerns in the determination of which cells are most beneficial for repairing injured tissue. This question along with determining the factors that induce the cells to migrate to the site of injury are of the utmost importance in furthering our efforts in not only finding reliable and functional cells for neural transplantation and repair, but also in being able to direct such cells to the site of injury.

CONTROL OF GRAFT REJECTION AND POTENTIAL FOR TUMORIGENICITY

Whether donor cells are xenogeneic, allogeneic or autogeneic to the host will be a key factor in the immune response. Allografts of neural fetal tissue used in both PD and HD patients demonstrate no graft rejection after immunosuppression treatments are withdrawn (Kordower et al., 1995; Kordower et al., 1997; Kordower et al., 1998; Freeman et al., 2000b; Hauser et al., 2002b), or in one study, where immunosuppression treatment was not used (Freed et al., 1992). Therefore, the use of immunosuppressants such as cyclosporin-A may not be necessary in allogeneic transplantation. In addition, as previously discussed the withdrawal of immunosuppressant therapy resulted in a loss of the transplantation benefits to the patient (Freeman et al., 2003). The safety and efficacy of non-immunosuppressed human recipients has not been addressed and should be carefully evaluated. For example, one PD patient who developed renal complications could not tolerate cyclosporin after the initial transplantation (Freeman et al., 1998). The autopsy of two PD patients that received multiple allografts with no immunosuppression for the last 12 months before they died (Kordower et al., 1996; Kordower et al., 1997a) displayed immune markers for macrophages, T and B cells, and microglia within the grafted sites (Kordower et al., 1997a). However, the grafts themselves were not rejected and it was not clear whether these findings were due to an ineffective immune response, trauma from the surgical procedure, or an early sign of possible graft rejection (Freeman et al., 1998).

Moreover, cyclosporin may have beneficial effects if inflammatory or autoimmune mechanisms contribute to the progression of diseases like PD and HD (Borlongan and Sanberg, 2002). The use of cyclosporin resulted in an increased survival of xenografts in rodent studies (Borlongan and Sanberg, 2002). One study demonstrated a sixfold increase in the number of surviving dopaminergic neurons when xenografts of human fetal ventral mesencephalic cells in parkinsonian rats were combined with long-term immunosuppressive therapy (Brundin et al., 1988). In addition, immunosuppressants have neurotrophic and neuroprotective properties (Borlongan et al., 1999). Therefore, immunosuppressant therapy may not only decrease the possibility of graft rejection, it may be beneficial in other ways to the patient. The issue of immunosuppression still remains to be delineated and will remain inconclusive until additional autopsies of allograft recipients have been evaluated.

One area of concern regarding the use of embryonic stem (ES) cells is that in addition to the ability to proliferate and differentiate upon transplantation in PD animal models. They also have the ability to form any fully differentiated cell type of the body and can develop rapidly into teratomas after transplantation (Bjorklund et al., 2002), thus retaining their mitotic ability after transplantation. Moreover, the embryonic carcinoma cell, stem cell of a germ-cell tumor, or teratocarcinoma have genetic anomalies. Thus, choosing to use stem cells that are further differentiated *in vitro* prior to transplantation (in order for the graft to contain only tissue-specific precursor cells) may be a wiser decision. The current methods for enhancing, selecting, and sorting a desired cell population has not yet been perfected; therefore, there is the chance that less specific/defined or undesired stem cells will be transplanted. Caution is encouraged in the use of ES cells for transplantation due to the fact that these cells retain their mitotic ability after transplantation, and therefore can form tissue masses or tumors in the graft recipient. Therefore, the more mature or adult neural stem cells may be more advantageous due to their limited plasticity, even though the quantities of cells are more limited (Borlongan and Sanberg, 2002).

ASSESSMENT OF FUNCTIONAL RECOVERY

Assessment of functional recovery for any disease depends not only on clinical rating scales, but also on neurophysiological measures that may be correlated to symptomatic recovery. Additionally, in order for comparisons to be made across clinical centers a unified clinical rating scale for each disorder or disease and standardization of neurophysiological measures must be employed. This lack of consensus hinders progression in the neural transplantation field.

In the case of PD, several standardized rating scales have been used to evaluate the results of grafts in patients, including the Unified PD Rating Scale (UPDRS) and the Schwab-England Disability Scores, as well as the Core Assessment Program for Intracerebral Transplantation (CAPIT), which incorporates the UPDRS. However, until recently there has been no agreement between centers on which test should be used. The CAPIT was developed to encourage uniformity and to standardize the rating scale to allow for congruent comparisons across clinical transplant studies and to serve as a registry (Freeman et al., 1998). The CAPIT includes patient inclusion criteria; rating scales for motor skills and dyskinesia; and testing time to be used for motor behaviors in relation to pharmacological challenge, brain imaging, and for baseline status and post-graft effects over time. This type of assessment program would allow for both intra-patient and inter-patient study design. Unfortunately, consensus has yet to be reached among clinical centers for the type of assessment program to use in all neural transplantation protocols.

A surrogate marker for graft survival is striatal f-dopa uptake measured by positron emission tomography (PET). Both striatal f-dopa uptake and cerebral blood flow of the supplementary motor area (SMA) and the dorsolateral pre-frontal cortex (DLPFC) can be measured and correlated to the functional recovery assessment by rating scales. PET scans demonstrate an increase in f-dopa occurs earlier than an increase in blood flow to the cerebral area after transplantation of fetal dopaminergic (DA) neurons in PD patients. This indicates that grafted DA neurons need to be integrated fully into these areas and have established "efferent and afferent" connections (Lindvall and Hagell, 2002). However, whether the increase in f-dopa is due to dopaminergic neurons of the graft or terminal sprouting of host DA neurons cannot be determined by PET analysis (Lindvall et al., 1990a; Kordower et al., 1995; Borlongan et al., 2000). The use of numerous measures allows for a clearer understanding of the patient's progress and response to treatment over time and they should be employed whenever possible.

Most centers conducting neural tissue transplantation in HD patients use either the CAPIT, which has been modified for HD (CAPIT-HD), or the Unified HD Rating Scale (UHDRS), or both for clinical evaluations. In some cases, the Schwab and England Disability Score, which defines the ability to perform daily living activities, is also employed. Similar to PD, there appears to be no consensus among the centers as to which scales should be used to permit easier comparison across trials. Neurophysiological measures include glucose uptake using a deoxyglucose PET scan within the striata and MRI scans of the transplanted areas to display graft growth. Both of these measures may be correlated to the clinical rating scale(s).

Briefly, the postmortem measures that are employed to study the effectiveness of cell transplantation must be mentioned. In HD, the Vonsattel Grades (0-4) is the standard scale often used to measure the degree of striatal neuropathology present at autopsy in the caudate and putamen (Grade 0: no or very little neuronal loss, Grade 1: up to 50% neuronal loss, Grade 2: striatal atrophy with neuronal loss and astrocytosis, Grade 3: extending atrophy, neuronal loss, and astrocytosis to globus pallidus; and Grade 4: severe atrophy and neuronal loss up to 95%). Currently, there is no equivalent scale to measure post-mortum the degree of degeneration or injury in other neurological conditions.

Unlike pharmaceutical trials, in which the regulation of new medications is extremely strict, neural transplantation is less regulated. Besides the two programs (CAPIT-PD and HD) mentioned and the updated CAPIT-PD, there are no approved assessment programs for other human transplantation studies. Although open-trial analysis (a study in which both investigator and patient are aware of the treatement condition) allows for quick access to critical information, there is a need for validated measures for clinical assessment. It is the belief of these authors that a program such as CAPIT would serve a useful function and that no transplantation study should be performed without the forethought of obtaining interpretable data that allows for comparisons across studies.

CONCLUSION

The development of NSCs to be used in cell-based therapies has certainly made progress in the last ten years. However, much more work lies ahead before any of the isolated NSCs are ready for clinical trials. This is evident from the number of questions that still remain to be answered. Among the most central are: what is the cellular requirement for the specific clinical conditions, should a heterogeneous or homogeneous population of cells be grafted, at what stage of differentiation should the cells be employed, will they integrate and function within the host environment, will culture conditions affect the transplantation or the recipient, and which donor source should be used to obtain the NSCs. These are a sample of the multitude of questions that remain unanswered. To date we know that NSCs isolated from different species will (1) survive in culture, the developing or adult brain, and the degenerating or injured brain, (2) migrate to the site of injury depending on their stage of differentiation, and (3) have the capability to differentiate into all neural cell types. Another major area that remains to be addressed is the identification of new surface or other markers that will aid in the isolation of NSCs and in characterizing the phenotype(s) of the cells both before and after transplantation (Gage et. al., 2000; see Chapter 3 in this volume).

For most of us in the clinical area there is an ever-present imperative to find a way to treat the injured or degenerating brain. However, our sense of urgency must be tempered with the practice of good science. Hopefully, we will learn from the past and apply the knowledge that has been obtained from previous clinical transplantation trials to the potential utilization of NSCs in repairing neurological deficits..

REFERENCES

Andrews PW, Damjanov I, Simon D, Banting GS, Carlin C, Dracopoli NC, Fogh J (1984) Pluripotent embryonal carcinoma clones derived from the human teratocarcinoma cell line Tera-2. Differentiation *in vivo* and *in vitro*. Lab Invest 50:147-162.

Akerud P, Canals JM, Snyder EY, Arenas E (2001) Neuroprotection through delivery of glial cell line-derived neurotrophic factor by neural stem cells in a mouse model of Parkinson's disease. J Neurosci 21:8108-8118.

Akiyama Y, Honmou O, Kato T, Uede T, Hashi K, Kocsis JD (2001) Transplantation of clonal neural precursor cells derived from adult human brain establishes functional peripheral myelin in the rat spinal cord. Exp Neurol 167:27-39.

Arenas E (2002) Stem cells in the treatment of Parkinson's disease. Brain Res Bull 57:795-808.

Bachoud-Levi AC, Bourdet C, Brugieres P, N'Guyen JP, Grandmougin T, Haddad B, Jeny R, Bartolomeo P, Boisse MF, Dalla Barba G, Degos JD, Ergis AM, Lefaucheur JP, Lisovoski F, Pailhous E, Remy P, Palfi S, Defer G, Cesaro P, Hantraye P, Peschanski M (1999) Safety and tolerability assessment of intrastriatal neural allografts in five patients with Huntington's disease. Experimental Neurology 161:194-202.

Bachoud-Levi AC, Remy P, Nguyen JP, Brugieres P, Lefaucheur JP, Bourdet C, Baudic S, Gaura V, Maison P, Haddad B, Boisse MF, Grandmougin T, Jeny R, Bartolomeo P, Dalla Barba G, Degos JD, Lisovoski F, Ergis AM, Pailhous E, Cesaro P, Hantraye P, Peschanski M (2000) Motor and cognitive improvements in patients with Huntington's disease after neural transplantation. Lancet 356:1975-1979.

Backlund EO, Granberg PO, Hamberger B, Knutsson E, Martensson A, Sedvall G, Seiger A, Olson L (1985) Transplantation of adrenal medullary tissue to striatum in parkinsonism. First clinical trials. J Neurosurg 62:169-173.

Bicknese AR, Goodwin HS, Quinn CO, Verneake CD (2002) Human umbilical cord blood cells can be induced to express markers for neurons and glia. Cell Transplantation 11:261-264.

Benoit BO, Savarese T, Joly M, Engstrom CM, Pang L, Reilly J, Recht LD, Ross AH, Quesenberry PJ (2001) Neurotrophin channeling of neural progenitor cell differentiation. J Neurobiol 46:265-280.

Bjorklund LM, Sanchez-Pernaute R, Chung S, Andersson T, Chen IY, McNaught KS, Brownell AL, Jenkins BG, Wahlestedt C, Kim KS, Isacson O (2002) Embryonic stem cells develop into functional dopaminergic neurons after transplantation in a Parkinson rat model. Proc Natl Acad Sci U S A 99:2344-2349.

Blau HM, Brazelton TR, Weimann JM (2001) The evolving concept of a stem cell: entity or function? Cell 105:829-841.

Borlongan CV, Sanberg PR (2002) Neural transplantation for treatment of Parkinson's disease. Drug Discov Today 7:674-682.

Borlongan CV, Freeman TB, Sanberg PR (2000) Reconstruction of the central nervous system by neural transplantation. In: Oxford Textbook of Surgery, Second Edition Edition (Morris SPJ, Wood WC, eds), pp 749-753: Oxford University Press.

Borlongan CV, Stahl CE, Fujisaki T, Sanberg PR, Watanabe S (1999) Cyclosporin A-induced hyperactivity in rats: is it mediated by immunosuppression, neurotrophism, or both? Cell Transplant 8:153-159.

Broxmeyer HA, ed (1998) Cellular characteristics of cord blood and cord blood transplantation. Bethesda: AABB Press.

Broxmeyer HE, Cooper S (1997) High-efficiency recovery of immature haematopoietic progenitor cells with extensive proliferative capacity from human cord blood cryopreserved for 10 years. Clin Exp Immunol 107 Suppl 1:45-53.

Broxmeyer HE (1996) Primitive hematopoietic stem and progenitor cells in human umbilical cord blood: an alternative source of transplantable cells. Cancer Treat Res 84:139-148.

Brundin P, Strecker RE, Widner H, Clarke DJ, Nilsson OG, Astedt B, Lindvall O, Bjorklund A (1988) Human fetal dopamine neurons grafted in a rat model of Parkinson's disease: immunological aspects, spontaneous and drug-induced behaviour, and dopamine release. Exp Brain Res 70:192-208.

Brundin P, Pogarell O, Hagell P, Piccini P, Widner H, Schrag A, Kupsch A, Crabb L, Odin P, Gustavii B, Bjorklund A, Brooks DJ, Marsden CD, Oertel WH, Quinn NP, Rehncrona S, Lindvall O (2000) Bilateral caudate and putamen grafts of embryonic mesencephalic tissue treated with lazaroids in Parkinson's disease. Brain 123 (Pt 7):1380-1390.

Brustle O, Choudhary K, Karram K, Huttner A, Murray K, Dubois-Dalcq M, McKay RD (1998) Chimeric brains generated by intraventricular transplantation of fetal human brain cells into embryonic rats. Nat Biotechnol 16:1040-1044.

Cardoso AA, Li ML, Batard P, Sansilvestri P, Hatzfeld A, Levesque JP, Lebkowski JS, Hatzfeld J (1993) Human umbilical cord blood CD34+ cell purification with high yield of early progenitors. J Hematother 2:275-279.

Chen J, Sanberg PR, Li Y, Wang L, Lu M, Willing AE, Sanchez-Ramos J, Chopp M (2001) Intravenous administration of human umbilical cord blood reduces behavioral deficits after stroke in rats. Stroke 32:2682-2688.

Buzanska L, Machaj EK, Zablocka B, Pojda Z, Domanska-Janik K (2002) Human cord blood-derived cells attain neuronal and glial features *in vitro*. J Cell Sci 115:2131-2138.

Cardoso AA, Li ML, Batard P, Sansilvestri P, Hatzfeld A, Levesque JP, Lebkowski JS, Hatzfeld J (1993) Human umbilical cord blood CD34+ cell purification with high yield of early progenitors. J Hematother 2:275-279.

Carpenter MK, Mattson M, Rao MS (2003) Sources of cells for CNS Therapy. In: Neural Stem Cells for Brain and Spinal Cord Repair. (Zigova T, Snyder EY, Sanberg PR, eds). New Jersey: Humana Press.

Carpenter MK, Inokuma MS, Denham J, Mujtaba T, Chiu CP, Rao MS (2001) Enrichment of neurons and neural precursors from human embryonic stem cells. Exp Neurol 172:383-397.

Carpenter MK, Cui X, Hu ZY, Jackson J, Sherman S, Seiger A, Wahlberg LU (1999) *In vitro* expansion of a multipotent population of human neural progenitor cells. Exp Neurol 158:265-278.

Carvey PM, Ling ZD, Sortwell CE, Pitzer MR, McGuire SO, Storch A, Collier TJ (2001) A clonal line of mesencephalic progenitor cells converted to dopamine neurons by hematopoietic cytokines: a source of cells for transplantation in Parkinson's disease. Exp Neurol 171:98-108.

Clarkson ED (2001) Fetal tissue transplantation for patients with Parkinson's disease: a database of published clinical results. Drugs Aging 18:773-785.

Crutcher KA (1990) Age-related decrease in sympathetic sprouting is primarily due to decreased target receptivity: implications for understanding brain aging. Neurobiol Aging 11:175-183.

Dietrich J, Noble M, Mayer-Proschel M (2002) Characterization of A2B5+ glial precursor cells from cryopreserved human fetal brain progenitor cells. Glia 40:65-77.

Englund U, Fricker-Gates RA, Lundberg C, Bjorklund A, Wictorin K (2002) Transplantation of human neural progenitor cells into the neonatal rat brain: extensive migration and differentiation with long-distance axonal projections. Exp Neurol 173:1-21.

Fink JS, Schumacher JM, Ellias SL, Palmer EP, Saint-Hilaire M, Shannon K, Penn R, Starr P, VanHorne C, Kott HS, Dempsey PK, Fischman AJ, Raineri R, Manhart C, Dinsmore J, Isacson O (2000) Porcine xenografts in Parkinson's disease and Huntington's disease patients: preliminary results. Cell Transplant 9:273-278.

Flax JD, Aurora S, Yang C, Simonin C, Wills AM, Billinghurst LL, Jendoubi M, Sidman RL, Wolfe JH, Kim SU, Snyder EY (1998) Engraftable human neural stem cells respond to developmental cues, replace neurons, and express foreign genes. Nat Biotechnol 16:1033-1039.

Freed CR (2002) Will embryonic stem cells be a useful source of dopamine neurons for transplant into patients with Parkinson's disease? Proc Natl Acad Sci U S A 99:1755-1757.

Freed CR, Breeze RE, Rosenberg NL, Schneck SA, Kriek E, Qi JX, Lone T, Zhang YB, Snyder JA, Wells TH, et al. (1992) Survival of implanted fetal dopamine cells and neurologic improvement 12 to 46 months after transplantation for Parkinson's disease. N Engl J Med 327:1549-1555.

Freed CR, Greene PE, Breeze RE, Tsai WY, DuMouchel W, Kao R, Dillon S, Winfield H, Culver S, Trojanowski JQ, Eidelberg D, Fahn S (2001) Transplantation of embryonic dopamine neurons for severe Parkinson's disease. N Engl J Med 344:710-719.

Freeman TB, Sanberg PR, Isacson O (1995a) Development of the human striatum: implications for fetal striatal transplantation in the treatment of Huntington's disease. Cell Transplant 4:539-545.

Freeman TB, Hauser RA, Sanberg PR, Saporta S (2000a) Neural transplantation for the treatment of Huntington's disease. Prog Brain Res 127:405-411.

Freeman TB, Hauser RA, Willing AE, Zigova T, Sanberg PR, Saporta S (2000b) Transplantation of human fetal striatal tissue in Huntington's disease: rationale for clinical studies. Novartis Found Symp 231:129-138; discussion 139-147.

Freeman TB, Olanow CW, Hauser RA, Kordower J, Holt DA, Borlongan CV, Sandberg PR (1998) Human fetal tissue transplantation. In: Neurosurgical Treatment of Movement Disorders (Germano IM, ed), pp 177-192. Park Ridge: The American Assoication of Neurological Surgeons.

Freeman TB, Vawter DE, Leaverton PE, Godbold JH, Hauser RA, Goetz CG, Olanow CW (1999) Use of placebo surgery in controlled trials of a cellular-based therapy for Parkinson's disease. N Engl J Med 341:988-992.

Freeman TB, Goetz CG, Kordower JH, Stoessl AJ, Brin MF, Shannon K, Perl DP, Godbold JH, Olanow CW (2003) Double blind controlled trial of bilateral fetal nigral transplantation in Parkinson's disease. Experimental Neurology Abstract 181:891.

Freeman TB, Olanow CW, Hauser RA, Nauert GM, Smith DA, Borlongan CV, Sanberg PR, Holt DA, Kordower JH, Vingerhoets FJ, et al. (1995b) Bilateral fetal nigral transplantation into the postcommissural putamen in Parkinson's disease. Ann Neurol 38:379-388.

Freeman TB, Cicchetti F, Hauser RA, Deacon TW, Li XJ, Hersch SM, Nauert GM, Sanberg PR, Kordower JH, Saporta S, Isacson O (2000c) Transplanted fetal striatum in Huntington's disease: phenotypic development and lack of pathology. Proc Natl Acad Sci U S A 97:13877-13882.

Fricker RA, Carpenter MK, Winkler C, Greco C, Gates MA, Bjorklund A (1999) Site-specific migration and neuronal differentiation of human neural progenitor cells after transplantation in the adult rat brain. J Neurosci 19:5990-6005.

Gage F, Christen Y (1997) Isolation, characterization, and utilization of CNS stem cells. Heidelberg ; New York: Springer Berlin.

Gage FH (1998) Cell therapy. Nature 392:18-24.

Gage FH (2000) Mammalian neural stem cells. Science 287:1433-1438.

Gage FH, Bjorklund A, Stenevi U, Dunnett SB (1983) Intracerebral grafting of neuronal cell suspensions. VIII. Survival and growth of implants of nigral and septal cell suspensions in intact brains of aged rats. Acta Physiol Scand Suppl 522:67-75.

Gritti A, Vescovi AL, Galli R (2002) Adult neural stem cells: plasticity and developmental potential. J Physiol Paris 96:81-90.

Gritti A, Frolichsthal-Schoeller P, Galli R, Parati EA, Cova L, Pagano SF, Bjornson CR, Vescovi AL (1999) Epidermal and fibroblast growth factors behave as mitogenic regulators for a single multipotent stem cell-like population from the subventricular region of the adult mouse forebrain. J Neurosci 19:3287-3297.

Ha Y, Choi JU, Yoon DH, Yeon DS, Lee JJ, Kim HO, Cho YE (2001) Neural phenotype expression of cultured human cord blood cells *in vitro*. Neuroreport 12:3523-3527.

Hagell P, Schrag A, Piccini P, Jahanshahi M, Brown R, Rehncrona S, Widner H, Brundin P, Rothwell JC, Odin P, Wenning GK, Morrish P, Gustavii B, Bjorklund A, Brooks DJ, Marsden

CD, Quinn NP, Lindvall O (1999) Sequential bilateral transplantation in Parkinson's disease: effects of the second graft. Brain 122 (Pt 6):1121-1132.

Hammang JP, Archer DR, Duncan ID (1997) Myelination following transplantation of EGF-responsive neural stem cells into a myelin-deficient environment. Exp Neurol 147:84-95.

Hauser RA, Sandberg PR, Freeman TB, Stoessl AJ (2002a) Bilateral human fetal striatal transplantation in Huntington's disease. Neurology 58:1704; author reply 1704.

Hauser RA, Freeman TB, Snow BJ, Nauert M, Gauger L, Kordower JH, Olanow CW (1999) Long-term evaluation of bilateral fetal nigral transplantation in Parkinson disease. Arch Neurol 56:179-187.

Hauser RA, Furtado S, Cimino CR, Delgado H, Eichler S, Schwartz S, Scott D, Nauert GM, Soety E, Sossi V, Holt DA, Sanberg PR, Stoessl AJ, Freeman TB (2002b) Bilateral human fetal striatal transplantation in Huntington's disease. Neurology 58:687-695.

Herrera DG, Garcia-Verdugo JM, Alvarez-Buylla A (1999) Adult-derived neural precursors transplanted into multiple regions in the adult brain. Ann Neurol 46:867-877.

Hoffer B, Olson L (1997) Treatment strategies for neurodegenerative diseases based on trophic factors and cell transplantation techniques. J Neural Transm Suppl 49:1-10.

Horuk R (2001) Chemokine receptors. Cytokine Growth Factor Rev 12:313-335.

Hows JM, Marsh JC, Bradley BA, Luft T, Coutinho L, Testa NG, Dexter TM (1992) Human cord blood: a source of transplantable stem cells? Bone Marrow Transplant 9 Suppl 1:105-108.

Huntington's Disease Collaborative Research Group (1993) A novel gene containing a trinucleotide repeat that is expanded and unstable on HD chromosomes. Cell 72:971-983.

Imai T, Chantry D, Raport CJ, Wood CL, Nishimura M, Godiska R, Yoshie O, Gray PW (1998) Macrophage-derived chemokine is a functional ligand for the CC chemokine receptor 4. J Biol Chem 273:1764-1768.

Johansson CB, Svensson M, Wallstedt L, Janson AM, Frisen J (1999) Neural stem cells in the adult human brain. Exp Cell Res 253:733-736.

Johe KK, Hazel TG, Muller T, Dugich-Djordjevic MM, McKay RD (1996) Single factors direct the differentiation of stem cells from the fetal and adult central nervous system. Genes Dev 10:3129-3140.

Jones RJ, Barber JP, Vala MS, Collector MI, Kaufmann SH, Ludeman SM, Colvin OM, Hilton J (1995) Assessment of aldehyde dehydrogenase in viable cells. Blood 85:2742-2746.

Karlsson J (2001) Survival of cultured and grafted embryonic dopaminergic neurones: effects of hypothermia and prevention of oxidative stress. In: Department of Physiological Sciences, p 152. Lund, Sweden: Neuroscience Center.

Kim CH, Broxmeyer HE (1998) *In vitro* behavior of hematopoietic progenitor cells under the influence of chemoattractants: stromal cell-derived factor-1, steel factor, and the bone marrow environment. Blood 91:100-110.

Klug MG, Soonpaa MH, Koh GY, Field LJ (1996) Genetically selected cardiomyocytes from differentiating embronic stem cells form stable intracardiac grafts. J Clin Invest 98:216-224.

Kondo T, Raff M (2000) Oligodendrocyte precursor cells reprogrammed to become multipotential CNS stem cells. Science 289:1754-1757.

Kopyov OV, Jacques S, Kurth M, Philpott LM, Lee A, Patterson M, Duma CM, Lieberman A, Eagle KS (1998) Fetal transplantation for huntington's disease: clinical studies. In: Cell Transplantation for Neurologica Disorders. (Freeman TB, Widiner H, eds), pp 95-134. New Jersey: Humana Press.

Kordower JH, Styren S, Clarke M, DeKosky ST, Olanow CW, Freeman TB (1997a) Fetal grafting for Parkinson's disease: expression of immune markers in two patients with functional fetal nigral implants. Cell Transplant 6:213-219.

Kordower JH, Freeman TB, Chen EY, Mufson EJ, Sanberg PR, Hauser RA, Snow B, Olanow CW (1998) Fetal nigral grafts survive and mediate clinical benefit in a patient with Parkinson's disease. Mov Disord 13:383-393.

Kordower JH, Freeman TB, Snow BJ, Vingerhoets FJ, Mufson EJ, Sanberg PR, Hauser RA, Smith DA, Nauert GM, Perl DP, et al. (1995) Neuropathological evidence of graft survival and striatal reinnervation after the transplantation of fetal mesencephalic tissue in a patient with Parkinson's disease. N Engl J Med 332:1118-1124.

Kordower JH, Rosenstein JM, Collier TJ, Burke MA, Chen EY, Li JM, Martel L, Levey AE, Mufson EJ, Freeman TB, Olanow CW (1996) Functional fetal nigral grafts in a patient with Parkinson's disease: chemoanatomic, ultrastructural, and metabolic studies. J Comp Neurol 370:203-230.

Kordower JH, Chen EY, Winkler C, Fricker R, Charles V, Messing A, Mufson EJ, Wong SC, Rosenstein JM, Bjorklund A, Emerich DF, Hammang J, Carpenter MK (1997b) Grafts of EGF-responsive neural stem cells derived from GFAP-hNGF transgenic mice: trophic and tropic effects in a rodent model of Huntington's disease. J Comp Neurol 387:96-113.

Kuehnle I, Goodell MA (2002) The therapeutic potential of stem cells from adults. Bmj 325:372-376.

Kuhn HG, Dickinson-Anson H, Gage FH (1996) Neurogenesis in the dentate gyrus of the adult rat: age-related decrease of neuronal progenitor proliferation. J Neurosci 16:2027-2033.

Kukekov VG, Laywell ED, Suslov O, Davies K, Scheffler B, Thomas LB, O'Brien TF, Kusakabe M, Steindler DA (1999) Multipotent stem/progenitor cells with similar properties arise from two neurogenic regions of adult human brain. Exp Neurol 156:333-344.

Kurimoto Y, Shibuki H, Kaneko Y, Ichikawa M, Kurokawa T, Takahashi M, Yoshimura N (2001) Transplantation of adult rat hippocampus-derived neural stem cells into retina injured by transient ischemia. Neurosci Lett 306:57-60.

Lee SH, Lumelsky N, Studer L, Auerbach JM, McKay RD (2000) Efficient generation of midbrain and hindbrain neurons from mouse embryonic stem cells. Nat Biotechnol 18:675-679.

Lendahl U, Zimmerman LB, McKay RD (1990) CNS stem cells express a new class of intermediate filament protein. Cell 60:585-595.

Li M, Pevny L, Lovell-Badge R, Smith A (1998) Generation of purified neural precursors from embryonic stem cells by lineage selection. Curr Biol 8:971-974.

Lindvall O, Hagell P (2001) Cell therapy and transplantation in Parkinson's disease. Clin Chem Lab Med 39:356-361.

Lindvall O, Hagell P (2002) Cell replacement therapy in human neurodegenerative disorders. Clinical Neuroscience Research 2:86-92.

Lindvall O, Backlund EO, Farde L, Sedvall G, Freedman R, Hoffer B, Nobin A, Seiger A, Olson L (1987) Transplantation in Parkinson's disease: two cases of adrenal medullary grafts to the putamen. Ann Neurol 22:457-468.

Lindvall O, Rehncrona S, Brundin P, Gustavii B, Astedt B, Widner H, Lindholm T, Bjorklund A, Leenders KL, Rothwell JC, et al. (1989) Human fetal dopamine neurons grafted into the striatum in two patients with severe Parkinson's disease. A detailed account of methodology and a 6-month follow-up. Arch Neurol 46:615-631.

Lindvall O, Brundin P, Widner H, Rehncrona S, Gustavii B, Frackowiak R, Leenders KL, Sawle G, Rothwell JC, Marsden CD, et al. (1990a) Grafts of fetal dopamine neurons survive and improve motor function in Parkinson's disease. Science 247:574-577.

Lindvall O, Rehncrona S, Brundin P, Gustavii B, Astedt B, Widner H, Lindholm T, Bjorklund A, Leenders KL, Rothwell JC, et al. (1990b) Neural transplantation in Parkinson's disease: the Swedish experience. Prog Brain Res 82:729-734.

Lindvall O, Sawle G, Widner H, Rothwell JC, Bjorklund A, Brooks D, Brundin P, Frackowiak R, Marsden CD, Odin P, et al. (1994) Evidence for long-term survival and function of dopaminergic grafts in progressive Parkinson's disease. Ann Neurol 35:172-180.

Liu J, Solway K, Messing RO, Sharp FR (1998) Increased neurogenesis in the dentate gyrus after transient global ischemia in gerbils. J Neurosci 18:7768-7778.

Lu D, Sanberg PR, Mahmood A, Li Y, Wang L, Sanchez-Ramos J, Chopp M (2002) Intravenous administration of human umbilical cord blood reduces neurological deficit in the rat after traumatic brain injury. Cell Transplant 11:275-281.

Lu L, Shen RN, Broxmeyer HE (1996) Stem cells from bone marrow, umbilical cord blood and peripheral blood for clinical application: current status and future application. Crit Rev Oncol Hematol 22:61-78.

Lundberg C, Martinez-Serrano A, Cattaneo E, McKay RD, Bjorklund A (1997) Survival, integration, and differentiation of neural stem cell lines after transplantation to the adult rat striatum. Exp Neurol 145:342-360.

Madrazo I, Drucker-Colin R, Diaz V, Martinez-Mata J, Torres C, Becerril JJ (1987) Open microsurgical autograft of adrenal medulla to the right caudate nucleus in two patients with intractable Parkinson's disease. N Engl J Med 316:831-834.

Martinez-Serrano A, Bjorklund A (1997) Immortalized neural progenitor cells for CNS gene transfer and repair. Trends Neurosci 20:530-538.

Martinez-Serrano A, Rubio FJ, Navarro B, Bueno C, Villa A (2001) Human neural stem and progenitor cells: *in vitro* and *in vivo* properties, and potential for gene therapy and cell replacement in the CNS. Curr Gene Ther 1:279-299.

Martinez-Serrano A, Lundberg C, Horellou P, Fischer W, Bentlage C, Campbell K, McKay RD, Mallet J, Bjorklund A (1995) CNS-derived neural progenitor cells for gene transfer of nerve growth factor to the adult rat brain: complete rescue of axotomized cholinergic neurons after transplantation into the septum. J Neurosci 15:5668-5680.

Mayani H, Lansdorp PM (1998) Biology of human umbilical cord blood-derived hematopoietic stem/progenitor cells. Stem Cells 16:153-165.

Mendez I, Baker KA, Hong M (2000a) Simultaneous intrastriatal and intranigral grafting (double grafts) in the rat model of Parkinson's disease. Brain Res Brain Res Rev 32:328-339.

Mendez I, Dagher A, Hong M, Gaudet P, Weerasinghe S, McAlister V, King D, Desrosiers J, Darvesh S, Acorn T, Robertson H (2002) Simultaneous intrastriatal and intranigral fetal dopaminergic grafts in patients with Parkinson disease: a pilot study. Report of three cases. J Neurosurg 96:589-596.

Mendez I, Dagher A, Hong M, Hebb A, Gaudet P, Law A, Weerasinghe S, King D, Desrosiers J, Darvesh S, Acorn T, Robertson H (2000b) Enhancement of survival of stored dopaminergic cells and promotion of graft survival by exposure of human fetal nigral tissue to glial cell line—derived neurotrophic factor in patients with Parkinson's disease. Report of two cases and technical considerations. J Neurosurg 92:863-869.

Meyer CH, Detta A, Kudoh C (1995) Hitchcock's experimental series of foetal implants for Parkinson's disease: co-grafting ventral mesencephalon and striatum. Acta Neurochir Suppl (Wien) 64:1-4.

Mokry J, Nemecek S (1998) Angiogenesis of extra- and intraembryonic blood vessels is associated with expression of nestin in endothelial cells. Folia Biol (Praha) 44:155-161.

Mujtaba T, Piper DR, Kalyani A, Groves AK, Lucero MT, Rao MS (1999) Lineage-restricted neural precursors can be isolated from both the mouse neural tube and cultured ES cells. Dev Biol 214:113-127.

Newman MB, Davis CD, Kuzmin-Nichols N, Sanberg PR (2003) Human umbilical cord blood (HUCB) cells for central nervous system repair. Neurotoxicity Research (in press).

Newman MB, Willing A, Cassady CJ, Manresa JJ, Kedziorek DA, Hart CD, Saporta S, Sanberg PR (2003) *In vitro* migration and phenotype identification of human umbilical cord blood (HUCB) cells to stroke brain. Experimental Neurology 181:101.

Newman MB, Zigova T, Willing A, Bickford PC, Saporta S, Sanchez R, Sanberg PR (2002) Migration behavior of human umbilical cord blood (HUCB) cells to normal and injured brain. Society for Neuroscience 423.17.

Ogawa Y, Sawamoto K, Miyata T, Miyao S, Watanabe M, Nakamura M, Bregman BS, Koike M, Uchiyama Y, Toyama Y, Okano H (2002) Transplantation of *in vitro*-expanded fetal neural progenitor cells results in neurogenesis and functional recovery after spinal cord contusion injury in adult rats. J Neurosci Res 69:925-933.

Okano HJ, Pfaff DW, Gibbs RB (1993) RB and Cdc2 expression in brain: correlations with 3H-thymidine incorporation and neurogenesis. J Neurosci 13:2930-2938.

Olanow CW, Kordower JH, Freeman TB (1996) Fetal nigral transplantation as a therapy for Parkinson's disease. Trends Neurosci 19:102-109.

Olanow CW and Brin MF (2001) Surgical therapies for Parkinson's disease. A physician's perspective. Adv Neurol 86:421-433.

Ookura T, Kawamoto K, Tsuzaki H, Mikami Y, Ito Y, Oh SH, Hino A (2002) Fibroblast and Epidermal Growth Factors Modulate Proliferation and Neural Cell Adhesion Molecule Expression in Epithelial Cells Derived from the Adult Mouse Tongue. *In vitro* Cell Dev Biol Anim 38:365-372.

Ostenfeld T, Caldwell MA, Prowse KR, Linskens MH, Jauniaux E, Svendsen CN (2000) Human neural precursor cells express low levels of telomerase *in vitro* and show diminishing cell proliferation with extensive axonal outgrowth following transplantation. Exp Neurol 164:215-226.

Palmer TD, Markakis EA, Willhoite AR, Safar F, Gage FH (1999) Fibroblast growth factor-2 activates a latent neurogenic program in neural stem cells from diverse regions of the adult CNS. J Neurosci 19:8487-8497.

Parent JM, Yu TW, Leibowitz RT, Geschwind DH, Sloviter RS, Lowenstein DH (1997) Dentate granule cell neurogenesis is increased by seizures and contributes to aberrant network reorganization in the adult rat hippocampus. J Neurosci 17:3727-3738.

Pelus LM, Horowitz D, Cooper SC, King AG (2003) Peripheral blood stem cell mobilization: A role for CXC chemokines. Oncology Hematology (in press).

Peretto P, Merighi A, Fasolo A, Bonfanti L (1999) The subependymal layer in rodents: a site of structural plasticity and cell migration in the adult mammalian brain. Brain Res Bull 49:221-243.

Philips MF, Mattiasson G, Wieloch T, Bjorklund A, Johansson BB, Tomasevic G, Martinez-Serrano A, Lenzlinger PM, Sinson G, Grady MS, McIntosh TK (2001) Neuroprotective and behavioral efficacy of nerve growth factor-transfected hippocampal progenitor cell transplants after experimental traumatic brain injury. J Neurosurg 94:765-774.

Pincus DW, Keyoung HM, Harrison-Restelli C, Goodman RR, Fraser RA, Edgar M, Sakakibara S, Okano H, Nedergaard M, Goldman SA (1998) Fibroblast growth factor-2/brain-derived neurotrophic factor-associated maturation of new neurons generated from adult human subependymal cells. Ann Neurol 43:576-585.

Pleasure SJ, Page C, Lee VM (1992) Pure, postmitotic, polarized human neurons derived from NTera 2 cells provide a system for expressing exogenous proteins in terminally differentiated neurons. J Neurosci 12:1802-1815.

Poltavtseva RA, Marey MV, Aleksandrova MA, Revishchin AV, Korochkin LI, Sukhikh GT (2002) Evaluation of progenitor cell cultures from human embryos for neurotransplantation. Brain Res Dev Brain Res 134:149-154.

Price J, Williams BP (2001) Neural stem cells. Curr Opin Neurobiol 11:564-567.

Redies C, Lendahl U, McKay RD (1991) Differentiation and heterogeneity in T-antigen immortalized precursor cell lines from mouse cerebellum. J Neurosci Res 30:601-615.

Reynolds BA, Weiss S (1992) Generation of neurons and astrocytes from isolated cells of the adult mammalian central nervous system. Science 255:1707-1710.

Rogister B, Ben-Hur T, Dubois-Dalcq M (1999a) From neural stem cells to myelinating oligodendrocytes. Mol Cell Neurosci 14:287-300.

Rogister B, Belachew S, Moonen G (1999b) Oligodendrocytes: from development to demyelinated lesion repair. Acta Neurol Belg 99:32-39.

Rubio FJ, Bueno C, Villa A, Navarro B, Martinez-Serrano A (2000) Genetically perpetuated human neural stem cells engraft and differentiate into the adult mammalian brain. Mol Cell Neurosci 16:1-13.

Sah DW, Ray J, Gage FH (1997) Regulation of voltage- and ligand-gated currents in rat hippocampal progenitor cells *in vitro*. J Neurobiol 32:95-110.

Sanberg PR, Willing AE, Cahill DW (2002) Novel cellular approaches to repair of neurodegenerative disease: from sertoli cells to umbilical cord blood stem cells. Neurotoxicity Research 4:95-101.

Sanchez-Ramos JR (2002) Neural cells derived from adult bone marrow and umbilical cord blood. J Neurosci Res 69:880-893.

Sanchez-Ramos JR, Song S, Kamath SG, Zigova T, Willing A, Cardozo-Pelaez F, Stedeford T, Chopp M, Sanberg PR (2001) Expression of neural markers in human umbilical cord blood. Exp Neurol 171:109-115.

Scheffler B, Horn M, Blumcke I, Laywell ED, Coomes D, Kukekov VG, Steindler DA (1999) Marrow-mindedness: a perspective on neuropoiesis. Trends Neurosci 22:348-357.

Shih CC, Weng Y, Mamelak A, LeBon T, Hu MC, Forman SJ (2001) Identification of a candidate human neurohematopoietic stem-cell population. Blood 98:2412-2422.

Siena S, Schiavo R, Pedrazzoli P, Carlo-Stella C (2000) Therapeutic relevance of CD34 cell dose in blood cell transplantation for cancer therapy. J Clin Oncol 18:1360-1377.

Snyder EY, Yoon C, Flax JD, Macklis JD (1997) Multipotent neural precursors can differentiate toward replacement of neurons undergoing targeted apoptotic degeneration in adult mouse neocortex. Proc Natl Acad Sci U S A 94:11663-11668.

Snyder EY, Deitcher DL, Walsh C, Arnold-Aldea S, Hartwieg EA, Cepko CL (1992) Multipotent neural cell lines can engraft and participate in development of mouse cerebellum. Cell 68:33-51.

Storms RW, Trujillo AP, Springer JB, Shah L, Colvin OM, Ludeman SM, Smith C (1999) Isolation of primitive human hematopoietic progenitors on the basis of aldehyde dehydrogenase activity. Proc Natl Acad Sci U S A 96:9118-9123.

Svendsen CN, Smith AG (1999) New prospects for human stem-cell therapy in the nervous system. Trends Neurosci 22:357-364.

Svendsen CN, Caldwell MA, Ostenfeld T (1999) Human neural stem cells: isolation, expansion and transplantation. Brain Pathol 9:499-513.

Svendsen CN, Clarke DJ, Rosser AE, Dunnett SB (1996) Survival and differentiation of rat and human epidermal growth factor-responsive precursor cells following grafting into the lesioned adult central nervous system. Exp Neurol 137:376-388.

Svendsen CN, Caldwell MA, Shen J, ter Borg MG, Rosser AE, Tyers P, Karmiol S, Dunnett SB (1997) Long-term survival of human central nervous system progenitor cells transplanted into a rat model of Parkinson's disease. Exp Neurol 148:135-146.

Tamaki S, Eckert K, He D, Sutton R, Doshe M, Jain G, Tushinski R, Reitsma M, Harris B, Tsukamoto A, Gage F, Weissman I, Uchida N (2002) Engraftment of sorted/expanded human central nervous system stem cells from fetal brain. J Neurosci Res 69:976-986.

Teng YD, Lavik EB, Qu X, Park KI, Ourednik J, Zurakowski D, Langer R, Snyder EY (2002) Functional recovery following traumatic spinal cord injury mediated by a unique polymer scaffold seeded with neural stem cells. Proc Natl Acad Sci U S A 99:3024-3029.

Thomson JA, Itskovitz-Eldor J, Shapiro SS, Waknitz MA, Swiergiel JJ, Marshall VS, Jones JM (1998) Embryonic stem cell lines derived from human blastocysts. Science 282:1145-1147.

To LB, Haylock DN, Simmons PJ, Juttner CA (1997) The biology and clinical uses of blood stem cells. Blood 89:2233-2258.

Tropepe V, Hitoshi S, Sirard C, Mak TW, Rossant J, van der Kooy D (2001) Direct neural fate specification from embryonic stem cells: a primitive mammalian neural stem cell stage acquired through a default mechanism. Neuron 30:65-78.

Uchida N, Buck DW, He D, Reitsma MJ, Masek M, Phan TV, Tsukamoto AS, Gage FH, Weissman IL (2000) Direct isolation of human central nervous system stem cells. Proc Natl Acad Sci U S A 97:14720-14725.

Vescovi AL, Gritti A, Galli R, Parati EA (1999a) Isolation and intracerebral grafting of nontransformed multipotential embryonic human CNS stem cells. J Neurotrauma 16:689-693.

Vescovi AL, Parati EA, Gritti A, Poulin P, Ferrario M, Wanke E, Frolichsthal-Schoeller P, Cova L, Arcellana-Panlilio M, Colombo A, Galli R (1999b) Isolation and cloning of multipotential stem cells from the embryonic human CNS and establishment of transplantable human neural stem cell lines by epigenetic stimulation. Exp Neurol 156:71-83.

Villa A, Navarro B, Martinez-Serrano A (2002) Genetic perpetuation of *in vitro* expanded human neural stem cells: cellular properties and therapeutic potential. Brain Res Bull 57:789-794.

Villa A, Snyder EY, Vescovi A, Martinez-Serrano A (2000) Establishment and properties of a growth factor-dependent, perpetual neural stem cell line from the human CNS. Exp Neurol 161:67-84.

Wagner JE, Kernan NA, Steinbuch M, Broxmeyer HE, Gluckman E (1995) Allogeneic sibling umbilical-cord-blood transplantation in children with malignant and non-malignant disease. Lancet 346:214-219.

Wagner JE, Broxmeyer HE, Byrd RL, Zehnbauer B, Schmeckpeper B, Shah N, Griffin C, Emanuel PD, Zuckerman KS, Cooper S, et al. (1992) Transplantation of umbilical cord blood after myeloablative therapy: analysis of engraftment. Blood 79:1874-1881.

Wenning GK, Odin P, Morrish P, Rehncrona S, Widner H, Brundin P, Rothwell JC, Brown R, Gustavii B, Hagell P, Jahanshahi M, Sawle G, Bjorklund A, Brooks DJ, Marsden CD, Quinn NP, Lindvall O (1997) Short- and long-term survival and function of unilateral intrastriatal dopaminergic grafts in Parkinson's disease. Ann Neurol 42:95-107.

Whittemore SR, Snyder EY (1996) Physiological relevance and functional potential of central nervous system-derived cell lines. Mol Neurobiol 12:13-38.

Willing AE, Othberg AI, Saporta S, Anton A, Sinibaldi S, Poulos SG, Cameron DF, Freeman TB, Sanberg PR (1999) Sertoli cells enhance the survival of co-transplanted dopamine neurons. Brain Res 822:246-250.

Yandava BD, Billinghurst LL, Snyder EY (1999) "Global" cell replacement is feasible via neural stem cell transplantation: evidence from the dysmyelinated shiverer mouse brain. Proc Natl Acad Sci U S A 96:7029-7034.

Yurek DM, Fletcher-Turner A (1999) GDNF partially protects grafted fetal dopaminergic neurons against 6-hydroxydopamine neurotoxicity. Brain Res 845:21-27.

Zigova T, Snyder EY, Sanberg PR (2003) Neural stem cells for brain and spinal cord repair. Totowa, N.J.: Humana Press.

Zigova T, Song S, Willing AE, Hudson JE, Newman MB, Saporta S, Sanchez-Ramos J, Sanberg PR (2002) Human umbilical cord blood cells express neural antigens after transplantation into the developing rat brain. Cell Transplant 11:265-274.

Zigova T, Barroso LF, Willing AE, Saporta S, McGrogan MP, Freeman TB, Sanberg PR (2000) Dopaminergic phenotype of hNT cells *in vitro*. Brain Res Dev Brain Res 122:87-90.

Zigova T, Pencea V, Betarbet R, Wiegand SJ, Alexander C, Bakay RA, Luskin MB (1998) Neuronal progenitor cells of the neonatal subventricular zone differentiate and disperse following transplantation into the adult rat striatum. Cell Transplant 7:137-156.

Chapter 14

Genetic Modification of Neural Stem/Progenitor Cells

Ping Wu and Weidong Xiao

INTRODUCTION

A major advance in neural stem/progenitor cell research is the ability to identify neural precursors and trace their differentiation lineages by manipulating gene expression, for example by expression of a visible marker in stem cells and then following the labeled cells as they develop into neurons or glia (review by Foster and Stringer, 1999). On the other hand, recent successes in isolation and culture of neural stem cells from embryonic, fetal or adult nervous system of rodent and human allow investigators to obtain large numbers of renewable cells that can then be genetically modified to explore biologic mechanisms underlying neural proliferation and differentiation (review by Temple, 1989; Stemple and Anderson, 1992; Reynolds et al., 1992; Davis and Temple, 1994; Johe et al., 1996; Kalyani et al., 1997; Fisher, 1997; Svendsen et al., 1998; Carpenter et al., 1999; Johansson et al., 1999; Foster and Stringer, 1999; Vescovi and Snyder, 1999; Villa et al., 2000; Uchida et al., 2000; Palmer et al., 2001). Moreover, genetically modified neural stem cells may be used to deliver therapeutic reagents into the nervous systems to treat various pathologic conditions (Uchida and Toya, 1996; Ourednik et al., 1999; Shihabuddin et al., 1999; Vescovi and Snyder, 1999; Martínez-Serrano et al., 2001; Blesch et al., 2002; Park et al., 2002).

Genetic modification of neural stem cells involves mainly two aspects: (1) overexpression of an endogenous gene or ectopic expression of a foreign gene (transgene expression) and (2) diminishing or completely blocking expression of an endogenous gene (gene knockout). The first aspect, transgene expression, is usually mediated by delivery of genetic materials into neural stem cells using either viral or nonviral gene transfer methods. The majority of genetic modification studies in neural stem/progenitor cells uses this strategy, and will be the focus of this review. On the other hand, even though such studies are yet comparatively few, gene knockout is achieved in neural stem cells (Aberg et al., 2001; Sakakibara et al., 2002) through application of antisense oligonucleotides or RNA that reduces specific gene expression by either steric blocking of

From: *Neural Stem Cells: Development and Transplantation*
Edited by: Jane E. Bottenstein © 2003 Kluwer Academic Publishers, Norwell, MA

translation or enzymatic degradation of a target mRNA through RNase H and/ or ribozyme mechanisms (Dagle and Weeks, 2001; Estibeiro and Godfray, 2001; Jaaskelainen and Urtti, 2002; Opalinska and Gewirtz, 2002). Furthermore, a small interfering RNA (siRNA) technology based on the phenomenon of RNA interference (RNAi; Fire, 1999; Bosher and Labouesse, 2000; Hammond et al., 2001; Zamore, 2001; Hannon, 2002) has recently emerged as a most advanced method to knock out specific gene expression (McManus and Sharp, 2002; Tuschl, 2002). The 22- to 25-nucleotide double-stranded RNA (dsRNA) induces efficiently and specifically the degradation of its homologous mRNA, which then causes silencing of the specific gene. In the foreseeable future, this powerful genetic tool is expected to have a great impact on neural stem cell research.

This chapter will focus on transgene expression mediated by viral or nonviral gene transfer methods, applications in neural stem cell research and potential clinical usage, as well as limitations and future directions. The cell types reviewed here include neural stem cells and neural progenitor cells. We use the term neural stem/progenitor cells to indicate those populations of cells that contain both stages of the lineage, since experimentally these cells are usually intermingled even though the starting population may consist of only neural stem cells.

VIRAL VECTOR-MEDIATED TRANSGENE EXPRESSION

Viruses, for their own survival, have gained an ability to efficiently enter host cells through millions of years of evolution. Upon binding, viruses penetrate the host cellular membrane, delivering their genetic materials into the cytoplasm and/or to the nuclei of host cells. Subsequently, the viral genome manages to escape intracellular degradation and initiates viral gene expression for replication and packaging. This natural capability to deliver exogenous genetic materials to the host makes viruses an efficient vehicle for gene transfer. So far, vectors based on viruses using recombinant DNA engineering are the most effective delivery system for inducing transgene expression in mammalian cells. The general rule in development of viral vectors is that an intact wild-type virus needs to be modified to remove unnecessary or toxic viral genes and to provide packaging capability for transgenes in order to achieve safe and effective gene transfer. To this end, only the essential elements for producing viral vectors are preserved. The deleted portions are then filled with an expression cassette usually consisting of a promoter, a gene of interest (the transgene) and a polyadenylation (polyA) signal. The missing genes that are critical for viral replication and packaging are provided in *trans* through helper plasmids, cosmids, viruses or packaging cells. Thus far, five types of recombinant viral vectors have been used to transfer exogenous genes into neural stem or neural progenitor/precursor cells.

Retroviral vectors

Simple retrovirus vectors usually refer to those originally derived from Moloney murine leukemia viruses (MoMLV), a type of oncoretrovirus. Retrovirus vectors were the first developed and most widely used viral vectors for gene delivery in mammalian cells. MoMLVs are enveloped viruses with two identical copies of single stranded RNA as their genome. Figure 1 shows that the 8-11 kb RNA genome consists of two long terminal repeats (LTRs) and four viral genes including *gag* (coding for internal structural proteins), *pro* and *pol* (coding for enzymes), and *env* (coding for envelope proteins). The LTRs are the essential *cis* elements for viral replication, integration and gene expression. Following infection, the viral genome is reverse transcribed into double stranded DNA, which then integrates randomly into the host genome. The LTRs, together with a packaging signal (Ψ) and a primer binding site (PBS), are the *cis*-acting sequences required in retrovirus expression vectors. All viral genes (*gag*, *pro*, *pol* and *env*) can be removed and replaced by genes of interest (transgenes). Recombinant retrovirus vectors have a capacity up to 8 kb. Expression of the

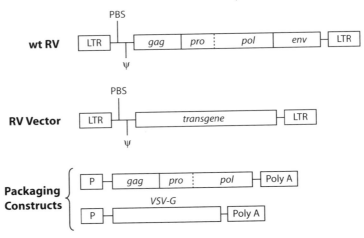

Figure 1. Schematic of the genome for a wild type retrovirus (wt RV), recombinant retroviral vector (RV vector) with a transgene of interest, and packaging constructs. Packaging cell lines are available with these packaging constructs integrated that allow stable expression of viral genes. LTR, long terminal repeat; PBS, primer binding site; Ψ, packaging signal; *gag-pro-pol* and *env*, viral genes; transgene, gene expression cassette containing the gene of interest; P, heterologous promoter; poly A, poly-adenylation signal; VSV-G, vesicular stomatitis virus G protein for pseudotyping.

transgenes can be driven by either an internal heterologous promoter or the viral promoter in the 5' LTR (Kim et al., 2000). To produce infectious retrovirus vectors, however, these viral genes need to be provided in *trans* in packaging cell lines (Cornetta et al., 1991; Kim et al., 2000; Palu et al., 2000). Particularly, vesicular stomatitis virus G protein (VSV-G) can be used to pseudotype retrovirus vectors, which broadens their host range (Friedmann and Yee, 1995). The higher stability of VSV-G also allows a higher titer viral production using ultracentrifugation (Kim et al., 2000).

MoMLV-based vectors were the first viral vectors to introduce foreign genes into neural progenitor/precursor cells. Recently, avian leukosis virus (ALV)-derived retrovirus vectors have also been used (Jungbluth et al., 1999; Dunn et al., 2000; Fults et al., 2002). Similar to their parental viruses, recombinant retrovirus vectors transduce only dividing cells, since their gene expression requires entry of the vector genome into the nucleus through the dissociated nuclear membrane, which only occurs during cell division. Taking advantage of this, retrovirus vectors containing specific markers were used to trace dividing neural progenitor/precursor cells in both developing and adult central nervous systems without labeling fully differentiated neurons or glial cells (Turner and Cepko, 1987; Holt et al., 1988; Wetts and Fraser, 1988; Luskin, 1993; Morshead et al., 1998; Morgan et al., 2000; Chambers et al., 2001; van Praag et al., 2002). Another property of retrovirus vectors is that they can efficiently integrate into host genomes. Therefore, they have been frequently used to transfer oncogenes to generate immortalized neural stem/progenitor cell lines (Evrard et al., 1990; Ryder et al., 1990; Redies et al., 1991; Snyder et al., 1992; Whittemore and White, 1993; Lundberg et al., 1997; Flax et al., 1998; Villa et al., 2000). Retrovirus vectors have also been used to introduce transgenes into primary cultured neural stem/progenitor cells (Williams and Price, 1995; Benedetti et al., 2000; Song et al., 2002; Owens, 2002). Earlier generations of retrovirus vectors are unable to transduce quiescent or slow dividing stem cells efficiently due to their dependency on active cell division and transcriptional silencing, a phenomenon of transgene shut off by silencer elements in LTRs through either *de novo* methylation or methylase-independent mechanisms (Daly and Chernajovsky, 2000; Pannell and Ellis, 2001). By modifying some of the silencer elements, a new retrovirus vector (GCDNsap) transduced more than 80% of primary cultured murine neural progenitor cells to express enhanced green fluorescent protein (GFP; Suzuki et al., 2002). However, these retrovirus vectors are unlikely to efficiently transduce quiescent neural stem cells that are mixed with more actively dividing progenitor cells in culture. Moreover, further studies are necessary to determine whether these vectors can drive long-term transgene expression without downregulation in neural progenitor cells.

Thus, retrovirus vectors may be used as a gene therapy tool for genetic modification of neural progenitor cells to deliver therapeutic reagents into the central nervous system (CNS). However, safety issues need to be thoroughly evaluated given that two cases of leukemia developed during a gene therapy clinical trial to treat severe combined immunodeficiency syndrome (SCID) using retrovirus vector-transduced hematopoietic stem cells (Fox, 2003). Particular concerns relate to possible generation of replication-competent viruses (RCVs) through homologous recombination during viral packaging and insertional mutagenesis when retrovirus vectors integrate into the host genome.

Lentiviral vectors

Lentiviruses are a subclass of retroviruses. Similar to an oncoretrovirus such as Moloney murine leukemia virus, lentiviruses are also enveloped RNA viruses containing two LTRs and four major viral genes (*gag, pro, pol* and *env*) (Figure 2). Their genomes are reverse transcribed into DNA and then integrate into the host genome. Unlike simple retroviruses, this complex retroviral subtype has additional regulatory (*tat* and *rev*) and accessory (*vif, vpr, vpu* and *nef*) genes that encode proteins involved in their life cycle (Naldini, 1998; Stevenson, 2002). The desirable feature of lentiviruses is that they are able to enter the nucleus without the nuclear membrane breakdown that occurs during mitosis (Stevenson, 2002). Therefore, lentiviruses infect both proliferating and non-proliferating cells. This is particularly attractive for creating lentivirus-based viral vectors to transfer genes into quiescent or slow-dividing neural stem cells.

The best characterized lentivirus vectors are derived from human immunodeficiency virus type 1 (HIV-1). Since the first attempt to develop an HIV-1-based lentivirus vector (Parolin et al., 1994) and the first successful use of the lentivirus vectors to express a transgene in rat brain neurons *in vivo* (Naldini et al., 1996a, b), many efforts have been made to improve both the efficiency and safety of lentivirus vectors through several generations of viral packaging systems and transgene vectors (review by Lever, 2000; Naldini and Verma, 2000; Kafri, 2001; Ailles and Naldini, 2002). Similar to the MLV oncoretroviral vectors (Figure 2), lentivirus transgene expression vectors include LTRs, a packaging signal (Ψ) and a primer binding site (PBS). Unlike those derived from MLV, lentivirus vectors require the *rev*-responsive element (RRE, a *cis* sequence inside the viral *env* gene), which is recognized and bound by the *Rev* protein to ensure transport of vector RNA from nucleus to cytoplasm (Ailles and Naldini, 2002). To minimize generation of RCVs through homologous recombination, all viral genes required for generating infectious lentivirus vectors are split into 3 separate packaging constructs: a *gag-pro-pol* vector, a *rev* vector, and a vector providing pseudotypic *env* protein (VSV-G). Recently, lentivirus vector gene

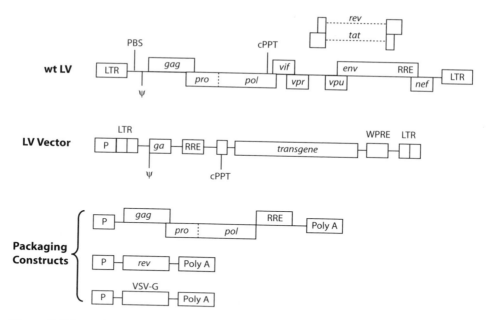

Figure 2. Schematic of the genome for a wild type lentivirus (wt LV), recombinant HIV-1 based lentiviral vector (LV vector), and packaging constructs. Viral packaging is achieved usually by cotransfection of 4 plasmids including the LV vector and 3 packaging constructs. LTR, long terminal repeat; PBS, primer binding site; Ψ, packaging signal; *gag-pro-pol* and *env*, viral genes; cPPT, central polypurine tract; *vif, vpr, vpu, nef, tat*, accessory or regulatory viral genes; *rev*, coding for a regulatory protein that binds to RRE, Rev-responsive element, required for vector RNA transport; P, heterologous promoter; transgene, gene expression cassette containing the gene of interest; WPRE, woodchuck hepatitis virus posttranscriptional regulatory element; poly A, poly-adenylation signal; VSV-G, vesicular stomatitis virus G protein for pseudotyping.

expression efficiency has been increased by including (1) an HIV-1 central DNA flap, containing the central polypurine tract (cPPT) and the central termination sequence (CTS) to enhance vector genome nuclear import (Zennou et al., 2001) or (2) a woodchuck hepatitis virus posttranscriptional regulatory element (WPRE) to stabilize mRNA transcripts and facilitate their nuclear export (Zufferey et al., 1999). Furthermore, the U3 regions of the LTRs may be modified to improve the safety of these lentivirus vectors (Ailles and Naldini, 2002; Galimi and Verma, 2002).

Given their capability to transduce genes in nondividing cells, lentivirus vectors have been tested recently in neural stem/progenitor cells. Most studies using HIV-1-based lentivirus vectors show that they introduce foreign genes efficiently (up to 90%) in either primary cultured or immortalized neural progenitor cells from both human and rodent (Englund et al., 2000; Buchet et al., 2002; Englund et al., 2002a,b; Ostenfeld et al., 2002). The transgene expression seems stable, in some cases for at least 6 months when lentivirus vector-treated cells are grafted into the brain (Buchet et al., 2002). Other studies, though, reported downregulation of gene expression by 6 weeks in grafted primary neural stem cells (Ostenfeld et al., 2002) or by 1 week in immortalized neural progenitor cells (Rosenqvist et al., 2002). The latter may indicate a transgene silencing effect of lentivirus vectors similar to that of MoMLV-based retrovirus vectors (Pannell and Ellis, 2001).

Safety concerns are particularly raised for HIV-1-based lentivirus vectors since their parental viruses are dangerously pathogenic to humans. Although efforts have been made to improve the design of HIV-1 vectors, possibilities of infectious RCVs are not completely eliminated (Delenda et al., 2002). Other non-primate lentivirus vectors have thus been developed as alternatives (review by Curran and Nolan, 2002). Specifically, Hughes et al. (2002) demonstrated that vectors based on feline immunodeficiency virus (FIV) transduced mouse primary neural progenitor cells efficiently. Additional concerns are held for lentivirus vectors, as they integrate randomly into host chromosomes. Consequently, whether undesirable incidents, similar to retrovirus vectors in SCID patients, occur or not remains to be carefully examined.

Adenoviral vectors

Adenoviruses have been extensively studied for vector development to transfer genetic materials into mammalian cells. Adenoviruses are non-enveloped viruses with a linear double stranded DNA genome of approximately 35 kb in length. About 50 serotypes of adenoviruses have been identified in humans. Since they can cause respiratory tract infections, they are often called cold viruses. Some of them are oncogenic. The commonly used adenovirus vectors are based on type 2 or type 5 adenoviruses. The adenovirus genome contains six early genes (E1A, E1B, E2A, E2B, E3 and E4) coding for proteins pertaining to regulatory functions and 5 late genes (L1-5) coding for structural proteins (Figure 3; Russell, 2000). Adenovirus vectors are constructed by deleting E1 and E3 (first generation); deleting E1, E2 and E4 (second generation); or deleting all viral genes (third generation; also called "gutless" vectors). To grow recombinant adenoviruses, the viral genes need to be supplied in *trans* by a helper virus, plasmid or integrated into a helper cell genome such as the human fetal kidney 293 cell line (review by Danthinne and Werth, 2000; Hitt and Graham, 2000; Russell, 2000; Nasz and Adam, 2001). The maximal transgene capacities of adenovirus vectors thus range between 8.2 to 37 kb, depending upon which generation is used for construction.

One of the most attractive features of adenovirus vectors is that they are very efficient at transducing both dividing and non-dividing cells. In addition,

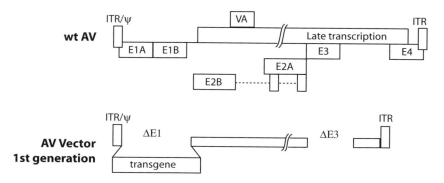

Figure 3. Schematic of the genome for a wild type adenovirus (wt AV) and the first generation of recombinant adenoviral vector (AV vector) that are still used widely. ITR, inverted terminal repeat; Ψ, packaging signal; E1-E4, early viral genes; Late transcription, late viral genes; VA, virus-associated RNA; transgene, gene expression cassette containing the gene of interest; ΔE1 or ΔE3, deleted E1 or E3 gene. The AV vector can be produced through a homologous recombination between a shuttle plasmid containing the transgene expression cassette with genomic AV DNA which has deletions in Ψ, E1 and/or E3. Another method is to clone the expression cassette into a large plasmid containing the whole adenovirus genome, and recombinant adenovirus can be rescued in 293 cells.

they can be produced at very high titers ($>10^{11}$ plaque forming units/ml). Adenovirus vectors used in neural stem/progenitor cell studies are mainly derived from the first generation of recombinant viruses. They have been shown to effectively transduce cultured neural stem/progenitor cells (Hughes et al., 2002; Falk et al., 2002). One of the applications of adenovirus gene delivery is to label neural stem/progenitor cells with a specific marker such as β-galactosidase (β-gal) or GFP and then trace these cells when grafted *in vivo* (Gage et al., 1995; Sabate et al., 1995; Chow et al., 2000; Mizumoto et al., 2001). In addition, adenovirus vectors have also been used directly to transduce ependymal or subependymal (includes neural stem cells) cells in the adult brain (Yoon et al., 1996; Benraiss et al., 2001). Although adenovirus vectors are more efficient in terms of transgene expression than most other viral vectors (Falk et al., 2002), cytotoxic effects and strong immune responses are often observed. Safety and toxicity of adenovirus vectors has been a big concern in clinical trials (NIH Report, 2002). Another drawback is that they alter the properties of neural stem/progenitor cells, e.g., inducing glial cell differentiation (Hughes et al., 2002). Since adenovirus vectors remain episomally after transduction, their transgene expression tends to be transient. This is a problem if a long-term gene expression is required in continuously dividing neural stem/progenitor cells. However, it has been reported that adenovirus-mediated gene expression is stable for up to 4 months in neural stem/progenitor cells after transplantation, although by then these cells have probably already stopped dividing (Chow et al., 2000).

On the other hand, no risk of insertional mutagenesis needs to be considered for this type of vector.

Adeno-associated viral vectors

Adeno-associated viruses (AAV) are one of the smallest human DNA viruses. They are considered to be "defective" since productive AAV replication depends on helper functions supplied by an unrelated helper virus such as an adenovirus, herpes simplex virus (HSV) types I and II, cytomegalovirus, or pseudorabies virus (Muzyczka and Berns, 2001). Under special circumstances, AAV may replicate autonomously in rare immortalized and transformed cell lines when treated with genotoxic agents or even in differentiating keratinocytes of a normal skin model without genotoxic agents (Yakobson et al., 1989; Yalkinoglu et al., 1991; Meyers et al., 2000). In the absence of a helper virus, more than 70% of wild type AAV can integrate site-specifically into the AAVS1 site of the host gene (19q-13-qTer; Kotin et al., 1990; Samulski et al., 1991). Additional attractive features for using AAV as a gene transfer vector are: (1) it infects both dividing and nondividing cells and (2) it is nonpathogenic in humans. Thus far, several AAV serotypes have been detected in human and non-human primates (Gao et al., 2002). Among all those serotypes, AAV2 has been characterized most extensively and is the first to be explored as a gene delivery vector (review by Muzyczka, 1992). This simple virus contains only an icosahedral protein capsid and a single stranded DNA molecule. The 4.8-kb DNA genome (Figure 4) contains two viral genes, *rep* and *cap*, flanked by two inverted terminal repeats (ITR, 145 bp). Four Rep proteins encoded by the *rep* gene control viral replication, integration, assembly, and regulation of viral structural gene expression. Three Cap proteins encoded by the *cap* gene form the capsid. Since the only viral sequence required for constructing an AAV transgene plasmid is the ITRs, all the viral genes (*rep* and *cap*) can be deleted and replaced by a transgene expression cassette. The viral genes are then provided in *trans* (AAV helper plasmid), along with a mini-adenovirus plasmid supplying the essential helper function, to produce recombinant AAV (rAAV; review by Monahan and Samulski, 2000; Smith-Arica and Bartlett, 2001). High titers of rAAV are achieved through modulating AAV *rep* gene expression (Xiao et al., 1998). The rAAV vectors are purified by either ultracentrifugation or affinity chromatography (Grimm et al., 1998) using the primary receptor for AAV-2, heparin sulfate proteoglycan (HSPG; Summerford et al., 1999; Bartlett et al., 2000). AAV vectors purified by chromatography have much higher purity than those purified by CsCl gradient and ultracentrifugation and may have higher infectivity (Zolotukhin et al., 1999).

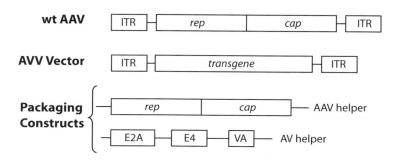

Figure 4. Schematic of a genome for a wild type adeno-associated virus (wtAAV), recombinant AAV vector, and packaging constructs. AAV vectors are packaged by cotransfection of an AAV vector plasmid with an AAV helper and an AV helper in the human embryonic kidney cell line 293. ITR, inverted terminal repeat; *rep* and *cap*, viral genes; transgene, gene expression cassette containing the gene of interest. The AAV helper contains the viral genes *rep* and *cap* but not the ITRs. The AV helper contains E2, E4 and VA RNA of AV that are essential to provide the helper function for AAV replication.

AAV vectors usually drive a long-term transgene expression (Carter and Samulski, 2000), which is desirable for many applications including delivery of genes to correct genetic defects. This longevity is thought to be due to their capability to integrate into the host genome (Wu et al., 1998) or the persistence of high molecular weight episomal circular concatamers (Yang et al., 1997; Duan et al., 1999). Another attractive feature of AAV vectors is that they are not cytotoxic, and induce minimal T-cell mediated immune responses (Monahan and Samulski, 2000; Rabinowitz and Samulski, 2000; Smith-Arica and Bartlett, 2001; Zhao et al., 2001). On the other hand, one limitation of rAAV is its small capacity (about 5 kb) for a transgene insert. Recent developments, however, make it possible to use AAV vectors for genes large than 5 kb using a split vector strategy (Duan et al., 2000). However, the effectiveness of this new technology remains to be studied. Although wild-type AAVs integrate into the host genome site-specifically, rAAVs, in the absence of Rep proteins, integrate into the transduced cell genome randomly. Application of Rep in *trans* may be one way to restore AAV's potential for site-specific integration (Monahan and Samulski, 2000; Owens et al., 2002).

Since AAV vectors introduce genes efficiently into both dividing and non-dividing cells, we hypothesized that AAV vectors would be an ideal vector to transduce human neural stem/progenitor cells (hNSC/NPC). Using an AAV2-derived CAGegfp vector containing a strong constitutive promoter, CAG, and a gene coding for enhanced green fluorescent protein (EGFP), we successfully transduced hNSC/NPCs (Figure 5a-b) in a dose-dependent manner (Wu et al,. 2002b). When used on dissociated cells, 100% of the transduced cells were

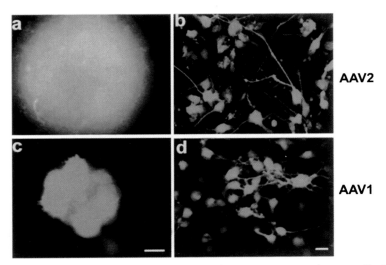

Figure 5. Recombinant AAV vectors transduce fetal human neural stem cells (hNSCs) efficiently. The K048 line is derived from the cortex of an 8-week aborted human embryo and was expanded *in vitro* with epidermal growth factor (EGF), fibroblast growth factor 2 (FGF2) and leukemia inhibitory factor (LIF). (**a**) and (**c**), neurospheres (passage 25) are transduced with AAV2gfp and AAV1gfp, respectively, at multiplicity of infections (MOIs) of 25. Live images are taken 2 weeks after treatment. (**b**) and (**d**), hNSCs transduced with both types of AAV vectors were further differentiated by withdrawal of growth factors. AAV2-transduced cells retain neuronal differentiation capability (analyzed by immunostaining with neuronal markers; Wu et al., 2002b). The differentiation phenotypes of AAV1-transduced cells remain to be determined. Scale bars: **a** and **c**, 100 μm; **b** and **d**, 20 μm.

green fluorescent protein labeled. GFP gene expression in hNSC/NPCs increases gradually and reaches a peak in about 2 weeks to 1 month depending on the amount of virus used. This is typical for AAV-mediated gene expression and is due to a slow conversion from a single stranded DNA genome to an expressionable double stranded form (Ferrari et al., 1996; Fisher et al., 1996; Sanlioglu et al., 2001). Furthermore, AAV-mediated GFP expression lasts for at least 3 months (longest time tested) *in vitro*. While GFP-expressing hNSC/ NPCs can differentiate into both neurons and glial cells, they gradually stop proliferating. It is suggested that GFP expression may cause cessation of proliferation (Martínez-Serrano et al., 2000; Martínez-Serrano et al., 2001). Alternatively, AAV capsid proteins or trace amounts of cellular proteins may contribute to this inhibition. More recently, we found that a vector derived from a type 1 AAV (AAV1) could also deliver a GFP transgene efficiently into hNSC/NPCs (Figure 5c-d). In contrast, Hughes et al. (2002) reported that vectors derived from different serotypes of AAV (including AAV2, AAV4 and AAV5) did not transduce mouse neural progenitor cells effectively. Cells from different species may be one of the explanations for the discrepant observations on AAV2-derived vectors, as it is unclear whether these mouse neural progenitor cells have receptors for AAV2, such as the HSPG primary receptor and/or coreceptors including fibroblast growth factor receptor (FGFR) or aVb5 integrin (Qing et al., 1999; Summerford et al., 1999).

Herpes simplex viral vectors

Herpes simplex virus type 1 (HSV-1), also developed as a gene delivery vector, is a large enveloped DNA virus with a natural neurotropism in humans (Burton et al., 2001; Roizman and Knipe 2003). The viral genome is a 152-kb linear double stranded DNA molecule containing two unique regions, long and short (termed U_L and U_S). Both U_L and U_S regions are flanked by inverted repeat sequences, internal repeat (IR) and terminal repeat (TR) (Figure 6). Over 80 genes have been identified and classified into three main groups: the immediate-early (IE or α) genes, early (E or β) genes, and late (L or γ) genes. The IE and E genes code proteins for regulation of viral gene expression and replication, while the L genes code for structural proteins.

Three types of HSV-1-based vectors have been developed (review by Simonato et al., 2000) including amplicon, replication-defective and replication-conditional vectors. The amplicon vectors are recombinant plasmids containing a transgene cassette and minimal HSV-1 viral sequences required for replication and packaging: an origin of DNA replication (*ori*) and a DNA packaging signal (*pac*; review by Fraefel et al., 2000). In this system, all the other viral regulatory and structural genes are supplied *in trans* by either helper viruses or helper cosmids to produce recombinant amplicon vectors. The replica-

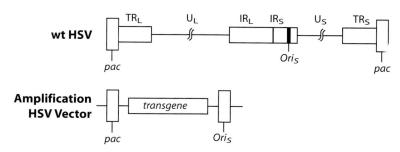

Figure 6. Schematic of a genome for a wild type herpes simplex virus (wt HSV) and amplicon HSV vector. HSV vectors are packaged by transfection of the amplicon HSV plasmid together with either HSV cosmid helpers or mutated HSV helper viruses. *pac*, packaging signal; ori$_S$, origin of DNA replication; U_L, unique long segment; U_S, unique short segment; TR$_L$, terminal repeat of the U_L segment; TR$_S$, terminal repeat of the U_S segment; IR$_L$, internal repeat of the U_L segment; IR$_S$, internal repeat of the U_S segment; transgene, gene expression cassette containing the gene of interest.

tion-defective vectors are derived from HSV-1 viruses by deletion or mutation of several essential viral genes, which are toxic and/or important for the initiation of viral replication. The deleted gene(s) are then replaced with a transgene cassette (Burton et al., 2001). To produce recombinant vectors, the essential viral genes are provided in *trans*. The third type of HSV-1 vector is the replication-conditional vector (review Jacobs et al., 1999; Burton et al., 2001). By deleting one or more genes that are essential for viral replication in nondividing cells only, they retain the capability to replicate in dividing cells. This type of vector has been developed specifically to selectively kill tumor cells either by their lytic replication in dividing cells or by delivery of anti-tumor agents (Andreansky et al., 1996; Jacobs et al., 1999).

As for most of the other viral vectors described above, HSV-1 vectors can also transduce both dividing and nondividing cells. Another attractive feature is their large capacity for transgene insertion, theoretically up to 150 kb, which could be beneficial for incorporation of a large gene or multiple genes when needed. Recently, a few groups tested the capability of HSV-1 vectors to transduce neural precursor/progenitor cells. Vicario and Schimmang (2003) reported using amplicon vectors to deliver foreign genes into neural progenitor cells that originated from mouse embryonic stem cells . However, the transgene expression was only transient, lasting for 5-6 days. This gene expression instability is often observed with HSV-1 vectors probably due to silencing of the viral promoters. Application of tissue-specific promoters may be one way to prolong transgene expression (Fraefel et al., 2000). Alternatively, episomal transduction without integration of HSV-1 vectors may also contribute in part to the transient gene expression, especially in actively dividing progenitor cells. The positive aspect of this episomal feature of HSV vectors is the lack of insertional mutagenesis associated with integration viral vectors, e.g., retroviral, lentiviral, or AAV vectors. Although not assessed in neural stem/progenitor cells, virion-related cytotoxicity and inflammation are other drawbacks of HSV-1 vectors when applied in the central nervous system. Further studies are required to improve the vector designs and packaging systems to reduce toxicity, as well as to eliminate possible production of replication competent viruses through homologous recombination between vectors and helpers or the existing latent HSV in host cells (Lachmann and Efstathiou, 1999).

In summary, five types of viral vectors have been tested in neural stem/progenitor cells. Their unique features as well as their strengths and weaknesses are summarized in Table 1 for comparison among different types of viral and non-viral vectors, and for their use in gene transfer to neural stem cells. Thus far, only limited studies have been carried out to compare the efficiency of different vectors for transferring genes into neural stem/progenitor cells. For example, Hughes et al. (2002) reported that a feline lentivirus vector or an aden-

Table 1. Comparison of viral and non-viral vectors for gene delivery into neural stem/progenitor cell

	Viral Vectors					Non-viral Vectors
	Retrovirus	Lentivirus	Adenovirus	AAV	HSV	
Target cells	dividing cells	dividing non-dividing	dividing non-dividing	dividing non-dividing	dividing non-dividing	dividing non-dividing
Gene transfer efficiency	++	++	+++	++	+	+
Capacity	≤8 kb	≤9 kb	≤8 kb (1st generation) ≤37 kb (Gutless vector)	≤4.5 kb	30 kb	Unlimited
Gene expression	long-term	long-term	transient	long-term	transient	transient
Integration	+	+	-	+	-	-
Titer	≥10^7 infectious units/ml	≥10^8 transducing units/ml	≥10^{10} plaque forming units/ml	≥10^9 transducing units/ml	≥10^7 infectious units/ml	N/A*
Cytotoxicity	-	-	+	-	+	+
Immune response	low	low	high	low	high	low
Large scale preparation of vectors	relatively easy	relatively difficult	1st generation: easy Gutless vector: very difficult	difficult	difficult	easy

*N/A: non-applicable

ovirus vector transduced embryonic mouse neural stem cells efficiently, while AAV type 2, 4 and 5 vectors failed to transduce these cells in culture. Luskin et al. (1988) observed a low efficiency when using retrovirus vectors to transfer foreign genes into neural progenitor cells. However, our group found that AAV type 1 and 2 vectors can efficiently deliver a reporter gene into primary cultured human neural stem cells (Wu et al., 2002a,b). In addition, Falk et al. (2000) reported several viral vectors transduce primary cultured neural stem cells from adult mouse lateral wall of lateral ventricles in a rank of AV vector > retrovirus vector > lentivirus vector. These discrepant findings may reflect the differences in the types of cells used among different groups, the different culture methods, and/or the promoters or other genetic components in the various vectors used.

NONVIRAL GENE DELIVERY SYSTEMS

Despite the high gene transfer efficiency mediated by viral vectors, most of them induce immune responses in addition to the safety risks related with replication competent viruses or insertional mutagenesis. Limited capacity for gene inserts and technical difficulties in production are other drawbacks. Thus, nonviral vectors such as cationic lipids or polymers have been used as alternative gene delivery vehicles (review by Audouy and Hoekstra, 2001; Pedroso de Lima et al., 2001; Merdan et al., 2002). Although in normal circumstances, non-viral vectors are not as efficient as viral vectors, they can carry very large DNA molecules, are easily modified, and are produced at relatively low cost.

Cationic lipids, often termed liposomes, are lipid bilayers with positive charges. They condense negatively charged DNA molecules and form a lipid-DNA complex called a lipoplex (review by Chesnoy and Huang, 2000; Audouy and Hoekstra, 2001). Lipoplexes bind to the negatively charged cell membrane through electrostatic interactions and are internalized via endocytosis (Felgner et al., 1994). Through a still unknown mechanism, DNA then escapes from endosomes and enters the cell nucleus.

Cationic polymers, bearing protonable amines, are also used to condense DNA molecules and mask their negative charges (De Smedt et al., 2000; Merdan et al., 2002). The polymer-DNA complex is referred as a polyplex, which electrostatically binds to the surface of cells and then is endocytosed. Although cationic polymers have been used for more than a decade in gene delivery, the underlying mechanisms for formation of polyplexes and the intracellular transport pathway from endosomes to nuclei remain unclear.

Both cationic lipids and polymers have been applied to facilitate gene delivery into neural stem/progenitor cells, although much less frequently than viral vectors. Using cationic lipids, foreign genes are introduced into either primary cultured neural stem/progenitor cells from embryonic and adult brain

(Wang et al., 1998, 2000; Falk et al., 2002; Kim et al., 2002) or immortalized cell lines (Corti et al., 1996; Eaton and Whittemore, 1996). The commercially available lipids include Lipofectin®, Lipofectamine™, Lipofectamine Plus™, Cellfectin®, and DMRIE-C (Invitrogen/Life Technologies) as well as Effectene (Qiagen). The reported efficiencies of cationic lipid-mediated transgene expression in neural stem/progenitor cells vary from 0.007% to 15%. This huge variation is probably due to differences in (1) types and dosage of liposomes or polymers, (2) the promoters in different plasmid constructs, (3) the types of cells, and (4) the different handling procedures. Although requiring further examination, DMRIE-C seems to give the highest gene transfer efficiency, up to 15% in one report (Falk et al., 2002). Cationic polymers such as SuperFect (Qiagen) and polyethyleimine (PEI) have also been used to transfer genes into primary cultured neural stem/progenitor cells *in vitro* (Falk et al., 2002; Kim et al., 2002). In addition, cells in the subventricular zone (SVZ) of adult mouse brain are reported to express foreign genes after using the PEI delivery method (Lemkine et al., 2002).

The most attractive features of the cationic lipids and polymers as gene delivery vehicles are convenient large scale production, targeting all cell types, unlimited DNA capacities, episomal gene expression with low risk of insertional mutagenesis, low immune responses and relatively easy modification for targeted gene delivery. However, the low transfection efficiency and considerable cytotoxicity are weaknesses of these gene delivery systems. A recent study by Kim et al. (2002) showed an apoptosis effect caused by SuperFect. In confirmation of this, an enhancement of SuperFect-mediated GFP gene expression was seen in primary fetal neural stem cells after cotransfection with plasmids containing anti-apoptotic genes *Bcl-2* or *Bcl-X$_L$*.

APPLICATIONS OF GENETICALLY MODIFIED NEURAL STEM/PROGENITOR CELLS

Genetic modification of neural stem/progenitor cells has and will continue to contribute to our knowledge of stem cell biology and to development of cell and gene therapy to treat diseases in the nervous system. It is not the intent of this review to include all previous studies in this exciting and growing field, but rather to point out potential applications of gene manipulation of neural stem/progenitor cells by selecting representative studies using either viral or non-viral vectors.

Genetic modifications for tracing or marking neural stem/progenitor cells.

Genetic tracers for identification of endogenous stem cell/progenitor cells and their progeny *in vivo* and *in vitro*

Based on the unique properties of retrovirus vectors to transfer genes into dividing cells but not fully differentiated neurons or glial cells, scientists have used these vectors to label endogenous dividing CNS neural progenitor/precursor cells with specific markers since the late 1980's. These cells and their progeny, including neurons, were then identified and traced *in vivo*. Clonal analyses using retrovirus-delivered exogenous reporter genes such as β-galactosidase provide strong evidence for the existence of a common multipotent progenitor/precursor cell that can differentiate into many types of neurons and glia in various regions of the developing CNS, including retina (Turner and Cepko, 1987; Holt et al., 1988; Hubener et al., 1995), cortex (Luskin, 1993; Williams and Price, 1995), and spinal cord (Leber et al., 1990) from rats or Xenopus. A similar strategy has been used to characterize the proliferation capabilities of a single neural stem/progenitor cell and to follow its migration pathway in adult mouse brain (Morshead et al., 1998). In addition, using a retrovirus vector containing enhanced green fluorescent protein (EGFP), van Praag et al. (2000) reported continuous neurogenesis in the adult mouse hippocampus. Furthermore, EGFP allows the newly generated neurons to be visualized and examined directly by electrophysiological recording. In addition, Song et al. (2002) applied a retrovirus vector containing EGFP to label neural stem cells isolated from adult rat hippocampus. They cocultured the retrovirus-EGFP-labeled stem cells with adult astrocytes and found that the astrocytes play a crucial role in directing neural stem cells to a neuronal fate. In conclusion, retrovirus-mediated genetic labeling of neural stem/progenitor cells contributes significantly to our growing knowledge of the properties of these cells in the CNS.

Genetic tracers for identification of grafted neural stem cell/progenitor cells

An obvious use of neural stem/progenitor cells is to replace lost neural cells in degenerated or injured CNS. Thus, much effort has been made to determine whether grafted cells survive, differentiate and integrate in animal recipients. One way to track and distinguish transplanted from host cells is to label the grafted cells with exogenous markers by genetic modification prior to transplantation. The genetic markers include an *E. coli* bacterial *lacZ* gene coding for β-galactosidase and a mutated jellyfish gene coding for GFP. Expression of the *lacZ* gene is detected through histochemical staining, while the GFP reporter expression can be visualized either directly under a fluorescent micro-scope or indirectly through immunostaining with antibodies specific against GFP.

The lacZ reporter was used first to genetically mark neural stem/progeni-tor cells derived from rodent and human CNS with the aid of retrovirus vectors (Flax et al., 1998), adenovirus vectors (Gage et al., 1995; Sabate et al., 1995; Chow et al., 2000), and lentivirus vectors (Hughes et al., 2002; Pluchino et al., 2003). Recently, preference has switched to GFP for labeling such cells before grafting. One of the reasons is that the natural fluorescence of GFP is readily detected without further processing. Vectors to transfer GFP into neural stem/progenitor cells include retrovirus (Young et al., 2000; Suzuki et al., 2002), lentivirus (Englund et al., 2002a, b; Ostenfeld et al., 2002), adenovirus (Mizumoto et al., 2001) and AAV vectors (Wu et al., 2002a, b). One particularly attractive feature of the GFP labeling is that it fills both the cell bodies and their processes without leaking. This allows one to track differentiated neurons derived from grafted human neural stem cells with a much better defined morphology, and to follow their integration into the host environment (Figure 7). Furthermore, GFP can be used as an indicator for guiding electrophysiological recording on live grafted cells in brain slices to further analyze functional maturation of grafted neural stem/progenitor cells.

Genetic modification to isolate and enrich neural stem/progenitor cells

Transfer of a marker gene can also be used to facilitate the isolation and enrichment of neural stem/progenitor cells from heterogeneous CNS tissues. This strategy is particularly useful for obtaining cells from adult CNS tissues that contain only few neural stem/progenitor cells, which makes it difficult to get a large number of relatively pure cells without a long period of clonal ex-pansion. Using a nonviral gene transfer technique with Lipofectin®, Roy et al, (2000) have selectively extracted neural stem/progenitor cells from adult hu-man hippocampus. The plasmid vector they used contains a GFP marker gene under the control of the enhancer element of nestin. Following fluorescence-activated cell sorting (FACS), neural stem/progenitor cells are selected based on their expression of the fluorescent GFP gene. With a similar strategy, this group also isolated and purified more restricted neuronal progenitor cells from either embryonic chick and rat forebrain (Wang et al., 1998) or adult rat lateral ventricular wall (Wang et al., 2000) using a Tα-1 tubulin promoter that is spe-cific for neuronal progenitors and young neurons. Although the GFP marking together with FACS is a relatively convenient method for isolation and enrich-ment of neural stem/progenitor cells, its use for further clonal expansion may be limited since continuous expression of GFP may affect the proliferation of

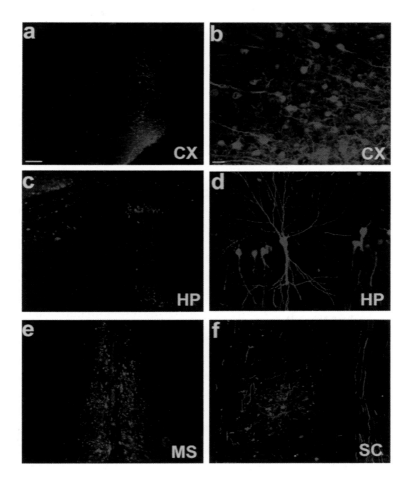

Figure 7. Recombinant AAV vectors containing GFP are used to trace grafted hNSCs in rat brain and spinal cord. hNSCs, originally derived from the cortex of an 8-week human embryo, are cultured with FGF2/heparin/laminin for 6-7 days *in vitro* before implantation into adult rats. Four days before grafting, cells are transduced with AAV2gfp. Animals are treated with the immunosuppressant cyclosporin A for 1 month and are then perfused with 4% paraformaldehyde.. Brain and spinal cord are cryosectioned and examined by confocal microscopy. (**a**) and (**b**) frontal cortex; (**c**) and (**d**) hippocampus; (**e**) medial septum; (**f**) spinal cord. Note, that grafted hNSCs integrated into the CNS and differentiated into neurons (determined by immunostaining with neuronal markers; Wu et al., 2002a). Scale bars: **a, c, e** and **f**, 250 μm; **b** and **d**, 25 μm.

these cells *in vitro* (Englund et al., 2000; Martínez-Serrano et al., 2000; Wu et al., 2002b).

Genetic modification to enhance proliferation of neural stem/progenitor cells

Proliferation of neural stem/progenitor cells requires either epigenetic or genetic stimulants. Without such signals, they quickly exit the cell cycle and begin differentiating. Over the past two and a half decades, retrovirus vectors have been used predominantly, due to their high integration efficiency, to create immortal cell lines by introducing oncogenes into neural stem/progenitor cells. The commonly used oncogenes include v-myc and the temperature-sensitive mutant of SV40 large T antigen. The earlier cell lines are mainly derived from mouse or rat CNS (Bartlett et al., 1988; Frederiksen et al., 1988; Evrard et al., 1990; Ryder et al., 1990; Redies et al., 1991; Snyder et al., 1992; Whittemore and White, 1993). Recently, similar immortal cell lines have been created from embryonic human brain tissue (Villa et al., 2000). Although it may not be desirable to use oncogene-overexpressing or proto-oncogene overexpresssing cells in clinical applications, these immortalized neural stem/progenitor cells, with their homogeneous phenotypes (if they are cloned lines), have contributed greatly to our knowledge of cell biology and the development of transplantation strategies to replace lost neural cells or to deliver therapeutic reagents into the CNS. For more detailed information, readers are referred to Chapter 8 in this volume and to several previous reviews (Fisher, 1997; Martínez-Serrano and Bjorklund, 1997; Foster and Stringer, 1999; Vescovi and Snyder, 1999).

Genetic modification to study and direct differentiation of neural stem/progenitor cells

Modification of specific gene expression in neural stem/progenitor cells not only helps us to dissect molecular mechanisms underlying neural development, but also provides us tools to guide these cells toward a desired phenotypic differentiation. For example, using a retrovirus vector to overexpress a Notch1 gene in either cortical or olfactory bulb (OB) precursors *in vivo* or *in vitro*, Chambers et al. (2001) observed an overall inhibitory effect of the Notch1 signal on neuronal differentiation. However, precursors derived from the two different areas of the mouse brain during early development respond differently to Notch1 overexpression. While the OB precursors respond to Notch1 by remaining quiescent without further differentiation, the cortical precursors cease proliferation and then differentiate rapidly toward a glial lineage.

Effort has also been made to apply genetic tools to induce neuronal differentiation from neural stem/progenitor cells. For example, Falk et al. (2002) applied a retrovirus-based vector to ectopically express neurogenin 2, a basic helix-loop-helix transcription factor (Anderson, 1999), in adult mouse neural stem cells derived from the lateral wall of the lateral ventricles. They observed that 90% of the transduced cells differentiated into neurons. Another neurogenic transcription factor, NeuroD, was introduced into an immortalized human neural stem cell line by retrovirus-mediated transduction (Cho et al., 2002). These cells then differentiated into neurons with TTX-sensitive Na^+ currents .

Generation of large quantities of cells with a specific neuronal phenotype from the multipotent neural stem/progenitor cells is a challenge, but it is an ultimate goal of cell therapy. Genetic modification may be one way to achieve this goal. For example, Wagner et al. (1991) stably transfected the mouse C17.2 neural stem cell line with Nurr1, an orphan nuclear receptor critical for dopaminergic neuron development. They demonstrated that more than 80% of the Nurr1-overexpressing cells differentiated into dopaminergic neurons when cocultured with astrocytes. The study indicated that both Nurr1-overexpression and unknown factors derived from type 1 astrocytes are required for dopaminergic neuron differentiation.

Genetic modification of neural stem cells to deliver therapeutic reagents into the CNS

Neural stem cells have great potential to replace lost neurons or glial cells in CNS degeneration or injury. In addition, these cells can be genetically modified to become vehicles or biological "minipumps" to deliver therapeutic reagents into the diseased brain or spinal cord. Their incredible plasticity and remarkable capabilities to migrate and integrate into neural circuitry make these cells extremely valuable for the development of combined cell and gene therapies to treat various neurological disorders.

Neurotrophic factors

Since lack of sufficient neurotrophic factors may be one of the reasons for limited regeneration following neurodegeneration or neurotrauma, neural stem/progenitor cells may be genetically modified to deliver desired neurotrophic factors locally to facilitate neural regeneration. Using neural progenitor cell lines that were derived from embryonic rat hippocampus and then genetically modified to produce nerve growth factor (NGF) or brain derived neurotrophic factor (BDNF), Martínez-Serrano and colleagues were the first to report these protective effects of genetically modified neural progenitor cells on damaged host neurons in several animal models. These include: (1) rescue of axotomized cholinergic neurons in the medial septum (Martínez-Serrano et al., 1995b), (2) reversal of cholinergic neuron atrophy in cognitively impaired aged rats (Martínez-Serrano et al., 1995a; Martínez-Serrano and Bjorklund, 1998), and (3) protection of striatal neurons against excitotoxic damage (Martínez-Serrano and Bjorklund, 1996). Later, Liu et al. (1999) demonstrated enhancement of axonal growth and regeneration in animals suffering from spinal cord injury by transplantation of a neural stem cell line derived from mouse cerebellum (C17.2) and genetically modified to secrete neurotrophin 3 (NT-3). In addition to the paracrine effect of neural stem cell-delivered trophic factors on host neurons, BDNF and glial-derived neurotrophic factor (GDNF) enhanced the survival and neuronal differentiation of the grafted cells in an autocrine fashion (Eaton and Whittemore, 1996; Ostenfeld et al., 2002).

Metabolic or neurotransmitter enzymes

Neural stem/progenitor cells have been tested for their ability to deliver therapeutic enzymes into the CNS to treat inherited genetic neurodegenerative diseases such as mucopolysaccharidosis type VII (MPS VII), lysosomal storage disease (Sly disease), or Tay-Sachs disease (Zlokovic and Apuzzo, 1997; Ourednik et al., 1999; Park et al., 2002). When grafted into neonatal mice, genetically modified neural progenitor cells derived from the C17.2 cell line spread widely throughout the brain and constitutively secreted enzymes such as β-glucuronidase or β-hexosaminidase. By doing so, they corrected the gene deficit in the animal model of Sly or Tay-Sachs disease, respectively (Snyder et al., 1995; Lacorazza et al., 1996). Another type of lysosomal storage disorder, Krabbe disease, is caused by mutation of the galactocerebrosidase gene and may be corrected by a similar strategy using genetically engineered neural progenitor cells to overexpress galactocerebrosidase (Torchiana et al., 1998). Recently, Buchet et al. (2002) successfully engineered human neural progenitor cells to secret β-glucuronidase, which may pave the way toward a clinical treatment for Sly disease.

One way to replace specific neuronal phenotypes that are damaged in diseases is to provide the same type of neurons by directed differentiation of neural stem cells. On the other hand, neural stem cells can be genetically modified to ectopically express a specific enzyme for a specific neurotransmitter. Along this line, Corti et al. (1999) reported using an adenovirus vector to modify human neural progenitor cells to express tyrosine hydroxylase, the rate-limiting enzyme in the production of the neurotransmitter dopamine.

Anti-tumor reagents

Brain tumors such as glioblastoma are difficult to treat due to the extensive infiltration of tumor cells into the normal brain parenchyma. Based on the fact that neural stem/progenitor cells migrate toward tumor cells (Aboody et al., 2000; Park et al., 2002; Hughes et al., 2002), these cells can be used to deliver anti-tumor reagents near the tumor cells. Along this line, Herrlinger et al. (2000) used a replication-conditional HSV-1 vector, which can replicate only in dividing cells, to infect the C17.2 neural precursor cell line without killing them, since the cells were growth-arrested by mimosine pre-treatment. The infected cells were then implanted into intracerebral gliomas. In the absence of mimosine *in vivo*, these neural precursor cells start dividing, which permits HSV-1 replication. Thus, acting as "Trojan horses", C17.2 cells delivered a large number of HSV-1 mutants near the tumor cells which then killed the tumor cells. However, concerns remain regarding the risk of endogenous neural stem/progenitor cells being infected and damaged by HSV-1 vectors. Benedetti et al. (2000) genetically engineered either primary or immortalized neural progenitor cells with a retrovirus vector containing an anti-tumor molecule, interleukin-4 (IL-4). They then grafted cells into established gliomas in mice, and found a progressive decrease in the size of large tumors with enhanced animal survival.

SUMMARY

Genetic modifications using viral or nonviral vectors have provided a powerful tool for identifying proliferating and differentiating neural stem cells. In addition to insights into basic developmental and cell biology, genetically modified neural stem cells have been explored for their potential to deliver therapeutic reagents into the CNS. Various gene delivery vehicles have been used in these studies. Generally speaking, vectors derived from viruses transfer genetic materials into neural stem cells more efficiently, while nonviral vectors, such as liposomes and polymers, are easier to prepare and probably safer to use. The choice of a gene delivery method for any given application needs to be carefully selected case by case. For example, if long-term and stable transgene expression is required, vectors originating from retrovirus, lentivirus, or AAV are usually the most favorable. On the other hand, adenovirus, HSV or liposome/polymers may be more appropriate for transient expression. Besides the choice of vectors, targeted gene expression using a cell-specific promoter should be considered and is desirable for studies that require turning on a specific gene expression at a specific developmental stage of the neural stem cells while turning it off during other stages. Another useful feature of genetic modification of neural stem cells is regulatable transgene expression. This is particularly important for neural stem cell-mediated delivery of therapeutic reagents, such as neurotrophic factors or enzymes, where expression levels need to be finely tuned or shut down completely to avoid unexpected deleterious effects. Thus far, regulatory systems based on tetracycline (tet-on or tet-off) have been tested in neural stem cells using retrovirus (Hoshimaru et al., 1996; Cho et al., 2002) or adenovirus vectors (Corti et al., 1999). Although these results are exciting, further optimization of the tet system as well as development of other systems are required to obtain tighter and safer control that is more appropriate for clinical applications.

Other issues also need be considered when using genetic tools to modify neural stem cells. For example, transgenes or other components in viral or non-viral vectors may directly or indirectly affect proliferation and/or differentiation of genetically modified neural stem cells. These may be virion proteins in the recombinant vectors, the reporter gene such as GFP, or therapeutic reagents such as neurotrophic factors or cytokines. Although not extensively studied, genetic modifications of neural stem cells with anti-immune, anti-inflammation or anti-apoptotic factors may help enhance the survival rate of grafted cells (Giannoukakis et al., 1999; Guillot et al., 2000). Furthermore, when choosing vectors with integration capability, we need to be cautious of insertional mutagenesis that may be harmful in clinical applications or complicate data interpretation. Along this line, development of better vectors that can either integrate into the host genome site specifically or retain a capability of long-term transgene expression without integration may be more desireable.

Genetic tools have helped us greatly in enhancing our knowledge of neural stem cell basic science and potential therapeutic application. There is no doubt that these studies will increase in number and stimulate rapid growth in the neural stem cell field, which in turn will facilitate efforts to understand and direct specific neural cell differentiation as well as to develop combined stem cell and gene therapy to treat various neurological disorders.

Acknowledgements

The authors are grateful to Dr. Richard E. Coggeshall for critical review and helpful discussion.

REFERENCES

Aberg MA, Ryttsen F, Hellgren G, Lindell K, Rosengren LE, MacLennan AJ, Carlsson B, Orwar O, Eriksson PS (2001) Selective introduction of antisense oligonucleotides into single adult CNS progenitor cells using electroporation demonstrates the requirement of STAT3 activation for CNTF-induced gliogenesis. Mol Cell Neurosci 17: 426-443.

Aboody KS, Brown A, Rainov NG, Bower KA, Liu S, Yang W, Small JE, Herrlinger U, Ourednik V, Black PM, Breakefield XO, Snyder EY (2000) Neural stem cells display extensive tropism for pathology in adult brain: evidence from intracranial gliomas. Proc Natl Acad Sci U S A 97: 12846-12851.

Ailles LE, Naldini L (2002) HIV-1-derived lentiviral vectors. Curr Top Microbiol Immunol 261: 31-52.

Anderson DJ (1999) Lineages and transcription factors in the specification of vertebrate primary sensory neurons. Curr Opin Neurobiol 9: 517-524.

Andreansky SS, He B, Gillespie GY, Soroceanu L, Markert J, Chou J, Roizman B, Whitley RJ (1996) The application of genetically engineered herpes simplex viruses to the treatment of experimental brain tumors. Proc Natl Acad Sci U S A 93: 11313-11318.

Audouy S, Hoekstra D (2001) Cationic lipid-mediated transfection *in vitro* and *in vivo* (review). Mol Membr Biol 18: 129-143.

Bartlett JS, Wilcher R, Samulski RJ (2000) Infectious entry pathway of adeno-associated virus and adeno-associated virus vectors. J Virol 74: 2777-2785.

Bartlett PF, Reid HH, Bailey KA, Bernard O (1988) Immortalization of mouse neural precursor cells by the c-myc oncogene. Proc Natl Acad Sci U S A 85: 3255-3259.

Benedetti S, Pirola B, Pollo B, Magrassi L, Bruzzone MG, Rigamonti D, Galli R, Selleri S, Di Meco F, De Fraja C, Vescovi A, Cattaneo E, Finocchiaro G (2000) Gene therapy of experimental brain tumors using neural progenitor cells. Nat Med 6: 447-450.

Benraiss A, Chmielnicki E, Lerner K, Roh D, Goldman SA (2001) Adenoviral brain-derived neurotrophic factor induces both neostriatal and olfactory neuronal recruitment from endogenous progenitor cells in the adult forebrain. J Neurosci 21: 6718-6731.

Blesch A, Lu P, Tuszynski MH (2002) Neurotrophic factors, gene therapy, and neural stem cells for spinal cord repair. Brain Res Bull 57: 833-838.

Bosher JM, Labouesse M (2000) RNA interference: genetic wand and genetic watchdog. Nat Cell Biol 2: E31-E36.

Buchet D, Serguera C, Zennou V, Charneau P, Mallet J (2002) Long-term expression of beta-glucuronidase by genetically modified human neural progenitor cells grafted into the mouse central nervous system. Mol Cell Neurosci 19: 389-401.

Burton EA, Bai Q, Goins WF, Glorioso JC (2001) Targeting gene expression using HSV vectors. Adv Drug Deliv Rev 53: 155-170.

Carpenter MK, Cui X, Hu ZY, Jackson J, Sherman S, Seiger A, Wahlberg LU (1999) *In vitro* expansion of a multipotent population of human neural progenitor cells. Exp Neurol 158: 265-278.

Carter PJ, Samulski RJ (2000) Adeno-associated viral vectors as gene delivery vehicles. Int J Mol Med 6: 17-27.

Chambers CB, Peng Y, Nguyen H, Gaiano N, Fishell G, Nye JS (2001) Spatiotemporal selectivity of response to Notch1 signals in mammalian forebrain precursors. Development 128: 689-702.

Chesnoy S, Huang L (2000) Structure and function of lipid-DNA complexes for gene delivery. Annu Rev Biophys Biomol Struct 29: 27-47.

Cho T, Bae JH, Choi HB, Kim SS, McLarnon JG, Suh-Kim H, Kim SU, Min CK (2002) Human neural stem cells: electrophysiological properties of voltage-gated ion channels. Neuroreport 13: 1447-1452.

Chow SY, Moul J, Tobias CA, Himes BT, Liu Y, Obrocka M, Hodge L, Tessler A, Fischer I (2000) Characterization and intraspinal grafting of EGF/bFGF-dependent neurospheres derived from embryonic rat spinal cord. Brain Res 874: 87-106.

Cornetta K, Morgan RA, Anderson WF (1991) Safety issues related to retroviral-mediated gene transfer in humans. Hum Gene Ther 2: 5-14.

Corti O, Horellou P, Colin P, Cattaneo E, Mallet J (1996) Intracerebral tetracycline-dependent regulation of gene expression in grafts of neural precursors. Neuroreport 7: 1655-1659.

Corti O, Sabate O, Horellou P, Colin P, Dumas S, Buchet D, Buc-Caron MH, Mallet J (1999) A single adenovirus vector mediates doxycycline-controlled expression of tyrosine hydroxylase in brain grafts of human neural progenitors. Nat Biotechnol 17: 349-354.

Curran MA, Nolan GP (2002) Nonprimate lentiviral vectors. Curr Top Microbiol Immunol 261: 75-105.

Dagle JM, Weeks DL (2001) Oligonucleotide-based strategies to reduce gene expression. Differentiation 69: 75-82.

Daly G, Chernajovsky Y (2000) Recent developments in retroviral-mediated gene transduction. Mol Ther 2: 423-434.

Danthinne X, Werth E (2000) New tools for the generation of E1- and/or E3-substituted adenoviral vectors. Gene Ther 7: 80-87.

Davis AA, Temple S (1994) A Self-Renewing Multipotential Stem-Cell in Embryonic Rat Cerebral-Cortex. Nature 372: 263-266.

De Smedt SC, Demeester J, Hennink WE (2000) Cationic polymer based gene delivery systems. Pharm Res 17: 113-126.

Delenda C, Audit M, Danos O (2002) Biosafety issues in lentivector production. Curr Top Microbiol Immunol 261: 123-141.

Duan D, Li Q, Kao AW, Yue Y, Pessin JE, Engelhardt JF (1999) Dynamin is required for recombinant adeno-associated virus type 2 infection. J Virol 73: 10371-10376.

Duan D, Yue Y, Yan Z, Engelhardt JF (2000) A new dual-vector approach to enhance recombinant adeno-associated virus-mediated gene expression through intermolecular cis activation. Nat Med 6: 595-598.

Dunn KJ, Williams BO, Li Y, Pavan WJ (2000) Neural crest-directed gene transfer demonstrates Wnt1 role in melanocyte expansion and differentiation during mouse development. Proc Natl Acad Sci U S A 97: 10050-10055.

Eaton MJ, Whittemore SR (1996) Autocrine BDNF secretion enhances the survival and serotonergic differentiation of raphe neuronal precursor cells grafted into the adult rat CNS. Exp Neurol 140: 105-114.

Englund U, Bjorklund A, Wictorin K, Lindvall O, Kokaia M (2002a) Grafted neural stem cells develop into functional pyramidal neurons and integrate into host cortical circuitry. Proc Natl Acad Sci U S A 99: 17089-17094.

Englund U, Ericson C, Rosenblad C, Mandel RJ, Trono D, Wictorin K, Lundberg C (2000) The use of a recombinant lentiviral vector for ex vivo gene transfer into the rat CNS. Neuroreport 11: 3973-3977.

Englund U, Fricker-Gates RA, Lundberg C, Bjorklund A, Wictorin K (2002b) Transplantation of human neural progenitor cells into the neonatal rat brain: extensive migration and differentiation with long-distance axonal projections. Exp Neurol 173: 1-21.

Estibeiro P, Godfray J (2001) Antisense as a neuroscience tool and therapeutic agent. Trends Neurosci 24: S56-S62.

Evrard C, Borde I, Marin P, Galiana E, Premont J, Gros F, Rouget P (1990) Immortalization of bipotential and plastic glio-neuronal precursor cells. Proc Natl Acad Sci U S A 87: 3062-3066.

Falk A, Holmstrom N, Carlen M, Cassidy R, Lundberg C, Frisen J (2002) Gene delivery to adult neural stem cells. Exp Cell Res 279: 34-39.

Felgner JH, Kumar R, Sridhar CN, Wheeler CJ, Tsai YJ, Border R, Ramsey P, Martin M, Felgner PL (1994) Enhanced gene delivery and mechanism studies with a novel series of cationic lipid formulations. J Biol Chem 269: 2550-2561.

Ferrari FK, Samulski T, Shenk T, Samulski RJ (1996) Second-strand synthesis is a rate-limiting step for efficient transduction by recombinant adeno-associated virus vectors. J Virol 70: 3227-3234.

Fire A (1999) RNA-triggered gene silencing. Trends Genet 15: 358-363.

Fisher KJ, Gao GP, Weitzman MD, DeMatteo R, Burda JF, Wilson JM (1996) Transduction with recombinant adeno-associated virus for gene therapy is limited by leading-strand synthesis. J Virol 70: 520-532.

Fisher LJ (1997) Neural precursor cells: applications for the study and repair of the central nervous system. Neurobiol Dis 4: 1-22.

Flax JD, Aurora S, Yang C, Simonin C, Wills AM, Billinghurst LL, Jendoubi M, Sidman RL, Wolfe JH, Kim SU, Snyder EY (1998) Engraftable human neural stem cells respond to developmental cues, replace neurons, and express foreign genes. Nat Biotechnol 16: 1033-1039.

Foster GA, Stringer BM (1999) Genetic regulatory elements introduced into neural stem and progenitor cell populations. Brain Pathol 9: 547-567.

Fox JL (2003) US authorities uphold suspension of SCID gene therapy. Nat Biotechnol 21: 217.

Fraefel C, Jacoby DR, Breakefield XO (2000) Herpes simplex virus type 1-based amplicon vector systems. Adv Virus Res 55: 425-451.

Frederiksen K, Jat PS, Valtz N, Levy D, McKay R (1988) Immortalization of precursor cells from the mammalian CNS. Neuron 1: 439-448.

Friedmann T, Yee JK (1995) Pseudotyped retroviral vectors for studies of human gene therapy. Nat Med 1: 275-277.

Fults D, Pedone C, Dai C, Holland EC (2002) MYC expression promotes the proliferation of neural progenitor cells in culture and *in vivo*. Neoplasia 4: 32-39.

Gage FH, Coates PW, Palmer TD, Kuhn HG, Fisher LJ, Suhonen JO, Peterson DA, Suhr ST, Ray J (1995) Survival and differentiation of adult neuronal progenitor cells transplanted to the adult brain. Proc Natl Acad Sci U S A 92: 11879-11883.

Galimi F, Verma IM (2002) Opportunities for the use of lentiviral vectors in human gene therapy. Curr Top Microbiol Immunol 261: 245-254.

Gao GP, Alvira MR, Wang L, Calcedo R, Johnston J, Wilson JM (2002) Novel adeno-associated viruses from rhesus monkeys as vectors for human gene therapy. Proc Natl Acad Sci U S A 99: 11854-11859.

Giannoukakis N, Thomson A, Robbins P (1999) Gene therapy in transplantation. Gene Ther 6: 1499-1511.

Grimm D, Kern A, Rittner K, Kleinschmidt JA (1998) Novel tools for production and purification of recombinant adenoassociated virus vectors. Hum Gene Ther 9: 2745-2760.

Guillot C, Le Mauff B, Cuturi MC, Anegon I (2000) Gene therapy in transplantation in the year 2000: moving towards clinical applications? Gene Ther 7: 14-19.

Hammond SM, Caudy AA, Hannon GJ (2001) Post-transcriptional gene silencing by double-stranded RNA. Nat Rev Genet 2: 110-119.

Hannon GJ (2002) RNA interference. Nature 418: 244-251.

Herrlinger U, Woiciechowski C, Sena-Esteves M, Aboody KS, Jacobs AH, Rainov NG, Snyder EY, Breakefield XO (2000) Neural precursor cells for delivery of replication-conditional HSV-1 vectors to intracerebral gliomas. Mol Ther 1: 347-357.

Hitt MM, Graham FL (2000) Adenovirus vectors for human gene therapy. Adv Virus Res 55: 479-505.

Holt CE, Bertsch TW, Ellis HM, Harris WA (1988) Cellular determination in the Xenopus retina is independent of lineage and birth date. Neuron 1: 15-26.

Hoshimaru M, Ray J, Sah DW, Gage FH (1996) Differentiation of the immortalized adult neuronal progenitor cell line HC2S2 into neurons by regulatable suppression of the v-myc oncogene. Proc Natl Acad Sci U S A 93: 1518-1523.

Hubener M, Gotz M, Klostermann S, Bolz J (1995) Guidance of thalamocortical axons by growth-promoting molecules in developing rat cerebral cortex. Eur J Neurosci 7: 1963-1972.

Hughes SM, Moussavi-Harami F, Sauter SL, Davidson BL (2002) Viral-mediated gene transfer to mouse primary neural progenitor cells. Mol Ther 5: 16-24.

Jaaskelainen I, Urtti A (2002) Cell membranes as barriers for the use of antisense therapeutic agents. Mini Rev Med Chem 2: 307-318.

Jacobs A, Breakefield XO, Fraefel C (1999) HSV-1-based vectors for gene therapy of neurological diseases and brain tumors: part II. Vector systems and applications. Neoplasia 1: 402-416.

Johansson CB, Svensson M, Wallstedt L, Janson AM, Frisen J (1999) Neural stem cells in the adult human brain. Exp Cell Res 253: 733-736.

Johe KK, Hazel TG, Muller T, Dugich-Djordjevic MM, McKay RD (1996) Single factors direct the differentiation of stem cells from the fetal and adult central nervous system. Genes Dev 10: 3129-3140.

Jungbluth S, Bell E, Lumsden A (1999) Specification of distinct motor neuron identities by the singular activities of individual Hox genes. Development 126: 2751-2758.

Kafri T (2001) Lentivirus vectors: difficulties and hopes before clinical trials. Curr Opin Mol Ther 3: 316-326.

Kalyani A, Hobson K, Rao MS (1997) Neuroepithelial stem cells from the embryonic spinal cord: Isolation, characterization, and clonal analysis. Dev Biol 186: 202-223.

Kim SH, Kim S, Robbins PD (2000) Retroviral vectors. Adv Virus Res 55: 545-563.

Kim YC, Shim JW, Oh YJ, Son H, Lee YS, Lee SH (2002) Co-transfection with cDNA encoding the Bcl family of anti-apoptotic proteins improves the efficiency of transfection in primary fetal neural stem cells. J Neurosci Methods 117: 153-158.

Kotin RM, Siniscalco M, Samulski RJ, Zhu XD, Hunter L, Laughlin CA, McLaughlin S, Muzyczka N, Rocchi M, Berns KI (1990) Site-specific integration by adeno-associated virus. Proc Natl Acad Sci U S A 87: 2211-2215.

Lachmann RH, Efstathiou S (1999) Gene transfer with herpes simplex vectors. Curr Opin Mol Ther 1: 622-632.

Lacorazza HD, Flax JD, Snyder EY, Jendoubi M (1996) Expression of human beta-hexosaminidase alpha-subunit gene (the gene defect of Tay-Sachs disease) in mouse brains upon engraftment of transduced progenitor cells. Nat Med 2: 424-429.

Leber SM, Breedlove SM, Sanes JR (1990) Lineage, arrangement, and death of clonally related motoneurons in chick spinal cord. J Neurosci 10: 2451-2462.

Lemkine GF, Mantero S, Migne C, Raji A, Goula D, Normandie P, Levi G, Demeneix BA (2002) Preferential transfection of adult mouse neural stem cells and their immediate progeny *in vivo* with polyethylenimine. Mol Cell Neurosci 19: 165-174.

Lever AM (2000) Lentiviral vectors: progress and potential. Curr Opin Mol Ther 2: 488-496.

Liu Y, Himes BT, Solowska J, Moul J, Chow SY, Park KI, Tessler A, Murray M, Snyder EY, Fischer I (1999) Intraspinal delivery of neurotrophin-3 using neural stem cells genetically modified by recombinant retrovirus. Exp Neurol 158: 9-26.

Lundberg C, Martínez-Serrano A, Cattaneo E, McKay RD, Bjorklund A (1997) Survival, integration, and differentiation of neural stem cell lines after transplantation to the adult rat striatum. Exp Neurol 145: 342-360.

Luskin MB (1993) Restricted proliferation and migration of postnatally generated neurons derived from the forebrain subventricular zone. Neuron 11: 173-189.

Luskin MB, Pearlman AL, Sanes JR (1988) Cell lineage in the cerebral cortex of the mouse studied *in vivo* and *in vitro* with a recombinant retrovirus. Neuron 1: 635-647.

Martínez-Serrano A, Bjorklund A (1996) Protection of the neostriatum against excitotoxic damage by neurotrophin-producing, genetically modified neural stem cells. J Neurosci 16: 4604-4616.

Martínez-Serrano A, Bjorklund A (1997) Immortalized neural progenitor cells for CNS gene transfer and repair. Trends Neurosci 20: 530-538.

Martínez-Serrano A, Bjorklund A (1998) Ex vivo nerve growth factor gene transfer to the basal forebrain in presymptomatic middle-aged rats prevents the development of cholinergic neuron atrophy and cognitive impairment during aging. Proc Natl Acad Sci U S A 95: 1858-1863.

Martínez-Serrano A, Fischer W, Bjorklund A (1995a) Reversal of age-dependent cognitive impairments and cholinergic neuron atrophy by NGF-secreting neural progenitors grafted to the basal forebrain. Neuron 15: 473-484.

Martínez-Serrano A, Lundberg C, Horellou P, Fischer W, Bentlage C, Campbell K, McKay RD, Mallet J, Bjorklund A (1995b) CNS-derived neural progenitor cells for gene transfer of nerve growth factor to the adult rat brain: complete rescue of axotomized cholinergic neurons after transplantation into the septum. J Neurosci 15: 5668-5680.

Martínez-Serrano A, Rubio FJ, Navarro B, Bueno C, Villa A (2001) Human neural stem and progenitor cells: *in vitro* and *in vivo* properties, and potential for gene therapy and cell replacement in the CNS. Curr Gene Ther 1: 279-299.

Martínez-Serrano A, Villa A, Navarro B, Rubio FJ, Bueno C (2000) Human neural progenitor cells: better blue than green? Nat Med 6: 483-484.

McManus MT, Sharp PA (2002) Gene silencing in mammals by small interfering RNAs. Nat Rev Genet 3: 737-747.

Merdan T, Kopecek J, Kissel T (2002) Prospects for cationic polymers in gene and oligonucleotide therapy against cancer. Adv Drug Deliv Rev 54: 715-758.

Meyers C, Mane M, Kokorina N, Alam S, Hermonat PL (2000) Ubiquitous human adeno-associated virus type 2 autonomously replicates in differentiating keratinocytes of a normal skin model. Virology 272: 338-346.

Mizumoto H, Mizumoto K, Whiteley SJ, Shatos M, Klassen H, Young MJ (2001) Transplantation of human neural progenitor cells to the vitreous cavity of the Royal College of Surgeons rat. Cell Transplant 10: 223-233.

Monahan PE, Samulski RJ (2000) Adeno-associated virus vectors for gene therapy: more pros than cons? Mol Med Today 6: 433-440.

Morgan JC, Majors JE, Galileo DS (2000) Wild-type and mutant forms of v-src differentially alter neuronal migration and differentiation *in vivo*. J Neurosci Res 59: 226-237.

Morshead CM, Craig CG, van der KD (1998) *In vivo* clonal analyses reveal the properties of endogenous neural stem cell proliferation in the adult mammalian forebrain. Development 125: 2251-2261.

Muzyczka N (1992) Use of adeno-associated virus as a general transduction vector for mammalian cells. Curr Top Microbiol Immunol 158: 97-129.

Muzyczka N, Berns KI (2001) Parvoviridae: The Viruses and Their Replication. In: Fundamental Virology (Knipe DM, Howley PM, Griffin D., eds), pp 1089-1121. Philadelphia: Lippincott Williams & Wilkins.

Naldini L (1998) Lentiviruses as gene transfer agents for delivery to non-dividing cells. Curr Opin Biotechnol 9: 457-463.

Naldini L, Blomer U, Gage FH, Trono D, Verma IM (1996a) Efficient transfer, integration, and sustained long-term expression of the transgene in adult rat brains injected with a lentiviral vector. Proc Natl Acad Sci U S A 93: 11382-11388.

Naldini L, Blomer U, Gallay P, Ory D, Mulligan R, Gage FH, Verma IM, Trono D (1996b) *In vivo* gene delivery and stable transduction of nondividing cells by a lentiviral vector. Science 272: 263-267.

Naldini L, Verma IM (2000) Lentiviral vectors. Adv Virus Res 55: 599-609.

Nasz I, Adam E (2001) Recombinant adenovirus vectors for gene therapy and clinical trials. Acta Microbiol Immunol Hung 48: 323-348.

NIH Report (2002) Assessment of adenoviral vector safety and toxicity: report of the National Institutes of Health Recombinant DNA Advisory Committee. Hum Gene Ther 13: 3-13.

Opalinska JB, Gewirtz AM (2002) Nucleic-acid therapeutics: basic principles and recent applications. Nat Rev Drug Discov 1: 503-514.

Ostenfeld T, Tai YT, Martin P, Deglon N, Aebischer P, Svendsen CN (2002) Neurospheres modified to produce glial cell line-derived neurotrophic factor increase the survival of transplanted dopamine neurons. J Neurosci Res 69: 955-965.

Ourednik V, Ourednik J, Park KI, Snyder EY (1999) Neural stem cells — a versatile tool for cell replacement and gene therapy in the central nervous system. Clin Genet 56: 267-278.

Owens GC, Mistry S, Edelman GM, Crossin KL (2002) Efficient marking of neural stem cell-derived neurons with a modified murine embryonic stem cell virus, MESV2. Gene Ther 9: 1044-1048.

Owens RA (2002) Second generation adeno-associated virus type 2-based gene therapy systems with the potential for preferential integration into AAVS1. Curr Gene Ther 2: 145-159.

Palmer TD, Schwartz PH, Taupin P, Kaspar B, Stein SA, Gage FH (2001) Cell culture. Progenitor cells from human brain after death. Nature 411: 42-43.

Palu G, Parolin C, Takeuchi Y, Pizzato M (2000) Progress with retroviral gene vectors. Rev Med Virol 10: 185-202.

Pannell D, Ellis J (2001) Silencing of gene expression: implications for design of retrovirus vectors. Rev Med Virol 11: 205-217.

Park KI, Ourednik J, Ourednik V, Taylor RM, Aboody KS, Auguste KI, Lachyankar MB, Redmond DE, Snyder EY (2002) Global gene and cell replacement strategies via stem cells. Gene Ther 9: 613-624.

Parolin C, Dorfman T, Palu G, Gottlinger H, Sodroski J (1994) Analysis in human immunodeficiency virus type 1 vectors of cis-acting sequences that affect gene transfer into human lymphocytes. J Virol 68: 3888-3895.

Pedroso de Lima MC, Simoes S, Pires P, Faneca H, Duzgunes N (2001) Cationic lipid-DNA complexes in gene delivery: from biophysics to biological applications. Adv Drug Deliv Rev 47: 277-294.

Pluchino S, Quattrini A, Brambilla E, Gritti A, Salani G, Dina G, Galli R, Del Carro U, Amadio S, Bergami A, Furlan R, Comi G, Vescovi AL, Martino G (2003) Injection of adult neurospheres induces recovery in a chronic model of multiple sclerosis. Nature 422: 688-694.

Qing K, Mah C, Hansen J, Zhou S, Dwarki V, Srivastava A (1999) Human fibroblast growth factor receptor 1 is a co-receptor for infection by adeno-associated virus 2. Nat Med 5: 71-77.

Rabinowitz JE, Samulski RJ (2000) Building a better vector: the manipulation of AAV virions. Virology 278: 301-308.

Redies C, Lendahl U, McKay RD (1991) Differentiation and heterogeneity in T-antigen immortalized precursor cell lines from mouse cerebellum. J Neurosci Res 30: 601-615.

Reynolds BA, Tetzlaff W, Weiss S (1992) A multipotent EGF-responsive striatal embryonic progenitor cell produces neurons and astrocytes. J Neurosci 12: 4565-4574.

Roizman B, Knipe DM (2003) Herpes Simplex Viruses and Their Replication. In: Fundamental Virology (Knipe DM, Howley PM, Griffin D., eds), pp 1123-1184. Philadelphia: Lippincott Williams & Wilkins.

Rosenqvist N, Hard Af SC, Samuelsson C, Johansen J, Lundberg C (2002) Activation of silenced transgene expression in neural precursor cell lines by inhibitors of histone deacetylation. J Gene Med 4: 248-257.

Roy NS, Wang S, Jiang L, Kang J, Benraiss A, Harrison-Restelli C, Fraser RA, Couldwell WT, Kawaguchi A, Okano H, Nedergaard M, Goldman SA (2000) *In vitro* neurogenesis by progenitor cells isolated from the adult human hippocampus. Nat Med 6: 271-277.

Russell WC (2000) Update on adenovirus and its vectors. J Gen Virol 81: 2573-2604.

Ryder EF, Snyder EY, Cepko CL (1990) Establishment and characterization of multipotent neural cell lines using retrovirus vector-mediated oncogene transfer. J Neurobiol 21: 356-375.

Sabate O, Horellou P, Vigne E, Colin P, Perricaudet M, Buc-Caron MH, Mallet J (1995) Transplantation to the rat brain of human neural progenitors that were genetically modified using adenoviruses. Nat Genet 9: 256-260.

Sakakibara S, Nakamura Y, Yoshida T, Shibata S, Koike M, Takano H, Ueda S, Uchiyama Y, Noda T, Okano H (2002) RNA-binding protein Musashi family: roles for CNS stem cells and a subpopulation of ependymal cells revealed by targeted disruption and antisense ablation. Proc Natl Acad Sci U S A 99: 15194-15199.

Samulski RJ, Zhu X, Xiao X, Brook JD, Housman DE, Epstein N, Hunter LA (1991) Targeted integration of adeno-associated virus (AAV) into human chromosome 19. EMBO J 10: 3941-3950.

Sanlioglu S, Monick MM, Luleci G, Hunninghake GW, Engelhardt JF (2001) Rate limiting steps of AAV transduction and implications for human gene therapy. Curr Gene Ther 1: 137-147.

Shihabuddin LS, Ray J, Gage FH (1999) Stem cell technology for basic science and clinical applications. Arch Neurol 56: 29-32.

Simonato M, Manservigi R, Marconi P, Glorioso J (2000) Gene transfer into neurones for the molecular analysis of behaviour: focus on herpes simplex vectors. Trends Neurosci 23: 183-190.

Smith-Arica JR, Bartlett JS (2001) Gene therapy: recombinant adeno-associated virus vectors. Curr Cardiol Rep 3: 43-49.

Snyder EY, Deitcher DL, Walsh C, Arnold-Aldea S, Hartwieg EA, Cepko CL (1992) Multipotent neural cell lines can engraft and participate in development of mouse cerebellum. Cell 68: 33-51.

Snyder EY, Taylor RM, Wolfe JH (1995) Neural progenitor cell engraftment corrects lysosomal storage throughout the MPS VII mouse brain. Nature 374: 367-370.

Song H, Stevens CF, Gage FH (2002) Astroglia induce neurogenesis from adult neural stem cells. Nature 417: 39-44.

Stemple DL, Anderson DJ (1992) Isolation of A Stem-Cell for Neurons and Glia from the Mammalian Neural Crest. Cell 71: 973-985.

Stevenson M (2002) Molecular biology of lentivirus-mediated gene transfer. Curr Top Microbiol Immunol 261: 1-30.

Summerford C, Bartlett JS, Samulski RJ (1999) AlphaVbeta5 integrin: a co-receptor for adeno-associated virus type 2 infection. Nat Med 5: 78-82.

Suzuki A, Obi K, Urabe T, Hayakawa H, Yamada M, Kaneko S, Onodera M, Mizuno Y, Mochizuki H (2002) Feasibility of ex vivo gene therapy for neurological disorders using the new retroviral vector GCDNsap packaged in the vesicular stomatitis virus G protein. J Neurochem 82: 953-960.

Svendsen CN, ter Borg MG, Armstrong RJE, Rosser AE, Chandran S, Ostenfeld T, Caldwell MA (1998) A new method for the rapid and long term growth of human neural precursor cells. J Neurosci Methods 85: 141-152.

Temple S (1989) Division and Differentiation of Isolated Cns Blast Cells in Microculture. Nature 340: 471-473.

Torchiana E, Lulli L, Cattaneo E, Invernizzi F, Orefice R, Bertagnolio B, Di Donato S, Finocchiaro G (1998) Retroviral-mediated transfer of the galactocerebrosidase gene in neural progenitor cells. Neuroreport 9: 3823-3827.

Turner DL, Cepko CL (1987) A common progenitor for neurons and glia persists in rat retina late in development. Nature 328: 131-136.

Tuschl T (2002) Expanding small RNA interference. Nat Biotechnol 20: 446-448.

Uchida K, Toya S (1996) Grafting of genetically manipulated cells into adult brain: toward graft-gene therapy. Keio J Med 45: 81-89.

Uchida N, Buck DW, He D, Reitsma MJ, Masek M, Phan TV, Tsukamoto AS, Gage FH, Weissman IL (2000) Direct isolation of human central nervous system stem cells. Proc Natl Acad Sci U S A 97: 14720-14725.

van Praag H, Schinder AF, Christie BR, Toni N, Palmer TD, Gage FH (2002) Functional neurogenesis in the adult hippocampus. Nature 415: 1030-1034.

Vescovi AL, Snyder EY (1999) Establishment and properties of neural stem cell clones: plasticity *in vitro* and *in vivo*. Brain Pathol 9: 569-598.

Vicario I, Schimmang T (2003) Transfer of FGF-2 via HSV-1-based amplicon vectors promotes efficient formation of neurons from embryonic stem cells. J Neurosci Methods 123: 55-60.

Villa A, Snyder EY, Vescovi A, Martínez-Serrano A (2000) Establishment and properties of a growth factor-dependent, perpetual neural stem cell line from the human CNS. Exp Neurol 161: 67-84.

Wagner J, Akerud P, Castro DS, Holm PC, Canals JM, Snyder EY, Perlmann T, Arenas E (1999) Induction of a midbrain dopaminergic phenotype in Nurr1-overexpressing neural stem cells by type 1 astrocytes. Nat Biotechnol 17: 653-659.

Wang S, Roy NS, Benraiss A, Goldman SA (2000) Promoter-based isolation and fluorescence-activated sorting of mitotic neuronal progenitor cells from the adult mammalian ependymal/subependymal zone. Dev Neurosci 22: 167-176.

Wang S, Wu H, Jiang J, Delohery TM, Isdell F, Goldman SA (1998) Isolation of neuronal precursors by sorting embryonic forebrain transfected with GFP regulated by the T alpha 1 tubulin promoter. Nat Biotechnol 16: 196-201.

Wetts R, Fraser SE (1988) Multipotent precursors can give rise to all major cell types of the frog retina. Science 239: 1142-1145.

Whittemore SR, White LA (1993) Target regulation of neuronal differentiation in a temperature-sensitive cell line derived from medullary raphe. Brain Res 615: 27-40.

Williams BP, Price J (1995) Evidence for multiple precursor cell types in the embryonic rat cerebral cortex. Neuron 14: 1181-1188.

Wu P, Phillips MI, Bui J, Terwilliger EF (1998) Adeno-associated virus vector-mediated transgene integration into neurons and other nondividing cell targets. J Virol 72: 5919-5926.

Wu P, Tarasenko YI, Gu Y, Huang LY, Coggeshall RE, Yu Y (2002a) Region-specific generation of cholinergic neurons from fetal human neural stem cells grafted in adult rat. Nat Neurosci 5: 1271-1278.

Wu P, Ye Y, Svendsen CN (2002b) Transduction of human neural progenitor cells using recombinant adeno-associated viral vectors. Gene Ther 9: 245-255.

Xiao X, Li J, Samulski RJ (1998) Production of high-titer recombinant adeno-associated virus vectors in the absence of helper adenovirus. J Virol 72: 2224-2232.

Yakobson B, Hrynko TA, Peak MJ, Winocour E (1989) Replication of adeno-associated virus in cells irradiated with UV light at 254 nm. J Virol 63: 1023-1030.

Yalkinoglu AO, Zentgraf H, Hubscher U (1991) Origin of adeno-associated virus DNA replication is a target of carcinogen-inducible DNA amplification. J Virol 65: 3175-3184.

Yang CC, Xiao X, Zhu X, Ansardi DC, Epstein ND, Frey MR, Matera AG, Samulski RJ (1997) Cellular recombination pathways and viral terminal repeat hairpin structures are sufficient for adeno-associated virus integration *in vivo* and *in vitro*. J Virol 71: 9231-9247.

Yoon SO, Lois C, Alvirez M, Alvarez-Buylla A, Falck-Pedersen E, Chao MV (1996) Adenovirus-mediated gene delivery into neuronal precursors of the adult mouse brain. Proc Natl Acad Sci U S A 93: 11974-11979.

Young MJ, Ray J, Whiteley SJ, Klassen H, Gage FH (2000) Neuronal differentiation and morphological integration of hippocampal progenitor cells transplanted to the retina of immature and mature dystrophic rats. Mol Cell Neurosci 16: 197-205.

Zamore PD (2001) RNA interference: listening to the sound of silence. Nat Struct Biol 8: 746-750.

Zennou V, Serguera C, Sarkis C, Colin P, Perret E, Mallet J, Charneau P (2001) The HIV-1 DNA flap stimulates HIV vector-mediated cell transduction in the brain. Nat Biotechnol 19: 446-450.

Zhao N, Liu DP, Liang CC (2001) Hot topics in adeno-associated virus as a gene transfer vector. Mol Biotechnol 19: 229-237.

Zlokovic BV, Apuzzo ML (1997) Cellular and molecular neurosurgery: pathways from concept to reality—part II: vector systems and delivery methodologies for gene therapy of the central nervous system. Neurosurgery 40: 805-812.

Zolotukhin S, Byrne BJ, Mason E, Zolotukhin I, Potter M, Chesnut K, Summerford C, Samulski RJ, Muzyczka N (1999) Recombinant adeno-associated virus purification using novel methods improves infectious titer and yield. Gene Ther 6: 973-985.

Zufferey R, Donello JE, Trono D, Hope TJ (1999) Woodchuck hepatitis virus posttranscriptional regulatory element enhances expression of transgenes delivered by retroviral vectors. J Virol 73: 2886-2892.

Index